Memoirs of the American Mathematical Society
Number 147

Daniel Gorenstein
and Koichiro Harada

Finite Groups Whose 2-Subgroups Are Generated by at Most 4 Elements

Published by the
American Mathematical Society
Providence, Rhode Island
1974

AMS (MOS) subject classifications (1970). Primary 20D05

Library of Congress Cataloging in Publication Data CIP

Gorenstein, Daniel.
Finite groups whose 2-subgroups are generated by at most 4 elements.

(Memoirs of the American Mathematical Society, no. 147)
Bibliography: p.
1. Finite groups. I. Harada, Koichiro, joint author. II. Title. III. Series: American Mathematical Society. Memoirs, no. 147.
QA3.A57 no. 147 [QA171] 510'.8s [512'.2] 74-11282
ISBN 0-8218-1847-3

Printed in the United States of America

TABLE OF CONTENTS

PART IV: A CHARACTERIZATION OF THE GROUP $D_4^2(3)$

PART V: CENTRAL INVOLUTIONS WITH NON 2-CONSTRAINED CENTRALIZERS

PART VI: A CHARACTERIZATION OF THE GROUP M_{12}

ABSTRACT

The object of this memoir is to determine all finite simple (and more generally fusion-simple) groups each of whose 2-subgroups can be generated by at most 4 elements. Using a result of MacWilliams, we obtain as a corollary the classification of all finite simple groups whose Sylow 2-subgroups do not possess an elementary abelian normal subgroups of order 8.

The general introduction below provides a fairly detailed outline of the over-all proof of our main classification theorem, including the methods employed. The proof itself is divided into six major parts; and the introductory section of each part gives a description of the principal results to be proved in that part.

Acknowledgement

We would like to thank Mrs. Lynne Edwards, Miss Dodie Huffman, Mrs. Gertrude Jackson, and Mrs. Angela Shelton for their patience and excellent typing. Also the National Science Foundation for their partial support under grants GP 32319 and GP 32835 .

June 17, 1973 Gorenstein
Harada

INTRODUCTION

In Parts I-VI of this paper we shall determine all finite simple groups each of whose 2-subgroups can be generated by at most 4 elements. For brevity, we say that such a group G has <u>sectional</u> 2-<u>rank</u> <u>at</u> <u>most</u> 4. The reason we use this terminology is that the given condition is equivalent to the assertion that every section of G has 2-rank at most 4. (By definition, a <u>section</u> of a group G is a homomorphic image of a subgroup of G, while the 2-<u>rank</u> of G is the maximum rank of an abelian 2-subgroup of G).

The advantage of considering groups of sectional 2-rank at most 4 rather than the wider class of groups of 2-rank at most 4 is that the first condition is inductive to all sections of G, while the second is not. Groups of sectional 2-rank at most 4 are an important class of groups since by a result of MacWilliams [35], it includes all groups in which $SCN_3(2)$ is empty; i.e. in which a Sylow 2-subgroup possesses no abelian normal subgroups of rank 3.

We shall prove

<u>Main Theorem</u>. If G is a finite simple group of sectional 2-rank at most 4, then G is isomorphic to one of the groups of the following list:

Received by the Editors February 11, 1974.

I. Odd characteristic:

$L_2(q)$, $L_3(q)$, $U_3(q)$, $G_2(q)$, $D_4^2(q)$, $PSp(4,q)$, $Re(q)^*$,q an odd power of 3, $L_4(q)$, $q \not\equiv 1 (\mathrm{mod}\ 8)$, $U_4(q)$, $q \not\equiv 7\ (\mathrm{mod}\ 8)$, $L_5(q)$,q $\equiv$ $\equiv -1 (\mathrm{mod}\ 4)$, or $U_5(q)$,$q \equiv 1\ (\mathrm{mod}\ 4)$;

II. Even characteristic:

$L_2(8)$, $L_2(16)$, $L_3(4)$, $U_3(4)$, or $Sz(8)$;

III. Alternating:

A_7, A_8, A_9, A_{10}, A_{11};

IV. Sporadic:

M_{11}, M_{12}, M_{22}, M_{23}, J_1, J_2, J_3, M^c, or L.

Here $D_4^2(q)$, $Re(q)^*$, J_1, J_2, J_3, M^c, and L denote respectively the triality twisted $D_4(q^3)$, the groups of Ree type of characteristic 3, Janko's group of order $2^3 \cdot 3 \cdot 5 \cdot 7 \cdot 11 \cdot 19$, the Hall-Janko group of order $2^7 \cdot 3^3 \cdot 5^2 \cdot 7$, the Higman-McKay-Janko group of order $2^7 \cdot 3^5 \cdot 5 \cdot 17 \cdot 19$, McLaughlin's group of order $2^7 \cdot 3^6 \cdot 5^3 \cdot 7 \cdot 11$, and the Lyons-Sims group of order $2^8 \cdot 3^7 \cdot 5^6 \cdot 7 \cdot 11 \cdot 31 \cdot 37 \cdot 67$.

Corollary A. If G is a simple group in which $SCN_3(2)$ is empty, then G is isomorphic to one of the groups of the following list:

I. 2-rank 2:

$L_2(q)$, $L_3(q)$, $U_3(q)$, q odd, $U_3(4)$, A_7, or M_{11};

II. 2-rank 3:

$$G_2(q) \quad \text{or} \quad D_4^2(q),\ q \equiv 1,\ 7 \pmod 8;$$

III. 2-rank 4:

$L_4(q)$, $q \equiv 7 \pmod 8$, $U_4(q)$, $q \equiv 1 \pmod 8$, $PSp(4,q)$, $q \equiv 1,7 \pmod 8$, $L_5(q)$, $q \equiv -1 \pmod 4$, $U_5(q)$, $q \equiv 1 \pmod 4$, J_2, J_3, or L.

All simple groups of 2-rank 2 have been determined in [3]. However, the nature of that proof does not yield a classification of all nonsolvable groups of 2-rank 2 and, in particular, of all such quasisimple groups -- i.e., perfect central extensions of simple groups. Clearly if G has 2-rank 2, $SCN_3(2)$ is empty in G and so G has sectional 2-rank at most 4. Hence as a further consequence of our Main Theorem, we have

Corollary B. If G is a quasisimple group of 2-rank with $O(G) = 1$, then either G is simple or G is isomorphic to $Sp(4,q)$, q odd.

Here $O(G)$ denotes the largest normal subgroup of G of odd order; it is called the core of G.

On the basis of our Main Theorem, it is very easy to determine all non 2-constrained fusion-simple groups of sectional 2-rank at most 4. (A group G with $O(G) = 1$ is 2-constrained if $C_G(O_2(G)) \subseteq O_2(G)$, where $O_2(G)$ denotes the largest normal 2-subgroup of G; and more generally G is 2-constrained if $G/O(G)$ is. Likewise a group G is fusion-simple if $O(G) = 1$, the center of G is trivial, and G has no normal subgroups of index 2). However, one of our principal subsidiary results (Theorem B of Part II) enables us to classify all such 2-constrained groups. Hence our analysis yields the following additional result:

Corollary C. If G is a fusion-simple group of sectional 2-rank at most 4, then one of the following holds for the derived group G' of G:

(i) G' is simple;

(ii) G' is the direct product of two simple groups of sectional 2-rank 2;

(iii) G' is the direct product of a simple group of sectional 2-rank 2 and $Z_{2^n} \times Z_{2^n}$ for some n; or

(iv) G' is a nontrivial extension of E_8 or E_{16} by A_5, A_6, A_7, $Z_3 \times A_5$, or $L_3(2)$.

Here Z_m denotes a cyclic group of order m and E_{2^n} denotes an elementary abelian group of order 2^n.

Although some of our individual results are independent of any induction assumption, the overall proof of the Main Theorem is carried out by induction. Thus a minimal counterexample G has the property that the nonsolvable composition factors of its proper subgroups satisfy the conclusion of the Main Theorem and so are of "known type". As a consequence, we are able to use specific properties of these groups in our analysis.

With the exception of a large portion of Part I, which entails considerations involving p-local subgroups for odd primes p, (the normalizer in G of a nonidentity p-subgroup is called a p-local subgroup of G) our arguments utilize primarily two principal techniques:

(I) 2-local analysis;

(II) Sylow 2-group fusion analysis.

By 2-local analysis, we mean a purely group-theoretic study of the 2-local structure of G. Partially this involves entirely local investigations, such as in Theorem B of Part II, which gives the possible **structures** of nonsolvable 2-constrained 2-local subgroups of G. However, the emphasis is on obtaining information about the normalizers in G of the elementary abelian 2-subgroups of maximal rank which is forced as a consequence of Glauberman's Z*-theorem. The effect of this analysis is to yield an approximate structure of a Sylow 2-subgroup S of G and partial information concerning the centralizers in G (modulo their cores) of the involutions of S. It is to reach this stage in Part I that we need to use the p-local subgroups for odd primes p.

At this point, one has just about enough information to carry out the same kind of detailed fusion analysis that is used in classifying known simple groups in terms of either the structure of their Sylow 2-groups or the centralizer of one of their involutions. For brevity, we refer to this as Sylow 2-group fusion analysis. In particular, Thompson's fusion lemma [42, Lemma 5.38] and its extension [27, Lemma 16] as well as Glauberman's Z*-theorem [11], of course, play a critical role. The effect of this analysis is to eliminate many of the possibilities for S on the basis of the (fusion) simplicity of G. Ultimately an exact list of possibilities for S is determined.

Once this stage is reached, we are in a position to invoke prior classification theorems to show that G is not a counterexample to the Main Theorem. Indeed, in the case that S is isomorphic to a Sylow 2-subgroup of one of the groups listed in the Main Theorem, one can apply the principal result of [3], [5], [16], [17], [18], [19], [21], [36], [37], or [43] to deduce that G is isomorphic to one of the groups of our Main Theorem.

However, our list of possibilities for S also includes the case that S is nonabelian of the form $S = S_1 \times S_2$ with S_i dihedral, quasi-dihedral, or homocyclic abelian of rank 2, $i = 1, 2$, as well as the case that S possesses a maximal cyclic subgroup of index 4. By the principal results of [9], [20], [40], [41], and [38], there are no simple groups with Sylow 2-subgroups of any of these forms.

In the course of our proof we also make use of the classification of simple groups of 2-rank 2 obtained in [3] to conclude that our minimal counterexample G has 2-rank at least 3. In addition, in Part I we use results of Janko [33], Janko-Thompson [34], and Thompson [42] concerning simple groups in which all 2-local subgroups are solvable.

Finally our argument also utilizes results of MacWilliams [35] and Harada [29] in certain cases in which $SCN_3(S)$ is empty and S has 2-rank at least 3. Under these assumptions, S possesses a unique normal four subgroup U. The principal result of [35] is a determination of the possibilities for S when the involutions of U are all conjugate in G. Included in this list is the split extension of a group $A \cong Z_{2^n} \times Z_{2^n}$, $n \geq 3$, by a four group $\langle b_1, b_2 \rangle$ with b_1 raising every element of A to the $2^{n-1} - 1$ power and b_2 interchanging a pair of generators of A. The principal theorem of [29] shows that there are no simple groups whose Sylow 2-subgroups are of this form. Hence by the main result of [35], S is necessarily of type $G_2(q)$, $q \equiv 1, 7 \pmod 8$, J_2, or L. In particular,

S is isomorphic to a Sylow 2-subgroup of one of the groups of the Main Theorem.

Throughout the paper, we shall denote the sectional 2-rank and the 2-rank of a group X by $r(X)$ and $m(X)$, respectively. In other regards, our notation is standard and includes the use of the bar convention for homomorphic images.

Finally because of the extreme length of the proof of the Main Theorem, we should like to give a "flow diagram" of its proof. Here L_t denotes the unique maximal normal semisimple subgroup of the group $C_G(t)/O(C_G(t))$, t an involution of G. By definition, a semisimple group is a central product of quasisimple groups, a quasisimple group being a perfect central extension of a (nonabelian) simple group. For convenience we set $L_t = 1$ if $C_G(t)/O(C_G(t))$ does not contain a normal semisimple subgroup.

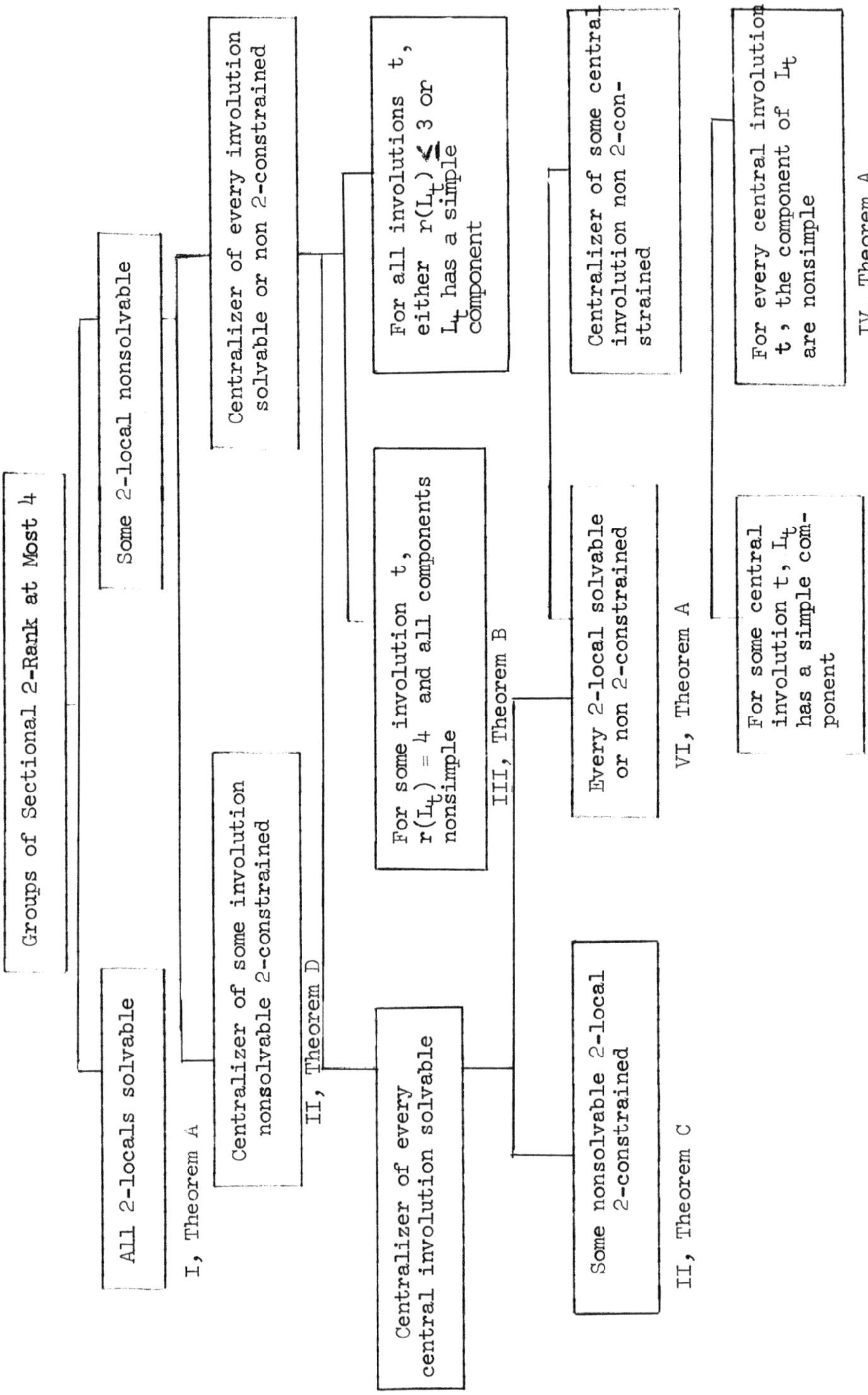
Groups of Sectional 2-Rank at Most 4
All 2-locals solvable
I, Theorem A
Some 2-local nonsolvable
Centralizer of some involution nonsolvable 2-constrained
II, Theorem D
Centralizer of every involution solvable or non 2-constrained
Centralizer of every central involution solvable
Some nonsolvable 2-local 2-constrained
II, Theorem C
Every 2-local solvable or non 2-constrained
VI, Theorem A
For some involution t, r(L_t) = 4 and all components nonsimple
III, Theorem B
For all involutions t, either r(L_t) ≤ 3 or L_t has a simple component
Centralizer of some central involution non 2-constrained
For some central involution t, L_t has a simple component
V, Theorem A
For every central involution t, the component of L_t are nonsimple
IV, Theorem A

PART I

SOLVABLE 2-LOCAL SUBGROUPS

1. Introduction. In this part we study groups of sectional 2-rank at most 4 in which all 2-local subgroups are solvable. Our main result is as follows:

Theorem A. If G is a nonsolvable group of sectional 2-rank at most 4 in which every 2-local subgroup is solvable, then $G'/O(G')$ is isomorphic to one of the following groups:

$L_2(q)$, q odd, $L_3(3)$, $U_3(3)$, $L_2(8)$, $L_2(16)$, $Sz(8)$, $U_3(4)$, A_7, or M_{11}.

Not only our assumption on the sectional 2-rank of G, but also the solvability of 2-local subgroups carries over to every section of G. Hence if G is a minimal counterexample to Theorem A, it follows that the nonsolvable composition factors of the proper subgroups of G satisfy the conclusion of Theorem A.

On the basis of a number of general theorems: Gorenstein-Walter [24], Gorenstein [15], Janko [33], Janko-Thompson [34] and Thompson [42], we are able to show, in addition, that $SCN_3(2)$ is nonempty in G, every 2-local subgroup of G has a trivial core, some local subgroup of G is nonsolvable, and some 2-local subgroup M of G has a noncyclic Sylow p-subgroup for some odd p (our assumption on the sectional 2-rank of G immediately forces $p = 3$ and the Sylow p-subgroup of M to be isomorphic to $Z_3 \times Z_3$).

The proof of Theorem A divides into two parts. First, using ideas of [42, Section 14], particularly a certain transitivity theorem of Thompson, we show that for some choice of M above, $O_2(M)$ is necessarily either abelian or of type $L_3(4)$. Secondly, by individual case analysis involving fusion arguments, we force the exact possibilities for the structure of a Sylow 2-subgroup of G. In each case, all simple groups with such a Sylow 2-subgroup have been previously determined and we conclude that G is not a counterexample to the theorem.

We remark here that Syskin [45] has determined all simple groups of 2-rank at most 3 in which all 2-local subgroups are solvable. Unfortunately use of his result would not shorten the analysis of Part I very much, since in almost all of the situations we must consider, we cannot exclude the possibility that G has 2-rank 4.

2. The minimal counterexample.

Henceforth G will denote a minimal counterexample to Theorem A.

Proposition 2.1. G is simple and every nonsolvable proper subgroup of G satisfies the conclusion of Theorem A.

Proof: The hypothesis of Theorem A clearly carries over to subgroups and homomorphic images of G. In particular, the minimality of G implies that every proper subgroup of G satisfies the conclusion of the theorem. Likewise, in proving that G is simple, we can assume that $O(G) = 1$.

If $O_2(G) \neq 1$, then $G = N_G(O_2(G))$ is a 2-local subgroup and so is solvable, which is not the case. Thus $O_2(G) = 1$. Let L be a minimal normal subgroup of G. Since $O(G) = O_2(G) = 1$, L is the direct product of isomorphic nonabelian simple groups L_i, $1 \leq i \leq m$. If $m > 1$, an involution x of L_1 would centralize L_2 and then $C_G(x)$ would be nonsolvable, contrary to hypothesis, so $m = 1$ and L is simple. For the same reason $|C_G(L)|$ is odd and, as $C_G(L) \lhd G$, it follows that

$C_G(L) = 1$. Hence G is isomorphic to a subgroup of $\mathrm{Aut}(L)$. If $L \subset G$, L would be isomorphic to one of the groups listed in the theorem, in which case G would satisfy the conclusion of the theorem. Hence we must have $G = L$ and so G is simple.

Combining our assumption on the 2-local subgroups of G with known classification theorems, we can derive several additional properties of G.

Proposition 2.2. $SCN_3(2)$ is nonempty in G.

Proof: Since the centralizer of every involution of G is solvable, the structure of G would be determined in the contrary case by a theorem of Janko and Thompson [34] and it would follow that G is isomorphic to one of the groups listed in Theorem A.

Proposition 2.3. $O(H) = 1$ for every 2-local subgroup H of G.

Proof: Since $SCN_3(2)$ is nonempty, G is connected in the sense of [15] or [24] and G has 2-rank at least 3. Since the centralizer of every involution of G is solvable, a theorem of Gorenstein and Walter [24, Theorem B] or [15] yields that $O(C_G(x)) = 1$ for every involution x of G.

But now [14, Theorem 5] implies that G is of characteristic 2 type in the sense of [14] and we conclude from [14, Theorem 1] that $O(H) = 1$ for every 2-local subgroup H of G.

Proposition 2.4. G possesses a nonsolvable p-local subgroup for some odd prime p.

Proof: Otherwise G would be an N-group and then the structure of G would be determined from Thompson's classification of such groups [42]. Again G would be isomorphic to one of the groups listed in Theorem A.

Proposition 2.5. Some 2-local subgroup of G has a noncyclic Sylow p-subgroup for some odd prime p.

Proof: Otherwise G would satisfy the hypotheses of a theorem of Janko [34] and again G would be isomorphic to one of the listed groups.

We establish one further property of G. In [42, Section 13], Thompson showed that a simple N-group G of characteristic 2 type does not possess a maximal 2-local subgroup M in which $O_2(M)$ is of symplectic type. In particular, in [42, Theorem 13.7] he treats the case that $O_2(M)$ is extra-special. Our interest here is limited to the single case that $O_2(M) \cong Q_8 * Q_8$ (and a Sylow 3-subgroup of M is elementary of order 9). An analysis of Thompson's proof in this case will reveal that it does not utilize the full assumption that G is an N-group, but only the fact that all 2-local subgroups are solvable. Hence his argument imples

Proposition 2.6. G does not possess a maximal 2-local subgroup M such that $O_2(M) \cong Q_8 * Q_8$.

One can also easily establish this particular case of Thompson's result in the following alternate manner. Assume G possesses a maximal 2-local subgroup M such that $R = O_2(M) \cong Q_8 * Q_8$. Clearly then a Sylow 2-subgroup S of M is a Sylow 2-subgroup of G. Since $SCN_3(S) \neq \emptyset$ by **Proposition** 2.2, one checks that $S/R \not\cong D_8$ or Z_4 and that $S/Z(S) \not\cong D_8 \times D_8$. Moreover, $S/Z(S)$ is not of type A_8 by a result of Harada [28] as G is simple. Likewise $S \supset R$, otherwise G has an isolated involution. It follows now that $|S/R| = 2$ and hence that $|S| = 64$. **Using the** extended form of Thompson's fusion lemma [27], **one argues** next that $S - R$ possesses an involution. This yields the following possibilities for S; namely, $S \cong QD_{16} * Q_8$ or $Q_{16} * Q_8$, or else S is of type A_8 or M_{12}. In the first two cases, a standard argument implies that G has an isolated involution . In the remaining cases, one invokes [17, Corollary A*] or [9, Theorem A] to obtain that $G \cong A_8, A_9, PSp(4,3), M_{12}, G_2(q)$, or $D_4^2(q)$, $q \equiv 3,5 \pmod 8$. However, this is a contradiction as each of these groups has a nonsolvable 2-local subgroup.

Remark. Recently Lundgren [44] has shown that there exists no simple group such that all 2-local subgroups are solvable, $SCN_3(2)$ nonempty, and having a maximal 2-local subgroup M with $O_2(M)$ of symplectic type.

Finally we can exclude a number of possibilities for the Sylow 2-subgroups of G. We first introduce some notation.

We denote by $L_3(4)^{(1)}$ the split extension of $L_3(4)$ by the transpose-inverse automorphism. We also denote by $Z_4 \cdot E_{16}$ the split extension of E_{16} by a cyclic group of order 4 acting freely on the E_{16} (i.e. $C_{E_{16}}(Z_4) \cong Z_2$).

Proposition 2.7. A Sylow 2-subgroup of G is not abelian, is not isomorphic to $Z_{2^n} \times Z_{2^n} \times D_8$, $D_8 \times D_8$, $D_8 \int Z_2$, $Z_2 \times Z_2 \int Z_2$, or $Z_4 \cdot E_{16}$ and is not of type $L_3(4)$, $L_3(4)^{(1)}$, A_8, A_{10}, $\hat{A}_8$, $\hat{A}_{10}$, or J_2.

Proof: Let S be a Sylow 2-subgroup of G. If S is abelian, then as $m(S) \geq 3$ by Proposition 2.2 and the centralizer of every involution of G is solvable, $G \cong L_2(8)$ or $L_2(16)$ by the main result of [5] or [43], so G is not a counterexample to Theorem A.

If S is of type $L_3(4)$, A_8, $\hat{A}_8$, $\hat{A}_{10}$, or J_2, then $G \cong L_3(4)$, A_8, A_9, $PSp(4,3)$, M^c, M_{22}, M_{23}, $U_4(3)$, L, J_2, or J_3 by

[16, Theorem C], [17, Corollary A*], [18, Corollary A], [18, Theorem B], and [16, Theorem A]. However, each of these groups possesses a non-solvable 2-local subgroup, contrary to the fact that all 2-local subgroups of G are solvable. Furthermore, if S is of type A_{10}, in which case also $S \cong D_8 \wr Z_2$, then again G has a **nonsolvable** 2-local subgroup by [18, Proposition 6.1 (iv)], giving the same contradiction.

If $S \cong Z_2 \times Z_2 \times D_8$ or $D_8 \times D_8$, then G is not simple by the main result of [20]. If $S \cong Z_2 \times Z_2 \wr Z_2$, then G has a normal subgroup of index 2 by [27, Lemma 18] and again G is not simple. If $S \cong Z_4 \cdot E_{16}$, then S is of exponent 8 and a result of Fong [9] implies that G is not simple. Finally if S is of type $L_3(4)^{(1)}$, then S is isomorphic to the group T_2 of [18, Lemma 7.1 (viii)], as is easily checked. We can therefore repeat the argument of [18, Lemma 9.4] to obtain that G has a normal subgroup of index 2. We note that the proof of this lemma applies to an arbitrary group H with such a Sylow 2-subgroup T_2. Indeed, since $T_2/Z(T_2) \cong Z_2 \times Z_2 \wr Z_2$, $N_H(Z(T_2))$ contains a normal subgroup of index 2 with Sylow 2-subgroup of type $L_3(4)$ and hence the argument of the lemma does not require H to be embedded as a subgroup of a fusion-simple group with Sylow 2-subgroups of type $\hat{A}_{10}$. Thus in each case we reach a contradiction.

Remark. Ultimately we shall force a Sylow 2-subgroup of G to be of one of the above structures. Proposition 2.7 will then yield a final contradiction.

3. Odd order groups acting on 2-groups.

If H is a 2-local subgroup of G, $O(H) = 1$ and so any subgroup of H of odd order acts faithfully on $O_2(H)$. Moreover, $O_2(H)$ has sectional 2-rank at most 4. Because of this, our analysis of the 2-local subgroup of G will depend upon several general results concerning a group B of odd order which acts faithfully on a 2-group R of sectional 2-rank at most 4. Most of these results concern the case that B is elementary abelian of order 9, but our final result concerns the case $|B| = 5$.

Throughout R will denote a 2-group of sectional 2-rank at most 4 and B a group of odd order which acts faithfully on R. We begin with the following preliminary result.

Lemma 3.1. If $|B| = 3$, $[R,B] = R$, and $|R| \leq 16$, then $R \cong Z_2 \times Z_2$, $Z_4 \times Z_4$, E_{16}, or Q_8.

Proof: If $|R| \leq 8$, our lemma is clear; so we may assume $|R| = 16$. If R is abelian, we must have one of the first three possibilities for R. For both assertions, the condition $[R,B] = R$ must be invoked.

Suppose then that $R' \neq 1$ and set $\bar{R} = R/R'$. Since $[\bar{R},B] = \bar{R}$, the only possibility is that $\bar{R}$ is a four group, whence $|R'| = 4$ and $R' = \mho(R)$. If B centralizes R', then so does $[B,R] = R$ and consequently $R' \subseteq Z(R)$ in this case. Suppose B does not centralize R'. Since $|R'| = 4$ and $Z = R' \cap Z(R) \neq 1$, B does not centralize Z. Since B leaves Z invariant, this forces $Z = R'$ and so $R' \subseteq Z(R)$ in this case as well. Since R/R' is elementary, it also follows that R' must be elementary. But $R = \langle Z(R),x,y\rangle$ as R/R' is a four group and $R' \subseteq Z(R)$. Hence R' is generated by $[x,y]$ and so R' is cyclic. This contradiction establishes the lemma.

In Lemmas 3.2-3.8, we assume that B is elementary of order 9.

Lemma 3.2. If $|R| \leq 64$, then either $R \cong E_{16}$, $Z_2 \times Z_2 \times Z_4 \times Z_4$, $Z_2 \times Z_2 \times Q_8$, $Q_8 * Q_8$, $Q_8 \times Q_8$, or R is of type $L_3(4)$.

Proof: Since R admits B faithfully and $r(R) \leq 4$, clearly $R \cong E_{16}$ or $Z_2 \times Z_2 \times Z_4 \times Z_4$ if R is abelian, so we can assume that R is nonabelian. Setting $\bar{R} = R/\Phi(R)$, we have that $C_B(\bar{R}) = 1$ and so $\bar{R}$ is elementary of order at least 16. Since $r(R) \leq 4$, we must have $\bar{R} \cong E_{16}$. Since $1 \neq R' \subseteq \Phi(R)$, we have $|R| \geq 32$.

We have $\bar{R} = C_{\bar{R}}(b_1) \times C_{\bar{R}}(b_2)$ for suitable elements b_1, b_2 of $B^{\#}$, where $C_{\bar{R}}(b_i)$ is a four group with $[\bar{R}, b_j] = C_{\bar{R}}(b_i)$, $i \neq j$, $1 \leq i, j \leq 2$. Setting $R_i = [R, b_i]$, it follows that $R_i < R$, that R_i is B-invariant, and that $\bar{R}_j = C_{\bar{R}}(b_i)$, $i \neq j$, $1 \leq i, j \leq 2$. Thus R_1R_2 is a group and covers $\bar{R}$, so $R = R_1R_2$. Moreover, $|\Phi(R)| = 2$ or 4 according as $|R| = 32$ or 64 and so $|R_i| \leq 4\,|\Phi(R)| \leq 16$, $i = 1,2$. Since $R_i = [R_i, b]$, we conclude from the preceding lemma that $R_i \cong Z_2 \times Z_2$, Q_8, E_{16}, or $Z_4 \times Z_4$. In particular, if $R = R_1 \times R_2$, then using the fact that $|R| \leq 64$ and $r(R) \leq 4$, we obtain the lemma at once. Since $R_i \triangleleft R$, $i = 1,2$, we can therefore suppose that $R_o = R_1 \cap R_2 \neq 1$.

Consider first the case that $R_i \not\cong E_{16}$ for both $i = 1$ and 2. Since b_j centralizes $\bar{R}_i$ and $R_i \cong Z_2 \times Z_2$, Q_8, or $Z_4 \times Z_4$, it follows that b_j centralizes R_i for $i \neq j$, $1 \leq i, j \leq 2$, whence B centralizes $R_o \neq 1$. Thus b_i has a nontrivial fixed point on R_i and so $R_i \cong Q_8$, $i = 1,2$. Since b_j centralizes R_i, so does $R_j = [b_j, R_j]$, $i \neq j$, $1 \leq 1, j \leq 2$. Hence R_1 centralizes R_2 and we conclude that $R \cong Q_8 \times Q_8$ or $Q_8 * Q_8$.

Assume finally that, say $R_1 \cong E_{16}$, in which case $|R| = 64$ and $C_{R_1}(b_1) = 1$. Since R_0 is B-invariant, the last condition implies that R_0 is not cyclic. Since $\bar{R}_1 \cap \bar{R}_2 = 1$, $R_0 \subseteq \Phi(R)$, whence $R_0 \cong Z_2 \times Z_2$ and $R_0 = \Phi(R)$. Clearly R_0 is a minimal B-invariant normal subgroup of R and consequently $R_0 \subseteq Z(R)$. Thus R is of class 2. If $R_2 \cong Q_8$, then b_1 centralizes $R_2' \cong Z_2$ and $R_2' \subseteq R_0$, contrary to the fact that $C_{R_0}(b_1) = 1$. Hence $R_2 \cong Z_2 \times Z_2$, $Z_4 \times Z_4$ or E_{16}. In the first two cases b_1 centralizes R_2. But $R_0 \subseteq R_2$ and $C_{R_0}(b_1) = 1$, a contradiction. Therefore, also $R_2 \cong E_{16}$.

Our conditions imply that R_2 possesses a B-invariant four subgroup T_2 disjoint from R_1, so R splits over R_1. Moreover, if $b \in B^{\#}$ centralized R_1, then $[b,T_2] = T_2$ as b does not centralize R and consequently T_2 would centralize R_1. But then R would be abelian, which is not the case. Thus $C_B(R_1) = 1$ and so $BT_2 \cong Z_3 \times A_4$ can be identified with a subgroup of $\mathrm{Aut}(R_1) \cong GL(4,2) \cong A_8$. If $B_0 = C_B(T_2)$, then B_0 corresponds to a 3-cycle in A_8 since its centralizer contains a subgroup of the form $Z_3 \times A_4$. Hence $C_{A_8}(B_0) \cong Z_3 \times A_5$. Thus a Sylow 2-subgroup of $C_{A_8}(B_0)$ is determined up to conjugacy in the holomorph $GL(4,2) \cdot E_{16}$. On the other hand, a 2-group S of type $L_3(4)$ possesses an automorphism group of type (3,3) which leaves invariant the two elementary subgroups of S of order 16. We conclude therefore that R is of type $L_3(4)$ and the lemma is proved.

Now let B_i denote the distinct cyclic subgroups of B of order 3 and set $R_i = C_R(B_i)$, $1 \leq i \leq 4$. We fix this notation for the next five lemmas.

Lemma 3.3. If $R_i = 1$ and $R_j, R_k \cong 1$ or $Z_2 \times Z_2$ for suitable $i \neq j \neq k \neq i$, $1 \leq i, j, k \leq 4$, then either $R \cong Z_2 \times Z_2 \times Z_{2^n} \times Z_{2^n}$, $n \geq 1$, or R is of type $L_3(4)$.

Proof: We proceed by induction on $|R|$ and note that our conditions carry over to subgroups and homomorphic images of R on which B acts faithfully. For definiteness, assume $i = 2$, $j = 3$, $k = 4$. Since $R_2 = 1$, R admits a fixed-point-free automorphism of order 3. As is well known, this implies that R has class at most 2. If R is abelian, our conditions force R to satisfy the first alternative of the lemma as $r(R) \leq 4$. Hence we can suppose that R is of class 2.

Let W be a minimal B-invariant subgroup of $R' \cap Z(R)$, so that $W \cong Z_2$ or $Z_2 \times Z_2$. However, the first case cannot occur as $C_R(B_2) = 1$, so W is a four group. Setting $\bar{R} = R/W$, we conclude by induction that $\bar{R} \cong Z_2 \times Z_2 \times Z_{2^n} \times Z_{2^n}$, $n \geq 1$, or $\bar{R}$ is of type $L_3(4)$.

Consider the latter case first, in which case $\bar{R}' = Z(\bar{R}) \cong Z_2 \times Z_2$. Since $W \subset R'$ and R has class 2, it follows that $R' \subset Z(R)$, $|R'| = 16$, and R' covers $Z(\bar{R})$. On the other hand, as $\tilde{R} = R/R' \cong E_{16}$, we see that R' is elementary. Thus $R' \cong E_{16}$. Since $r(R) \leq 4$, this in turn implies that $R\text{-}R'$ contains no involutions. But by assumption, R_i is elementary, $i = 3,4$ and consequently $R_i \subset R'$, $i = 3,4$. Since $R_2 = 1$, this yields that $\tilde{R} = \tilde{R}_1$. Thus B_1 centralizes $\tilde{R} = R/R' = R/\Phi(R)$ and so B_1 centralizes R, which is not the case.

We conclude that $\bar{R} \cong Z_2 \times Z_2 \times Z_{2^n} \times Z_{2^n}$, $n \geq 1$. If $n = 1$, then $|R| = 64$ and it follows from the preceding lemma that R is of type $L_3(4)$. Hence we can suppose that $n \geq 2$. Again by our hypothesis on

R_2, R_3, R_4, we must have $\bar{R}_1 \cong Z_{2^n} \times Z_{2^n}$ and $\bar{R}_i \cong Z_2 \times Z_2$, $i = 3$ or 4, say $i = 3$. Let T_1 be the inverse image of $\bar{R}_1$ in R, so that T_1 is B-invariant, normal in R, and $R = T_1R_3$. Since $R_3 \cong Z_2 \times Z_2$ in this case, we also have that $T_1 \cap R_3 = 1$. Since $\bar{T}_1 = T_1/W = \bar{R}_1 \cong Z_{2^n} \times Z_{2^n}$ and $W \subset Z(T_1)$, we see that $|T_1'| = 1$ or 2. In either case, it follows that B centralizes T_1', whence $T_1' \subset R_2 = 1$. We conclude therefore that T_1 is abelian.

Finally by the structure of $\bar{T}_1$ and the fact that B acts on T_1, either $T_1 \cong Z_{2^{n+1}} \times Z_{2^{n+1}}$ or $T_1 = W \times U_1$, where U_1 is is B-invariant and $U_1 \cong Z_{2^n} \times Z_{2^n}$. In the first case, B_1 centralizes T_1 as it centralizes $T_1/W = \bar{R}_1$. But $[R_3, B_1] = R_3$, as $R_3^{B_1} = R_3$ and $[\bar{R}_3, B_1] = \bar{R}_3$. Since T_1 is normal in R, it follows that R_3 centralizes T_1. Thus $R = T_1 \times R_3 \cong Z_{2^{n+1}} \times Z_{2^{n+1}} \times Z_2 \times Z_2$ and so R satisfies the conclusion of the lemma.

Suppose then that $T_1 = W \times U_1$. Since U_1 is B-invariant and $\bar{U}_1 = \bar{R}_1$, we have that $U_1 \subset R_1$. But $V_1 = \mho^1(U_1) \lhd R$ as $T_1 \lhd R$ and $W \cong Z_2 \times Z_2$. Since $[R_3, B_1] = R_3$, R_3 centralizes V_1. Since $W \subset Z(R)$ and $T_1 \cap R_3 = 1$, it follows that $WV_1R_3 = W \times V_1 \times R_3$ and consequently $r(R) \geq 6$, which is not the case.

<u>Lemma 3.4</u>. If $R_i = 1$ for at least two values of i, $i \leq i \leq 4$, then R is abelian.

<u>Proof</u>: For definiteness, assume $R_1 = R_2 = 1$. In particular, we have $C_R(B) = 1$. We proceed by induction on $|R|$. We can assume R is not elementary abelian, otherwise the lemma holds trivially. Let Z be a minimal B-invariant subgroup of $Z(R) \cap \Phi(R)$. Then $Z \cong Z_2 \times Z_2$ as

as $C_R(B) = 1$. By induction, $\bar{R} = R/Z$ is abelian and so $\bar{R} \cong Z_{2^m} \times Z_{2^m} \times Z_{2^n} \times Z_{2^n}$. Since B acts faithfully on $\bar{R}$, $\bar{R}$ must have rank 4 and so $m \geq 1$, $n \geq 1$. Since $\bar{R}_1 = \bar{R}_2 = 1$, we also have $\bar{R} = \bar{R}_3 \times \bar{R}_4$ with $\bar{R}_3 = [\bar{R}, B_4] \cong Z_{2^m} \times Z_{2^m}$ $\bar{R}_4 = [\bar{R}, B_3] \cong Z_{2^n} \times Z_{2^n}$.

We claim that $R_3' = R_4' = 1$. Suppose, say, $R_3' \neq 1$. Since R_3' is B-invariant and $\bar{R}$ is abelian, the only possibility is that $Z = R_3'$. But as $Z \subseteq Z(R_3)$ and $\bar{R}_3$ is abelian of rank 2, it is clear that R_3' must be cyclic. This proves our assertion.

Without loss we may assume that $Z \subseteq R_3$. Since $C_R(B) = 1$, we have $Z \cap R_4 = 1$. Since $\bar{R}_3 \cap \bar{R}_4 = 1$, it follows that $R_3 \cap R_4 = 1$. In addition, we have that $R_3 \lhd R$ and $R = R_3R_4$. Since B_3 centralizes R_3, so does $[R, B_3]$. But $R_4 \subseteq [R, B_3]$ as $\bar{R}_4 = [\bar{R}, B_3]$. Hence R_4 centralizes R_3 and consequently $R = R_3 \times R_4$. Since R_3 and R_4 are abelian, so therefore is R.

We next prove

<u>Lemma 3.5</u>. If $\Phi(R) \cong E_{16}$ and B acts faithfully on $\Phi(R)$, then $R \cong Z_4 \times Z_4 \times Z_4 \times Z_4$.

<u>Proof</u>: Set $T = \Phi(R)$ and $\bar{R} = R/T$. Since B acts faithfully on R, it acts faithfully on $\bar{R}$, so $\bar{R} \cong E_{16}$ and consequently $\bar{R} = \bar{R}_i\bar{R}_j$ for suitable i,j, $i \neq j$, say $i = 1$, $j = 2$. Then $\bar{R}_k \cong Z_2 \times Z_2$ and $C_{\bar{R}_k}(B) = 1$, $k = 1,2$. Thus $R = TR_1R_2$ and $U_k = TR_k$ is invariant under B and of order 2^6, $k = 1,2$. Since B acts faithfully on $T \cong E_{16}$, we have $C_T(B) = 1$ and consequently $C_{U_k}(B) = 1$, $k = 1,2$. We conclude therefore from Lemma 3.2 that $U_k \cong Z_2 \times Z_2 \times Z_4 \times Z_4$ or U_k is of type $L_3(4)$, $k = 1,2$.

Consider first the case that U_1 and U_2 are both of type $L_3(4)$. Since $C_{\bar{U}_k}(B_k) = \bar{R}_k \cong Z_2 \times Z_2$, $k = 1,2$, it follows in this case that $C_T(B_k) = 1$ for $k = 1$ and 2. This implies that $T = R_3R_4$ with $R_3 \cong R_4 \cong Z_2 \times Z_2$. Since $|R| = 2^8$, this in turn implies that each $R_i \cong Z_2 \times Z_2$, $1 \leq i \leq 4$. Furthermore, either R_3 or $R_4 \subset Z(R)$, say $R_4 \subset Z(R)$. Setting $\tilde{R} = R/R_4$, we have that $|\tilde{R}| = 2^6$ and that $\tilde{R}_i \cong Z_2 \times Z_2$, $i = 1,2,3$. Applying Lemma 3.2 once again, we conclude that also $\tilde{R}$ is of type $L_3(4)$ with $\tilde{R}_3 = Z(\tilde{R})$.

Now choose x_1 in $R_1^{\#}$ and x_2 in $R_2^{\#}$. By the structure of $\tilde{R}$, we have $(\tilde{x}_1\tilde{x}_2)^2 \in \tilde{R}_3^{\#}$, whence $x_1x_2x_1x_2 = x_3x_4$, where $x_3 \in R_3^{\#}$ and $x_4 \in R_4$. Hence $x_2x_1x_2x_1 = (x_3x_4)^{x_1}$. But $x_2x_1x_2x_1 = (x_1x_2x_1x_2)^{-1}$ and consequently $(x_3x_4)^{x_1} = (x_3x_4)^{-1} = x_3x_4$, whence $x_3^{x_1} = x_3$. On the other hand, U_1 is of type $L_3(4)$ and x_1, x_3 are noncentral involutions of U_1. Moreover, $T = R_3R_4$ and R_1R_4 are the two elementary subgroups of U_1 of order 16, so x_1, x_3 lie in distinct such subgroups. However, any two such involutions of U_1 do not commute.

Next assume $U_1 \cong Z_2 \times Z_2 \times Z_4 \times Z_4$ and U_2 is of type $L_3(4)$. The first isomorphism imples that $R_1 \cong Z_4 \times Z_4$. Since $R_h \subset T$ for $h = 3,4$, it follows that $U_1 = R_h \times R_1$ for some $h = 3$ or 4, say $h = 3$. We see then that $T = R_3\Omega_1(R_1)$ and $U_2 = R_3\Omega_1(R_1)R_2$ with R_3 or $\Omega_1(R_1) \subseteq Z(U_2)$, but not both. On the other hand, as $U_1 \lhd R$, $\mho^1(U_1) = \Omega_1(R_1) \subseteq Z(R)$, so, in fact, $\Omega_1(R_1) = Z(U_2)$. Since R_2 and R_3 are not contained in $Z(U_2)$, this yields that $Z(R) = \Omega_1(R_1)$ or R_1.

Now we use the fact that $R_4 = 1$. Thus B_4 acts regularly on R and so R must be of class 2. Hence $R/Z(R)$ is abelian. If $Z(R) = R_1$, then clearly $R/Z(R)$ is elementary, so in either case $R' = \Omega_1(R_1)$. Now setting $\tilde{R} = R/\Omega_1(R_1)$, we have that $\tilde{R} = \tilde{R}_1\tilde{R}_2\tilde{R}_3$ with $\tilde{R}$ abelian and $\tilde{R}_i \cong Z_2 \times Z_2$. Thus $\tilde{R}$ is of rank 6, contrary to $r(R) \leq 4$.

There remains the case that $U_k \cong Z_2 \times Z_2 \times Z_4 \times Z_4$ for both $k = 1$ and 2. This time we obtain $R_k \cong Z_4 \times Z_4$, $k = 1,2$. Since $\bar{R} = R/T = \bar{R}_1\bar{R}_2$, we have $\bar{R}_2 = [\bar{R},B_1]$, so B_1 has no fixed points on $\bar{R}_2$. Since B_1 leaves R_2 invariant, it follows that $C_{R_2}(B_1) = 1$. Thus $R_1 \cap R_2 = 1$. Since $|R_k| = 2^4$ and $|R| = 2^8$, we conclude that $R = R_1R_2$. This in turn yields that $R_2 = [R,B_1]$ and consequently $R_2 \lhd R$. By symmetry, also $R_1 \lhd R$. Since $R_1 \cap R_2 = 1$, we conclude that $R = R_1 \times R_2 \cong Z_4 \times Z_4 \times Z_4 \times Z_4$ and the lemma is proved.

As a corollary, we prove

Lemma 3.6. If B acts faithfully on $\pi(R)$ and each $|R_i| \leq 2^5$, $1 \leq i \leq 4$, **then** $R \cong Z_4 \times Z_4 \times Z_4 \times Z_4$.

Proof: Set $T = \Phi(R)$. If $\Phi(T) = 1$, then T is elementary and the faithful action of B on T together with $r(R) \leq 4$ implies that $T \cong E_{16}$. Hence the desired conclusion follows from the preceding lemma. We can therefore suppose that $\Phi(T) \neq 1$.

We have that $\Phi(T) \cap Z(R) \neq 1$ as $\Phi(T) \lhd R$. Let W be a minimal B-invariant subgroup of $\Phi(T) \cap Z(R)$ and set $\bar{R} = R/W$. Then $\bar{T} = \Phi(\bar{R})$ and as $W \subseteq \Phi(T)$, B acts faithfully on $\bar{T}$. Moreover, B acts faithfully on $\bar{R}$. Each $|\bar{R}_i| \leq 2^5$, $1 \leq i \leq 4$, and $r(\bar{R}) \leq 4$. Hence all our conditions carry over to B and $\bar{R}$. We conclude therefore by induction on $|R|$ that $\bar{R} \cong Z_4 \times Z_4 \times Z_4 \times Z_4$. Without loss we can suppose that $\bar{R} = \bar{R}_1 \times \bar{R}_2$, where $\bar{R}_k \cong Z_4 \times Z_4$, $k = 1,2$.

By the minimality of W, we have that $W \cong Z_2$ or $Z_2 \times Z_2$. In particular, $R' \subseteq W \subseteq Z(R)$ and R' is elementary. It follows directly from this that $\mho^1(R) \subseteq Z(R)$. Consider first the case $W \cong Z_2$, whence B centralizes W. Thus $W \subseteq R_k$, $k = 1,2$. Since $\bar{R}_k \cong Z_4 \times Z_4$, with $[B,\bar{R}_k] = \bar{R}_k$, we see that $\mho^1(R_k)$ possesses a B-invariant four subgroup W_k with $[W_k,B] = W_k$, $k = 1,2$. Then each $W_k \subset Z(R)$ and now our conditions imply that $WW_1W_2 = W \times W_1 \times W_2 \cong E_{32}$, contrary to $r(R) \leq 4$.

On the other hand, if $W \cong Z_2 \times Z_2$, then $C_W(B) = 1$ by the minimality of W and so $W \subset R_i$ for some i. Since $|R_k| \leq 2^5$ by hypothesis and since $|\bar{R}_k| = 2^4$, $k = 1,2$, we have $i \neq 1$ or 2, so $i = 3$ or 4, say $i = 3$. Moreover, $R_k \cong \bar{R}_k$ and so $W_k = \mho^1(R_k)$ is a four group, $k = 1,2$. Thus $WW_1W_2 = R_3W_1W_2 = R_3 \times W_1 \times W_2 \cong E_{64}$, again contradicting $r(R) \leq 4$.

Before stating our next result, we introduce the following notation. We write $D^*_{2^{2n+1}}$ for the 2-group given by the generators a,b,c subject to the relations

$$a^{2^n} = b^{2^n} = c^2 = 1,\ [a,b] = 1,\ a^c = a^{-1},\ b^c = b^{-1},\ n \geq 2.$$

Thus $\langle a,b\rangle \cong Z_{2^n} \times Z_{2^n}$ and is inverted by the involution c.

<u>Lemma 3.7</u>. Suppose that for each i, $1 \leq i \leq 4$, one of the following holds:

(a) R_i is cyclic or generalized quaternion;

(b) R_i is elementary of order at most 8;

(c) $R_i \cong Z_4 \times Z_4$ or $Z_4 * Q_8$; or

(d) $R_i \cong D^*_{32}$.

Under these conditions, R has one of the following **structures**:

(i) $R \cong Z_{2^m} \times Z_{2^m} \times Z_{2^n} \times Z_{2^n}$, $m = 1$ or 2, $n = 1$ or 2;

(ii) $R \cong Z_2 \times Z_2 \times Q_8$;

(iii) $R \cong Q_8 * Q_8$;

(iv) R is of type $L_3(4)$;

(v) R is a nonabelian special 2-group of order 2^7 with $Z(R) \cong Z_2 \times Z_2 \times Z_2$, $R_i \cong Q_8$, and $[Z(R), B_i] \cong Z_2 \times Z_2$ for some i, $1 \leq i \leq 4$.

Proof: We first note that if $R_i \cong Z_{2^m}$ or Q_{2^n} for some i, then $m \leq 2$ and $n = 3$. Indeed, suppose false for some i, say $i = 1$. Since B leaves R_1 invariant, B then centralizes R_1 and consequently $R_1 \subset R_j$ for all j, $1 \leq j \leq 4$. But then by the assumed structure of R_j, we see that each R_j is also cyclic or generalized quaternion of order at least 16. Hence B centralizes each R_j and so centralizes R, which is not the case. In particular, we conclude from this result that $|R_i| \leq 2^5$ for all i, $1 \leq i \leq 4$, and so, in any event, $|R| \leq 2^{20}$.

If $|R| \leq 2^6$, Lemma 3.2 implies that either R has one of the asserted forms or $R \cong Q_8 \times Q_8$. However, in the latter case, some $R_i \cong Q_8 \times Z_2$, contrary to hypothesis. Hence we can suppose that $|R| \geq 2^7$. If B acts faithfully on $T = \mathfrak{F}(R)$, the lemma is a consequence of the preceding lemma inasmuch as $|R_i| \leq 2^5$ for all i, $1 \leq i \leq 4$. Hence we can also assume that B does not act faithfully on T. Thus B_i centralizers T for some i, say $i = 1$, whence $T \subset R_1$. In particular, $|T| \leq 2^5$. Furthermore, as usual, we have $R/T \cong E_{16}$ and consequenctly $|R| \leq 2^9$.

Consider first the case that $|T| = 2^5$, in which case $T = R_1$. By the first paragraph of the proof, the only possibility is that $T = D_{32}^*$. T possesses a unique subgroup $W \cong Z_4 \times Z_4$ and we have that $W \lhd R$ and B-invariant. Setting $\bar{R} = R/W$, we have that $|\bar{R}| = 32$, whence $\bar{R} \cong Z_2 \times Z_2 \times Q_8$ or $Q_8 * Q_8$ by Lemma 3.2. In either case, $\bar{R}_i \cong Q_8$ and $\bar{T} = Z(\bar{R}_i)$ for some $i = 2,3,4$. Considering the action of $\bar{R}_i$ on $W \cong Z_4 \times Z_4$ and using the structure of $\mathrm{Aut}(W)$, we see that $\bar{T}$ must centralize W. However, $\bar{T}$ inverts W as $T \cong D_{32}^*$. Hence $|T| \leq 2^4$.

Assume next that $|T| = 2^4$. Since B leaves invariant some four subgroup of R/T, there exists a B-invariant subgroup U of R of order 2^6 containing T. Since $|R_i| \leq 2^5$, no B_i centralizes U, $1 \leq i \leq 4$, and so B acts faithfully on U. Hence $U \cong Z_2 \times Z_2 \times Z_4 \times Z_4$, $Q_8 \times Q_8$, or U is of type $L_3(4)$ by Lemma 3.2. However, $T \subset U$, $T \subset R_1$, and $|T| = 2^4$. These conditions rule out the third possibility. In the second case $C_U(B_i) \cong Q_8 \times Z_2$ for some $i = 2,3,4$. Since $|R_i| \leq 2^5$, B_i acts regularly on R/U and so $R_i = C_U(B_i)$. However, by assumption, no R_i is of this form. Thus $U \cong Z_2 \times Z_2 \times Z_4 \times Z_4$. Since $T \subset R_1$ and $|T| = 2^4$, we have $T \cong Z_4 \times Z_4$. Now setting $V = \Omega_1(T)$ and $\bar{R} = R/V$, we have that $|\bar{R}| = 64$ and we conclude easily from Lemma 3.2 that $\bar{R}$ is of type $L_3(4)$.

On the other hand, B_1 acts regularly on R/T as $|R_1| \leq 2^5$ and so $[R,B_1]$ covers R/T. But $[R,B_1]$ centralizes T as B_1 does, so $R = [R,B_1]T$ centralizes T. Thus $T \subset Z(R)$. Since R/T is elementary, we conclude now that $R' \subset T$ and R' is elementary. Hence $R' = V$ and consequently $\bar{R} = R/V$ is abelian, which is a contradiction. Thus $|T| \leq 2^3$.

Since we are assuming that $|R| \geq 2^7$, it follows now that $|T| = 2^3$ and $|R| = 2^7$. We claim first that $T = R_1$. Indeed, if $T \subset R_1$, then $|R_1| = 2^5$, $R_1 \cong D_{32}^*$, and $R_1 \triangleleft R$. Let W be the subgroup of R_1 with $W \cong Z_4 \times Z_4$. As above, R/W is not isomorphic to Q_8. Hence $R/W \cong Z_2 \times Z_2 \times Z_2$. This implies that $T = \mho(R) \subset W$ and that B has a non-trivial fixed point on R/T as it has one on R/W. Since B is faithfully represented on R/T, it follows that R/T must have rank at least 5, which is a contradiction. Thus $T = R_1$, as asserted.

Setting $\bar{R} = R/T$ once again, we have $\bar{R} = \bar{R}_i \times \bar{R}_j$ for suitable $i,j = 2,3$, or 4 with $i \neq j$ and $\bar{R}_i = \bar{R}_j \cong Z_2 \times Z_2$; say $i = 2$, $j = 3$. Setting $U_k = TR_k$, $k = 2,3$, it follows again from Lemma 3.2 that $U_k \cong Z_2 \times Z_2 \times Q_8$ or $Q_8 * Q_8$, $k = 2,3$. Furthermore, if we set $W = C_R(B)$, we see that $|W| = 2$, $W \subset T$, and $W = U_k'$, $k = 2$ and 3. Since each $U_k \triangleleft R$, also $W \triangleleft R$. We set $\tilde{R} = R/W$, so that $|\tilde{R}| = 2^6$. Our conditions imply that $|R_i| = 8$ and $R_i \supset C_R(B)$, $i = 1,2,3$, whence $\tilde{R}_i \cong Z_2 \times Z_2$, $i = 1,2,3$. Another application of Lemma 3.2 now yields that R is of type $L_3(4)$. In addition, $\tilde{T} = \Phi(\tilde{R}) = \mho^1(\tilde{R})$ and this implies that T splits over W. Thus we have that $T \cong Z_2 \times Z_2 \times Z_2$. Since B_1 acts regularly on R/T and centralizes T, we conclude easily from the 3-subgroup lemma that $T \subset Z(R)$, whence R is of class 2. Since T is elementary and $T = \mho(R)$, we must have $T = R'$. Thus R is, in fact, a special 2-group. Furthermore, by the structure of U_2, we have $R_2 \cong Q_8$. Likewise $[T,B_2] = [Z(R), B_2] \cong Z_2 \times Z_2$. Hence R has the structure asserted in (v) and the proof is complete

In view of Lemmas 3.3, 3.4, and 3.7, it will be convenient to say that a 2-group R is of <u>restricted type</u> if R admits an elementary abelian subgroup B of order 9 acting faithfully on it and one of the following holds:

(a) $R \cong Z_{2^m} \times Z_{2^m} \times Z_{2^n} \times Z_{2^n}$, $m = 1,2$, $n \geq 1$;

(b) $R \cong Z_2 \times Z_2 \times Q_8$;

(c) $R \cong Q_8 * Q_8$;

(d) R is of type $L_3(4)$; or

(e) R is special of order 2^7 with $Z(R) \cong Z_2 \times Z_2 \times Z_2$ with $C_R(B_i) \cong Q_8$ and $[Z(R), B_i] \cong Z_2 \times Z_2$ for some i, $1 \leq i \leq 4$.

We need one final result:

Lemma 3.8. We have $R_i \not\cong Q_8 * Q_8$ for any i, $1 \leq i \leq 4$.

Proof. Suppose, say, that $R_1 \cong Q_8 * Q_8$. Since B acts faithfully on R, $R \supset R_1$ and so $T = N_R(R_1) \supset R_1$. Since B centralizes R_1, so therefore does $T_1 = [T, B_1]$, whence $T_1 \cap R_1 \subseteq Z(R_1)$. Thus $\bar{T} = T/Z(R_1) = \bar{R}_1 \times \bar{T}_1$ with $\bar{R}_1 \cong E_{16}$. However, $T = R_1T_1$ and so $\bar{T}_1 \neq 1$. Hence $r(T) \geq 5$, contrary to $r(R) \leq 4$.

We conclude with a result in the case $|B| = 5$.

Lemma 3.9. If $|B| = 5$ and $C_R(B) \neq 1$, then we have

(i) R is nonabelian special, B centralizes R', $[R,B] = R$, $|R'| \leq 2^4$, and $|R/R'| = 2^4$;

(ii) If $|R'| \leq 4$, then either $R \cong Q_8 * D_8$ or R is of type $U_3(4)$.

Proof: We proceed by induction on $|R|$ to establish (i). Setting $R_0 = C_R(B)$, we have $R_0 \neq 1$ by hypothesis. Let Z be a minimal B-invariant normal subgroup of R. Then $Z \subseteq Z(R)$, Z is elementary, and B acts irreducibly on Z. Hence either $|Z| = 2$ or 2^4. However, in the latter case, $Z \cap R_0 = 1$, whence $ZR_0 = Z \times R_0$ and $r(R) \geq 5$, which is not the case. Hence $|Z| = 2$ and $Z \subseteq R_0$.

Set $\bar{R} = R/Z$ and suppose first that $Z = R_0$. Then $C_{\bar{R}}(B) = 1$ and so if $\bar{Q}$

is a minimal B-invariant normal subgroup of $\bar{R}$, we have that $\bar{Q} \subset Z(\bar{R})$ and $\bar{Q}$ is elementary of order 2^4. If Q denotes the inverse image of $\bar{Q}$ in R, then Q is B-invariant of order 2^5. Since B acts irreducibly on $\bar{Q}$, Q would be elementary if it were abelian, contrary to $r(R) \leq 4$. Hence Q is extra-special. Since R induces trivial automorphisms of Q/Z, it follows from [13, Lemma 5.4.6] that $R = QT$, where $T = C_R(Q)$. Since $Q \cap T = Z$, we have $\bar{Q} \cap \bar{T} = 1$, whence $\overline{QT} = \bar{Q} \times \bar{T}$ as $\bar{Q} \subset Z(\bar{R})$. Since $r(R) \leq 4$, this forces $\bar{T} = 1$. Thus $R = Q$ and (i) holds.

We can therefore assume that $R_0 \supset Z$, in which case $\bar{R}_0 = C_{\bar{R}}(B) \neq 1$. Hence the lemma holds by induction for $\bar{R}$. In particular, $\bar{R} = [\bar{R},B]$, $\bar{R}_0 = \bar{R}' = Z(\bar{R})$, and $|\bar{R}/\bar{R}'| = 2^4$. Setting $R_1 = [R,B]$, we see that R_1 covers $\bar{R}$ and that either $R_1 = R$ or else $R = R_1Z = R_1 \times Z$. However, the latter case is excluded as $r(R) \leq 4$. Thus $R_1 = R$. Since $\bar{R}_0 \lhd \bar{R}$, we have that $R_0 \lhd R$, so $[R_0,R,B] = 1$. Likewise $[R_0,B,R] = 1$, so $[R,B,R_0] = [R,R_0] = 1$ by the three-subgroup lemma. Hence R_0 centralizes R and so $R_0 \subset Z(R)$. Since R/R_0 is elementary of order 2^4, we also have that R' is elementary. This in turn forces $R_0 = R' = Z(R)$, otherwise $r(R) \geq 5$. We conclude that R is nonabelian special. For the same reason, $|R'| \leq 2^4$ and so (i) holds in this case as well.

Suppose finally that $|R'| \leq 4$. Since $R' = R_0 \neq 1$, either $|R'| = 2$ or 4. In the first case, R is extra-special, so $R \cong Q_8 * D_8$ inasmuch as $Q_8 * Q_8$ does not possess an automorphism of order 5. Moreover, in the latter case, it follows from a result of Thompson [42, Lemma 5.26] that R is of type $U_3(4)$. Thus (ii) also holds.

4. The local subgroups of G.

Using our assumption that $r(G) \leq 4$, we shall now analyze the local subgroups of G. The analysis will depend on the following property of the subgroup structure of $GL(4,2) \cong A_8$, whose proof we leave to the reader.

Lemma 4.1. If X is a solvable subgroup of A_8 with $O_2(X) = 1$, then X is isomorphic to either a subgroup of a Sylow 3-normalizer, Sylow 5-normalizer, or Sylow 7-normalizer, which have the respective orders $3^2 \cdot 2^3$, $5 \cdot 3 \cdot 2^2$, and $7 \cdot 3$.

We note also that a Sylow 3-subgroup of A_8 is elementary of order 9 and a Sylow 7-normalizer of A_8 is nonabelian.

As a consequence, we have

Proposition 4.2. Every 2-local subgroup H of G has order $2^a h$, where $h = 1,3,5,7,9,15$, or 21. Moreover, H does not contain cyclic subgroups of order 9 or 21.

Proof: Set $Q = O_2(H)$. Since H is solvable with $O(H) = 1$, $\bar{H} = H/Q$ is faithfully represented on $\bar{Q} = Q/\Phi(Q)$ as a vector space over $GF(2)$. But $|\bar{Q}| \leq 16$ as $r(G) \leq 4$ and so $\bar{H}$ is isomorphic to a solvable subgroup of $GL(4,2) \cong A_8$. Now the proposition follows from the preceding lemma.

Combined with Proposition 2.5, this yields

Proposition 4.3. There exists a 2-local subgroup of G having elementary abelian Sylow 3-subgroups of order 9.

Since a Sylow 7-normalizer in A_8 is of odd order, Lemma 4.1 yields the following further restriction on the 2-local subgroups of G.

Lemma 4.4. If H is a 2-local subgroup of G whose order is divisible by 7, then $O_2(H)$ is a Sylow 2-subgroup of H.

Since the centralizer of a Sylow 5-subgroup of A_8 is of odd order, Lemma 4.1 similarly yields

Lemma 4.5. If H is a 2-local subgroup of G with nontrivial Sylow 5-subgroup P, then we have

(i) A Sylow 2-subgroup of $C_H(P)$ is contained in $O_2(H)$;

(ii) $O_2(H)P$ is normal in H.

We turn now to the p-local subgroups of G for p odd.

Lemma 4.6. If H is a nonsolvable p-local subgroup of G, then we have

(i) H contains a four subgroup T;

(ii) $p = 3, 5$, or 7;

(iii) If $p = 7$, then $|O_p(H)| = 7$;

(iv) If $p = 5$, then $|O_p(H)| \leq 5^3$;

(v) If $p = 3$ and $C_{O_p(H)}(T) \neq 1$, then $|O_p(H)| \leq 3^4$.

Proof: Let S be a Sylow 2-subgroup of H and set $P = O_p(H)$. Since H is nonsolvable, it does not have a normal 2-complement, so S is not cyclic. If S were generalized quaternion, the Brauer-Suzuki Theorem [7] would imply that $H = O(H)C_H(x)$ for any involution of x of H. Since $C_G(x)$ is solvable, so is H, which is not the case. We conclude that S contains a four subgroup T. Thus (i) holds.

If $t \in T^{\#}$, $C_P(t) \subseteq C_G(t)$, whence $|C_P(t)| = 1,3,5,7,9$ by Proposition 4.2. Since $P \neq 1$, $C_P(t) \neq 1$ for some t in $T^{\#}$ and so we must have $p = 3, 5$, or 7, proving (ii). Likewise, the Brauer-Wielandt formula implies that $|P| \leq p^3$ if $p = 5$ or 7 and, if $C_P(T) \neq 1$, that $|P| \leq 3^4$ if $p = 3$. In particular, (iv) and (v) hold.

Finally if $p = 7$ and $|P| > 7$, T does not centralize P, so $[C_P(t), T] \neq 1$ for some t in $T^{\#}$. However, $[C_P(t), T] \subseteq N = C_G(t)$ and so 7 divides $|N|$. But then $O_2(N)$ is a Sylow 2-subgroup of N by the preceding lemma and so $T \subseteq O_2(N)$, whence $[C_P(t),T] \subseteq P \cap O_2(N) = 1$. Thus T centralizers $C_P(t)$, which is not the case. Thus (iii) also holds.

As a consequence, we have

<u>Proposition 4.7.</u> The p-local subgroups of G are solvable for all $p \geq 7$.

<u>Proof:</u> For $p > 7$, this follows from Lemma 4.6 (ii). Suppose then that G possessed a nonsolvable 7-local subgroup H. By Lemma 4.6 (iii), $P = O_7(H)$ is of order 7. This implies that $C = C_H(P)$ is also nonsolvable. In particular, C does not have a normal 2-complement and so by Frobenius' theorem [13, Theorem 7.4.5], some 2-subgroup $T \neq 1$ of C is normalized, but not centralized by an element x of odd order. As P centralizes T and x, it follows that $x \notin P$, that $\langle P,x\rangle \subset N_G(T)$, and that $\langle P,x\rangle$ is abelian of odd order at least 21, contrary to Proposition 4.2.

Our next objective is to show that also all 5-local subgroups of G are solvable. We first prove

<u>Lemma 4.8.</u> If some 5-local subgroup of G is nonsolvable, then we have

(i) A Sylow 2-subgroup S of G has order at most 2^7;

(ii) For some involution z of $Z(S)$, a Sylow 5-subgroup P of $C_G(z)$ is nontrivial and a Sylow 2-subgroup of $N_G(P)$ is isomorphic to a subgroup of D_8.

Proof: Consider first the case that $N = N_G(P)$ is nonsolvable for some subgroup P of G of order 5. Then so is $C = C_G(P)$ and hence P centralizes an involution t of G. Choose an involution z of G such that $H = C_G(z)$ contains both P and a Sylow 2-subgroup R of $C_G(t)$, and such that a Sylow 2-subgroup S of H containing R has maximal order. Set $Q = O_2(H)$. Since $O(H) = 1$ and H is solvable, P acts faithfully on Q. Since $z \in C_Q(P)$, we have $C_Q(P) \neq 1$. Since $r(Q) \leq 4$, Lemma 3.9 (i) applies and yields that Q is nonabelian special with P centralizing Q' and $|Q/Q'| = 16$. If S^* is a Sylow 2-subgroup of G containing S, then $Z(S^*) \subset H$ and as $Q \subset S$ and H is 2-constrained with $O(H) = 1$, it follows that $Z(S^*) \subseteq Q'$. But then if we let z^* be an involution of $Z(S^*)$, it follows that $\langle P, S^* \rangle \subset C_G(z^*)$. Hence by our maximal choice of H, we have $S = S^*$. Thus S is a Sylow 2-subgroup of G and $z \in Z(S)$.

By Lemma 4.5 (i), $C_S(P) \subset Q$, whence $C_S(P) = Q'$ is elementary. Since $R \subset S$, it follows that $C_R(P)$ is also elementary. But t was an arbitrary involution of C and R an arbitrary Sylow 2-subgroup of $C_G(t)$. Hence if we choose t in the center of a Sylow 2-subgroup T of C and take R to contain T, then $T \subset C_R(P)$ and our argument shows that T is elementary abelian. Thus the structure of C is determined from [5] or [43] (or by induction). We conclude that either $\bar{C} = C/O(C)$ contains a normal subgroup of odd index isomorphic to $L_2(q)$, $q \equiv 3,5 \pmod 8$, or else $N_{\bar{C}}(\bar{T})$ contains a cyclic subgroup of order at least 7. Since T centralizes P, it would follow in the latter case that $N_G(T)$ contained a subgroup of odd order $d \geq 35$, contrary to Proposition 4.2. Thus the latter case cannot occur and we see therefore that T is a four group.

But $Q' \subset C$ as P centralizes Q'. Hence $|Q'| \leq 4$ and so $|Q| \leq 2^6$. Furthermore, since the centralizer of every involution of G is solvable, we conclude now from the structure of $P\Gamma L(2,q)$ for $q \equiv 3,5 \pmod 8$ that $N/O(C)$ contains a normal subgroup of odd index isomorphic to $L_2(q)$ or $PGL(2,q)$. Correspondingly a Sylow 2-subgroup of N is dihedral of order 4 or 8. Since $N_S(P) \subset N$, it also follows that $|N_S(P):Q'| \leq 2$. But $S = QN_S(P)$ by Lemma 4.5 (ii) and as $|Q| \leq 2^6$, we conclude that $|S| \leq 2^7$. All parts of the lemma are therefore proved in this case.

We can therefore suppose that $N_G(P)$ is solvable for every subgroup P of G of order 5. By assumption, $K = N_G(V)$ is nonsolvable for some 5-subgroup $V \neq 1$ of G. We can clearly suppose that $V = O_5(K)$. By Lemma 4.6 (iv), $|V| \leq 5^3$. Since $C_G(V) \subseteq C_G(P)$ for any subgroup P of V of order 5, we have that $C = C_G(V)$ is solvable and that $|V| > 5$. Proposition 4.2 implies that $|C|$ is odd. Setting $\bar{K} = K/CV$, we conclude that $\bar{K}$ is isomorphic to a subgroup of $GL(3,5)$. Since K satisfies the conclusion of Theorem A, $\bar{K}$ does not contain a subgroup isomorphic to $SL(3,5)$. We conclude at once that $\bar{K} \cong L_2(5)$ or $PGL(2,5)$ and that V is elementary of order 5^3.

Now the representation of $\bar{K}$ on V is 5-stable as $\bar{K}$ does not involve $SL(2,5)$. Hence no 5-element of $\bar{K}$ has a quadratic minimal polynomial on V. This implies that if V^* is a Sylow 5-subgroup of K, then V is the unique elementary subgroup of K of order 5^3. In particular, V is characteristic in V^*. This in turn yields that V^* is a Sylow 5-subgroup of G and that V is characteristic in any 5-subgroup of G in which it lies.

If t is an involution of K', we have $P = C_V(t)$ is of order 5. As in the first part of the proof, there exists an involution z which centralizes P such that $H = C_G(z)$ contains a Sylow 2-subgroup S of G. Again with $Q = O_2(H)$, we have that Q' centralizes P, and that $|Q/Q'| = 2^4$. We need only show that a Sylow 2-subgroup of $N = N_G(P)$ is isomorphic to a subgroup of D_8 to conclude, as in the preceding case, that $|S| \leq 2^7$ and thus establish the lemma.

Setting $N = N_G(P)$, we have that $V \subset N$ and that N is solvable. Since $|V| = 5^3$, the Frattini argument together with Proposition 4.2 implies that $O_{5'}(N)$ is of odd order. Since N is solvable, it is 5-stable and as V is characteristic in a Sylow 5-subgroup of N, we conclude at once that $V \subseteq O_{5',5}(N)$. Moreover, V is characteristic in a Sylow 5-subgroup of $O_{5',5}(N)$. Hence by the Frattini argument $K = N_G(V)$ contains a Sylow 2-subgroup T of N. Since $\bar{K} \cong PGL(2,5)$ or $L_2(5)$, T is thus isomorphic to a subgroup of D_8 and the lemma is proved.

Now we prove

<u>Proposition 4.9</u>. Every 5-local subgroup of G is solvable.

<u>Proof</u>: Suppose false. Then G possesses an involution z satisfying the conditions of the preceding lemma. We set $H = C_G(z)$, and let S, P be a Sylow 2-subgroup and 5-subgroup of H respectively. Then $|S| \leq 2^7$, $|P| = 5$, and a Sylow 2-subgroup of $N_G(P)$ is isomorphic to a subgroup of D_8. Moreover, $Q = Q_2(H)$ is nonabelian special with P centralizing Q' and $|Q/Q'| = 16$. But $C_G(P)$ contains no elementary subgroups of order 8 and so $|Q'| = 2$ or 4. We conclude therefore from Lemma 3.9 (ii) that $Q \cong D_8 * Q_8$ or that Q is of type $U_3(4)$. Moreover, by

Lemma 4.5 $S = QN_S(P)$ and $C_S(P) = Q'$. Since $N_S(P)/C_S(P)$ is cyclic of order 1,2, or 4 and since $N_S(P)$ is isomorphic to a subgroup of D_8, it follows in either case that $|S:Q| \leq 2$.

On the other hand, by Proposition 4.3, there exists a maximal 2-local subgroup M of G of order divisible by 9. Setting $R = O_2(M)$, we may assume that $R \subset S$. We claim first that $|R| \leq 2^6$. Indeed, either this is the case or $R = S$, $|S:Q| = 2$, Q is of type $U_3(4)$ and $N_S(P) \cong D_8$. However, in this case $|Z(S)| = 2$ and $Z(S) \subset Q'$. Moreover, we check that Q' is precisely the second center of S and that $Q = C_S(Q')$. Hence Q is characteristic in $S = R$ and so $M \subset N_G(Q)$. Maximality of M forces $M = N_G(Q)$ and as P normalizes Q, it follows that $P \subset M$. Thus $|M|$ is divisible by $9 \cdot 5$, contrary to Proposition 4.2. Thus $|R| \leq 2^6$, as asserted.

We can therefore invoke Lemma 3.3 to conclude that $R \cong Z_2 \times Z_2 \times Z_2 \times Z_2$, $Z_2 \times Z_2 \times Z_4 \times Z_4$, $Z_2 \times Z_2 \times Q_8$, $Q_8 * Q_8$, $Q_8 \times Q_8$, or that R is of type $L_3(4)$. Clearly Q contains no subgroup of any of these forms and consequently $S \supset Q$. Suppose first that $Q \cong Q_8 * D_8$. Then Q has 2-rank 2 and so S has 2-rank at most 3. This implies that $R \cong Z_2 \times Z_2 \times Q_8$, $Q_8 * Q_8$, or $Q_8 \times Q_8$. Furthermore, we can write $S = \langle Q,x \rangle$, where $x \in N_S(P)$. Since $\mathrm{Aut}(Q)/\mathrm{Inn}(Q) \cong S_5$, we know the action of x on Q and if we set $\bar{S} = S/Z(S)$, we have that $\bar{S} \cong Z_2 \times Z_2 \int Z_2$. In particular, $\bar{S}$ has no quaternion subgroups and $\bar{Q}$ is the unique elementary abelian subgroup of $\bar{S}$ of order 16. Since $Z(S) \subset Z(R)$, these conditions force $\bar{R}$ to lie in $\bar{Q}$, whence $R \subset Q$, which is clearly impossible.

Thus Q is a 2-group of type $U_3(4)$. We argue that S has 2-rank 2. Indeed, suppose S contains an elementary subgroup A of order 8. Since $|S:Q| \leq 2$ and all involutions of Q lie in Q', it follows that $Q' \subseteq A$, that $|S:Q| = 2$, and that $S = AQ$. Since Q' centralizes both Q and A, $Q' \subseteq Z(S)$. On the other hand, as $S \supset Q$ and Q' centralizes P, $Q' \subseteq N_S(P)$ and $N_S(P) \cong D_8$, so Q' is not contained in $Z(S)$, a contradiction. Hence S has 2-rank 2. But this contradicts Proposition 2.2.

Combining Propositions 4.7 and 4.9 with Proposition 2.4, we obtain

Proposition 4.10. G possesses a nonsolvable 3-local subgroup.

We shall now derive some further properties of the 3-local subgroups of G.

Lemma 4.11. If P is a subgroup of G of order 3 such that $N_G(P)$ is 3-constrained, then $N_G(P)$ is solvable.

Proof: Set $N = N_G(P)$ and $C = C_G(P)$, so that $|N/C| \leq 2$. Since N is 3-constrained, so also is C. Hence it will suffice to prove that C is solvable, so assume false.

Set $K = O_{3'}(C)$ and suppose first that K is nonsolvable. Since Theorem A holds for K and K is a 3'-group, the only possibility is that $K/O(K) \cong Sz(8)$. Hence a Sylow 2-subgroup R of K is normalized by a 7-element $x \neq 1$. Then $\langle P,x\rangle \subseteq N_G(R)$ and $\langle P,x\rangle$ is abelian of odd order at least 21, contrary to Proposition 4.2. Thus K is solvable.

This in turn implies that $|K|$ is odd. Indeed, otherwise a Sylow 2-subgroup R of K would be nontrivial and $C = KN_C(R)$ by the Frattini argument. Since $N_G(R)$ is solvable, C is thus solvable, contrary to assumption.

As in the proof of Lemma 4.6 (i), C contains a four subgroup T. Since $|K|$ is odd, it follows that T normalizes some Sylow 3-subgroup Q of $O_{3',3}(C)$. We have $P \subset Q$ and so $C_Q(T) \neq 1$. Since $T \subset N_G(Q)$ and $Q \subseteq O_3(N_G(Q))$, we conclude now from Lemma 4.6 (v) that $|Q| \leq 3^4$.

Finally set $\bar{C} = C/K$, so that $\bar{Q} = O_3(\bar{C})$ and $|\bar{Q}| \leq 3^4$. Since C is 3-constrained, we have $C_{\bar{C}}(\bar{Q}) \subseteq \bar{Q}$. But $\bar{C}$ is nonsolvable and $\bar{C}$ centralizes $\bar{P}$, which is of order 3 in $\bar{Q}$. The only possibility is then that $\bar{Q}/\bar{P}$ is elementary of order 3^3 and that $\tilde{C} = \bar{C}/\bar{Q} \cong SL(3,3)$. ($GL(3,3)$ is excluded as the centralizer of every involution of $\tilde{C}$ must be solvable.) But then $|C_{\tilde{C}}(\tilde{t})|$ is divisible by 3 for each $\tilde{t}$ in $\tilde{T}^{\#}$. Since also $|C_{\bar{Q}}(\bar{t})| = 9$ for all $\bar{t}$ in $\bar{T}^{\#}$, we conclude that $|C_C(t)|$ is divisible by 27 for all t in $T^{\#}$, contrary to Proposition 4.2.

Before stating our next lemma, we recall some well-known facts about $\mathrm{Aut}(L)$, where $L = L_2(q)$, $L_3(3)$, or $U_3(3)$. In the first case, $\mathrm{Aut}(L) = P\Gamma L(2,q)$. Moreover, if q is an odd square, then $P\Gamma L(2,q)$ possesses exactly three distinct subgroups which contain L as a subgroup of index 2. One of these is $PGL(2,q)$, which has dihedral Sylow 2-subgroups; a second is $PGL^*(2,q)$, which has quasi-dihedral Sylow 2-subgroups, and the third has the form $L\langle t\rangle$, where t is an involution whose action on L is induced from an automorphism of $GF(q)$. Moreover, if $q = 9$, $L\langle t\rangle \cong S_6$. We note also that all involutions of $PGL^*(2,q)$ lie in $L_2(q)$. On the other hand, if q is odd, but not a square, then $PGL(2,q)$ is the only subgroup of $P\Gamma L(2,q)$ which contains $L_2(q)$ as a subgroup of index 2.

Similarly if $L = U_3(3)$ or $L_3(3)$, then $\mathrm{Aut}(L)$ possesses exactly one subgroup which contains L as a subgroup of index a positive power of 2; namely, $P\Gamma U(3,3)$ and $L_3(3)\langle t\rangle$, where t is an involution induced from the transpose-inverse map. We shall denote this latter group by $L_3^*(3)$.

Lemma 4.12. If P is a subgroup of G of order 3 such that $N = N_G(P)$ is nonsolvable, then we have

(i) P is a Sylow 3-subgroup of $O(N)$;

(ii) $|N/C_G(P)| \leq 2$;

(iii) $N/O(N) \cong L_2(q)$, $PGL(2,q)$, $PGL^*(2,q)$, q odd, $q \geq 5$, $P\Gamma L(2,9)$, S_6, $L_3(3)$, $L_3^*(3)$, $U_3(3)$, $P\Gamma U(3,3)$, or M_{11}.

Proof: Set $C = C_G(P)$. Then $|N/C| \leq 2$, so (ii) holds. Moreover, C is nonsolvable and $O(C) = O(N)$. Again as in Lemma 4.6 (i), C contains a four subgroup T. We let Q be a T-invariant Sylow 3-subgroup of $O_{3',3}(O(N))$, so that $Q \supseteq P$ and $Q \subseteq C$. To establish (i), we need only show that $Q = P$, for then as $O(N)$ is 3-constrained, it will follow that P is a Sylow 3-subgroup of $O(N)$. Assume then that $Q \supset P$, whence $|Q| \geq 9$.

Since N is nonsolvable, N is not 3-constrained by the preceding lemma. Clearly $Q \subseteq O_{3',3}(N)$ and so it follows at once from the definition that $K = C_G(Q)$ is nonsolvable. Applying Frobenius' theorem once again, we see that K contains a 2-subgroup V which is normalized, but not centralized by an element x of odd order. Then $\langle Q,x\rangle \subseteq N_G(V)$ and $\langle Q,x\rangle$ has order $9d$, where d is odd and $d > 1$, contrary to Proposition 4.2. Thus (i) holds.

Now set $\bar{N} = N/O(N)$. Since G is simple, $N \subset G$ and so N satisfies the conclusion of Theorem A. Hence $\bar{N}$ contains a normal subgroup

$\bar{L} \cong L_2(q)$, q odd, $q \geq 5$, A_7, $L_3(3)$, $U_3(3)$, M_{11}, $L_2(8)$, $L_2(16)$, $Sz(8)$, or $U_3(4)$ such that $C_{\bar{N}}(\bar{L}) = 1$. Clearly $\bar{L} \subseteq \bar{C}$. In each of the last four cases, if V is a 2-subgroup of C whose image is a Sylow 2-subgroup of $\bar{L}$, it is immediate, as C centralizes P, that $N_C(V)$ contains an abelian subgroup of odd order at least 21, contrary to Proposition 4.2. Likewise if $\bar{L} \cong A_7$, $\bar{L}$ contains a four subgroup $\bar{V}$ such that $|N_{\bar{L}}(\bar{V})|$ is divisible by 9. But then if V is a four subgroup of C which maps on $\bar{V}$, $|N_C(V)|$ is divisible by 27, again contradicting Proposition 4.2. Similarly if $|\bar{N}/\bar{L}|$ is not a power of 2, it follows easily that $|N_{\bar{N}}(\bar{V})|$ is divisible by an odd integer $d \geq 9$, where $\bar{V}$ is a four subgroup of $\bar{L}$, and we reach the same contradiction as in the preceding case. We thus conclude from the known structure of $\mathrm{Aut}(\bar{L})$ that either $\bar{L} \cong L_2(q)$ or else $\bar{N} \cong L_3(3)$, $U_3(3)$, $L_3^*(3)$, $P\Gamma U(3,3)$, or M_{11}.

Hence it remains to treat the case that $\bar{L} \cong L_2(q)$. Since $|\bar{N}/\bar{L}|$ is a power of 2, it follows that either $\bar{N} \cong L_2(q)$, $PGL(2,q)$, $PGL^*(2,q)$, or else $\bar{N}-\bar{L}$ possesses an involution $\bar{t}$ whose action on $\bar{L}$ is induced from an automorphism of $GF(q)$. As is well-known, $C_{\bar{L}}(\bar{t}) \cong PGL(2,r)$, where $r^2 = q$. If $r > 3$, $C_{\bar{L}}(\bar{t})$ is nonsolvable and hence so is $C_G(t)$ for some involution t of N, a contradiction. Thus $q = 9$, and $\bar{L}\langle\bar{t}\rangle \cong S_6$. Hence either $\bar{N} = \bar{L}\langle\bar{t}\rangle \cong S_6$ or $\bar{N} \supset \bar{L}\langle\bar{t}\rangle$. In the latter case, $|\bar{N}/\bar{L}| = 4$ and $\bar{N} \cong \mathrm{Aut}(\bar{L}) \cong P\Gamma L(2,9)$. This establishes (iii) and completes the proof.

By Proposition 4.3, G possesses an elementary abelian subgroup of order 9. We let B be an arbitrary such subgroup of G.

<u>Lemma 4.13</u>. $N_G(B)$ is solvable.

Proof: Since $N_G(B)/C_G(B)$ is isomorphic to a subgroup of $GL(2,3)$, which is solvable, we need only prove that $C = C_G(B)$ is solvable. Suppose false. Then C does not have a normal 2-complement, so again by Frobenius' theorem, some 2-subgroup $T \neq 1$ of C is normalized, but not centralized by an element x of C of odd order. Then $\langle B,x\rangle \subset N_G(T)$ and $\langle B,x\rangle$ is abelian of order $9m$ for some $m > 1$, contrary to Proposition 4.2.

Lemma 4.14. If P is subgroup of B of order 3 such that $N_G(P)$ is nonsolvable and R is a B-invariant 2-subgroup of $N_G(P)$, then $R \cong Z_{2^n}$, $n \geq 0, Z_2 \times Z_2, Z_2 \times Z_2 \times Z_2, Z_4 \times Z_4, Q_8, Q_8 * Z_4, Q_8 * Q_8$, or D_{32}^*.

Proof: Setting $N = N_G(P)$, Lemma 4.12 implies that P is a Sylow 3-subgroup of $O(N)$ and that $\bar{N} = N/O(N) \cong L_2(q)$, $PGL(2,q)$, $PGL^*(2,q)$, q odd, $q \geq 5$, S_6, $P\Gamma L(2,9)$, $L_3(3)$, $L_3^*(3)$, M_{11}, $U_3(3)$, or $P\Gamma U(3,3)$. Then $|\bar{B}| = 3$, $\bar{R} \cong R$, and $\bar{R}$ is $\bar{B}$-invariant. We conclude readily now for each possible structure of $\bar{N}$ that $\bar{R}$ has one of the forms listed and the lemma follows.

Lemma 4.15. If B is a Sylow 3-subgroup of G, then B is inverted by a 2-element of G.

Proof: Since G is simple and B is elementary of order 9, it follows in this case that $B \subseteq N_G(B)'$. Since $\bar{N} = N_G(B)/C_G(B)$ is isomorphic to a 3'-subgroup of $GL(2,3)$, it is immediate that B is inverted by an involution of $\bar{N}$ and the lemma follows.

Lemma 4.16. If B is a Sylow 3-subgroup of G and N is a nonsolvable 3-local subgroup of G, then $N/O(N) \cong PGL(2,q)$ or $PGL^*(2,q)$, q odd, $q \geq 5$.

Proof: We may assume that $N = N_G(P)$ for some nontrivial subgroup P of B. Since $N_G(B)$ is solvable by Lemma 4.13, $|P| = 3$. Setting $C = C_G(P)$, it follows from the preceding lemma that $|N/C| = 2$. Thus $\bar{N} = N/O(N)$ has a normal subgroup of index 2. Moreover, $B \subseteq N$ and P is a Sylow 3-subgroup of $O(N)$ by Lemma 4.12. Hence $|\bar{B}| = 3$ and so a Sylow 3-subgroup of $\bar{N}$ has order 3. The possibilities for $\bar{N}$ are again given by Lemma 4.12. Our present conditions force $\bar{N} \cong PGL(2,q)$ or $PGL^*(2,q)$ for some odd $q \geq 5$.

5. The structure of $O_2(M)$.

By Proposition 4.3, there exists a maximal 2-local subgroup M of G such that a Sylow 3-subgroup B of M is elementary of order 9. There may, of course, exist more than one maximal 2-local subgroup of G containing B. Our aim in this section will be to show that M can always be chosen so that $O_2(M)$ is of restricted type.

We first let M be an arbitrary such maximal 2-local subgroup and set $R = O_2(M)$. We have $O(M) = 1$ by Proposition 2.3 and as M is solvable, $C_M(R) \subseteq R$. Note also that M is a $\{2,3\}$-group by Proposition 4.2. Furthermore, the maximality of M implies that $M = N_G(R)$ and that R is a maximal B-invariant 2-subgroup of G.

We let B_i denote the distinct subgroups of B of order 3 and we set $R_i = C_R(B)$, $1 \leq i \leq 4$. We fix this notation.

We shall first establish a useful transitivity theorem, analogous to [42, Lemma 6.2]. We begin with two preliminary lemmas.

Lemma 5.1. If H is a solvable subgroup of G containing B, then $O_{3'}(H)$ has a normal 2-complement.

Proof: Set $K = O_{3'}(H)$ and let T be a B-invariant Sylow 2-subgroup of $O_{2',2}(K)$. Since K is solvable, either $|K|$ is odd, in which case the lemma is clear, or $T \neq 1$. However, in the latter case, Proposition 4.2 implies that $N_G(T)$ is a $\{2,3\}$-group as $B \subseteq N_G(T)$. Hence $N_K(T)$ is a 2-group. Since $K = O_{2',2}(K)N_K(T)$ by the Frattini argument, it follows that $K = O_{2',2}(K)$. Thus K has a normal 2-complement in this case as well.

Lemma 5.2. If $C = C_G(B_i)$ is solvable for some i and B is a Sylow 3-subgroup of $C_G(B)$, then any B-invariant 2-subgroup of C lies in $O_{3'}(C)$.

Proof: We have $B \subseteq C$. Clearly the lemma holds if B is a Sylow 3-subgroup of C as C is solvable; so we may assume that this is not the case. If V is a Sylow 2-subgroup of $O_{3'}(C)$, it follows now by the Frattini argument that $|N_C(V)|$ is divisible by 27. Hence $V = 1$ by Proposition 4.2 and so $O_{3'}(C)$ is of odd order.

Set $\bar{C} = C/O_{3'}(C)$ and $\bar{Q} = O_3(\bar{C})$. We assume the lemma is false, in which case $\bar{B}$ normalizes some 2-group $\bar{T} \neq 1$ of $\bar{C}$. We choose such a $\bar{T}$ of least possible order. Since $\bar{B}_i \subseteq \bar{Q}$, $\bar{T}$ centralizes $\bar{B}_i$. Since $|\bar{B}/\bar{B}_i| = 3$, the minimality of $\bar{T}$ implies that either $|\bar{T}| = 2$ and $\bar{B}$ centralizes $\bar{T}$ or $\bar{T}$ is a four group and $[\bar{T},\bar{B}] = \bar{T}$.

By assumption, B is a Sylow 3-subgroup of $C_G(B)$ and consequently $C_{\bar{Q}}(\bar{B}) \subseteq \bar{B}$. But then if $\bar{B}$ centralizes $\bar{T}$, $\bar{T}$ centralizes $C_{\bar{Q}}(\bar{B})$ and so $\bar{T}$ centralizes $\bar{Q}$ by Thompson's $A \times B$-lemma, contrary to the fact that $C_{\bar{C}}(\bar{Q}) \subseteq \bar{Q}$ as $\bar{C}$ is solvable. Thus $\bar{T}$ is a four group and $[\bar{T},\bar{B}] = \bar{T}$. As usual, Lemma 4.6 now yields that $|\bar{Q}| \leq 3^4$, whence, in fact, $\bar{Q}/\bar{B}_i$ is elementary of order 3^3 and $|\bar{Q}| = 3^4$. This forces $\bar{Q}$ to be abelian and hence elementary abelian with $\bar{Q} = \bar{Q}_0 \times \bar{B}_i$, where $\bar{Q}_0 = [\bar{Q},\bar{T}]$ is of order 3^3. But $\bar{Q}_0$ is $\bar{B}$-invariant and consequently $C_{\bar{Q}_0}(\bar{B}) \neq 1$. Thus

$C_{\bar{Q}}(\bar{B}) \not\leq \bar{B}$, which is not the case.

This enables us to establish the desired transitivity theorem.

Lemma 5.3. Suppose $C_G(B_i)$ is solvable for some i and B is a Sylow 3-subgroup of $C_G(B)$. If T_1 and T_2 are two maximal B-invariant 2-subgroups of G such that $C_{T_j}(B_i) \neq 1$, $j = 1,2$, then $T_2 = T_1^x$ for some x in $O(C_G(B))$.

Proof: Suppose false and choose T_1, T_2 to violate the conclusion in such a way that $T_1 \cap T_2 = D$ has maximal order. Set $S_j = C_{T_j}(B_i)$, $j = 1,2$ and $C = C_G(B_i)$. Since C is solvable, the preceding lemma implies that $S_j \subseteq K = O_{3'}(C)$, $j = 1,2$. But K has a normal 2-complement by Lemma 5.1, again as C is solvable. Hence $\langle S_1, S_2^y \rangle$ is a 2-group for some y in $C_{O(K)}(B)$. But as $C_G(B) \subseteq C_G(B_i) = C$, we see that $C_{O(K)}(B) \subseteq O(C_G(B))$. We conclude now by a standard argument that we must have $D \neq 1$.

Setting $H = N_G(D)$, we have that H is a 2-local subgroup of G containing B. By Proposition 4.2, B is a Sylow 3-subgroup of H, so $N_{T_i}(D) \subseteq O_2(H)$, $i = 1,2$. Since $N_{T_i}(D) \supset D$ for $i = 1,2$, we reach a contradiction by a standard argument.

We shall now treat the case that B is not a Sylow 3-subgroup of G. We shall prove

Proposition 5.4. If B is not a Sylow 3-subgroup of G, then $R = O_2(M)$ is of restricted type.

We carry out the proof by contradiction. Thus we assume R is not of restricted type. We first prove

Lemma 5.5. For some i, R_i is not isomorphic to

Z_{2^n}, $n \geq 0$, $Z_2 \times Z_2$, $Z_2 \times Z_2 \times Z_2$, $Z_4 \times Z_4$, Q_{2^m}, $m \geq 3$, $Z_4 * Q_8$, or D^*_{32}.

Proof: In the contrary case, R would be of restricted type by Lemma 3.7.

For definiteness, we assume that Lemma 5.5 holds for $i = 1$.

On the other hand, we have

Lemma 5.6. If $N_G(B_i)$ is nonsolvable for some i, then $R_i \cong Z_{2^n}$, $n \geq 0$, $Z_2 \times Z_2$, $Z_2 \times Z_2 \times Z_2$, $Z_4 \times Z_4$, Q_8, $Q_8 * Z_4$, or D^*_{32}.

Proof: Since R_i is a B-invariant 2-subgroup of $N_G(B_i)$, this lemma follows from Lemmas 4.14 and 3.8.

Since R_1 is not of any of these forms, we have

Lemma 5.7. $N_G(B_1)$ is solvable.

We next prove

Lemma 5.8. B is a Sylow 3-subgroup of $C_G(B_1)$.

Proof: Set $C = C_G(B_1)$ and $K = O_{3'}(C)$. Then $BR_1 \subset C$ and C is solvable by the preceding lemma. Suppose B is not a Sylow 3-subgroup of C. Then if T is a Sylow 2-subgroup of K, $|N_C(T)|$ is divisible by 27, so $T = 1$ by Proposition 4.2. Thus $|K|$ is odd. Hence if we set $\bar{C} = C/K$, we have that $\bar{R}_1 \cong R_1$. Moreover, if $\bar{Q} = O_3(\bar{C})$, we have $C_{\bar{R}_1}(\bar{Q}) = 1$ as C is 3-constrained. Also $\bar{B}_1 \subseteq \bar{Q}$.

Since R_1 contains a four subgroup T_1, it follows now as in Lemma 4.6 that $|\bar{Q}| \leq 3^4$ and that $C_{\bar{Q}}(\bar{T}_1) = \bar{B}_1$, otherwise $|C_G(t_1)|$ would be divisible by 27 for some t_1 in $T_1^{\#}$, contrary to Proposition 4.2. Since $[\bar{T}_1, \bar{B}] \subseteq \bar{R}_1$, this implies that $\bar{B} \cap \bar{Q} = \bar{B}_1$. We conclude therefore that $\tilde{C} = \bar{C}/\bar{Q}$ is isomorphic to a subgroup of $GL(3,3)$ and that $|\tilde{B}| = 3$. Since $\tilde{R}_1$ is $\tilde{B}$-invariant and $\tilde{R}_1$ does not have one of the forms listed in Lemma 5.3, we check that the only possibility is that $\tilde{R}_1 \cong Q_8 \times Z_2$.

But then $\tilde{B}$ centralizes a four subgroup of $\tilde{R}_1$ and so B centralizes a four subgroup of R_1, which we can take as T_1. But T_1 contains an involution t_1 such that $C_{\bar{Q}}(\bar{t}_1) \supset \bar{B}_1$ and we conclude that $|C_G(t_1)|$ is divisible by 27, giving the same contradiction.

As a consequence, we have

Lemma 5.9. If $N_G(B_i)$ is **solvable** and $R_i \neq 1$, then B is a Sylow 3-subgroup of $C_G(B_i)$.

Proof: Suppose false, in which case $C_G(B_i)$ contains a 3-subgroup Q of order 3^3 with $B \lhd Q$. Choose y in $Q-B$. Then

$$R_i^y = C_R(B_i)^y = C_{R^y}(B_i) \neq 1.$$

Thus R^y is a B-invariant 2-subgroup of G with $C_{R^y}(B_i) \neq 1$. Moreover, $R_i = C_R(B_i) \neq 1$ and R is a maximal B-invariant 2-subgroup of G. Since B is a Sylow 3-subgroup of $C_G(B)$ by Lemma 5.8, we can therefore apply Lemma 5.3 to conclude that $R^{yx} = R$ for some x in $C_G(B)$. Then $z = yx \in M$ and z acts on B in the same way as y. But B is a Sylow 3-subgroup of $C_G(B)$. Since $y \in Q-B$ and Q is a 3-group, z thus induces an automorphism of B of order 3, which implies that B is not a Sylow 3-subgroup of M. However, this contradicts Proposition 4.2 and establishes the lemma.

We can now reach a final contradiction. Since B is not a Sylow 3-subgroup of G, we have $B \lhd Q$ for some 3-subgroup Q of G of order 3^3. By Lemma 5.8, $B \nsubseteq Z(Q)$ and so $Z(Q) = B_i$ for some i. Since $Q \subseteq C_G(B_i)$, we conclude from the preceding lemma that either $R_i = 1$ or $N_G(B_i)$ is nonsolvable. In particular, $i \neq 1$. Assume, for definiteness, that $i = 4$.

Then there exists y in $Q-B$ with y cyclically permuting B_1, B_2, B_3 under conjugation. Hence $C_G(B_i)$ is solvable for $i = 2,3$ as well as for $i = 1$. Suppose $R_i \neq 1$ for $i = 2$ or 3, say $i = 3$. Then

$$R_3^y = C_R(B_3)^y = C_{R^y}(B_1) \neq 1.$$

Applying Lemma 5.3 as in the preceding lemma, it follows that $R^{yx} = R$ for some x in $C_G(B)$. Then $z = yx \in M$ and we again conclude that B is not a Sylow 3-subgroup of M, a contradiction. Thus $R_2 = R_3 = 1$.

Lemma 3.4 now yields that R is abelian. Since clearly $C_R(B) = 1$, we have, in fact, $R = R_1 \times R_4$. If $R_4 = 1$, then B_1 centralizes $R = R_1$, contrary to the fact that B acts faithfully on R. Thus $R_4 \neq 1$ and so, by what we have shown above, $N_G(B_4)$ is nonsolvable. The possible structures of R_4 are now given by Lemma 5.6. Since $C_{R_4}(B) = 1$, the only possibilities are that $R_4 \cong Z_2 \times Z_2$ or $Z_4 \times Z_4$. Similarly $R_1 \neq 1$ and as $r(R) \leq 4$, the only possibilities for R_1 are $R_1 \cong Z_{2^n} \times Z_{2^n}$ for some $n \geq 1$. Thus R is, in fact, of restricted type. This establishes Proposition 5.4.

We now treat the case that B is a Sylow 3-subgroup of G. In this case, we prove.

<u>Proposition 5.10</u>. If B is a Sylow 3-subgroup of G, we can choose M so that $R = O_2(M)$ is of restricted type and $C_R(B) = 1$.

We carry out the proof in a sequence of lemmas. To begin with, we let M be an arbitrary maximal 2-local subgroup of G containing B. By Proposition 4.10, G possesses a nonsolvable 3-local subgroup. Moreover, $N_G(B)$ is solvable by Lemma 4.13. Since B is a Sylow 3-subgroup of G, we thus obtain the following additional result:

Lemma 5.11. $N_G(B_i)$ is nonsolvable for some i.

For definiteness, assume $N_G(B_4)$ is nonsolvable. We set $N = N_G(B_4)$ and fix this notation. We next prove

Lemma 5.12. We can choose M so that R_4 is a four group and $[R_4,B] = R_4$.

Proof: Since B is a Sylow 3-subgroup of G, $\bar{N} = N/O(N) \cong PGL(2,q)$ or $PGL^*(2,q)$, q odd, by Lemma 4.15. Moreover, B_4 is a Sylow 3-subgroup of $O(N)$ by Lemma 4.12. Hence $|\bar{B}| = 3$ and so $\bar{B}$ normalizes some four subgroup $\bar{R}_4^*$ of $\bar{N}$. Moreover, $\bar{R}_4^*$ is a maximal $\bar{B}$-invariant 2-subgroup of $\bar{N}$ by the structure of $PGL(2,q)$ and $PGL^*(2,q)$. Since $B_4 \lhd N$, it follows that N contains a B-invariant four subgroup R_4^* which maps on $\bar{R}_4^*$. Our conditions imply that $[R_4^*,B] = R_4^*$, that R_4^* centralizes B_4, and that R_4^* is a maximal B-invariant 2-subgroup of N.

Since $B \subseteq N_G(R_4^*)$, there exists a maximal 2-local subgroup M^* of G containing R_4^*B. We set $R^* = O_2(M^*)$. Since B is a Sylow 3-subgroup of M^*, which is solvable, and since R_4^* is B-invariant, we have $R_4^* \subseteq R^*$. Since $C_{R^*}(B_4) \subseteq N$, we conclude from the preceding paragraph that $R_4^* = C_{R^*}(B_4)$.

Thus replacing M by M^*, if necessary, we see that the conditions of this lemma will be satisfied.

Clearly $C_R(B) \subseteq R_4$. Since R_4 is a four group with $[R_4,B] = R_4$, it follows at once that $C_R(B) = 1$. Hence the proposition will hold if R is of restricted type; so we may assume the contrary. Arguing as in Lemmas 5.5 and 5.6, and again using Lemma 3.7, we conclude now that $N_G(B_i)$ is solvable and $R_i \neq 1$ for some i. We have $i \neq 4$ as $N = N_G(B_4)$ is nonsolvable. For definiteness, we assume $i = 1$.

We next prove

Lemma 5.13. The following conditions hold:

(i) A Sylow 2-subgroup of N is dihedral or quasi-dihedral;

(ii) If T is a Sylow 2-subgroup of $N_N(B)$, then $T/C_T(B)$ is of order 4.

Proof: The first statement is immediate from the fact that $\bar{N} = N/O(N) \cong PGL(2,q)$ or $PGL^*(2,q)$, q odd. Next let T be a Sylow 2-subgroup of $N_N(B)$. By Lemma 4.14, some 2-element t inverts B. Then t inverts B_4 and so $t \in N$, which implies that $T \not\subseteq C = C_G(B_4)$. Thus $T_1 = T \cap C$ is of index 2 in T. Set $T_0 = C_T(B)$. We have $\bar{C} \cong L_2(q)$, $|\bar{B}| = 3$, and $\bar{T}_1$ is a Sylow 2-subgroup of $N_{\bar{C}}(\bar{B})$. Since $\bar{B}$ is a Sylow 3-subgroup of $\bar{C}$, q is not a power of 3. It follows therefore from the structure of $L_2(q)$ that $N_{\bar{C}}(\bar{B})$ is a dihedral group. Hence $|\bar{T}_1/\bar{T}_0| = 2$ and consequently $|T/T_0| = 4$, so (ii) also holds.

We set $K = N_N(B)$ and let T be a Sylow 2-subgroup of K. We fix this notation as well. We first prove

Lemma 5.14. $|K \cap M|$ is not divisible by 4.

Proof: Let V be a Sylow 2-subgroup of $K \cap M$ and suppose by way of contradiction that $|V| \geq 4$. Since R_4 is B-invariant, B centralizes $R_4 \cap V$. But $C_R(B) = 1$, as we have noted above, and consequently $R_4 \cap V = 1$.

On the other hand, as $V \subseteq K \cap M \subseteq N \cap M$, V normalizes $N \cap R = R_4$ as $R \lhd M$. Thus $R_4 \lhd R_4V$ and $|R_4V| \geq 16$ as $|R_4| = 4$, $|V| \geq 4$, and $R_4 \cap V = 1$. Since R_4 is a four subgroup of R_4V and $R_4V \subseteq N$, R_4V is dihedral or quasi-dihedral by Lemma 5.13(i). However, this is impossible as a dihedral or quasi-dihedral group of order at least 16 does not possess a normal four subgroup.

This enables us to prove.

Lemma 5.15. T does not normalize B_1.

Proof: Suppose false. Then for any t in $T^{\#}$, we have

$$R_1^t = C_R(B_1)^t = C_{R^t}(B_1) \neq 1,$$

as $R_1 \neq 1$. Also $N_G(B_1)$ is solvable. Obviously B is a Sylow 3-subgroup of $C_G(B)$ in the present case. It follows therefore from Lemma 5.3 that $R^{ty_t} = R$ for some y_t in $O(C_G(B))$. Thus $ty_t \in M = N_G(R)$. Since $C_G(B) \subseteq N$ and $T \subseteq K = N_N(B)$, it also follows that each $ty_t \in K$. Set $U = \langle ty_t \mid t \in T^{\#}\rangle$, so that $U \subseteq K \cap M$. Since each $y_t \in O(C_G(B))$, we conclude at once that a Sylow 2-subgroup T^* of U is, in fact, a Sylow 2-subgroup of K. Hence $|T^*| \geq 4$ by Lemma 5.13(ii). Since $T^* \subseteq K \cap M$, the preceding lemma is thereby contradicted.

Since T does not normalize B_1, B_1 has exactly two conjugates in B under the action of T, say B_1 and B_2. (Since T leaves B_4 invariant, B_1 and B_4 are not conjugate under T.) As our final lemma we prove

Lemma 5.16. We have $R_2 = 1$ and either $R_3 = 1$ or R_3 is a four group.

Proof: Suppose $R_2 \neq 1$ and again let $t \in T^{\#}$. If t leaves B_1 invariant, it follows as in the preceding lemma that $ty_t \in M$ for some y_t in $O(C_G(B))$. In the contrary case, we have

$$R_2^t = C_R(B_2)^t = C_{R^t}(B_1) \neq 1.$$

Again Lemma 5.3 yields that $ty_t \in M$ for some y_t in $O(C_G(B))$ in this case as well. But now we reach a contradiction exactly as in the preceding lemma. We conclude that $R_2 = 1$.

Since T leaves B_4 invariant and permutes B_1, B_2, T must leave B_3 invariant. If $L = N_G(B_3)$ were solvable and $R_3 \neq 1$, we could argue on R_3, B_3 exactly as we did in Lemma 5.15 (using Lemma 5.3 for B_3) to obtain that $K \cap M$ contained a Sylow 2-subgroup of K, again contradicting Lemma 5.14. Hence either $R_3 = 1$ or L is nonsolvable. Thus if $R_3 \neq 1$, it follows that $R_3 \cong Z_{2^n}$, $n \geq 1$, or $Z_2 \times Z_2$ as R_3 is a B-invariant 2-subgroup of L and $L/O(L) \cong PGL(2,q)$ or $PGL^*(2,q)$, q odd, by Lemma 4.15. However, as $C_R(B) = 1$, we have $C_{R_3}(B) = 1$. The only possibility is then that $R_3 \cong Z_2 \times Z_2$. Thus R_3 has the asserted structure and the lemma is proved.

By Lemmas 5.12 and 5.16, we have that $R_2 = 1$, that $R_3 = 1$ or a four group, and that R_4 is a four group. Lemma 3.3 now yields that R is of restricted type. Since $C_R(B) = 1$, this establishes Proposition 5.10.

6. The case $C_R(B) \neq 1$.

The next step in the proof is to reduce to the case that $C_R(B) = 1$. We shall prove

Proposition 6.1. We can choose B, M, and R so that either $R \cong Z_{2^m} \times Z_{2^m} \times Z_{2^n} \times Z_{2^n}$, $m = 1,2$, $n \geq 1$, or R is of type $L_3(4)$.

If a Sylow 3-subgroup of G has order 9, the desired conclusion follows from Proposition 5.10; so we may assume this is not the case. We choose B, M, and R arbitrary with $R = O_2(M)$ of restricted type. We can suppose that R is not of one of the stated forms, in which case $R \cong Q_8 * Q_8$, $Z_2 \times Z_2 \times Q_8$, or R is special of order 2^7. We shall derive a contradiction in each case in a sequence of lemmas.

First of all, we have

Lemma 6.2. R is not isomorphic to $Q_8 * Q_8$.

Proof: This is an immediate consequence of Proposition 2.6.

Thus $R \cong Z_2 \times Z_2 \times Q_8$ or R is special of order 2^7. With the notation B_i, R_i as before, $1 \leq i \leq 4$, we clearly have in the first case $R_i \cong Q_8$ for some i and the same conclusion holds in the second case by Lemma 3.7 (v). For definiteness, we assume that $R_1 \cong Q_8$. We set $N = N_G(B_1)$ and $C = C_G(B_1)$ and fix this notation.

Lemma 6.3. If N is nonsolvable, then

(i) $C/O(C) \cong L_3(3)$ or M_{11};

(ii) $N_G(R_1) \subseteq M$.

Proof: In either case $R_1' = C_R(B)$ and $R_1' \subset Z(R)$. Since $M = RN_M(B)$ by the Frattini argument, it follows that $R_1' \triangleleft M$, whence $M = N_G(R_1')$ by the maximality of M. Since $N_G(R_1) \subset N_G(R_1')$ we obtain (ii).

Now set $\bar{N} = N/O(N)$, so that by Lemma 4.12, $\bar{N}' \cong L_2(q)$, q odd, $L_3(3)$, $U_3(3)$, or M_{11} with $\bar{B}$ of order 3 and $|\bar{N}: \bar{N}'| \leq 2^2$. Since $[\bar{R}_1, \bar{B}] \cong Q_8$ and lies in $\bar{N}'$, the first possibility is excluded. We also have that $O(C) = O(N)$ and $\bar{N}' \subseteq \bar{C}$. We conclude now from our conditions that either (i) holds or else $\bar{C} \cong L_3^*(3)$, $P\Gamma U(3,3)$, or $U_3(3)$. However, in each of these cases, $\bar{B}$ normalizes a subgroup $\bar{T}_1$ of $\bar{C}$ with $\bar{T}_1 \supset \bar{R}_1$ and $\bar{T}_1 \cong Q_8 * Z_4$. Since C centralizes B_1, it follows that B normalizes a subgroup T_1 of C with $T_1 \supset R_1$ and $T_1 \cong Q_8 * Z_4$. Then $T_1 \subset N_G(R_1) \subset M$, by (ii). However, as M is a $\{2,3\}$-group with Sylow 3-subgroup B and $N_{T_1}(R_1)$ is B-invariant, we have $T_1 \subseteq R = O_2(M)$. But

$R_1 = C_R(B_1)$, so $T_1 = R_1$, which is not the case.

We can now prove

Lemma 6.4. N is solvable.

Proof: Suppose false. By the structure of C given in the preceding lemma, a Sylow 2-subgroup T_1 of C is quasi-dihedral of order 16 and we can take T_1 to contain R_1. Then $T_1 \subseteq N_G(R_1) \subseteq M$. Moreover, $T_1 = R_1\langle z^*\rangle$ for some involution z^* which centralizes B_1.

We have $Q_1 = [Z(R), B_1] \cong Z_2 \times Z_2$ in either case, so that $Z(R) = Q_1 \times \langle z\rangle$, where $\langle z\rangle = C_R(B) = R_1'$ is normal in M. Since $T_1 \supset R_1$, we also have $z \neq z^*$. Now $z^* \in M$ and z^* centralizes B_1, so z^* acts on Q_1. Since $[Q_1, B_1] = Q_1$, Thompson's $A \times B$-lemma implies that z^* centralizes Q_1. Thus $Q_1 \subseteq M^* = C_G(z^*)$.

Now we use the fact that C has only one conjugacy class of involutions. Thus $z^* \sim z$ in C and so M^* and M have the same structures. Since $B_1Q_1 \subseteq M^*$ and $[B_1, Q_1] = Q_1$, we have that $Q_1 \subset R^* = O_2(M^*)$. For the same reason, either $Q_1 \subset Z(R^*)$ or $Q_1 \cap Z(R^*) = 1$. However, in the latter case, $Q_1Z(R^*) \cong E_{32}$, contrary to $r(R^*) \leq 4$. Thus, in fact, $Q_1 \subset Z(R^*)$.

Finally we have $Q_1 = [Z(R), B_1] = [Z(R), B]$ and as $M = RN_M(B)$, it follows that also $Q_1 \triangleleft M$. Maximality of M forces $M = N_G(Q_1)$. Similarly if B^* is a Sylow 3-subgroup of M^* containing B_1, we have $Q_1 = [Z(R^*), B_1] = [Z(R^*), B^*]$ and we conclude by the same argument that $M^* = N_G(Q_1)$. Thus $M = M^* = N_G(Q_1)$. Since $M = C_G(z)$ and $M^* = C_G(z^*)$, this forces $z = z^*$, which is not the case.

We next prove

Lemma 6.5. B is not a Sylow 3-subgroup of $C_G(B)$.

Proof: Since B is not a Sylow 3-subgroup of G, we have $B \lhd Q$ for some 3-subgroup Q of G of order 27. Choose y in $Q-B$. Then $B_i^y = B_1$ for some i, $1 \leq i \leq 4$. Since $C_R(B) \neq 1$, $R_i \neq 1$ and so

$$R_i^y = C_R(B_i)^y = C_{R^y}(B_1) \neq 1.$$

Likewise also $R_1 \neq 1$. Hence if B is a Sylow 3-subgroup of $C_G(B)$, we can apply Lemma 5.3 to obtain, as usual, that $R^{yx} = R$ for some x in $C_G(B)$. Thus $yx \in M$. Also y does not centralize B if B is a Sylow 3-subgroup of $C_G(B)$. It follows at once that $|B\langle yx\rangle|$ is divisible by 27; so $|M|$ is as well, contrary to Proposition 4.2.

In either case, $R_i = Z(R) \cong Z_2 \times Z_2 \times Z_2$ for some i, say $i = 2$. We set $K = C_G(B_2)$ and next prove

Lemma 6.6. K is nonsolvable.

Proof: Suppose that K is solvable. We have $R_2 \subseteq K$ and $R_2 \cong Z_2 \times Z_2 \times Z_2$. Since $C_G(B) \subset K$, the preceding lemma implies that $|K|$ is divisible by 27. As usual, the Frattini argument plus Proposition 4.2 yields now that $|O_{3'}(K)|$ is odd. Setting $\overline{K} = K/O_{3'}(K)$ and $\overline{P} = O_3(\overline{K})$, we have that $\overline{R}_2$ acts faithfully on $\overline{P}$ and $\overline{R}_2$ centralizes $\overline{B}_2 \subset \overline{P}$. Since $\overline{R}_2 \cong Z_2 \times Z_2 \times Z_2$, it follows at once that $|C_{\overline{P}}(\overline{t})| \geq 27$ for some $\overline{t}$ in $\overline{R}_2^{\#}$. Once again Proposition 4.2 yields a contradiction.

Lemma 6.7. $K/O(K) \cong P\Gamma L(2,9)$ or S_6.

Proof: We know that K is nonsolvable and that K contains an elementary subgroup of order 8. Applying Lemma 4.12, we see that either the lemma holds or else $\overline{K} = K/O(K) \cong PSL^*(3,3)$ or $P\Gamma U(3,3)$. In both cases $\overline{K}$ contains a subgroup $\overline{T} \cong Q_8 * Z_4$ which is invariant under a subgroup $\overline{B}^*$ of $\overline{K}$ of order 3. Hence there exist subgroups T, B^* of K

with $T \cong Q_8 * Z_4$, $B^* \cong Z_3 \times Z_3$, $B_2 \subset B^*$, and B^* normalizing T. In particular, $C_T(B^*) \cong Z_4$.

We let M^* be a maximal 2-local subgroup of G containing $N_G(T)$. Then $TB^* \subseteq M$ and as T is B^*-invariant, we have $T \subseteq R^* = O_2(M^*)$, as usual. On the other hand, as a Sylow 3-subgroup of G is not of order 9, R^* is of restricted type by Proposition 5.4. In particular, this implies that $|C_{R^*}(B^*)| \leq 2$. Since $C_T(B^*) \subseteq C_{R^*}(B^*)$ and $|C_T(B^*)| = 4$, this yields a contradiction and establishes the lemma.

It remains to eliminate this single configuration. We set $Z = \langle z \rangle = C_R(B)$, so that $Z = R_1'$ is of order 2 and $Z \subset K$. Since $C_G(B) \subseteq C_G(B_2) = K$ and B is not a Sylow 3-subgroup of $C_G(B)$, we easily obtain that a Sylow 3-subgroup F of K is abelian of order 27 with $F \cap O(K) = B_2$. Furthermore, by the structure of a Sylow 3-normalizer in A_6, F is elementary and can be chosen to contain B and to be Z-invariant. Moreover, $N_K(F)$ contains a Z-invariant cyclic subgroup Y of order 4 such that $YZ \cong D_8$. We let y be the involution of Y. We also set $H = N_G(F)$ and fix this notation for our final argument.

Lemma 6.8. We have B_2 normal in H.

Proof: By Proposition 4.2, $|C_H(F)|$ is odd. Hence setting $\bar{H} = H/C_H(F)$, we have that $\overline{YZ} \cong D_8$. Moreover, we can identify $\bar{H}$ with a subgroup of $GL(3,3)$. Since z centralizes B, $\bar{z} \notin SL(3,3)$. Since $SL(3,3)$ is minimal simple, it follows that either $\bar{H}$ is solvable or $\bar{H} \cong GL(3,3)$. However, in the latter case, $|C_{\bar{H}}(\bar{z})|$ is divisible by 3, whence $|C_G(z)|$ is divisible by 27, which is not the case. Thus $\bar{H}$ is solvable.

We shall argue that $\bar{y} \in Z(\bar{H})$. Since $C_F(y) = B_2$, this will clearly imply that $B_2 < H$.

Set $\bar{Q} = O_2(\bar{H})$. We first prove that $\bar{y} \in \bar{Q}$. Suppose that $O(\bar{H}) \neq 1$. Since 9 divides $|C_F(z)|$, $\bar{z}$ must invert $O(\bar{H})$. Since $\bar{z}$ is noncentral in $\bar{Y}\bar{Z} \cong D_8$, this forces $\bar{y}$ to centralize $O(\bar{H})$. Since $\bar{H}$ is solvable, we conclude that $\bar{y} \in \bar{Q}$. On the other hand, if $O(\bar{H}) = 1$, then $\bar{Q} \neq 1$ and $\bar{H}/\bar{Q}$ is either of odd order or isomorphic to S_3 as $\bar{H}$ is isomorphic to a solvable subgroup of $GL(3,3)$. Again we conclude that $\bar{y} \in \bar{Q}$.

If $\bar{Q}$ is not elementary, then $\langle\bar{y}\rangle = \Omega_1(\mho^1(\bar{Q}))$ as $Z_2 \times QD_{16}$ is a Sylow 2-subgroup of $GL(3,3)$. Hence $\bar{y} \in Z(\bar{H})$ in this case. Thus we can suppose that $\bar{Q}$ is elementary. If $\bar{H}_0$ denotes the subgroup of $\bar{H}$ which maps into $SL(3,3)$, then $|\bar{H} : \bar{H}_0| \leq 2$ and so $\bar{y} \in \bar{Q}_0 = \bar{Q} \cap \bar{H}_0$. If $\langle\bar{y}\rangle = \bar{Q}_0$, we are again done. Since QD_{16} is a Sylow 2-subgroup of $SL(3,3)$, the only other possibility is that $\bar{Q}_0$ is a four group. Since the normalizer in $GL(3,3)$ of a four subgroup of $SL(3,3)$ is isomorphic to $Z_2 \times S_4$, $\bar{H}$ is isomorphic to a subgroup of $Z_2 \times S_4$. We therefore conclude that either (a) $\bar{H} \cong S_4$ with $\bar{Q} \subset \bar{Y}\bar{Z}$, or (b) $\bar{H} \cong Z_2 \times S_4$.

Case (a) cannot arise here inasmuch as $\bar{Q}$ would then have no nontrivial fixed points on F, contrary to the fact that $\bar{Y}\bar{Z}$ centralizes B_2. Consider case (b). Then $Z(\bar{H}) \cong Z_2$ must invert F and consequently $K^* = N_G(B_2)$ contains $K = C_G(B_2)$ as a subgroup of index 2. But then Lemma 6.7 implies that $K^*/O(K^*) \cong P\Gamma L(2,9)$ and so contains a subgroup of index 2 isomorphic to $PGL(2,9)$. We conclude at once now that $N_{K^*}(F)$ contains a cyclic subgroup Y^* of order 8. But then $Y^* \subset H$ and $\bar{Y}^*$ is also cyclic of order 8, contrary to the fact that $\bar{H} \cong Z_2 \times S_4$.
Hence case (b) is also excluded and we conclude finally that $\bar{y} \in Z(\bar{H})$, as

asserted.

We need one additional fact.

Lemma 6.9. F is a Sylow 3-subgroup of $N_G(B)$.

Proof: We have $C_G(B) \subseteq K$ and by the structure of K, Z is a Sylow 2-subgroup of $C_K(B)$. Hence Z is a Sylow 2-subgroup of $C_G(B)$. Since $M = N_G(Z)$, it follows now by the Frattini argument that $N_G(B) = C_G(B)N_M(B)$. But F is a Sylow 3-subgroup of $C_G(B)$ as $C_G(B) \subseteq K$. Since B is a Sylow 3-subgroup of M, the lemma follows.

Now we can derive a final contradiction. We reexamine $C = C_G(B_1)$, which contains $\langle R_1, F\rangle$ and is solvable by Lemma 6.4. As usual, $|O_{3'}(C)|$ is odd by Proposition 4.2. We set $\bar{C} = C/O_{3'}(C)B_1$ and $\bar{P} = O_3(\bar{C})$. We have $|\bar{B}| = 3$ and $\bar{R}_1\bar{B} \cong SL(2,3)$, so $\bar{B} \not\subseteq \bar{P}$. Again $\bar{z}$ must invert $\bar{P}$ and so $\bar{P}$ is abelian. Moreover, $\bar{P} \neq 1$, as C is solvable. In particular, $\bar{P}_1 = C_{\bar{P}}(\bar{B}) \neq 1$. Since $\bar{P}_1$ is $\bar{F}$-invariant, there thus exists in C an F-invariant 3-subgroup $P_1 \supseteq B_1$ which maps on $\bar{P}_1$. But clearly P_1 normalizes B. Hence $P_1 \subseteq F$ by the preceding lemma. Thus $\bar{P}_1 \subseteq \bar{F}$. Since $|\bar{F}| = 9$ with $\bar{B} \subset \bar{F}$ and $\bar{B} \not\subseteq \bar{P}$, we conclude that $|\bar{P}_1| = 3$ and that $\bar{F} = \overline{BP_1}$.

Now consider the action of $\bar{R}_1\bar{B}$ on $\bar{Q} = \Omega_1(\bar{P})$. Since $\bar{B}$ and $C_{\bar{Q}}(\bar{B}) = \bar{P}_1$ have order 3, clearly $\bar{Q}$ is elementary of order at most 27. Since $\bar{R}_1\bar{B} \cong SL(2,3)$ and $\bar{z}$ inverts $\bar{Q}$, the only possibility is that $|\bar{Q}| = 9$. Hence $\bar{F}\bar{Q} = \bar{B}\bar{Q}$ is of order 27 and so $\bar{Q}$ normalizes $\bar{F}$. We can therefore find a 3-group Q in C which normalizes F and maps on $\bar{Q}$. Then $Q \subseteq H = N_G(F)$ and it follows now from Lemma 6.8 that Q centralizes B_2. Thus $\bar{Q}$ centralizes $\bar{B}_2 = \bar{B}$. Since $[\bar{R}_1, \bar{B}] = \bar{R}_1$, this forces $\bar{R}_1$ to centralizes $\bar{Q}$, contrary to the fact that $\bar{z} \in \bar{R}_1$ and $\bar{z}$

inverts $\bar{Q}$.

7. Proof of Theorem A.

By Proposition 6.1, we can choose B, M, and $R = O_2(M)$ so that R is of restricted type with $C_R(B) = 1$. It follows therefore from Lemma 3.7 that either

$R \cong Z_{2^m} \times Z_{2^m} \times Z_{2^n} \times Z_{2^n}$, $m = 1,2$, $n \geq 1$, or R is of type $L_3(4)$.

We shall now eliminate each of these possibilities. We first consider the elementary abelian case. However, for application in a subsequent part of our analysis we make a portion of our argument into a general lemma.

Let then X be a group which contains a subgroup $A \cong E_{16}$, set $C = C_G(A)$, and $N = N_G(A)$ and suppose the following conditions hold:

(a) $C = O(N)A$;

(b) N/C is isomorphic to a subgroup of a Sylow 3-normalizer in A_8 and has order divisible by 9;

(c) A Sylow 2-subgroup S of N has order at least 2^6;

(d) If **T** is a Sylow 2-subgroup of G containing S and U is a normal four subgroup of $N_T(S)$, then $C_X(u)$ is solvable for each u in $U^{\#}$.

Under these assumptions, we prove

Lemma 7.1. The following conditions hold:

(i) S is isomorphic to a subgroup of $D_8 \int Z_2$;

(ii) S is a Sylow 2-subgroup of X.

Proof. Let P be a Sylow 3-subgroup of N and let V be a Sylow 2-subgroup of $N_N(P)$. Since $PC/C \lhd N/C$, $N = CN_N(P)$ by the Frattini argument and so AV is a Sylow 2-subgroup of N. Without loss we can assume that $S = AV$. Since P normalizes A and $C_A(P) = 1$, we see that $A \cap V = 1$. But if $\overline{N} = N/O(N)$, $\overline{N}$ can be identified with a subgroup of the holomorph $\overline{Y}$ of $\overline{A}$ and we have $\overline{Y}/\overline{A} \cong GL(4,2) \cong A_8$. It follows from the structure of A_8 that $N_{\overline{Y}}(\overline{P}) \cong D_8$ and that $\overline{P}$ is a Sylow 3-subgroup of $\overline{Y}$. In particular, $\overline{P}$ is uniquely determined up to conjugacy in $\overline{Y}$ and so the structure of $\overline{A}N_{\overline{Y}}(\overline{P})$ is also uniquely determined. In fact, its Sylow 2-subgroups are of type A_{10}. We conclude therefore that $\overline{S} = \overline{AV}$ is isomorphic to a subgroup of a 2-group of type A_{10}. Since a 2-group of type A_{10} is isomorphic to $D_8 \int Z_2$, it follows that (i) holds.

We argue now that S is a Sylow 2-subgroup of X. We have that V is isomorphic to a subgroup of D_8. Since $|S| \geq 2^6$ by hypothesis, we check that either $S \cong D_8 \times D_8$, $D_8 \int Z_2$, S is of type A_8, or $V \cong Z_4$. In the latter two cases, A is clearly characteristic in S and consequently S is a Sylow 2-subgroup of X. We can therefore suppose that $S \cong D_8 \times D_8$ or $D_8 \int Z_2$. In particular, S contains a characteristic subgroup D of index at most 2 with $D \cong D_8 \times D_8$. We have $D = D_1 \times D_2$, where $D_i = \langle a_i, b_i \rangle$ with a_i, b_i involutions, $i = 1,2$. We set $\langle z_i \rangle = Z(D_i)$, $i = 1,2$. Since every elementary abelian normal subgroup of S of order 16 lies in D, we can suppose without loss that $A = \langle a_1, z_1 \rangle \times \langle a_2, z_2 \rangle$.

We assume, by way of contradiction, that S is not a Sylow 2-subgroup of G, in which case $N_G(S)$ contains a 2-element x with $x \notin S$ and $x^2 \in S$. Since D is characteristic in S, we have $D^x = D$. However, $A^x \neq A$, since otherwise $x \in N = N_G(A)$, contrary to the fact that $\langle S, x \rangle$ is a 2-group containing S properly and S is a Sylow 2-subgroup of N. We can assume the notation is so chosen that $A^x = \langle b_1, z_1 \rangle \times \langle c, z_2 \rangle$, where

$c = a_2$ or b_2.

Now set $H = C_G(z_2)$ and $\tilde{H} = H/\langle z_2\rangle$. Since $\langle z_1, z_2\rangle$ is normal in a Sylow 2-subgroup of $N_G(S)$, condition (d) implies that H is solvable. We have that $D_1 \subset H$ and that $\tilde{D}_1 \cong D_1 \cong D_8$. Since H is solvable, every chief factor of $\tilde{H}$ is elementary abelian. Hence $\tilde{H}$ possesses a normal subgroup $\tilde{K}$ such that $\tilde{D}_1 \cap \tilde{K} \neq 1$ and $\tilde{D}_1 \not\subseteq \tilde{K}$. Since $\tilde{D}_1$ is generated by its involutions, we conclude from this that the involutions of $\tilde{D}_1$ are not all conjugate in $\tilde{H}$. We shall now contradict this fact.

First of all, by the structure of $\bar{N} = N/O(N)$, we have that $\overline{DP} = \bar{F}_1 \times \bar{F}_2$, where $\bar{F}_i \cong S_4$, $i = 1,2$. Without loss we can assume that $\bar{D}_i$ is a Sylow 2-subgroup $\bar{F}_i$, $i = 1,2$. Thus $\langle \bar{a}_i, \bar{z}_i\rangle = \bar{A} \cap \bar{F}_i$, $i = 1,2$. Since $C_N(D_2)$ covers F_1 and $z_2 \in D$, it follows, in particular, that the involution of $\langle a_1, z_1\rangle$ are conjugate in $C_G(D_2)$ and hence in H. Therefore the involutions of $\langle \tilde{a}_1, \tilde{z}_1\rangle$ are conjugate in $\tilde{H}$.

Next set $E_i = D_i^x$, $i = 1,2$. Since D is characteristic in S, x normalizes D and so $D = E_1 \times E_2$. Note that $D/\langle z_1 z_2\rangle \cong D_8 * D_8$, a group which does not possess a direct factor isomorphic to D_8. This implies that $z_1 z_2 \notin E_i$, $i = 1$ or 2. Since $Z(D) = \langle z_1, z_2\rangle$ and $Z(D) \cap E_i \neq 1$, $i = 1,2$, it follows that $z_1 \in E_i$ for some i. For such a value of i, we see that E_i must have the form $\langle a_1 d, b_1 e\rangle$ for suitable d,e in $\langle z_2\rangle$. Since $A^x = \langle b_1, z_1, c, z_2\rangle$ with $c = a_2$ or b_2, we conclude that $A^x \cap E_1 = \langle b_1 e, z_1\rangle$. But now reasoning on $N^x/O(N^x)$ as we just done with $\bar{N}$ and using the fact that $z_2 \in E_2$, it follows that the involutions of $\langle b_1 e, z_1\rangle$ are all conjugate in $C_{N^x}(E_2)$ and hence in H. Since $e \in \langle z_2\rangle$, this yields that the involutions of $\langle \tilde{b}_1, \tilde{z}_1\rangle$ are all conjugate in $\tilde{H}$.

Thus all involutions of $\tilde{D}_1$ are conjugate in $\tilde{H}$, which is the desired contradiction.

With the aid of this last result, we next prove

Proposition 7.2. R is not elementary abelian of order 16.

Proof. Suppose false. We have that $O(M) = 1$, that $R = O_2(M) \cong E_{16}$, and that $B \cong Z_3 \times Z_3$. Moreover, M and the centralizer of every involution of G is solvable by the basic hypothesis of Theorem A. It is immediate therefore that conditions (a), (b), and (d) of the previous lemma hold with G, R, and M in the roles of X, A, and N respectively. Hence if a Sylow 2-subgroup S of M has order at least 2^6, then the preceding lemma yields that S is a Sylow 2-subgroup of G and is isomorphic to a subgroup of $D_8 \int Z_2$. Hence in this case $S \cong D_8 \times D_8$, $D_8 \int Z_2$, or $Z_4 \cdot E_{16}$, or S is of type A_8. In each case, Proposition 2.7 yields a contradiction.

We conclude therefore that $|S| \leq 2^5$. If $S = R$ or $S \cong Z_2 \times Z_2 \int Z_2$, then R is characteristic in S and as S is a Sylow 2-subgroup of $M = N_G(R)$, S is a Sylow 2-subgroup of G. Again Proposition 2.7 yields a contradiction. The only other possibility is that $|S| = 2^5$ and $S \cong Z_2 \times Z_2 \times D_8$, in which case $M \cong A_4 \times S_4$. We can write $S = D_1 \times D_2$, where $D_1 \cong Z_2 \times Z_2$ and $D_2 = \langle a_2, b_2 \rangle \cong D_8$ with $[D_1, B_i] = 1$ and $D_2B_i \cong S_4$ for some i, $1 \leq i \leq 4$. We also set $\langle z_2 \rangle = Z(D_2)$. Without loss we can assume that $R = D_1 \times \langle a_2, z_2 \rangle$ and that b_2 inverts B_i. However, S is not a Sylow 2-subgroup of G, otherwise G would not be simple by Proposition 2.7. Hence $N_G(S)$ contains a 2-element x with $x \notin S$ and $x^2 \in S$.

Since $Z(S) = D_1 \times \langle z_2 \rangle \cong E_8$, we have $|C_{Z(S)}(x)| \geq 4$ and consequently $C_{D_1}(x) \neq 1$. We let a_1 be an involution of $C_{D_1}(x)$ and we set $H = C_G(a_1)$. Then H is a 2-local subgroup of G containing SB_i. Since b_2 inverts B_i and D_1 centralizes B_i, it follows that no element of $D_1\langle b_2\rangle - D_1$ lies in $O_2(H)$. We shall now contradict this fact.

Since $x \notin S$ and $M = N_G(R)$, $R^x \neq R$. Since $R^x \subseteq S$, the only possibility is that $R^x = D_1 \times \langle z_2, b_2\rangle$. Now consider M^x which also has S as a Sylow 2-subgroup and contains B_i^x which is inverted by an involution of $S-R^x$. Moreover, B_i^x normalizes, but does not centralize a four subgroup U of R^x. Since $SB_i^x \cong Z_2 \times Z_2 \times S_4$, it follows that $\langle z_2\rangle = S' \subseteq U$. Furthermore, since $R^x = D_1 \times \langle z_2, b_2\rangle = O_2(SB^x)$ and since $b_2 \in R^x - Z(S)$, we see that, in fact, $U = \langle z_2, b_2d\rangle$ for some d in D_1.

But $B_i^x \subseteq H$ as x centralizes a_1. Since H is a 2-local subgroup of G, it follows now from Proposition 4.2 that every four subgroup of H which is normalized, but not centralized by B_i^x lies in $O_2(H)$. In particular, $U = \langle z_2, b_2d\rangle \subseteq O_2(H)$. Since b_2d is in $D_1\langle b_2\rangle - D_1$, we have the desired contradiction.

Proposition 7.3. R is not abelian.

Proof. Assume false, in which case $R \cong Z_{2^m} \times Z_{2^m} \times Z_{2^n} \times Z_{2^n}$, $m = 1$ or 2 and $n \geq 2$ by the preceding proposition. Let T be a Sylow 2-subgroup of $N_M(B)$ and set $S = RT$. Then S is a Sylow 2-subgroup of M and $R \cap T = 1$ (as $C_R(B) = 1$). If R is homocyclic, T is isomorphic to a subgroup of D_8. However, in the contrary case T does not contain a cyclic subgroup of order 4, otherwise B would be forced to centralize $\mho^{n-1}(R)$, which is a four group. Hence T is isomorphic to a subgroup of $Z_2 \times Z_2$ in this case. Furthermore, in either case, $T \neq 1$, since otherwise

$R = S$ would be a Sylow 2-subgroup of $M = N_G(R)$ and hence of G. Thus G would have abelian Sylow 2-subgroups, contrary to Proposition 2.7.

Observe now that the wreathed product $W_k = Z_{2^k} \int Z_2$ contains a subgroup of index 2 isomorphic to $Q_{2^{k+1}} * Z_{2^k}$ (cf. [2, Lemma 2.1.2]) and so W_k has sectional 2-rank 3. This means that S does not contain a subgroup isomorphic to $Z_2 \times Z_2 \times W_k$ for any $k \geq 2$ as $r(S) \leq 4$.

We first rule out the case in which $m = 2$ $(n \geq 2)$. If T contains an involution t with $R\langle t\rangle/\Phi(R) \cong Z_2 \times Z_2 \int Z_2$, then we can write $R = R_1 \times_\Phi R_2$ with $R_1\langle t\rangle \cong W_2$ and $R_2\langle t\rangle \cong W_n$. Setting $V = \langle t\rangle$, $Z_1 = \Omega_1(Z(R_1V))$, and $\overline{RV} = RV/Z_1$, we check that $\overline{C}_1 = C_{\overline{R}_1}(\overline{V}) \cong Z_2 \times Z_2$ and consequently $\overline{C}_1 \times \overline{R}_2\overline{V} \cong Z_2 \times Z_2 \times W_n$, contrary to $r(S) \leq 4$. Hence $R\langle t\rangle/\Phi(R) \cong Z_2 \times Z_2 \times D_8$ for any involution t of T. Now let $R = R_1 \times R_2$ be a B-invariant decomposition of R into B-invariant subgroups with $R_1 \cong Z_{2^m} \times Z_{2^m}$ and $R_2 \cong Z_{2^n} \times Z_{2^n}$. Because of the structure of $R\langle t\rangle/\Phi(R)$, t cannot interchange R_1 and R_2 and so t must normalize both R_1 and R_2. Our conditions imply that $R_i\langle t\rangle/\Phi(R_i) \cong E_8$ and $R_j\langle t\rangle \cong W_m$ or W_n, $j = 1,2$ $i \neq j$. But then $R\langle t\rangle/\Phi(R_i) \cong Z_2 \times Z_2 \times W_k$, where $k = m$ or n, giving the same contradiction.

Therefore $m = 1$. In particular, R is not homocyclic. Again let t be an involution of T and let $R = R_1 \times R_2$ be as above, so that $R_1 \cong Z_2 \times Z_2$. Clearly t must normalize both R_1 and R_2. If $R_1\langle t\rangle \cong E_8$, then $R_2\langle t\rangle \cong W_n$ and so $R\langle t\rangle \cong Z_2 \times Z_2 \times W_n$, giving the same contradiction. Hence $R_1\langle t\rangle \cong D_8$. Now as with t, T must also leave both R_1 and R_2 invariant, and T is elementary. If $|T| = 4$, we can find an involution t_1 in T such that $R_1\langle t_1\rangle \cong E_8$, contrary to what we have just shown. We conclude therefore that $T = \langle t \rangle \cong Z_2$.

Suppose next that t centralizes R_2, in which case $S \cong D_8 \times Z_{2^n} \times Z_{2^n}$. We shall prove that S is a Sylow 2-subgroup of G, which will contradict Proposition 2.7. Indeed, S possesses two subgroups isomorphic to R, each of which has $\mho^1(R) = \Phi(R)$ as its Frattini subgroup. Hence if $R^x \subset S$ for some x in G, we have $\Phi(R)^x = \Phi(R)$. But $M = N_G(\Phi(R))$ by the maximality of M and so $x \in M$. Hence $R^x = R$ and consequently R is weakly closed in S, which yields the desired conclusion that S is a Sylow 2-subgroup of G.

Therefore t does not centralize R_2. Clearly then R is the unique abelian subgroup of its order in $S = R\langle t \rangle$ and hence S is a Sylow 2-subgroup of G. We can write $R_i = \langle a_i, b_i \rangle$ with $a_i^t = b_i$, $i = 1,2$. Since G has no normal subgroups of index 2, the transfer lemma [27, Lemma 16] is applicable to ta_1 and the maximal subgroup R of S. We then conclude that either ta_1 is conjugate in G to an element y of R or else $(ta_1)^2 = a_1b_1$ is conjugate in G to an element s of $S - R$. In the first case, $(ta_1)^2 = a_1b_1$ is conjugate to $y^2 \in \langle a_2^2, b_2^2 \rangle$. However, as R is abelian and weakly closed in S, two elements of R conjugate in G are conjugate in $M = N_G(R)$. But $a_1b_1 \in R_1 \lhd M$, while $y^2 \in R_2$ and $R_1 \cap R_2 = 1$, so clearly these elements are not conjugate in M. Hence necessarily $a_1b_1 \sim s$ in G for some s in $S - R$.

As usual, we can find an element g in G such that

$$s^g = a_1b_1 \quad \text{and} \quad C_S(s)^g \subseteq S = C_S(a_1b_1).$$

On the other hand, s and t clearly induce the same action on R and consequently $C_{R_2}(s) \cong Z_{2^n}$. It follows now that the involution w of

$C_{R_2}(s)$ is mapped by g into R. Again by the weak closure of R in S, this in turn implies $w^g \in \Omega_1(R_2)$. Since $C_B(R_1) \cong Z_3$ and acts transitively on $\Omega_1(R_2)^{\#}$, we can therefore find an element b in $C_B(R_1)^{\#}$ such that $w^{gb} = w$. But then $s \sim a_1 b_1$ in $C_G(w)$. However, $C_B(w) \cong Z_3$ acts transitively on $R^{\#}$. We thus conclude that all involutions of $R\langle s\rangle = \langle a_1, b_1, s\rangle \cong D_8$ are conjugate in the group $C_G(w)$. But this is clearly impossible as $C_G(w)$ is solvable. This completes the proof of the proposition.

Thus R is necessarily of type $L_3(4)$. To treat this case we need some information about the automorphism group of a 2-group of this form.

<u>Lemma 7.4</u>. Let A be a 2-group of type $L_3(4)$, set $B = \mathrm{Aut}(A)$, $C = O_2(B)$, let D be a Sylow 3-subgroup of B, and let T be a Sylow 2-subgroup of $N_B(D)$. Then we have

(i) $|B| = 2^{10} \cdot 3^2$, $O(B) = 1$, $C \cong E_{2^8}$, and C is the stabilizer of the chain $A \supset Z(A) \supset 1$;

(ii) $B/C \cong N_B(D) \cong S_3 \times S_3$,

(iii) $C_B(D) = D$ and $T \cong Z_2 \times Z_2$;

(iv) The semi-direct product TA is of type $\hat{A}_{10}$;

(v) If t is an involution of T, then $\langle t\rangle A$ is of type $L_3(4)^{(1)}$, $\hat{A}_8$, or J_2.

<u>Proof</u>. A can be generated by involutions $z_1, z_2, a_1, a_2, b_1, b_2$ satisfying

$$[a_1, b_1] = [a_2, b_2] = z_1,\ [a_2, b_1] = z_2, \text{ and } [a_1, b_2] = z_1 z_2$$

with all remaining commutators of pairs of generators being trivial (cf. [16, Lemma 4.7]).

Clearly A is generated by a_1, a_2, b_1, b_2, $Z(A) = A' = \langle z_1, z_2\rangle \cong Z_2 \times Z_2$, and $A/Z(A) \cong E_{16}$. One checks directly that any mapping α of A into A determined by setting $a_i^{\alpha} = a_i d_i$ and $b_i^{\alpha} = b_i e_i$ for some d_i, e_i in $\langle z_1, z_2\rangle$, $i = 1,2$, is an automorphism of A of order at most 2 which stabilizes the chain $A \supset Z(A) \supset 1$. Moreover, any two such automorphisms centralize each other. Hence the set of all such automorphisms of A generate a subgroup C_0 of B with $C_0 \cong E_{2^8}$. Since C_0 stabilizes the chain $A \supset Z(A) \supset 1$, it follows that $C_0 \subseteq C = O_2(B)$. We note that as $Z(A) \cong Z_2 \times Z_2$ and $A/Z(A) \cong E_{16}$, C_0 includes every automorphism which stabilizes this chain. In particular, $C_0 \lhd B$. Furthermore, if $\beta \in B$ acts trivially on $A/Z(A)$, we conclude from the above relations that B also acts trivially on $Z(A)$, whence $\beta \in C_0$. Thus we also have that $\bar{B} = B/C_0$ is faithfully represented on $A/Z(A)$.

On the other hand, as in [16, Lemma 4.7], one sees that B is a $\{2,3\}$-group and $D \cong Z_3 \times Z_3$. In particular, our argument thus yields that $\bar{B}$ is isomorphic to a solvable subgroup of A_8. But by the structure of $\mathrm{Aut}(L_3(4))$ itself, we check that $N_{\bar{B}}(\bar{D})$ contains a subgroup isomorphic to $S_3 \times S_3$. Since $E_1 = \langle z_1, z_2, a_1, a_2\rangle$ and $E_2 = \langle z_1, z_2, b_1, b_2\rangle$ are the only elementary abelian subgroups of A of order 16, D leaves both of them invariant and hence so does $[Q,D]$, where $Q=O_2(B)$. Then our conditions force $[Q,D]$ to centralize $E_i/Z(A)$ for both i=1 and 2 and so also must centralize $A/Z(A)$. But this is a contradiction and so $\bar{D}$ centralizes $\bar{Q}$. Since no involution of A_8 centralizes a Sylow 3-subgroup of A_8, we conclude that $\bar{Q}=O_2(\bar{B})=1$ and hence that $\bar{D} \lhd \bar{B}$. But then also $C=C_0$ and C is the stabilizer of the chain $A \supset Z(A) \supset 1$.

Furthermore, we have that $\bar{T} \cong Z_2 \times Z_2$ or D_8 and that $C_{\bar{T}}(\bar{D}) = 1$. Hence either $C_B(D) = D$ or $C_C(D) \neq 1$. But C acts trivially on both

$Z(A)$ and $A/Z(A)$, so C leaves $E_1 = \langle z_1, z_2, a_1, a_2\rangle$ and $E_2 = \langle z_1, z_2, b_1, b_2\rangle$ invariant. We know that D leaves E_1 and E_2 invariant. But D acts faithfully on E_i by [16, Lemma 4.7(iii), (iv)]. Hence if we set $T_0 = C_C(D) = T \cap C$, it follows from Thompson's $A \times B$ - lemma that D acts faithfully on $C_{E_i}(T_0)$, $i = 1,2$. However, as $D \cong Z_3 \times Z_3$ and $E_i \cong E_{16}$, this forces $E_i = C_{E_i}(T_0)$, $i = 1,2$. Thus T_0 centralizes E_1E_2. But $A = E_1E_2$, forcing $T_0 = 1$, and we conclude that $C_B(D) = D$.

In particular, $T \cong \bar{T} \cong Z_2 \times Z_2$ or D_8. Suppose $T \cong D_8$. Then $T_1 = C_T(Z(A))$ has order at least 4. But then by the structure of $N_B(D)$, we have $[T_1, D] = D$ and consequently D also centralizes $Z(A)$. However, this contradicts [16, Lemma 4.7 (iv)]. Therefore $T \cong Z_2 \times Z_2$ and so $N_B(D) \cong B/C \cong S_3 \times S_3$. Thus (i), (ii), and (iii) hold.

The proof of [18, Proposition 9.2] show that there exists a nontrivial split extension of a 2-group of type $L_3(4)$ by $S_3 \times S_3$ having Sylow 2-subgroups of type $\hat{A}_{10}$. However, as D is a Sylow 2-subgroup of $B = \mathrm{Aut}(A)$ and T is a Sylow 2-subgroup of $N_B(D)$, the preceding analysis shows that any two nontrivial split extensions of a 2-group of type $L_3(4)$ by $S_3 \times S_3$ are necessarily isomorphic. Hence TA is of type $\hat{A}_{10}$, proving (iv), the structure of such a 2-group is described in [18, Lemma 7.1] and now (v) follows at once from [18, Lemma 7.1 (viii)].

We can now prove

Proposition 7.5. A Sylow 2-subgroup S of G is of type $L_3(4)$, $L_3(4)^{(1)}$, $\hat{A}_8$, $\hat{A}_{10}$, or J_2.

Proof. Again let T be a Sylow 2-subgroup of $N_M(B)$ and set $S = TR$, so that S is a Sylow 2-subgroup of M. Since R is of type $L_3(4)$ and $B \cong Z_3 \times Z_3$ acts faithfully on R, we can apply the preceding lemma to determine the possible structures of S. If $T \neq 1$, Lemma 7.4 (iv) and (v) yields that S is of type $L_3(4)^{(1)}$, $\hat{A}_8$, $\hat{A}_{10}$, or J_2. However, if $T = 1$, then $S = R$ is of type $L_3(4)$. On the other hand, in each case it is easily verified that R is the unique subgroup of type $L_3(4)$ in S. Hence R is characteristic in S and consequently S is a Sylow 2-subgroup of G, thus establishing the proposition.

Since Proposition 7.5 is in contradiction to Proposition 2.7, we conclude that no counterexample to Theorem A exists and the proof of the theorem is complete.

PART II

2-CONSTRAINED 2-LOCAL SUBGROUPS

1. Introduction. In Part I, we have determined all simple groups G of sectional 2-rank at most 4 with solvable 2-local subgroups. With the aid of that result we shall here determine the possibilities for G under the assumption that its 2-local subgroups are 2-constrained.

Our principal result is as follows:

Theorem A. If G is a simple group of sectional 2-rank at most 4 in which every 2-local subgroup is 2-constrained, then G is isomorphic to one of the following groups:

I. $L_2(q)$, q odd, $L_3(3)$, $U_3(3)$, $U_4(3)$, $PSp(4,3)$, or $G_2(3)$;

II. $L_2(8)$, $L_2(16)$, $L_3(4)$, $U_3(4)$, or $Sz(8)$;

III. A_7 or A_8;

IV. M_{11}, M_{22}, M_{23}, J_3 .

In view of the main result of Part I, we can assume in proving Theorem A that G possesses a nonsolvable 2-constrained 2-local subgroup M. In contrast to the analysis of Part I, our argument in this case does not require any induction assumption. This occurs because we are able to determine the structure of $M/O(M)$ as a purely local problem, independent of the embedding of M in G. Thus approximately half of the paper is taken up with the proof of the following theorem:

Theorem B. If X is a nonsolvable 2-constrained group of sectional 2-rank at most 4 with $O(X) = 1$, then one of the following holds:

I. (a) $X/O_2(X) \cong GL(3,2)$;

(b) $O_2(X) \cong E_8$, E_{16}, or $Z_4 \times E_8$; or

II. (a) $X/O_2(X) \cong A_5$, A_6, A_7, S_5, or $X/O_2(X)$ contains a subgroup of index at most 2 isomorphic to $Z_3 \times A_5$;

(b) $O_2(X) \cong E_{16}$ or $Q_8 * D_8$.

We wish to thank J.L. Alperin who made the important observation that a group X with $O_2(X) \cong Z_{2^n} \times Z_{2^n} \times Z_{2^n}$, $n \geq 2$, and $X/O_2(X) \cong GL(3,2)$ has sectional 2-rank 5.

Theorem B gives rather precise information concerning the structure of $M/O(M)$. Using this knowledge, we are able to pin down the structure of a Sylow 2-subgroup of our group G of Theorem A by a careful analysis of the fusion of involutions. Once this is accomplished, known classification theorems will again enable us to complete the proof of Theorem A. In fact, we shall be able to treat a slightly more general situation which will be important for our subsequent work in the non 2-constrained case.

We shall prove

Theorem C. If G is a simple group of sectional 2-rank at most 4 in which the centralizer of every central involution is solvable and some

2-local subgroup of G is nonsolvable and 2-constrained, then G is isomorphic to $L_4(3)$, $U_4(3)$, $PSp(4,3)$, $G_2(3)$, $L_3(4)$, A_8, A_9, A_{10}, A_{11}, or M_{22}.

Theorem D. If G is a simple group of sectional 2-rank at most 4 in which the centralizer of some involution is nonsolvable and 2-constrained, then G is isomorphic to J_2, J_3, or M_{23}.

Clearly if G is a simple group of sectional 2-rank at most 4 in which all 2-local subgroups are 2-constrained and some 2-local subgroup is nonsolvable, then the hypotheses of either Theorem C or D holds in G and so G satisfies the conclusion of Theorem A.

In conclusion we remark that recently Sehgal [39] has classified all simple groups G in which $SCN_3(2)$ is empty and $C_G(t)$ is 2-constrained for every involution t of G for which a Sylow 2-subgroup of $C_G(t)$ has index at most 2 in a Sylow 2-subgroup of G (generalizing the corresponding result of Janko-Thompson [34] for the case that $C_G(t)$ is solvable). Since such a group has sectional 2-rank at most 4, we see that effectively our Theorem D includes Sehgal's result as a special case.

Our notation follows that of Part I.

2. The automorphism groups of certain 2-groups. We collect here a number of properties of the automorphism groups of certain specific 2-groups, which we require for the proofs of Theorems B, C, and D.

Lemma 2.1. Let $A \cong Z_4 \times Z_4$, set $B = \mathrm{Aut}(A)$, $C = O_2(B)$, and let D,E denote respectively a Sylow 2-subgroup and a Sylow 3-subgroup of B. Then we have

(i) $|B| = 2^5 \cdot 3$ and $O(B) = 1$;

(ii) $C \cong E_{16}$ and C is the stabilizer of the chain $A \supset \Phi(A) \supset 1$;

(iii) $D \cong Z_2 \times Z_2 \int Z_2$;

(iv) $C_C(E)$ is a four group and $N_B(E)$ has dihedral Sylow 2-subgroups of order 8;

(v) A Sylow 2-subgroup S of the semi-direct product $N_B(E)A$ is given by generators $\alpha,\beta,\delta,\epsilon$ satisfying the following conditions:

(a) $\alpha^4 = \beta^4 = \delta^2 = \epsilon^2 = 1$ with $A = \langle\alpha,\beta\rangle \cong Z_4 \times Z_4$, $\langle\delta,\epsilon\rangle \cong D_8$, $\gamma = (\delta\epsilon)^2$ inverting A, ϵ interchanging α and β, and $\alpha^\delta = \alpha^{-1}\beta^2, \beta^\delta = \beta\alpha^2$;

(b) $\langle \gamma,\delta \rangle = C_C(E)$;

(c) $\langle \delta,A \rangle \cong \langle \delta\gamma,A \rangle$ and $\langle \epsilon,A \rangle \cong \langle \epsilon\gamma,A \rangle \cong Z_4 \int Z_2$;

(d) $\langle \gamma,\epsilon,E \rangle \cong Z_2 \times S_3$;

(e) $\langle \gamma,\epsilon,A \rangle$ is of type M_{12};

(f) Every involution of $S - \langle \gamma,\epsilon,A \rangle$ is conjugate in S to δ;

(vi) The semi-direct product of A with any subgroup of B isomorphic to $Z_2 \times S_3$ has a Sylow 2-subgroup of type M_{12}.

Proof: If one represents B as a matrix group

$$\left\langle \begin{pmatrix} a & b \\ c & d \end{pmatrix} \;\middle|\; a,b,c,d \in Z/4Z \text{ with } ad-bc \not\equiv 0 \pmod 4 \right\rangle ,$$

the various statements can be straightforwardly verified. Some details are given in [35, p. 364,365].

Lemma 2.2. Let $A \cong E_{16}$ and set $B = \mathrm{Aut}(A)$. Then we have

(i) $B \cong GL(4,2) \cong A_8$;

(ii) An element of B of order 3 which corresponds to a 3-cycle (123) in A_8 acts regularly on A;

(iii) A central involution of B centralizes a subgroup of A of order 8;

(iv) If S is a Sylow 2-subgroup of the semi-direct product of A with a subgroup L of B isomorphic to A_4, then one of the following holds;

(a) $|Z(S)| \geq 4$ and either $r(S) \geq 5$ or S is of type $L_3(4)$; or

(b) $|Z(S)| = 2$ and S is of type A_8;

(v) Under the conditions of (iv), if Z(S) has order 2, then $(AL)' \cong Q_8 * Q_8$ and a Sylow 3-subgroup of L acts regularly on the Frattini factor group of $(AL)'$;

(vi) A Sylow 2-subgroup of the semi-direct product of A with a subgroup L of B isomorphic to S_5 is of type $\hat{A}_8$ or A_{10} according as L acts transitively or intransitively on $A^{\#}$;

(vii) A Sylow 2-subgroup of the semi-direct product of A with a subgroup of B isomorphic to a Sylow 3-normalizer in A_8 is of type A_{10}.

(viii) Let K be a subgroup of B which corresponds to $N_{A_8}(\langle(123)\rangle)$. Then the semi-direct product KA has Sylow 2-subgroups of type $\hat{A}_8$.

<u>Proof</u>: First, (i) is obvious and one can verify (ii) using the isomorphism of A_8 with GL(4,2) (cf. Huppert [32, p. 157]). Likewise (iii) is also a consequence of the same reference as (12)(34)(56)(78) is a central involution of A_8.

Now let L be a subgroup of B isomorphic to A_4 and set M equal to the semi-direct product of L and A. Let $T = \langle t_1, t_2 \rangle \cong Z_2 \times Z_2$ be a Sylow 2-subgroup of L and set $S = TA$, so that S is a Sylow 2-subgroup of M. In proving (iv), we can assume that $r(S) \leq 4$. If $A_1 = C_A(t_1) \cong E_8$, then $A_2 = C_A(t_2) \cong E_8$ as $t_1 \sim t_2$ in L. If $A_1 = A_2$, then T centralizes A_1. Hence $TA_1 \cong E_{32}$, contrary to $r(S) \leq 4$. Thus $A_1 \neq A_2$ and so $A_1 \cap A_2 = Z(S) \cong Z_2 \times Z_2$. Clearly then $A = \langle A_1 \cap A_2, a_1, a_2 \rangle$, where $a_i \in A_i$ and t_j does not centralize a_i, $i \neq j$, $1 \leq i, j \leq 2$. Since $t_1 \sim t_1t_2$ in L also $C_A(t_1t_2) \cong E_8$. Since t_1t_2 centralizes $A_1 \cap A_2$ and does not centralize a_1 or a_2, this forces t_1t_2 to centralize a_1a_2. Hence $a_1^{t_2} a_2^{t_1} = a_1a_2$, which implies that $[a_1, t_2] = [a_2, t_1]$. Since L acts on $A/Z(S) \cong Z_2 \times Z_2$, T centralizes this section. It follows that $S' = \langle [a_1, t_2] \rangle \cong Z_2$ and consequently $S/S' = AT/S' \cong E_{32}$, again contradicting $r(S) \leq 4$.

We conclude therefore that $C_A(t_1) \cong Z_2 \times Z_2$. By (iii), t_1 corresponds to a noncentral involution of A_8 and hence so does each involution of T. We now count the number of conjugacy classes of four groups consisting of three noncentral involutions of A_8. Let U be such a four group. Since all noncentral involutions of A_8 are conjugate to $(12)(34)$, we may assume that $u = (12)(34) \in U$. Setting $C = C_{A_8}(u)$, we have that C contains a normal subgroup C_1 of index 2 isomorphic to $Z_2 \times Z_2 \times A_4$. Furthermore, we check that a Sylow 2-subgroup of C is isomorphic to $Z_2 \times Z_2 \int Z_2$.

We conclude from the structure of C that all four subgroups of C which contain u and do not lie in C_1 are conjugate in C. Hence either U is a member of this conjugacy class or $U \subseteq C_1$. But by the fusion pattern of involution in A_8, we check that $O_2(C_1) \cong E_{16}$ contains 6 noncentral involutions and 9 central involutions and, in fact, that $O_2(C_1)$ possesses a unique four subgroup containing u and consisting of noncentral involutions. We thus conclude that A_8 possesses at most two conjugacy classes of four groups of the required type.

On the other hand, B contains two subgroups L_1 and L_2 such that $L_1A \cong A_5 \cdot E_{16}^{(1)}$ and $L_2A \cong A_5 \cdot E_{16}^{(2)}$. Moreover, four subgroups T_i of L_i correspond to four subgroups of A_8 consisting of noncentral involutions, $i = 1,2$, again by (iii). We know that T_1A is of type A_8 and $Z(T_1A) \cong Z_2$, while T_2A is of type $L_3(4)$ and $Z(T_2A) \cong Z_2 \times Z_2$. In view of the conclusion of the preceding paragraph, this establishes (iv).

Next assume that $Z(S) \cong Z_2$. Then by (iv), S is of type A_8. Hence S contains a characteristic subgroup $Q = Q_1 * Q_2$, $Q_1 \cong Q_2 \cong Q_8$. Since $S \lhd M$, we have $Q \lhd M$. Hence $M/Q \cong Z_2 \times Z_3$. Let x be a 3-element of M. Then $Q_1^x = Q_1$ and $Q_2^x = Q_2$. Since x centralizes a 2-element of S which interchanges Q_1 and Q_2, we have that $[Q_1,x] = Q_1$ and $[Q_2,x] = Q_2$. This clearly implies (v).

As for (vi), we first note that M_{23} has Sylow 2-subgroups of type $\hat{A}_8$ and contains a split extension of S_5 by E_{16} in which S_5 acts transitively on $E_{16}^{\#}$ (see [18, Proposition 4.7.]). On the other hand, A_{10} contains a split extension of S_5 by E_{16} in which S_5 acts intransitively on $E_{16}^{\#}$ (see [17, Proposition 6.1. (iv)]). Hence to prove (vi), it is enough to show that A_8 possesses exactly two conjugacy classes of subgroups isomorphic to S_5. However, as is easily verified, A_8 has exactly two conjugacy classes of A_5's. One is expressed as a premutation group on six letters and the other on five letters. Correspondingly the normalizers of these groups in A_8 are isomorphic to S_5 and to $(A_5 \times Z_3)Y$, where $Y \cong Z_2$, $A_5Y \cong S_5$, and $Z_3Y \cong S_3$. From this, our desired assertion follows.

Finally we prove (vii) and (viii). Since there is only one conjugacy class of Sylow 3-normalizers in A_8, the structure of S is uniquely determined in this case. On the other hand, one sees easily that A_{10} contains such an extension of E_{16} by a Sylow 3-normalizer in A_{10} (see [17, Proposition 6.1 (v)]). This proves (vii). Since $N_{A_7}((<123>))$ contains a Sylow 2-subgroup of $N_{A_8}((<123>))$, KA has Sylow 2-subgroups of type $A_7 \cdot E_{16}$. On the other hand, M_{23} contains a split extension of E_{16} by A_7 and has Sylow 2-subgroups of type $\hat{A}_8$. This proves (viii).

<u>Lemma 2.3</u>. Let $A \cong E_{16}$ and let B be a nonsolvable subgroup of Aut(A) with $O_2(B) = 1$. Then we have

(i) If B acts transitively on $A^{\#}$, then B possesses a subgroup $L \cong A_5$ which acts transitively on $A^{\#}$;

(ii) If B acts intransitively on $A^{\#}$, then either $B \cong GL(3,2)$, A_5, or S_5.

<u>Proof</u>: We have that $\mathrm{Aut}(A) \cong A_8 \cong GL(4,2)$. One checks that B must be isomorphic to A_8, A_7, S_6, A_6, $GL(3,2)$, S_5, A_5, $Z_3 \times A_5$, or else that B contains a subgroup of index 2 isomorphic to $Z_3 \times A_5$. Considering the action of B on A in each case, we obtain (i) and (ii).

In [Lemma 7.4, Part I] we have studied the automorphism group B of a 2-group A of type $L_3(4)$. If $C = O_2(B)$, D is a Sylow 3-subgroup of B, and T is a Sylow 2-subgroup of $N_B(D)$, we have established all of the following facts: $|B| = 2^{10} \cdot 3^2$, $O(B) = 1$, $C \cong E_{2^8}$, C is the stabilizer of the chain $A \supset Z(A) \supset 1$, $B/C \cong N_B(D) \cong S_3 \times S_3$, $C_B(D) = 1$, $T \cong Z_2 \times Z_2$, the semi-direct product TA is of type $\hat{A}_{10}$, and if t is an involution of T, then $\langle t\rangle A$ is of type $L_3(4)^{(1)}$, $\hat{A}_8$, or J_2. Here $L_3(4)^{(1)}$ again denotes the split extension of $L_3(4)$ by an automorphism of order 2 induced from transpose-inverse map of $GL(3,4)$.

We shall need additional detailed information about the structure of B. Using the same notation, we prove

Lemma 2.4. The following conditions hold:

(i) If D_1 is a subgroup of D of order 3 which acts regularly on A, then $D_1 \triangleleft N_B(D)$ and $N_B(D_1)$ has a Sylow 2-subgroup of type A_8;

(ii) If D_2 is a subgroup of D of order 3 which centralizes $Z(A)$, then D_2 acts regularly on $A/Z(A)$ and $N_B(D) = N_B(D_2)$.

(iii) If U is a four subgroup of $N_B(D_1)$ or $N_B(D_2)$ disjoint from C, then UA is of type $\hat{A}_{10}$;

(iv) If u is an involution of $N_B(D_1)$ or $N_B(D_2)$ not in C, then $A\langle u\rangle$ is of type $\hat{A}_8$, J_2, or $L_3(4)^{(1)}$;

(v) If u is a 2-element of B which normalizes each of the two elementary subgroups of A of order 16, then $u \in \langle t\rangle C$ for some t in $N_B(D)^{\#}$ and $\langle t\rangle A$ is of type $\hat{A}_8$.

Proof: We shall treat only the case of D_1, the corresponding results for D_2 can be proved similarly; and (v) follows from the fact that

2-groups of type J_2 or $L_3(4)^{(1)}$ have no normal elementary subgroups of order 16, while those of type $\hat{A}_8$ do.

D possesses a unique subgroup of order 3 which acts regularly on A and consequently $D_1 \lhd N_B(D)$. By [16, Lemma 4.7] there exist D-invariant four subgroups $W_1 = \langle a_1, a_2 \rangle$, $W_2 = \langle b_1, b_2 \rangle$ such that $A = \langle a_1, a_2, b_1, b_2 \rangle$. An element σ of C centralizes D_1 if and only if for some d_1 in $D_1^{\#}$ (which we can assume cyclically permutes a_1, a_2, a_1a_2 and b_1, b_2, b_1b_2 respectively), we have

$$a_1^{\sigma} = a_1u,\ a_2^{\sigma} = a_2u^{d_1},\ (a_1a_2)^{\sigma} = a_1a_2u^{d_1^2}$$

$$b_1^{\sigma} = b_1v,\ b_2^{\sigma} = b_2v^{d_1},\ (b_1b_2)^{\sigma} = b_1b_2v^{d_1^2}$$

for suitable u,v in $Z(A)$. Since the elements u,v can be chosen independently in $Z(A) \cong Z_2 \times Z_2$, it follows that $C_1 = C_C(D_1) \cong E_{16}$.

By [Lemma 7.4 (ii), Part I], $N_B(D)/D_1 \cong Z_2 \times S_3$ and this group acts on C_1. By [Lemma 7.4 (iii), Part I], D/D_1 acts regularly on C_1. We claim that $N_B(D)/D_1$ acts faithfully on C_1. If not, the kernel $\langle t \rangle$ is of order 2 and we have $\langle D_1, t \rangle \cong S_3$. Since $Z(S)$, W_1, W_2 are all the D-invariant four subgroups of A, either t normalizes or interchanges W_1, W_2. If $W_1^t = W_2$, we can suppose without loss that $a_1^t = b_1$. Then from $a_1^{t\sigma} = a_1^{\sigma t}$ for σ in C_1, we get $b_1^{\sigma} = b_1u^t$. But $b_1^{\sigma} = b_1v$ with u and v independent in $Z(A)$, so we have a contradiction. Thus $W_i^t = W_i$, $i = 1,2$. Without loss, we may assume that $a_1^t = a_2$. Again from $a_1^{\sigma t} = a_1^{t\sigma}$, we get $a_2u^{d_1} = a_2u^t$. Since u is arbitrary in $Z(A)$, it follows that d_1t centralizes $Z(A)$. But then $(d_1t)(d_1t)^t = d_1td_1^{-1}t = d^2$

centralizes $Z(A)$, contrary to the regularity of D_1 on A. Thus $N_B(D)/D_1$ acts faithfully on C_1, as asserted.

We have already noted that D/D_1 acts regularly on C_1. Hence identifying $N_B(D)/D_1$ in its action on C_1 with a subgroup of A_8, we can identify D/D_1 with $X = \langle(123)\rangle$ and so $N_B(D)/D_1$ can be identified with a subgroup L of $N_{A_8}(X) = (X \times F)\langle(12)(56)\rangle$, where $F \cong A_5$. Since L does not centralize X, we see that a Sylow 2-subgroup of L, which is isomorphic to $Z_2 \times Z_2$, is uniquely determined up to conjugacy. Hence the structure of a Sylow 2-subgroup of LC_1 is uniquely determined and is isomorphic to that of $N_B(D_1)$. One checks that it is, in fact, of type A_8.

Finally if S is a 2-group of type A_8, then S contains a unique elementary subgroup E of order 16, $S = EV$, where V is a four subgroup disjoint from E, and every complement of E in S is conjugate to V. Likewise every involution of $S - E$ is conjugate to one in V. Now (iii) and (iv) follow for D_1 from [Lemma 7.4 (iv) and (v), Part I].

We shall argue next that every nontrivial extension of E_{16} by A_5 or S_5 splits. The proof depends on the following preliminary result which we shall also need in a later part of the paper.

<u>Lemma 2.5</u>. Suppose $A \lhd B$ with $A \cong E_{16}$ and $B/A \cong Z_2 \times Z_2$. If $Z(B) \cong Z_2$ and $B - A$ contains involutions b_1, b_2 such that $A\langle b_i\rangle \cong Z_2 \times Z_2 \int Z_2$, $i = 1,2$ and $B = \langle A, b_1, b_2\rangle$, then B splits over A.

<u>Proof:</u> Set $Z(A\langle b_1\rangle) = \langle z_1, z_2\rangle \cong Z_2 \times Z_2$. Since $A\langle b_1\rangle \lhd B$, b_2 normalizes $\langle z_1, z_2\rangle$. Since $Z(B) \cong Z_2$, we may assume the notation is chosen so that $z_1^{b_2} = z_2$ and $(z_1 z_2)^{b_2} = z_1 z_2$. If $[b_1, b_2] = 1$, then

clearly B splits over A, so we can assume that $[b_1, b_2] \neq 1$. By the structure of $A\langle b_1\rangle$, the only involutions of $A\langle b_1\rangle - A$ are z_1b_1, z_2b_1, $z_1z_2b_1$, and b_1. It follows that $b_1^{b_2} = z_1b_1$, z_2b_1, or $z_1z_2b_1$. However, in the first case $b_1 = (b_1^{b_2})^{b_2} = z_1z_2b_1$, a contradiction. Similarly $b_1^{b_2} \neq z_2b_1$ and so the only possibility is that $b_1^{b_2} = z_1z_2b_1$. But this implies that $(z_1b_1)^{b_2} = z_2(z_1z_2b_1) = z_1b_1$. Thus $\langle z_1b_1, b_2 \rangle$ is a four group disjoint form A and so B splits over A.

Lemma 2.6. Every nontrivial extension of E_{16} by A_5 or S_5 splits.

Proof: Assume $A \cong E_{16}$, $A \lhd B$, and $B/A \cong A_5$ or S_5 with B acting nontrivially on A. First consider the A_5 case. Let D be a Sylow 5-subgroup of B. Then $N_A(D) = 1$ and $N_B(D)$ is dihedral of order 10. Hence if S is a Sylow 2-subgroup of B, it follows that $S - A$ contains an involution t.

If B acts transitively on $A^{\#}$, then $N_B(S)$ contains an element v of order 3 which acts regularly on A and hence on S. This implies first that $Z(S) \cong Z_2 \times Z_2$. Since $S - A$ contains an involution and v acts regularly on $S/Z(S)$, it also follows that $S/Z(S) \cong E_{16}$ and that $W = \langle t, t^v, Z(S) \rangle$ is v-invariant with $W/Z(S) \cong Z_2 \times Z_2$. This in turn forces $W \cong E_{16}$. Clearly $Z(S) \subseteq A$ and $W \cap A = Z(S)$. Hence W contains a four group disjoint from A and so S splits over A. By Gaschütz' theorem, we conclude that B splits over A.

Suppose then that B acts intransitively on $A^{\#}$. Since $A_5 \cdot E_{16}^{(1)}$ has Sylow 2-subgroups of type A_8 (cf. [17, Theorem A^*]), we have $Z(S) \cong Z_2$ whether or not B splits over A. Again let v be an element of $N_B(S)$ of order 3. Then the involutions t and $t' = t^v$ lie in distinct cosets of A in S and so $S = A\langle t,t'\rangle$. Furthermore, we have $A\langle t\rangle \cong A\langle t'\rangle \cong Z_2 \times Z_2 \int Z_2$ as this is the case in $A_5 \cdot E_{16}{}^{(1)}$. Hence S splits over A by the preceding lemma and again B splits over A by Gaschütz' theorem.

Assume next that $B/A \cong S_5$. By the preceding argument, B contains a subgroup $L \cong A_5$. If L acts transitively on A, then a Sylow 2-subgroup T of LA is of type $L_3(4)$ and L contains a subgroup P of order 3 which acts regularly on T. Set $K = N_B(T)$. Since $B/A \cong S_5$, we see that $K/T \cong S_3$. Since $C_T(P) = 1$, it follows by the Frattini argument that $N_K(P)$ contains an involution u. Thus $S = \langle T,u\rangle$ is a Sylow 2-subgroup of B. Moreover, u does not stabilize the chain $T \supset \Phi(T) \supset 1$ as u inverts P. It follows therefore from Lemma 2.4(v) that S must be of type $\hat{A}_8$. Hence S splits over A and now Gaschütz' theorem implies that B splits over A.

On the other hand, if L acts intransitively on $A^{\#}$, a Sylow 2-subgroup T of LA is of type A_8. But then all complements of A in T are conjugate. Hence all complements to A in LA are conjugate to L (cf. [32, p. 121]). It follows therefore by the Frattini argument that $B = AN_B(L)$. Since $N_B(L) \cap A = 1$, $N_B(L)$ is thus a complement to A in B and so B splits over A in this case as well.

Lemma 2.7. Let $A \cong Q_8 * D_8$, set $B = \mathrm{Aut}(A)$, and $C = O_2(B)$. Then we have

(i) $C \cong E_{16}$, $B/C \cong S_5$, and B splits over C;

(ii) B has Sylow 2-subgroups of type A_{10};

(iii) Any two subgroups of B isomorphic to A_5 are conjugate in B;

(iv) If L is a subgroup of B isomorphic to A_5, then LA has Sylow 2-subgroups of type J_2.

Proof C: As is well known, $\mathrm{Inn}(A) \cong E_{16}$ and $\mathrm{Aut}(A)/\mathrm{Inn}(A) \cong S_5$. Hence $C \cong E_{16}$ and $B/C \cong S_5$. Moreover, B splits over C by the preceding lemma. Since $A - Z(A)$ contains 10 involutions and 20 elements of order 4, B does not act transitively on $C^{\#}$. Thus $B \cong S_5 \cdot E_{16}^{(1)}$, which has Sylow 2-subgroups of type A_{10} by Lemma 2.2(vi) and so (i) and (ii) hold.

Next let L_1, L_2 be two subgroups of B isomorphic to A_5 and let T_1, T_2 be Sylow 2-subgroups of L_1, L_2 respectively. Since T_1C, T_2C are each Sylow 2-subgroups of B', we can assume without loss in proving (iii) that $T_1C = T_2C$. But $L_1C \cong A_5 \cdot E_{16}^{(1)}$ and so T_1C is of type A_8. One checks now that any two complements of C in T_1C are conjugate in T_1C, whence $T_1 \sim T_2$ in B. It follows again from [32, p. 121] that also L_1, L_2 are conjugate in B, proving (iii).

By (iii), the structure of a Sylow 2-subgroup of LA is uniquely determined, where L is any subgroup of B isomorphic to A_5. On the other hand, by [16, Lemma 6.4], there exists a group of this form with Sylow 2-subgroups of type J_2. Thus (iv) also holds.

Finally we have

Lemma 2.8. If A is a 2-group of type $Sz(8)$ or $Sz(8) \times Z_2$, then $\mathrm{Aut}(A)$ does not possess a 2-element which acts nontrivially on $A/Z(A)$.

Proof: We have $A = A_1 \times Z$, where A_1 is of type $Sz(8)$ and Z is of order 1 or 2. A_1 is generated by six elements $a_1, a_2, a_3, a_4, a_5, a_6$ satisfying

$$a_1^2 = a_2^2 = a_3^2 = a_4^4 = a_5^4 = a_6^4 = 1, \ [a_4,a_5] = a_1, \ [a_4,a_6] = a_2,$$

(2.1)

$$[a_5,a_6] = a_3, \ a_4^2 = a_1a_3, a_5^2 = a_2a_3, a_6^2 = a_1a_2 ,$$

with all other commutators of pairs of generators being trivial.

We have $Z(A_1) = \Phi(A_1) = \langle a_1,a_2,a_3\rangle$ and $Z(A) = Z(A_1) \times Z \cong E_8$ or E_{16}. As is well known, A_1 admits a nonabelian group of order 21 as an automorphism group. Hence so does A. Let D be such an automorphism group of A. Then D acts faithfully on $\bar{A} = A/Z(A) \cong E_8$. Suppose that our lemma is false, in which case A also admits an automorphism which induces an involution of $\bar{A}$. This means that $B = \mathrm{Aut}(A)$ induces $\mathrm{Aut}(\bar{A}) \cong GL(3,2)$. Hence if C denotes the stabilizer of the chain $A \supset Z(A) \supset \Phi(A) \supset 1$, we have $B/C \cong GL(3,2)$.

Considering the action of $GL(3,2)$ on E_8, we see that any four subgroup of E_8 is normalized by a subgroup of $GL(3,2)$ isomorphic to S_4. It follows from this that B possesses a 2-element x with the property:

$$\bar{a}_4^{\,x} = \bar{a}_4, \ \bar{a}_5^{\,x} = \bar{a}_6, \text{ and } \bar{a}_6^{\,x} = \bar{a}_5 .$$

Hence

(2.2) $a_4^x = a_4\epsilon_1$, $a_5^x = a_6\epsilon_2$, and $a_6^x = a_5\epsilon_3$

for suitable elements ϵ_1, ϵ_2, ϵ_3 in $Z(A)$. But now using the commutator relations of (2.1), we obtain from (2.2) that

(2.3) $a_1^x = a_2$ and $a_3^x = a_3$.

In particular, this yields

(2.4) $(a_1a_3)^x = a_2a_3$.

On the other hand, using the fact that $a_4^2 = a_1a_3$, (2.2) yields

(2.5) $(a_2a_3)^x = a_2a_3$.

However, (2.4) and (2.5) are inconsistent as $a_1 \neq a_2$.

3. Theorem B, the GL(3,2) case. Let X be a nonsolvable 2-constrained group with $O(X) = 1$ and $r(X) \leq 4$. Setting $Y = O_2(X)$, $\bar{X} = X/Y$, and $\bar{Y} = Y/\Phi(Y)$, we have that $\bar{Y}$ is elementary abelian of order at most 16 and that $\bar{X}$ acts faithfully on $\bar{Y}$. Thus $\bar{X}$ is isomorphic to a nonsolvable subgroup of $GL(4,2) \cong A_8$ with $O_2(\bar{X}) = 1$ and so one of the following holds:

$\bar{X} \cong GL(3,2)$, $A_5, S_5, A_6, S_6, A_7, A_8, Z_3 \times A_5$, or $\bar{X}$ contains a subgroup of index 2 isomorphic to $Z_3 \times A_5$.

In particular, either $\bar{X} \cong GL(3,2)$ or $\bar{X}$ contains a subgroup isomorphic to A_5. We shall treat these two possibilities separately. In this section, we prove

Proposition 3.1. If X is a nonsolvable 2-constrained group with $O(X) = 1$ and $r(X) \leq 4$ such that $X/O_2(X) \cong GL(3,2)$, then

$$O_2(X) \cong E_8,\ E_{16},\ \text{or}\ Z_4 \times E_8.$$

We shall carry out the proof in a long sequence of lemmas. To begin with, we let X denote an arbitrary group with the following properties:

(a) $O(X) = 1$;

(b) $C_X(Y) \subseteq Y$ for some normal 2-subgroup Y of X.

We also denote by S a Sylow 2-subgroup of X. Thus $Y \subseteq S$ as $Y \subseteq O_2(X)$.

Lemma 3.2. If $Y \cong E_8$ and $X/Y \cong A_4$, then

(i) If $Z(X) \neq 1$, then $S \cong Q_8 * Q_8$ and X splits over Y;

(ii) If $Z(X) = 1$, then X normalizes a four subgroup of Y and either

(a) $S \cong Z_2 \times Z_2 \int Z_2$;

(b) $S \cong \langle \alpha, \beta, \gamma \mid \langle \alpha, \beta \rangle \cong Z_4 \times Z_4,\ \gamma^2 = 1,\ \alpha^\gamma = \alpha^{-1},\ \beta^\gamma = \beta^{-1} \rangle$; or

(c) $S \cong \langle \alpha, \beta, \gamma \mid \langle \alpha, \beta \rangle \cong Z_4 \times Z_4,\ \gamma^2 = 1,\ \alpha^\gamma = \alpha^{-1}\beta^2,\ \beta^\gamma = \beta\alpha^2 \rangle$.

Proof: Set $Z = Z(S)$ and let B be a Sylow 3-subgroup of X. Since $C_X(Y) \subseteq Y$, we have $Z \subset Y$, whence $Z \cong Z_2$ or $Z_2 \times Z_2$. Since $S \triangleleft X$, B leaves Z invariant. If $Z \cong Z_2 \times Z_2$, B does not centralize Z as B does not centralize Y. Hence in this case, $Z(X) = 1$. On the other hand, if $Z \cong Z_2$, B centralizes Z, $Z = Z(X)$, and $Z \subseteq \Phi(S)$.

Consider the second case first. B does not centralize Y/Z and so B acts regularly on S/Z which is of order 16. Hence $S/Z \cong Z_4 \times Z_4$ or E_{16}. In the first case, it follows at once that $Y = \Phi(S)$ and that $Y \subseteq Z(S)$, which is a contradiction. Thus $S/Z \cong E_{16}$ and so S is extra-special. Since $Y \in SCN_3(S)$, the only possibility is $S \cong Q_8 * Q_8$. Hence (i) holds.

Suppose next that $Z \cong Z_2 \times Z_2$. Since B acts nontrivially on S/Z, we have $S/Z \cong Q_8$ or E_8. In the first case, we again obtain the contradiction $Y \subseteq Z(S)$; so $S/Z \cong E_8$. It follows that $R = [B,S]$ is a group of order 16 on which B acts regularly, so $R \cong E_{16}$ or $Z_4 \times Z_4$. In addition, we have $C_S(B) = \langle \gamma \rangle$, where $|\gamma| = 2$. Since S is nonabelian, γ does not centralize R. Since B acts faithfully on both $C_R(\gamma)$ and $R/C_R(\gamma)$, we must have $C_R(\gamma) \cong Z_2 \times Z_2$. Moreover, if $R \cong E_{16}$, it is immediate that $S \cong Z_2 \times Z_2 \int Z_2$. On the other hand, if $R \cong Z_4 \times Z_4$, Lemma 2.1 (v) implies that S satisfies (ii) (b) or (c). Thus (ii) holds in this case.

Lemma 3.3. Assume the following conditions hold:

(a) Y is of type $L_3(4)$;

(b) $X/Y \cong S_3$;

(c) A Sylow 3-subgroup of X centralizes $\Phi(Y)$ and acts regularly on $Y/\Phi(Y)$;

(d) A 2-element of $X-Y$ normalizes an elementary group of Y of order 16.

Under these conditions, S is of type $\hat{A}_8$.

Proof: By Lemma 2.4 (iv) and (v), it will be enough to show that $N_X(B)-Y$ contains an involution, where B is a Sylow 3-subgroup of X. Setting $N = N_X(B)$, we have $N \cap Y = \Phi(Y) \cong Z_2 \times Z_2$. Hence it will suffice to prove that a Sylow 2-subgroup of N is isomorphic to D_8. But if u is a 2-element of $N-\Phi(Y)$, then $\hat{u} \in \langle t\rangle C$, where $\hat{u}$ is the automorphism of Y induced by u and t, C have the same meanings as in Lemma 2.4 (iv) (with Y as A). By the structure of $\langle t\rangle A$, t does not centralize $\Phi(A)$, so neither does u. Hence $\langle u,\Phi(Y)\rangle \cong D_8$ and the lemma is proved.

Lemma 3.4. If $Y \cong E_8$ and $X/Y \cong GL(3,2)$, then one of the following holds:

(i) S is of type A_8 and X splits over Y;

(ii) S is of type M_{12} and X does not split over Y.

Proof: Let y be an involution of $Z(S) \cap Y$. Then $W = C_X(y)$ contains S and $W/Y \cong S_4$, as is immediate from the action of $GL(3,2)$ on Y. Applying Lemma 3.2 (i), it follows now that W contains a subgroup $L \cong A_4$ such that $L \cap Y = 1$, $S_1 = LY \cap S \cong Q_8 * Q_8$, $S_1 \lhd W$, and $W/S_1 \cong S_3$. This implies that S has a coset $aY \neq Y$ which contains an involution. Since $GL(3,2)$ has only one conjugacy class of involutions, we conclude that every coset of Y whose image in X/Y is of order 2 contains an involution. In particular, $S - S_1$ contains an involution t. Moreover, a 3-element of

$W - S_1$ acts regularly on $S_1/\langle y\rangle$. These conditions force t to either interchange or induce an outer automorphism on the two factors of S_1 that are isomorphic to Q_8. Correspondingly we see that S is of type A_8 or M_{12}. Since correspondingly S does or does not split over Y, so also X does or does not split over Y.

The following lemma will be crucial.

Lemma 3.5. If $Y \cong Z_4 \times Z_4 \times Z_4$ and $X/Y \cong GL(3,2)$, then $r(X) \geq 5$.

Proof: Set $\bar{X} = X/\Omega_1(Y)$ and $\tilde{X} = \bar{X}/\bar{Y} \cong X/Y \cong GL(3,2)$. Considering the action of $\tilde{X}$ on $\bar{Y} \cong E_8$, we see that $\tilde{X}$ contains a subgroup $\tilde{L} \cong A_4$ such that $\bar{Y}_1 = C_{\bar{Y}}(\tilde{L}) \cong Z_2$. Hence by Lemma 3.2(i), the inverse image of $\tilde{L}$ in $\bar{X}$ has the form $\overline{YL}$, where $\bar{L} \cong A_4$, $\bar{Y} \cap \bar{L} = 1$, and $\bar{Y}_1$ centralizes $\bar{L}$. Now considering the action of $\tilde{X}$ on $V = \Omega_1(Y)$, it follows likewise from Lemma 3.2(i) that the inverse image of $\bar{L}$ in X has the form VL, where $L \cong A_4$ and $V \cap L = 1$. Thus if Y_1 denotes the inverse image of $\bar{Y}_1$ in Y, we have that $Y_1 \cong Z_4 \times Z_2 \times Z_2$ with L normalizing Y_1 and $Y_1 \cap L = 1$.

Set $X_1 = YL$, let T be a Sylow 2-subgroup of L, and set $S_1 = YT$, so that S_1 is a Sylow 2-subgroup of X_1. We shall prove that $S_1/\mho^1(Y_1)$ has 2-rank at least 5 and this will establish the lemma. We set $\bar{X}_1 = X_1/\mho^1(Y_1)$. Then $\bar{S}_1 = \overline{YT}$ with $\bar{Y}_1 \cap \bar{T} = 1$ and $\bar{Y}_1, \bar{T}$ elementary abelian of order 8 and 4 respectively. We need only show that $\bar{Y}_1 \subseteq Z(\bar{S}_1)$ as then $\bar{Y}_1\bar{T} = \bar{Y}_1 \times \bar{T} \cong E_{32}$

whence $\bar{S}_1$ has 2-rank at least 5.

First of all, L contains a 3-element u which acts nontrivially on V. Since $\mho^1(Y_1) \cong Z_2$ and is contained in V, it follows that $\bar{u}$ acts regularly on $\bar{V}$ which is a four group. Since $\bar{u}$ acts regularly on the four group $\bar{T}$ and $\bar{T}$ normalizes $\bar{V}$, the only possibility is that $\bar{T}$ centralizes $\bar{V}$. Since $\bar{S}_1 = \overline{YT}$ with $\bar{Y}$ abelian, we conclude that $\bar{V} \subseteq Z(\bar{S}_1)$.

Finally set $\tilde{S}_1 = \bar{S}_1/\bar{V}$. Then $\tilde{Y}_1 \cong Z_2$ and $\tilde{Y}_1 \subseteq Z(\tilde{S}_1)$. But by Lemma 3.2 (i), $S_1/V \cong \bar{S}_1/\bar{V} = \tilde{S}_1 \cong Q_8 * Q_8$. In particular, $\tilde{Y}_1 = Z(\tilde{S}_1)$. Let $\bar{y}_1$ be a fixed element of $\bar{Y}_1 - \bar{V}$. If $\bar{a}$ is any element $\bar{S}_1$ such that $\tilde{a}$ has order 4, we have $\tilde{a}^2 = \tilde{y}_1$ by the structure of $\tilde{S}_1$. Hence $\bar{a}^2 = \bar{y}_1\bar{v}$ for some $\bar{v}$ in $\bar{V}$. But $\bar{V} \subseteq Z(\bar{S}_1)$ and so $\bar{a}$ centralizes $\bar{a}^2\bar{v}^{-1} = \bar{y}_1$. However, $\tilde{S}_1$ is generated by the set of its elements $\tilde{a}$ with the property that $\tilde{a}$ is of order 4. Hence $C_{\bar{S}_1}(\bar{y}_1)$ covers $\tilde{S}_1 = \bar{S}_1/\bar{V}$ and we conclude at once that $\bar{y}_1 \in Z(\bar{S}_1)$, whence $\bar{Y}_1 \subseteq Z(\bar{S}_1)$, as required.

The following lemma will be needed in analyzing the case that $Y \cong E_{16}$, $X/Y \cong GL(3,2)$, and $Z(X) \neq 1$.

Lemma 3.6. Assume the following conditions hold:

(a) Y is special of order 2^6 with Z(Y) of order 4;

(b) $X/Y \cong Z_3$ and an element of X of order 3 acts regularly on $Y/\Phi(Y)$;

(c) Y contains an elementary subgroup E of order 16 which is normal in X.

Under these conditions, we have either

(i) Y splits over E and Y is of type $L_3(4)$;

(ii) Y does not split over E, Y - E contains no involutions, and if $z \in Z(Y)^{\#}$ is such that $Y/\langle z\rangle \cong Q_8 * Q_8$, then Y possesses a quaternion subgroup with center $\langle z\rangle$.

Proof: We first note that our conditions imply that $r(X) \leq 4$. Let P be a Sylow 3-subgroup of X. If Y splits over E, then by Gaschütz' theorem $X = YP = EL$, where $L \cong A_4$ and $E \cap L = 1$. Since $Z(Y) \cong Z_2 \times Z_2$ and $r(Y) = 4$, Lemma 2.2 (iv) now yields that Y is of type $L_3(4)$.

Suppose then that Y does not split over E. If t is an involution of $Y - E$, then for u in $P^{\#}$, $T = \langle t,t^u,t^{u^2}\rangle$ is P-invariant and $T/T \cap Z(Y)$ is a four group on which P acts regularly. Since $|TZ(Y)| = 16$, the only possibility is that $TZ(Y) \cong E_{16}$. Hence we can choose t so that T is a four group. But then $T \cap E = 1$ and so Y splits over E, which is not the case. Thus $Y - E$ contains no involution.

Suppose finally that $z \in Z(Y)^{\#}$ is such that $\bar{Y} = Y/\langle z\rangle \cong Q_8 * Q_8$. Then $\bar{Y}$ contains an involution $\bar{t}$ not in $\bar{E}$. By the preceding paragraph, a representative t of $\bar{t}$ has order 4. Setting $T = \langle t,t^u,t^{u^2}\rangle$ we again have that $V = TZ(Y)$ is P-invariant of order 16 with P acting regularly on $V/Z(Y)$. But as $\bar{t}$ is an involution, we also see that $\bar{V} \cong E_8$. We conclude at once that $V \cong Q_8 \times Z_2$ and that a quaternion subgroup of V has center $\langle z\rangle$. Thus all parts of (ii) hold in this case.

Lemma 3.7. If $Y \cong E_{16}$ and $X/Y \cong GL(3,2)$, then one of the following holds:

(i) X/Y acts decomposably on Y with one three-dimensional invariant subspaces;

(ii) X/Y act indecomposably on Y with a one-dimensional invariant subspace; or

(iii) X/Y acts indecomposably on Y with a three-dimensional invariant subspace.

Proof: Clearly there exist at least three different actions of X/Y on Y as described above. Hence it will be enough to show that $GL(4,2) \cong A_8$ has at most three conjugacy classes of subgroups isomorphic to $GL(3,2) \cong L_2(7)$.

Examining the structure of $L_2(7)$, we see that it has no subgroups of index less than 7, that it has only one conjugacy class of subgroups of index 8 and that it has two of index 7. On the other hand, all primitive permutation representations of A_7 are known and from that result, one concludes that A_7 has exactly two conjugacy classes of subgroups isomorphic to $L_2(7)$. Since $N_{S_7}(L_2(7)) = N_{A_7}(L_2(7))$, it follows that S_7 has only one conjugacy class of such subgroups. Consequently S_8 has exactly two conjugacy classes of subgroups isomorphic to $L_2(7)$, represented by G_1 and G_2, with G_1, but not G_2, a subgroup of S_7.

Now consider their conjugacy in A_8. Since $N_{S_8}(G_2) \cong PGL(2,7)$ and $N_{A_8}(G_1) = G_1$, the conjugacy class containing G_2 does not split in A_8. We conclude therefore that $GL(4,2) \cong A_8$ has at most three conjugacy classes of subgroups isomorphic to $L_2(7)$.

Remark. Our argument shows that in each of the three cases the action of X/Y on Y is uniquely determined. Moreover, if $Z(S)$ is noncyclic, X must act decomposably on Y. Indeed, in cases (ii) and (iii), the orbit lengths of X on $Y^{\#}$ are 1,14 and 7,8 respectively, which implies in either case that $Z(S) \cong Z_2$. Furthermore, we note that $Z(X) \neq 1$ in cases (i) and (ii), while $Z(X) = 1$ in case (iii). Finally in cases (ii) and (iii), X/Y can be identified with a subgroup of $A_7 \subset GL(4,2)$.

We consider separately the cases $Z(X) \neq 1$ and $Z(X) = 1$.

Lemma 3.8. If $Y \cong E_{16}$, $X/Y \cong GL(3,2)$, and $Z(X) \neq 1$, then one of the following holds:

(i) X is indecomposable on Y, S is of type $\hat{A}_8$, and X splits over Y;

(ii) X is decomposable on Y, $Y_1 = [X,Y] \cong E_8$, $X/Y_1 \cong SL(2,7)$, $S - Y$ contains no involutions, $Z(S) \cong Z_2 \times Z_2$, and S possesses a quaternion group with $Z(X)$ as its center; or

(iii) $X = Z(X) \times O^2(X)$.

Proof: Clearly $Z(X) = \langle z\rangle \cong Z_2$. Setting $\bar{X} = X/\langle z\rangle$, we have that $\bar{Y} \cong E_8$ and $\bar{X}/\bar{Y} \cong GL(3,2)$. But now the proof of Lemma 3.4 shows that X possesses a subgroup T containing Y such that $\bar{T} \cong Q_8 * Q_8$ and $T \lhd S$.

Consider first the case that $z \notin \Phi(T)$; then $T \cong Z_2 \times (Q_8 * Q_8)$. In particular, $\Phi(T) \neq 1$, whence $\Phi(T) \cap Z(S) \neq 1$. It follows that $Z(S)$ is noncyclic. Hence by the preceding remark, X acts decomposably on Y. Setting $Y_1 = [X,Y]$, we have that $Y_1 \cong E_8$ and that X/Y_1 is isomorphic to an extension of $X/Y \cong GL(3,2)$ by $Y_1\langle z\rangle/Y_1 \cong Z_2$. By Schur's results, it follows that either $X/Y_1 \cong GL(3,2) \times Z_2$ or $SL(2,7)$. However, the latter case cannot occur under our present assumption as TY_1/Y_1 is a four group by the structure of T. We conclude at once that (iii) holds in this case.

We can therefore assume that $z \in \Phi(T)$. By the structure of $\bar{T}$, $|\Phi(T)| = 4$ and $\Phi(T) = Z(T)$. Thus since $Y \subseteq T$ and $C_X(Y) \subseteq Y$, we have $Z(T) \subseteq Y \cong E_{16}$ and so $\Phi(T) \cong Z_2 \times Z_2$. It also follows from the proof of Lemma 3.4 that if $\bar{N} = N_{\bar{X}}(\bar{T})$ then $\bar{N}/\bar{T} \cong S_3$ and a Sylow 3-subgroup of $\bar{N}$ acts regularly on $\bar{T}/\Phi(\bar{T})$. Hence if $N = N_X(T)$, we have that $N/T \cong S_3$ and a Sylow 3-subgroup P of N acts regularly on $T/\Phi(T)$.

Using P, we argue now that T is special. To prove this, we need only show that $T' = \Phi(T)$. Assume false. As $\Phi(T) \cong Z_2 \times Z_2$ and $T/\langle z\rangle \cong Q_8 * Q_8$, the only possibility is that $T' \cong Z_2$ and $z \notin T'$. Since $z \in \Phi(T)$, it follows that $\Phi(T) = \langle z, T'\rangle$. Hence if we set $\tilde{T} = T/T'$, we have that $\tilde{T}$ is abelian and that $\langle\tilde{z}\rangle = \Phi(\tilde{T})$. But then we see that P has a nontrivial fixed point on $\tilde{T}/\Phi(\tilde{T})$, contrary to the fact that P acts regularly on $T/\Phi(T)$. Thus T is special, as asserted.

The hypotheses of Lemma 3.6 are therefore all satisfied with TP, T, and Y in the roles of X, Y, and E respectively. We conclude that either T is of type $L_3(4)$ or T satisfies the various conditions of Lemma 3.6 (ii).

In the first case, it follows from Lemma 3.3 that S is of type $\hat{A}_8$. (Since P centralizes z, P centralizes $\Phi(T)$). Hence S splits over Y and consequently so does X. Since $|Z(S)| = 2$, X acts indecomposably on Y by our remark above. Thus all parts of (i) hold in this case.

Now consider the second alternative. Then $T - Y$ contains no involutions. Since $X/Y \cong GL(3,2)$, any involution of $S - T$ is conjugate to an element of T and hence of $T - Y$. It follows therefore that $S - T$ also contains no involutions. Since $N_T(P) = \Phi(T)$ and $N_N(P)/N_T(P) \cong S_3$, the only possibility is that a Sylow 2-subgroup of $N_N(P)$ is isomorphic to $Z_4 \times Z_2$, whence $\Phi(T) \subseteq Z(N_N(P))$. Since $N = TN_N(P)$ by the Frattini argument and $S \subseteq N$, we conclude that $\Phi(T) = Z(S)$. Thus $Z(S)$ is noncyclic and so X acts decomposably on Y. Since $z \in \Phi(T)$, we have that $X/Y_1 \cong SL(2,7)$ in this case, where again $Y_1 = [X,Y] \cong E_8$. Since $\bar{T} = T/\langle z\rangle \cong Q_8 * Q_8$, Lemma 3.6(ii) implies that T possesses a quaternion subgroup with center $\langle z\rangle$. Hence all parts of (ii) hold in this case.

Lemma 3.9. If $Y \cong E_{16}$, $X/Y \cong GL(3,2)$, and $Z(X) = 1$, then we have

(i) If $Y_1 = [X,Y]$, then $Y_1 \cong E_8$ and $X/Y_1 \cong Z_2 \times GL(3,2)$;

(ii) $O^2(X)$ contains a subgroup $L \cong A_4$ and disjoint from Y such that LY has Sylow 2-subgroups of type A_8;

(iii) All involutions of $S - (S \cap O^2(X))$ are conjugate in S;

(iv) If X splits over Y, then S is of type $\hat{A}_8$;

(v) If S contains at least two elementary abelian subgroups of order 16, then X splits over Y.

Proof: In this case, Lemma 3.7 (iii) must hold and it is immediate that $Y_1 = [X,Y] \cong E_8$. Moreover, all 8 involutions of $Y - Y_1$ are conjugate in X and hence in S. In particular, $C_S(y) = Y$ for any element y in $Y - Y_1$. This implies that $S/Y_1 \not\cong Q_{16}$, whence $S/Y_1 \not\cong SL(2,7)$. The only possibility therefore is that $X/Y_1 \cong Z_2 \times GL(3,2)$, so (i) holds.

By (i), $X' = O^2(X)$ is isomorphic to an extension of $GL(3,2)$ by E_8. Again the proof of Lemma 3.4 shows that X' contains a subgroup $T \cong Q_8 * Q_8$ and if $N = N_{X'}(T)$, then $N/T \cong S_3$ and a Sylow 3-subgroup P of N acts regularly on $T/\Phi(T)$. These conditions imply that $Y_1 \subseteq T$ and that $TP = Y_1 L$ with $L \cong A_4$, $Y_1 \cap L = 1$, and $Z(Y_1 L) \cong Z_2$. Using the fact that $C_S(y) = Y$ for every y in $Y - Y_1$, it follows now that $Z(TY) \cong Z_2$. Furthermore, $Y \cap L = 1$, since otherwise $Y \cap L \cong Z_2$ and $Y \cap L \lhd L$, contradicting $L \cong A_4$. We can therefore apply Lemma 2.2 (iv) to conclude that LY has Sylow 2-subgroups of type A_8, proving (ii).

By the remark following Lemma 3.7, if Aut (Y) is identified with A_8, then X/Y can be identified with a subgroup of A_7. Hence if X splits over Y, it follows that X can be identified with a subgroup of $A_7 \cdot E_{16}$ the unique nontrivial split extension of A_7 by E_{16}. Thus S is of type $A_7 \cdot E_{16}$ and so S is of type $\hat{A}_8$ (cf. [18]). This establishes (iv).

To prove (iii), consider first the case that X splits over Y, whence S is of type $\hat{A}_8$ by (iv) and $S_1 = S \cap X'$ is of type A_8 by Lemma 3.4(i). If one examines the properties of S described in [18, Lemma 3.1], one checks easily that all involutions of $S - S_1$ are conjugate in S, as asserted.

Suppose then that X does not split over Y. Then obviously X' does not split over Y_1 and consequently $S_1 = S \cap X'$ is of type M_{12} by Lemma 3.4 (ii). We let W be the normalizer in X' of a four subgroup of Y_1. Then we have that $W/Y_1 \cong S_4$. By the structure of S_1, we conclude easily that $W = N_{X'}(A)$, where $A \cong Z_4 \times Z_4$ and that $W = AN_W(B)$, where B is a Sylow 3-subgroup of W, $A \cap N_W(B) = 1$, and $N_W(B) \cong Z_2 \times S_3$. Since B normalizes, but does not centralize Y_1, we have $C_{Y_1}(B) \cong Z_2$ and so $C_Y(B) \cong Z_2 \times Z_2$. Hence B centralizes an element y of $Y - Y_1$. Then y normalizes S_1 and as A is characteristic in S_1, it follows that y normalizes A. Thus $N_X(A)$ is a split extension of A by $N_X(B) \cap N_X(A)$ which has order $2^3 \cdot 3$ and, moreover, $N_X(B) \cap N_X(A)$ acts faithfully on A. We conclude now from Lemma 2.1 (v), first, that the structure of S is uniquely determined and, secondly, that all involutions of $S - S_1$ are conjugate in S. Thus (iii) holds in this case as well.

Finally assume that S contains $E \cong E_{16}$ with $E \neq Y$. If $E \not\subseteq S_1$, then some involution of E would be conjugate to an element y of $Y - Y_1$ by (iii). But E is not conjugate to Y in S as $Y \triangleleft S$. Since $Y = C_S(y)$, we reach a contradiction. Hence $E \subseteq S_1$. But now S_1 is not of type M_{12}

as M_{12} has 2-rank 3. Thus S_1 is of type A_8 by Lemma 3.4 and consequently X' splits over Y_1 by that lemma, whence also X aplits over Y. This proves (v) and establishes the lemma.

The following lemma together with Lemma 3.5 will be the principal tools needed to establish Proposition 3.1.

Lemma 3.10. If $Y \cong Z_4 \times Z_2 \times Z_2 \times Z_2$, $X/Y \cong A_4$, and X splits over Y, then $r(X) \geq 5$.

Proof: Let L be a subgroup of X isomorphic to A_4 such that $X = YL$ with $Y \cap L = 1$. Without loss we can assume that $T = S \cap L$ is a Sylow 2-subgroup of L, so that $S = YT$. Since $C_X(Y) \subseteq Y$, $Z(S) \subseteq Y$. Since $TZ(S) = T \times Z(S)$ and T is a four group, the lemma holds if $Z(S)$ has rank at least 3, so we can suppose $Z(S)$ has rank at most 2 and $r(X) \leq 4$.

We let P be a Sylow 3-subgroup of L and set $S_1 = \Omega_1(Y)T$, so that P normalizes both S and S_1. Note that by the structure of Y, $Z(S_1) = \Omega_1(Z(S))$ and so $Z(S_1)$ has the same 2-rank as $Z(S)$. Furthermore, $\mho^1(Y)$ is clearly of order 2 and lies in $Z(S_1)$. Hence P centralizes $\mho^1(Y)$ and it follows that P centralizes both $Z(S)$ and $Z(S_1)$.

Consider first the case that $Z(S_1)$ is of rank 2. Then by Lemma 2.2 (iv), S_1 is of type $L_3(4)$. Since P centralizes $Z(S_1)$, it follows that P act regularly on $S_1/Z(S_1)$. We shall apply Lemma 2.4(iv) with S_1, P in the roles of A, D_2 respectively. Since $Y/\Omega_1(Y)$ is of order 2 and all elements of $Y - \Omega_1(Y)$ are of order 4, P centralizes an element of y of Y of order 4 and $y \notin S_1$. Hence $S = S_1\langle y\rangle$ and $y^2 \in Z(S)$. Setting $B = \mathrm{Aut}\,(S_1)$ and $C = O_2(B)$, we have that y induces an element $\bar{y}$ of $N_B(P)$ of order at most 2. If $y \notin C$, Lemma 2.4 (iv) implies that the semi-direct product $S_1\langle\bar{y}\rangle$ is of type $\hat{A}_8, J_2$, or $L_3(4)^{(1)}$. However, none of these

2-groups possesses a subgroup isomorphic to $\Omega_1(Y)\langle\bar{y}\rangle \cong E_{32}$. Hence $\bar{y} \in C$. But the elements of C stabilize $S_1 \supset \Phi(S_1) \supset 1$. Since $C_S(S_1) = Z(S_1) = \Phi(S_1)$, we conclude that $S/Z(S_1) \cong E_{32}$, whence $r(S) \geq 5$, a contradiction.

Suppose then that $Z(S_1) \cong Z_2$, whence S_1 is of type A_8 by Lemma 2.2 (iv). Moreover, by Lemma 2.2(v), $Q = (L\Omega_1(Y))' \cong Q_8 * Q_8$ and P acts regularly on $Q/Z(Q)$. Note that $X/\Omega_1(Y) \cong Z_2 \times A_4$, so that the image of L is normal in this factor group. Hence $\Omega_1(Y)L \lhd X$ and so also $Q \lhd X$. We can assume $C_S(Q) \subseteq Q$, since otherwise clearly $r(S) \geq 5$. Hence X/Q is isomorphic to a subgroup of $\mathrm{Out}(Q)$. But then $C_X(P)$ has a Sylow 2-subgroup of order at most 4. However, again P centralizes an element y of $Y - \Omega_1(Y)$. Since P centralizes $Q' \subseteq \Omega_1(Y)$, P centralizes a four subgroup of $\Omega_1(Y)$ and so $C_Y(P) \cong Z_2 \times Z_4$, a contradiction. This completes the proof.

Lemma 3.11. If $Y \cong Z_2 \times Z_4 \times Z_4 \times Z_4$ and $X/Y \cong GL(3,2)$, then $r(X) \geq 5$.

Proof: Clearly either $Z(X) = 1$ or $Z(X) \cong Z_2$ and $Y/Z(X) \cong Z_4 \times Z_4 \times Z_4$. In the latter case, Lemma 3.5 implies that $r(X/Z(X)) \geq 5$, whence $r(X) \geq 5$. Hence we can suppose $Z(X) = 1$. This implies that X acts indecomposably on $\Omega_1(Y)$ with $Y^* = \mho^1(Y) \cong E_8$ as a 3-dimensional invariant subspace.

Set $\bar{X} = X/Y^*$, so that $\bar{Y} \cong E_{16}$ and $\bar{X}$ leaves $\overline{\Omega_1(Y)} \cong Z_2$ invariant. Consider first the case that $\bar{X}$ acts decomposably on $\bar{Y}$, in which case $\bar{Y}_1 = [\bar{X},\bar{Y}] \cong E_8$ and the inverse image Y_1 of $\bar{Y}_1$ in Y is isomorphic to $Z_4 \times Z_4 \times Z_4$. If $X_1 = O^2(X) \subset X$, then $\bar{Y}_1 \subset \bar{X}_1$ and $\bar{X}_1/\bar{Y}_1 \cong GL(3,2)$. Hence Lemma 3.5 applies to the inverse image X_1 of $\bar{X}_1$ and yields that $r(X_1) \geq 5$, whence $r(X) \geq 5$. On the other hand, if $\bar{X}_1 = \bar{X}$, then $\bar{X}/\bar{Y}_1 \cong SL(2,7)$. It follows in this case from Lemma 3.8 (ii) that S contains an element t such that $\langle\bar{t}^2\rangle = \overline{\Omega_1(Y)} = Z(\bar{X})$. Setting $y = t^2$, we have that $y \in \Omega_1(Y) - Y^*$ and $C_S(y) \supset Y$. However, it follows as in Lemma 3.9 that

$C_S(y) = Y$ under the present assumptions concerning the action of X on $\Omega_1(Y)$. Thus the lemma holds if $\bar{X}$ acts decomposably on $\bar{Y}$ and so we can assume that this action is indecomposable.

By Lemma 3.8 (i), we have that $\bar{X}$ splits over $\bar{Y}$. Hence X possesses a subgroup X^* containing Y^* such that $X^*/Y^* \cong GL(3,2)$ and $X = X^*Y$. Now consider the group $X_0 = X^*\Omega_1(Y)$. Since $Z(X) = 1$, it follows that also $Z(X_0) = 1$ and consequently Lemma 3.9 (ii) yields that X_0 contains a subgroup $L \cong A_4$ such that $L \cap \Omega_1(Y) = 1$ and $L\Omega_1(Y)$ has Sylow 2-subgroups of type A_8. If T is a Sylow 2-subgroup of L, this implies, in particular, that $C_{\Omega_1(Y)}(T) \cong Z_2$.

Next we set $\tilde{X} = X/\Omega_1(Y)$ and consider the action of $\tilde{L}$ on $\tilde{Y} \cong E_8$. If $\tilde{L}$ normalizes a four subgroup $\tilde{U}$ of $\tilde{Y}$, then L acts on the inverse image U of $\tilde{U}$ in Y and $U \cong Z_4 \times Z_4 \times Z_2 \times Z_2$. Hence L also acts on $\mho^1(U) \cong Z_2 \times Z_2$. The only possibility then is that T centralizes $\mho^1(U)$, contrary to the conclusion of the preceding paragraph. Hence $C_{\tilde{Y}}(\tilde{L}) = \tilde{Y}_0 \cong Z_2$ and L normalizes the inverse image Y_0 of $\tilde{Y}_0$, which is isomorphic to $Z_4 \times Z_2 \times Z_2 \times Z_2$. But now all the assumptions of Lemma 3.10 are satisfied with LY_0 and Y_0 in the roles of X,Y respectively. Thus $r(LY_0) \geq 5$ and so $r(X) \geq 5$, as asserted.

To obtain a further consequence of the preceding result, we first prove

<u>Lemma 3.12</u>. If $Y \cong Z_4 \times Z_2 \times Z_2 \times Z_2$ and $X/Y \cong GL(3,2)$, then $X/\Omega_1(Y) \cong Z_2 \times GL(3,2)$.

<u>Proof</u>: Suppose false, in which case $X/\Omega_1(Y) \cong SL(2,7)$. Setting $\bar{X} = X/\mho^1(Y)$, we see that $\bar{X}$ satisfies the conditions of Lemma 3.8 (ii). In particular, $\bar{S}$ contains a quaternion subgroup $\bar{Q}$ with $Z(\bar{Q}) = Z(\bar{X})$. Let Y_1, Q be the inverse images of $Z(\bar{Q})$ and $\bar{Q}$ respectively in X. Since

$|Y_1| = 4$ and $Y_1 \triangleleft X$, the only possibility is that Y_1 is cyclic. Thus Q is isomorphic to a nonsplit extension of Q_8 by Z_2, contrary to the fact that the Schur multiplier of Q_8 is trivial.

We now prove

Lemma 3.13. If $Y \cong Z_4 \times Z_4 \times Z_4 \times Z_4$ and $X/Y \cong GL(3,2)$, then $r(X) \geq 5$.

Proof: Set $\bar{X} = X/\Omega_1(Y)$. Suppose first that $\bar{X}$ leaves invariant a 3-dimensional subspace $\bar{Y}_1$ of $\bar{Y}$. Then the inverse image Y_1 of $\bar{Y}_1$ is normal in X and $Y_1 \cong Z_4 \times Z_4 \times Z_4 \times Z_2$. Setting $\tilde{X} = X/\mho^1(Y_1)$, it follows that $\tilde{Y} \cong Z_2 \times Z_2 \times Z_2 \times Z_4$ and that $\tilde{Y}_1 \cong Z_2 \times Z_2 \times Z_2$ is a direct factor of $\tilde{Y}$. Hence by the preceding lemma, $\tilde{X}$ possesses a subgroup $\tilde{X}_1$ of index 2 containing $\tilde{Y}_1$ such that $\tilde{X}_1/\tilde{Y}_1 \cong GL(3,2)$. But now if X_1 denotes the inverse image of $\tilde{X}_1$ in X, we have that $X_1/Y_1 \cong GL(3,2)$ and so we conclude from Lemma 3.11 that $r(X_1) \geq 5$, whence $r(X) \geq 5$.

On the other hand, if $\bar{X}$ leaves invariant a one-dimensional subspace $\bar{Y}_1$ of $\bar{Y}$, then the inverse image Y_1 of $\bar{Y}_1$ is isomorphic to $Z_4 \times Z_2 \times Z_2 \times Z_2$. Hence if we again set $\tilde{X} = X/\mho^1(Y_1)$, we see that $\tilde{Y} \cong Z_4 \times Z_4 \times Z_4 \times Z_2$ and that $\tilde{X}/\tilde{Y} \cong GL(3,2)$. Again Lemma 3.11 yields that $r(\tilde{X}) \geq 5$ and so $r(X) \geq 5$ in this case as well.

Using Lemma 3.12, we can also prove the following result:

Lemma 3.14. If $Y \cong Z_8 \times Z_2 \times Z_2 \times Z_2$ and $X/Y \cong GL(3,2)$, then $r(X) \geq 5$.

Proof: By two applications of Lemma 3.12, we obtain that $X/\Omega_1(Y) \cong Z_4 \times GL(3,2)$. Moreover, $Y_1 = \mho^1(Y)$ is cyclic of order 4 and $Y_1 \subseteq Z(X)$. Setting $X_1 = O^2(X)$ and $\bar{X} = X/\mho^2(Y)$, our conditions imply that $\bar{Y}_1\bar{X}_1$

$= \bar{Y}_1 \times \bar{X}_1$ with $\bar{Y}_1 \cong Z_2$ and $\bar{X}_1$ isomorphic to a nontrivial extension of $GL(3,2)$ by E_8. By Lemma 3.4, a Sylow 2-subgroup of $\bar{X}_1$ is of type A_8 or M_{12}. In either case, we have $r(\bar{X}_1) = 4$, whence $r(\bar{X}) = 5$ and so $r(X) \geq 5$, as asserted.

We need one additional property of X when $X/Y \cong GL(3,2)$ and $r(X) \leq 4$ before we shall be in a position to establish Proposition 3.1. It depends on the following preliminary result.

Lemma 3.15. If $r(X) \leq 4$ and $X = YB$, where B is nonabelian of order 21, then we have

(i) $[B,Y] \cong Z_{2^n} \times Z_{2^n} \times Z_{2^n}$ for some n or of type $Sz(8)$;

(ii) $C_Y(B)$ is cyclic and disjoint from $[B,Y]$.

Proof: In proving (i), it clearly suffices to treat the case that $Y = [B,Y]$. If Y is abelian, it is the direct product of homocyclic subgroups Y_i, $1 \leq i \leq m$, on which B acts indecomposably. Since $Y = [B,Y]$, B centralizes no Y_i and so each Y_i has rank at least 2. Since $C_B(Y)=1$, $P = B'$ does not centralize some Y_i, say Y_1, and so Y_1 has rank at least 3. Since $r(Y) \leq 4$, the only possibility is that $Y = Y_1$ and Y_1 has rank 3 or 4. However, in the latter case, $C_{Y_1}(P)$ would be a nontrivial cyclic group. Since $C_{Y_1}(P)$ is B-invariant, this would imply that $C_{Y_1}(B) \neq 1$, which clearly contradicts the indecomposable action of B on Y_1. Hence $Y = Y_1 \cong Z_{2^n} \times Z_{2^n} \times Z_{2^n}$ for some n and (i) holds.

Suppose then that Y is nonabelian and let Z be a minimal B-invariant subgroup of $Z(Y)$, so that Z is elementary. If P centralizes Z, then $B/P \cong Z_3$ acts on Z and so clearly $Z \cong Z_2$ or $Z_2 \times Z_2$. In the contrary case, $C_B(Z) = 1$, whence $[Z,B] = Z$ and so $Z \cong Z_2 \times Z_2 \times Z_2$ by the argument of the preceding paragraph. Furthermore, if we set $\bar{X} = X/Z$, we have that

$\bar{Y} = [\bar{Y},\bar{B}]$ and so by induction $Y \cong Z_{2^n} \times Z_{2^n} \times Z_{2^n}$ for some n or $\bar{Y}$ is a 2-group of type $Sz(8)$.

We first rule out the case $Z \cong Z_2 \times Z_2$. Indeed, let W be the inverse image of $\Omega_1(\bar{Y})$ in Y, so that $\bar{W} = W/Z \cong E_8$ and P acts regularly on $\bar{W}$. By [13, Theorem 5.6.5] W is necessarily abelian, whence, in fact, W is elementary. Since $Z \cong Z_2 \times Z_2$, we have $W \cong E_{32}$, contrary to the fact that $r(X) \leq 4$.

Next consider the case that $Z \cong E_8$. In this case, $|W| = 2^6$ and P acts regularly on both Z and $\bar{W}$. Since $r(W) \leq 4$, W is not elementary abelian. This implies that $W - Z$ contains no involution. Indeed, otherwise the transitive action of P on $\bar{W}$ would imply that each coset of Z in W contained an involution. But then as $Z \subseteq Z(W)$, each coset would consist entirely of involutions, so W would be elementary abelian which is not the case. We conclude that $Z = \Omega_1(W)$ and hence that $Z = \Omega_1(Y)$. Since P acts transitively on $Z^{\#}$, Y is thus a Suzuki 2-group. Since Y is nonabelian, it follows now from [30] that Y is of type $Sz(8)$ and (ii) holds in this case.

Assume next that $Z \cong Z_2$ and $\bar{Y}$ is of type $Sz(8)$. Then YB is a nontrivial central extension of the normalizer of a Sylow 2-subgroup of $Aut(Sz(8))$. If Y splits over Z, then clearly $Y = Z \times Y_1$, where $Y_1 = [Y,P] \cong \bar{Y}$. Since $Z \cong Z_2$, B centralizes Z and consequently $Y_1 = [B,Y]$, contrary to our assumption $Y = [B,Y]$. Hence Y does not split over Z. Since a Sylow 2-subgroup of $Aut(Sz(8))$ is disjoint from its conjugates, the argument of [4] now yields that the Schur multiplier of $Aut(Sz(8))$ must be divisible by 2. However, this contradicts [4, Theorem 2].

Finally we consider the case $Z \cong Z_2$ and $\bar{Y} \cong Z_{2^n} \times Z_{2^n} \times Z_{2^n}$. Again

W is elementary. As Y is nonabelian, we have $W \subset Y$ and so $n \geq 2$. Since $Z \cong Z_2$ and $\bar{Y}$ is abelian, we also have that $V = \mho^1(Y) \subseteq Z(Y)$. Thus V is abelian and P does not centralize V. It follows at once that $\Omega_1(V)$ contains a B-invariant subgroup $Z^* \cong E_8$ on which P acts regularly. In particular, Z^* is a minimal B-invariant subgroup of $Z(Y)$ and so we are reduced to one of the cases treated above. This establishes (i).

In proving (ii), we now drop the assumption that $Y = [B,Y]$. We set $Y_1 = [Y,B]$ and $Y_0 = C_Y(B)$, so that $Y = Y_1Y_0$ and $Y_1 \lhd X$. Moreover, Y_1 has the structure given in (i). In particular, P acts regularly on Y_1, so $Y_1 \cap Y_0 = 1$. Setting $\bar{X} = X/\Phi(Y_1)$, we have that $\bar{Y} = \bar{Y}_1\bar{Y}_0$ with $\bar{Y}_1 \cong Z_2 \times Z_2 \times Z_2$ and $\bar{Y}_0 = C_{\bar{Y}}(\bar{B}) \cong Y_0$. Since B acts irreducibly on $\bar{Y}_1$, Thompson's $A \times B$-lemma implies that $\bar{Y}_0$ centralizes $\bar{Y}_1$, whence $\bar{Y} = \bar{Y}_1 \times \bar{Y}_0$. Since $r(Y) \leq 4$, we must have $r(\bar{Y}_0) \leq 1$ and so $\bar{Y}_0$ must be cyclic. Hence Y_0 is cyclic and (ii) holds.

<u>Lemma 3.16</u>. If $X/Y \cong GL(3,2)$ and $r(X) \leq 4$, then $[B,Y] \cong Z_{2^n} \times Z_{2^n} \times Z_{2^n}$ for some n for any subgroup B of X of order 21.

<u>Proof</u>: B exists and is nonabelian of order 21. Hence by the preceding lemma, applied to YB, either our conclusion holds or else $Y_1 = [B,Y]$ is of type $Sz(8)$. Since X/Y contains an involution which acts nontrivially on $Y/Z(Y)$, it follows from Lemma 2.8 that $Y \neq Y_1$. Since $Y = Y_1Y_0$, where $Y_0 = C_Y(B)$, we conclude that $Y_0 \neq 1$. Moreover, by the preceding lemma, Y_0 is cyclic and $Y_1 \cap Y_0 = 1$.

We argue first that $Y = Y_1 \times Y_0$. Let $Y_0 = \langle y \rangle$ and let b be an element of B of order 3. We have that $C_{Y_1}(b) \cong Z_4$ and we set $C_{Y_1}(b) = \langle y_1 \rangle$. Clearly y leaves $C_{Y_1}(b)$ invariant and so y either centralizes or inverts y_1. But B possesses a 7-element which transitively permutes the cosets of $\Phi(Y_1)$ in Y_1, other than $\Phi(Y_1)$ itself, and which centralizes y. It follows therefore that either y centralizes or inverts Y_1. However, the latter case is impossible as then Y_1 would be forced to be abelian. Hence $Y_0 = \langle y \rangle$ centralizes Y_1 and so $Y = Y_1 \times Y_0$, as asserted.

In particular, $Z(Y) = Z(Y_1) \times Y_0$ and so $\mho^1(Y_0) \lhd X$. Setting $\bar{X} = X/\mho^1(Y_0)$, we have now that $\bar{Y} = \bar{Y}_1 \times \bar{Y}_0$ with $\bar{Y}_1$ of type $Sz(8)$ and $\bar{Y}_0 \cong Z_2$. Then $\bar{X}/\bar{Y} \cong GL(3,2)$ and $\bar{X}/\bar{Y}$ acts faithfully on $\bar{Y}/Z(\bar{Y}) \cong E_8$. Hence $\bar{Y}$ admits an automorphism which induces an automorphism of $\bar{Y}/Z(\bar{Y})$ of order 2, again contradicting Lemma 2.8.

Remark. The argument that proved that Y_0 either centralizes or inverts Y_1 is valid equally well when $Y_1 \cong Z_4 \times Z_4 \times Z_4$.

We are at last in a position to establish Proposition 3.1. Let then X be a nonsolvable 2-constrained group with $O(X) = 1$ and $r(X) \leq 4$ such that $X/O_2(X) \cong GL(3,2)$. We set $Y = O_2(X)$ and we have $C_X(Y) \subseteq Y$, so the preceding results apply to X and Y. We argue by induction on $|X|$.

If Y is elementary abelian, then Y has rank at most 4, so clearly $Y \cong E_8$ or E_{16} and the proposition holds. Hence we can suppose that Y is not elementary, in which case $\Phi(Y) \neq 1$. We let Z be a minimal normal subgroup of X contained in $Z(Y) \cap \Phi(Y)$. Then $Z \cong Z_2$ or E_8. We set $\bar{X} = X/Z$. Since $Z \subseteq \Phi(Y)$, we have that $C_{\bar{X}}(\bar{Y}) \subseteq \bar{Y}$. Thus $\bar{X}$ is 2-constrained with $O(\bar{X}) = 1$ and $\bar{X}/\bar{Y} \cong GL(3,2)$. We conclude therefore by induction that $\bar{Y} \cong E_8$, E_{16}, or $Z_4 \times E_8$.

We let B be a subgroup of X of order 21 and note that in all cases $[\bar{B},\bar{Y}] \cong E_8$. Consider first the case that $Z \cong Z_2$. Then B centralizes Z and it follows from Lemma 3.15 that $Y_1 = [B,Y] \cong E_8$ and that $Y_0 = C_Y(B)$ is cyclic. Our present conditions force $Y_0 \cong Z_2$, Z_4, or Z_8. Moreover, Y_0

centralizes Y_1 by Thompson's $A \times B$-lemma, so $Y = Y_1 \times Y_0$. As Y is not elementary under our present assumptions, we have $Y_0 \cong Z_4$ or Z_8. However, as $r(X) \leq 4$, $Y \not\cong Z_8 \times Z_2 \times Z_2 \times Z_2$ by Lemma 3.14 and so $Y_0 \not\cong Z_8$. Hence $Y_0 \cong Z_4$ and consequently $Y \cong Z_4 \times E_8$. Thus the proposition holds when $Z \cong Z_2$.

Suppose then that $Z \cong E_8$, in which case X/Y acts faithfully on Z. Hence $Z \subseteq Y_1 = [B,Y]$. This time using Lemma 3.15 together with Lemma 3.16, we conclude that $Y_1 \cong Z_4 \times Z_4 \times Z_4$ and $Y_0 = C_Y(B) \cong 1$, Z_2, or Z_4. Since $r(X) \leq 4$, Lemma 3.5 implies that $Y \neq Y_1$. Hence $Y \supset Y_1$ and as $Y = Y_0Y_1$, it follows that $Y_0 \neq 1$. Thus $Y_0 \cong Z_2$ or Z_4.

We treat these two cases in succession; so assume first that $Y_0 \cong Z_2$. By Lemma 3.11, $Y \not\cong Z_2 \times Z_4 \times Z_4 \times Z_4$, so Y_0 does not centralize Y_1. By the remark following Lemma 3.16, we have then that Y_0 inverts Y_1. In particular, Y_1 is thus the unique abelian subgroup of index 2 in Y and so is characteristic in Y. Hence $\bar{Y}_1 \lhd \bar{X}$. If $\bar{X}/\bar{Y}_1 \cong Z_2 \times GL(3,2)$, we see that X possesses a subgroup X_1 of index 2 with $Y_1 \lhd X_1$ and $X_1/Y_1 \cong GL(3,2)$. Since $Y_1 \cong Z_4 \times Z_4 \times Z_4$, Lemma 3.5 implies that $r(X_1) \geq 5$, contrary to $r(X) \leq 4$. Thus $\tilde{X} = \bar{X}/\bar{Y}_1 \cong SL(2,7)$ with $\tilde{Y}_0 = Z(\tilde{X})$. This means that if $Y_0 = \langle y \rangle$, then y induces an automorphism of Y_1 which lies in the commutator subgroup of $\mathrm{Aut}\,(Y_1)$. However, as y inverts Y_1, a matrix representation (over Z_4) of y as an automorphism of Y_1 has determinant -1, which implies that this automorphism does not lie in $(\mathrm{Aut}\,(Y_1))'$. Thus when $Y_0 \cong Z_2$, we reach a contradiction whether Y_0 centralizes or inverts Y_1.

Suppose finally that $Y_0 \cong Z_4$, in which case $\bar{Y} \cong Z_4 \times Z_2 \times Z_2 \times Z_2$. Applying Lemma 3.12, it follows now that $\bar{X}/\Omega_1(\bar{Y}) \cong Z_2 \times GL(3,2)$. Hence $\bar{X}$

possesses a subgroup $\bar{X}_1$ of index 2 with $\bar{Y} \cap \bar{X}_1 = \Omega_1(\bar{Y})$ and $\bar{X}_1/\Omega_1(\bar{Y}) \cong GL(3,2)$. If X_1 denotes the inverse image of $\bar{X}_1$ in X, it follows that $Y \cap X_1 = Y_1(Y_0 \cap X_1)$ with $Y_0 \cap X_1 \cong Z_2$ and $X_1/Y \cap X_1 \cong GL(3,2)$. Moreover, $B \subseteq X_1$ and so $C_{Y \cap X_1}(B) = Y_0 \cap X_1 \cong Z_2$. Since also $r(X_1) \leq 4$, we see that X_1 satisfies the same conditions that X satisfied when $Y_0 \cong Z_2$. Hence the argument of the preceding paragraph can be repeated for X_1 to yield a contradiction. Thus Proposition 3.1 is finally proved.

4. Theorem B, the A_5 case. In this section we shall treat the case in which $X/O_2(X)$ contains a subgroup isomorphic to A_5.

Proposition 4.1. If X is a nonsolvable 2-constrained group with $O(X) = 1$ and $r(X) \leq 4$ such that $X/O_2(X)$ contains a subgroup isomorphic to A_5, then $O_2(X) \cong E_{16}$ or $Q_8 * D_8$.

Proof: In contrast to Proposition 3.1, the proof of Proposition 4.1 is quite short. Again we put $Y = O_2(X)$. Since the proposition is simply an assertion concerning the structure of Y, we can assume without loss that X/Y itself is isomorphic to A_5.

We argue by induction on $|Y|$. We let S be a fixed Sylow 2-subgroup of X. If Y is elementary abelian, then $Y \cong E_{16}$ as $r(X) \leq 4$ and X/Y acts faithfully on Y. Thus the proposition holds in this case and so we can assume Y is not elementary abelian.

Let Z be a minimal normal subgroup of X contained in $Z(Y) \cap \Phi(Y)$. If X/Y acts faithfully on Z, then $Z \cong E_{16}$, again as $r(X) \leq 4$. In the contrary case, clearly $Z \cong Z_2$. Setting $\bar{X} = X/Z$, our hypotheses carries over to $\bar{X}$ and so we conclude by induction that $\bar{Y} \cong E_{16}$ or $Q_8 * D_8$.

Consider first the case that $\bar{Y} \cong Q_8 * D_8$. We claim that $Z \cong Z_2$. If not, $Z \cong E_{16}$ and a subgroup P of X of order 5 acts regularly on Z.

But $\bar{P}$ centralizes $Z(\bar{Y})$ and consequently $Y_0 = C_Y(P) \neq 1$. Hence $ZY_0 = Z \times Y_0$ and so ZY_0 has 2-rank at least 5, contrary to $r(X) \leq 4$. Thus $Z \cong Z_2$, as asserted. Thus $|Y| = 2^6$. Since P acts faithfully on Y and centralizes Z, we conclude now from [Lemma 3.9, Part I] that Y is necessarily of type $U_3(4)$. However, one checks easily that the automorphism group of a 2-group of type $U_3(4)$ is solvable and so does not involve A_5.

Our argument shows that necessarily $\bar{Y} \cong E_{16}$. If $Z \cong Z_2$, then $|Y| = 2^5$. Since Y is not elementary abelian, clearly Y cannot be abelian and so Y is extra-special. Since $\mathrm{Aut}(Q_8 * Q_8)$ is solvable, the only possibility is that $Y \cong Q_8 * D_8$, in which case the proposition holds.

Assume finally that $Z \cong E_{16}$. Since $Z \subseteq Z(Y)$ and $r(X) \leq 4$, $Y - Z$ contains no involutions. We apply Lemma 2.6 twice to conclude that X splits over Y. Thus $X = LY$ with $L \cong A_5$ and $L \cap Y = 1$. Let v be an element of L of order 3. Since Y is not elementary abelian, it follows from [31, Theorem 8.2] that v does not act regularly on Y. Hence $C_Z(v) \neq 1$ and consequently $LZ \cong A_5 \cdot E_{16}^{(1)}$. In particular, LZ has Sylow 2-subgroups of type A_8.

Suppose that also $\overline{LY} \cong A_5 \cdot E_{16}^{(1)}$. Then L possesses a subgroup $L_1 \cong A_4$ such that $C_{\bar{Y}}(\bar{L}_1) = \bar{Y}_1 \cong Z_2$. But then L_1 acts on the inverse image Y_1 of $\bar{Y}_1$ and $Y_1 \cong Z_4 \times Z_2 \times Z_2 \times Z_2$. However, Lemma 3.10 can now be applied to L_1Y_1 and we conclude that $r(L_1Y_1) \geq 5$, contrary to $r(X) \leq 4$. Thus $\overline{LY} \cong A_5 \cdot E_{16}^{(2)}$ and $\bar{L}$ acts transitively on $\bar{Y}^{\#}$.

Let $Y_0 = C_Y(v)$. We shall argue in this final case that $r(Y) \geq 5$, again contradicting $r(X) \leq 4$, and this will complete the proof of the proposition. Observe first that $\bar{v}$ acts regularly on $\bar{Y}$ and hence $Y_0 \subseteq Z$.

Since $LZ \cong A_5 \cdot E_{16}^{(1)}$, we have, in fact, that $Y_0 \cong Z_2 \times Z_2$. Furthermore, as L acts on $Z^{\#}$ in orbits of length 5 and 10, Y_0 necessarily intersects each orbit nontrivially. This implies that the set of squares of elements of elements of $Y - Z$ also intersects Y_0 nontrivially. Hence if we set $\tilde{Y} = Y/Y_0$, it follows that $\tilde{Y} - \tilde{Z}$ contains an involution $\tilde{t}$. But v acts regularly on both Z/Y_0 and Y/Z and so v also acts regularly on $\tilde{Y}$. We conclude therefore that $\tilde{W} = \langle \tilde{Z}, \tilde{t}, \tilde{t}^v \rangle \cong E_{16}$.

Now let W be the inverse image of $\tilde{W}$ in Y. Then $Z \subseteq Z(W)$ and $W/Z \cong Z_2 \times Z_2$. Hence $|W'| = 1$ or 2 and consequently $\bar{W} = W/W'$ is abelian of order 32 or 64. Furthermore, $C_{\bar{W}}(v) = \bar{Y}_0$ is of order 2 or 4 (according as $|W'| = 2$ or 1) and $\bar{W} = \bar{Y}_0 \times \bar{W}_1$, where $\bar{W}_1 = [\bar{W}, v]$. Since $\bar{W}/\bar{Y}_0 \cong W/Y_0 = \tilde{W} \cong E_{16}$, we have $\bar{W}_1 \cong E_{16}$ and consequently $\bar{W}$ has rank at least 5. Thus $r(Y) \geq 5$, as asserted, and Proposition 4.1 is proved.

We now complete the proof of Theorem B. Since $\mathrm{Out}(Q_8 * D_8) \cong S_5$, it suffices to eliminate the cases $X/Y \cong S_6$ and $X/Y \cong A_8$, where $Y \cong E_{16}$. Suppose first that $X/Y \cong S_6$. By Lemmas 2.3 and 2.6, X contains a subgroup $L \cong A_5$ which is transitive on $Y^{\#}$. Let $L_1 \cong A_4$ be a subgroup of L. Then a Sylow 2-subgroup T of L_1Y is of type $L_3(4)$ by Lemma 2.2(iv) and a Sylow 3-subgroup B of L_1Y acts fixed-point-free on T. Furthermore, $N/Y \cong Z_2 \times S_4$, where $N = N_X(L_1Y)$, and so N is a split extension of T by $N_X(B) \cong Z_2 \times S_3$. As $r(X) \leq 4$, no 2-element of $N_X(B)$ stabilizes the chain $T \supset Z(T) \supset 1$. However, Lemma 7.4 of Part I yields that a Sylow 2-subgroup S of X is of type $\hat{A}_{10}$. This conflicts with $SCN_3(S) \neq \emptyset$. Thus the case $X/Y \cong S_6$ does not occur. Since $A_8 \supset S_6$, the latter case can not occur either. This completes the proof of Theorem B.

5. <u>Theorems C and D, initial reductions</u>. In the next three sections we

shall establish Theorems C and D.

Proposition 5.1. Theorems C and D hold if a Sylow 2-subgroup of G is of type $PSp(4,q)$, q odd, $L_3(4)$, $L_3(4)^{(1)}$, J_2, A_8, $\hat{A}_8$, A_{10}, or M_{12}.

Proof: Let G be a simple group satisfying the hypotheses of the Theorem C or D, let S be a Sylow 2-subgroup of G, and suppose S is one of the specified types.

If S is of type $L_3(4)^{(1)}$, then as noted in the proof of Proposition 2.7 of part I, the argument of [16, Lemma 9.4] implies that G is not simple, so this case cannot occur.

If S is of one of the remaining types [19, Theorem A], [16, Theorems A and C], [17, Theorem A], [21, Theorem A], [18, Theorem A], [36, Theorem 1.1], [17, Theorem B*], [37, Theorem 1.1], [18, Theorem B], and [6, Theorem A] together imply that G is isomorphic to any of the following groups: $PSp(4,q)$, q odd, $L_4(q)$, $q \equiv 3,5 (\mathrm{mod}\ 8)$, $U_4(q)$, $q \equiv 3,5(\mathrm{mod}\ 8)$, $G_2(q)$, $q \equiv 3,5(\mathrm{mod}\ 8)$, $D_4^2(q)$, $q \equiv 3,5(\mathrm{mod}\ 8)$, $L_3(4)$, A_8, A_9, A_{10}, A_{11}, J_2, J_3, M_{12}, M_{22}, M_{23}, M^c, or L.

Using the known structure of the centralizers of involutions and the normalizers of maximal elementary abelian 2-subgroups in each of these groups, we check easily that the only ones which satisfy the hypothesis of Theorem C or D are the groups listed in those theorems.

In proving Theorems C and D in the $SCN_3(2)$ empty case, we need the following additional result.

Proposition 5.2. Let G be a simple group with Sylow 2-subgroup S such that $SCN_3(S)$ is empty and the involutions of an element of $U(S)$ are conjugate in G. If S contains a subgroup of type $\hat{A}_8$ or a normal subgroup isomorphic to $Q_8 * D_8$, then S is of type J_2 or $\hat{A}_{10}$.

Proof: By the results of MacWilliams [35] and Harada [29], either the derived conclusion holds or S is of type $G_2(q)$, $q \equiv 1,7 \pmod 8$. However, one checks directly in these cases that S does not possess a normal subgroup isomorphic to $Q_8 * D_8$ or a subgroup of type $\hat{A}_8$.

In the next section as well as in Part VI, we shall encounter a certain configuration which it will be preferable to analyze abstractly. Thus we consider a group X which satisfies the following conditions:

(a) X has sectional 2-rank at most 4;

(b) The centralizer of every central involution of X is solvable;

(c) X contains a subgroup $Q \cong Q_8 * Q_8$;

(d) If $N = N_X(Q)$, then N contains a 3-element which acts regularly on $Q/Z(Q)$.

We conclude this section by establishing some general properties of such a group X. We first prove

Lemma 5.3. The following conditions hold:

(i) $N/O(N)Q$ is isomorphic to a subgroup of a Sylow 3-normalizer in A_8;

(ii) $Z(Q)$ is the center of any Sylow 2-subgroup of X containing Q and N covers $C_X(Z(Q))/O(C_X(Z(Q)))$.

Proof: Set $C = C_X(Q)$. Since $r(X) \leq 4$, it is immediate that Q is a Sylow 2-subgroup of QC. Hence if Y is a Sylow 2-subgroup of X containing Q, it follows that $Z(Y) = Z(Q)$. In addition, we have that $QC = Q \times O(N)$ and that $\bar{N} = N/QC$ is isomorphic to a subgroup of $\mathrm{Aut}(Q)/\mathrm{Inn}(Q)$. Thus $\bar{N}$ is isomorphic to a subgroup of a Sylow 3-normalizer in A_8 and so (i) and the first assertion of (ii) hold.

Next set $K = C_X(Z(Q))$. Since $Z(Y) = Z(Q)$, K is solvable by our hypothesis. Moreover, $N \subseteq K$ as $Z(Q)$ is characteristic in Q. Next let W be a Sylow 2-subgroup $O_{2',2}(K)$ and set $\tilde{K} = K/O(K)$, so that $\tilde{W} = O_2(\tilde{K})$. Since $r(\tilde{K}) \leq 4$ and K is solvable, $\tilde{K}/\tilde{W}$ is isomorphic to a solvable subgroup of A_8 with no nontrivial normal 2-subgroups. All such subgroups are described in [Lemma 4.1, Part I]. On the other hand, by assumption (b), $\tilde{K}$ contains a 3-element $\tilde{x}$ which acts regularly on $\tilde{Q}/Z(\tilde{Q})$. Since $[\tilde{Q},\tilde{x}]=\tilde{Q}$, we conclude at once that $\tilde{Q} \subseteq \tilde{W}$. Hence without loss we can assume that $Q \subseteq W \subseteq Y$.

If $Q = W$, then by the Frattini argument N covers $K/O(K)$ and (ii) holds. Hence we may suppose that $W \supset Q$, whence $W_1 = N_W(Q) = N \cap W \supset Q$. Since $N \subseteq K$, clearly $W_1 \subseteq O_{2',2}(N)$, whence $\bar{W}_1 \subseteq O_2(\bar{N})$. It follows therefore from the structure of $\bar{N}$ that $\bar{N} \cong Z_2 \times Z_3$ or $Z_2 \times S_3$, whence $\bar{W}_1 = O_2(\bar{N}) \cong Z_2$. Since $O_3(\bar{N})$ acts regularly on $Q/Z(Q)$, we see that $W_1/Z(Q) \cong Z_2 \times Z_2 \int Z_2$. Thus Q is characteristic in W_1 and consequently $W_1 = W$. Since Q is now characteristic in W, we conclude once again by the Frattini argument that N covers $K/O(K)$.

We now prove

Lemma 5.4. Assume that a Sylow 2-subgroup Y of X containing Q satisfies the following conditions:

(a) $|Y| \geq 2^8$;

(b) Q is normal in a subgroup V of Y of type A_{10} which splits over Q.

Then we have

(i) $Y = \langle Y_1, Y_2, t\rangle$, where $Y_1 \cong Y_2 \cong Q_{16}$ or QD_{16}, $Y_1Y_2 = Y_1 * Y_2$, and t is an involution which interchanges Y_1 and Y_2;

(ii) $|Y| = 2^8$ and $SCN_3(Y)$ is empty.

<u>Proof</u>: Let K and W be as in the preceding lemma. Since N covers $K/O(K)$ and $Y \subseteq K$, we have that N contains a Sylow 2-subgroup of K, which without loss we can take to be Y. Since $|Y| \geq 2^8$ by assumption, $|\bar{Y}| \geq 2^3$ and hence by the structure of $\bar{N} = N/O(N)Q$, we have, in fact, $|\bar{Y}| = 2^3$, whence $\bar{N}$ is isomorphic to a full Sylow 3-normalizer in A_8 and $|Y| = 2^8$. We have that $\bar{Y} \cong D_8$ and $Y/Z(Y)$ splits over $Q/Z(Y)$. Considering the action of $\bar{N}$ on Q, we see that $Y = \langle Q, u_1, u_2, t\rangle$, where $Q_1 * Q_2$ with $Q_1 \cong Q_2 \cong Q_8$, $u_1^2 \equiv u_2^2 \equiv t^2 \equiv 1 \pmod{Z(Y)}$, $\langle Q_1, u_1\rangle \cong \langle Q_2, u_2\rangle \cong Q_{16}$ or QD_{16}, $Q_1^t = Q_2$, and $u_1^t = u_2$.

Now we use the fact that $Q \lhd V \subseteq Y$, where V is of type A_{10} with V splitting over Q. One checks from the preceding relations that there is only one possibility for V; namely, $V = \langle Q, u_1u_2, t\rangle$ with $u_1^2 = u_2^2$ and $(u_1u_2)^2 = t_2 = 1$. Using the fact that u_1u_2 is an involution, we argue next that u_1 centralizes u_2. If $u_1^2 = u_2^2 = 1$, the desired conclusion is immediate. In the contrary case, $u_1^2 = u_2^2 = z$, where $\langle z\rangle = Z(Y)$. Then $\langle u_1, u_2\rangle$ is of order 8 and is clearly not isomorphic to D_8 or Q_8. Hence $\langle u_1, u_2\rangle$ is abelian and so u_1 centralizes u_2 in this case as well.

We conclude now that $Y_i = \langle Q_i, u_i\rangle \cong Q_{16}$ or QD_{16}, $i = 1,2$, and that $Y_1Y_2 = Y_1 * Y_2$ with t interchanging Y_1 and Y_2. Finally the last statement of the lemma can be checked directly from the structure of Y.

If $Y_1 \cong Q_{16}$ in Lemma 5.4, one checks that Y is of type $PSp(4,q)$ for suitable $q \equiv 1$ or $7 \pmod 8$ (cf. [21]) and so the possibilities for X are determined from [21]. We need a comparable classification theorem in the case that $Y_1 \cong QD_{16}$. Moreover, because of later applications in Part VI, it will be preferable to treat this case in general, independent of the special assumption satisfied by the group X above.

Thus we shall prove

Proposition 5.5. There exists no fusion-simple group in which a Sylow 2-subgroup has the form $Q\langle t\rangle$, where $Q = Q_1 * Q_2$ with $Q_1 \cong Q_2 \cong QD_{16}$ and t is an involution which interchanges Q_1 and Q_2.

We assume G is a fusion simple group with Sylow 2-subgroup S of the given form $Q\langle t\rangle$ and proceed in a sequence of lemmas to derive a contradiction. We let $Q_i = \langle a_i, b_i \mid a_i^2 = b_i^4 = 1\rangle$, so that $a_i b_i$ is of order 8, $i = 1,2$. We also put $z = b_i^2$, so that $\langle z\rangle = Z(Q_i)$, $i = 1,2$. Without loss we can assume that $a_1^t = a_2$ and $b_1^t = b_2$. Finally we put $u_i = (a_i b_i)^2$, $i = 1,2$, and $u = u_1 u_2$.

We begin with the following omnibus lemma concerning the structure of S, the various assertions of which can be directly checked.

Lemma 5.6. The following conditions hold:

(i) $S/\langle z\rangle$ is of type A_{10};

(ii) $S' = \langle z,u,a_1a_2,b_1b_2,u_1\rangle$ and $\Phi(S') = \langle z,u\rangle$;

(iii) S has 8 conjugacy classes of involutions represented by z,u, b_1u, a_1a_2, a_1, b_1b_2, t, and tz;

(iv) $C_S(u) = \langle z,u,a_1a_2,b_1b_2,a_1b_1,t\rangle$ is of order 2^7;

(v) $C_S(b_1u) = \langle z,b_1u_1,a_2b_2,b_1b_2\rangle$ is of order 2^5 and $C_S(b_1u)' = \langle u_2\rangle \cong Z_4$;

(vi) $C_S(a_1a_2) = \langle z,u,a_1,a_2,t\rangle \cong Z_2 \times Z_2 \int Z_2$ is of order 2^5 and $C_S(a_1a_2)' = \langle z, a_1a_2\rangle \cong Z_2 \times Z_2$;

(vii) $C_S(a_1) = \langle a_1\rangle \times \langle a_2,b_2\rangle$ is of order 2^5 and $C_S(a_1)' = \langle u_2\rangle \cong Z_4$;

(viii) $C_S(b_1b_2) = \langle z,u,b_1,b_2,t\rangle \cong Z_2 \times Z_2 \int Z_2$ is of order 2^5 and $C_S(b_1b_2)' = \langle z, b_1b_2\rangle \cong Z_2 \times Z_2$;

(ix) $C_S(t) = C_S(tz) = \langle z, a_1a_2, b_1b_2, t\rangle \cong Z_2 \times Z_2 \times D_8$ is of order 2^5 and $C_S(t)' = C_S(tz)' = \langle u\rangle$;

(x) $SCN_3(S)$ is empty;

(xi) $U = \langle z,u\rangle$ is the unique element of $\mathcal{U}(S)$;

(xii) Q is the normal closure of $\langle a_1, b_1u\rangle$ in S.

We now prove

Lemma 5.7. The following fusion relations hold:

(i) z is not conjugate to u, b_1u, or a_1 in G;

(ii) a_1, b_1u, and u are conjugate in $C_G(z)$.

Proof: If $z \sim u$ in G, then all involutions of U are conjugate in G and so by the main results of [35] and [29], S is either of type $G_2(q)$, $q \equiv 1,7 \pmod 8$, J_2 or L. However, in none of these cases is a Sylow 2-subgroup of the form of S. Hence $z \not\sim u$ in G.

Suppose $b_1u \sim z$ in G, whence $(b_1u)^g = z$ with $C_S(b_1u)^g \subseteq S$ for some g in G. Since $C_S(b_1u)' = \langle u_2\rangle$ and $u_2^2 = z$, it follows that $z^g \in \Phi(C_S(b_1u)') \subseteq \Phi(S') = U$. Since $u \sim zu$ in S and $z \not\sim u$ in G, this forces $z^g = z$, which is clearly impossible. Since also $z \in \Phi(C_S(a_1)')$, we conclude similarly that $z \not\sim a_1$ in G. Thus (i) holds.

Now set $T = \langle z, u, a_1b_1, a_1a_2, b_1b_2, t\rangle$, so that T is a maximal subgroup of S and every involution of $S - T$ is conjugate in S to b_1u or a_1. By Thompson's fusion lemma, $b_1u \sim y$ in G for some y in T. But as $|C_S(b_1u)| = |C_S(a_1)| = 2^5$, this implies that every involution of T has a centralizer in S of order at least that of any involution of $S - T$. Hence we can assume that the involution y is extremal. Thus we have $a_1^g = y$ and $C_S(a_1)^g \subseteq C_S(y)$. We can also suppose that $y = z$, u, a_1a_2, b_1b_2, t, or tz. However, the first possibility is excluded by (i). On the other hand, in the last 4 cases, $|C_S(y)| = |C_S(a_1)|$, but $C_S(y) \not\cong C_S(a_1)$ by the preceding lemma. Hence the only possibility is that $y = u$. Since $z \in \Phi(C_S(a_1)')$, it follows as in the proof of (i) that $z^g = z$. Thus $a_1 \sim u$ in $C_G(z)$. Similarly $b_1u \sim u$ in $C_G(z)$ and so (ii) also holds.

Now set $M = C_G(z)$ and $\bar{M} = M/O(M)$. We shall prove

Lemma 5.8. $\bar{M}$ possesses a normal subgroup with Sylow 2-subgroup $\bar{Q}$ of the form $\bar{L}_1 * \bar{L}_2$, where $\bar{L}_i \cong SL(2,q_i)$, q_i odd, or $\hat{A}_7$, $i = 1,2$. In particular, $Q \cong Q_{16} * Q_{16}$.

Proof: Set $\tilde{M} = \bar{M}/\langle\bar{z}\rangle$. By Lemma 5.7 (ii) $\bar{M}$ does not have a normal 2-complement and hence neither does $\tilde{M}$. Since $\langle\tilde{u}\rangle = \tilde{U} = Z(\tilde{S}) \cong Z_2$, this implies that $\tilde{u} \in \tilde{K} = O^2(\tilde{M})$. Since $u \sim a_1 \sim b_1u$ in M it follows that also $\tilde{a}_1$ and $\tilde{b}_1\tilde{u}$ are in $\tilde{K}$. But $Q = \langle a_1, b_1u\rangle^S$ by Lemma 5.6 (xii) and consequently $\tilde{Q} \subseteq \tilde{K}$. Hence either $\tilde{Q}$ is a Sylow 2-subgroup of $\tilde{K}$ or $\tilde{S} \subseteq \tilde{K}$.

Consider the latter possibility. Since $\tilde{u}$ is not isolated in $\tilde{K}$ and $\tilde{K} = O^2(\tilde{M})$ has no normal subgroups of index 2 and since $O(\tilde{K}) = 1$, $\tilde{K}$ is therefore fusion-simple. But $\tilde{S}$ is of type A_{10} by Lemma 5.6(i). We can therefore apply [17, Theorem B*] and [36] to conclude that $\tilde{K}$ possesses a normal subgroup $\tilde{L}$ of odd index with $\tilde{L} \cong A_{10}$, A_{11}, $L_4(q)$, $q \equiv 3 \pmod 8$ or $U_4(q)$, $q \equiv 5 \pmod 8$. Since S does not split over $\langle z\rangle$, the inverse image $\bar{L}$ of $\tilde{L}$ in $\bar{M}$ is a nontrivial central extension of $\tilde{L}$ by Z_2. However, the Schur multipliers of A_{10}, A_{11}, $L_4(q)$, or $U_4(q)$ are well known [8,25] and it follows that $\bar{L} \cong \hat{A}_{10}$, $\hat{A}_{11}$, $SL(4,q)$, or $SU(4,q)$. But it is easily verified that none of these groups has a Sylow 2-subgroup of the form of S.

Our argument yields that $\tilde{Q}$ is a Sylow 2-subgroup of $\tilde{K}$. But $\tilde{Q} \cong D_8 \times D_8$ and $\tilde{K}$ is fusion simple. Hence by the main result of [20], $\tilde{K}$

possesses a normal subgroup $\tilde{L}$ of odd index of the form $\tilde{L} = \tilde{L}_1 \times \tilde{L}_2$, where $\tilde{L}_i \cong L_2(q_i)$, q_i odd, or A_7 and $\tilde{Q} \cap \tilde{L}_i \cong D_8$, $i = 1,2$. Let $\bar{L}$, $\bar{L}_i$ be the inverse images in $\bar{M}$ of $\tilde{L}$, $\tilde{L}_i$ respectively, $i = 1,2$. Since $\langle z\rangle = Z(Q)$, $\bar{L}_i$ does not split over $\langle z\rangle$ and so $\bar{L}_i \cong SL(2,q_i)$ or $\hat{A}_7$, $i = 1,2$, and $\bar{L} = \bar{L}_1 * \bar{L}_2$. Furthermore, $\bar{Q}$ is a Sylow 2-subgroup of $\bar{L}$ and $\bar{Q} \cap \bar{L}_i \cong Q_{16}$, $i = 1,2$, so $Q \cong \bar{Q} \cong Q_{16} * Q_{16}$. Hence all Parts of the lemma hold.

Now we can derive an immediate contradiction. Indeed, by the lemma $Q \cong Q_{16} * Q_{16}$. However, by assumption, $Q \cong QD_{16} * QD_{16}$. But the groups $Q_{16} * Q_{16}$ and $QD_{16} * QD_{16}$ are not isomorphic, since the latter contains an involution whose centralizer is isomorphic to $Z_2 \times QD_{16}$, while the former does not. This completes the proof of Proposition 5.5.

6. Theorems C and D, the A_5 case. Henceforth G will denote a simple group of sectional 2-rank at most 4 which possesses a nonsolvable 2-constrained 2-local subgroup M. We let R be a Sylow 2-subgroup of $O_{2',2}(M)$ and set $\bar{M} = M/O(M)$. Then $O(\bar{M}) = 1$, $\bar{R} = O_2(\bar{M})$, $C_{\bar{M}}(\bar{R}) \subseteq \bar{R}$, $r(\bar{M}) \leq 4$, and $\bar{M}$ is nonsolvable. Thus $\bar{M}$ satisfies the hypotheses of Theorem B(I) or B(II).

It will be convenient to say correspondingly that M is of type GL(3,2) or A_5.

In view of Proposition 5.1, Theorems C and D will be proved provided we can show that a Sylow 2-subgroup of G is of type $L_3(4)$, $L_3(4)^{(1)}$, J_2, A_8, $\hat{A}_8$, A_{10}, $\hat{A}_{10}$, or M_{12}. This we shall do in Sections 6 and 7. We shall treat separately the cases that M is of type GL(3,2) and of type A_5. However, as we shall see, one of our results in the A_5 case will be needed in the analysis of the $GL(3,2)$ case. Hence it will be preferable to treat the case that M is of type A_5 first. This we shall do in the present section.

We shall divide the analysis into three parts, as follows:

(I) M is the centralizer of some involution of G;

(II) $\overline{M}$ acts transitively on $\overline{R}^{\#}$;

(III) $\overline{M}$ acts intransitively on $\overline{R}^{\#}$ and the centralizer of every central involution of G is solvable.

Clearly case (I) corresponds to Theorem D. We note, however, that in case (II), we need not impose the condition that the centralizer of every central involution is solvable.

We let T be a Sylow 2-subgroup of M containing R and S a Sylow 2-subgroup of G containing T and fix all this notation. We first prove

Proposition 6.1. If M is the centralizer of some involution of G, then S is of type J_2 or $\hat{A}_{10}$.

Clearly we have $R \cong Q_8 * D_8$ in this case. Since M is a 2-constrained 2-local subgroup, we have $Z(R) = Z(S) \cong Z_2$, $M = C_G(z)$ for some involution z of $Z(S)$, and $S = T$. We break up the proof into a few lemmas.

Lemma 6.2. $SCN_3(S)$ is empty and S possesses a unique element U in $U(S)$.

Proof: Suppose false and let $A \triangleleft S$ with $A \cong E_8$. Setting $\tilde{M} = M/\langle z\rangle$, it follows from Lemma 2.7 (ii) that $\tilde{S}$ is either of type A_{10} or A_8. One checks that correspondingly $U(\tilde{S})$ consists of exactly one or three elements and in either case these lie in the unique elementary abelian normal subgroup $\tilde{R}$ of order 16 in $\tilde{S}$. But $\tilde{A} \in U(\tilde{S})$, whence $\tilde{A} \subseteq \tilde{R}$ and consequently $A \subseteq R$. However, this is impossible as $R \cong Q_8 * D_8$ and so has 2-rank 2.

Since $|Z(\tilde{S})| = 2$ in either case, we see that $U(S)$ consists of a unique element U and the lemma is proved.

If the involutions of U are all conjugate in G, then S is of type J_2 or $\hat{A}_{10}$ by Proposition 5.2. We can therefore assume for the balance of the proof that the involutions of U are not all conjugate in G. Let $U = \langle z,u\rangle$. Since $u \sim zu$ in S, we have that z is not conjugate to u in G. Since $\overline{M}'/\overline{R} \cong A_5$, we see that the 10 noncentral involutions of R are all conjugate to u in M. Hence z is not conjugate in G to any involution of $R - \langle z\rangle$.

We next prove

Lemma 6.3. If $\overline{M}/\overline{R} \cong A_5$, then S is of type J_2.

Proof: By Glauberman's Z*-theorem, $z \sim a$ in G for some involution a in S - R as G is simple. Since $N_M(S)$ contains a 3-element acting regularly on $S/R \cong Z_2 \times Z_2$, it follows, in particular, that each coset of R in S - R contains an involution.

Again set $\tilde{M} = M/\langle z\rangle$. Since $\tilde{M}/O(\tilde{M})$ is an extension of A_5 by E_{16}, $\tilde{M}/O(\tilde{M})$ splits over $\tilde{R}O(\tilde{M})/O(\tilde{M})$ by Lemma 2.6. Let $\tilde{L}$ be a complement to $\tilde{R}O(\tilde{M})/O(\tilde{M})$. Then the inverse image $\overline{L}$ of $\tilde{L}$ in $\overline{M}$ is isomorphic to either $Z_2 \times A_5$ or to SL(2,5). In the first case $\overline{M}$ is isomorphic to a split extension of $Q_8 * D_8$ by A_5 and so $\overline{S}$ is of type J_2 by Lemma 2.7 (iv). Hence the lemma holds in this case. In the contrary case, we see that $S = RQ$, where $Q \cong Q_8$ and Q is normalized, but not centralized by a 3-element which can be taken to normalize R. We shall derive a contradiction by using the conclusion of the preceding paragraph.

We have that R can be expressed in exactly 10 distinct ways as the central product of Q_8 and D_8. Hence for some x in $Q - \langle z\rangle$, we can write $R = R_1 * R_2$ with $R_1 \cong Q_8$, $R_2 \cong D_8$ and x leaving both R_1 and R_2 invariant. But $\tilde{S}$ is of type A_8 in the present case and so $\langle\tilde{R},\tilde{x}\rangle \cong$

$Z_2 \times Z_2 \int Z_2$. This implies that $\tilde{x}$ does not centralize $\tilde{R}_1$ or $\tilde{R}_2$. Hence $\langle x,R_1\rangle \cong Q_{16}$ or QD_{16} and $\langle x,R_2\rangle \cong D_{16}$ or QD_{16}. However, as $|x| = 4$, we see that the only possibility is that, in fact, $\langle x,R_1\rangle \cong Q_{16}$ and $\langle x,R_2\rangle \cong QD_{16}$. One checks directly now that all involutions of $\langle x,R\rangle$ lie in R, thus contradicting the first paragraph of the proof.

We can therefore assume for the balance of the proof that $\bar{M}/\bar{R} \cong S_5$. We set $S_1 = S \cap M'$, so that $\bar{S}_1/\bar{R} \cong Z_2 \times Z_2$ is a Sylow 2-subgroup of $\bar{M}'/\bar{R} \cong A_5$. We first prove

Lemma 6.4. The following conditions hold:

(i) z is conjugate in G to an involution of $S_1 - R$;

(ii) $\bar{M}'$ splits over $\bar{R}$;

(iii) S_1 is of type J_2.

Proof: It will suffice to prove (i), for then the argument of the preceding lemma can be repeated with M' in place of M to yield (ii) and (iii).

Suppose then that (i) is false, in which case $z \sim a$ in G for some involution a in $S - S_1$, again by Glauberman's Z*-theorem. Again setting $\tilde{M} = M/\langle z\rangle$, it follows from Lemma 2.7(ii) that $\tilde{S}$ is of type A_{10} (whence $\tilde{S} \cong D_8 \int Z_2$). Likewise we have that $\tilde{S}_1$ is of type A_8. A description of $\tilde{S}$ is given at the beginning of [17, section 6] in terms of generators z_1, z_2, a_1, a_2, b_1, b_2, u. With this identification, we can identify $\tilde{R}$ with $B = \langle z_1,z_2,b_1,b_2\rangle$ and $\tilde{S}_1$ with $\langle B,a_1a_2,u\rangle$. We see then that $\tilde{a}$ will correspond to a noncentral involution of the subgroup $T = \langle a_1,b_1\rangle \times \langle a_2b_2\rangle \cong D_8 \times D_8$ which lies in one of the two factors. Hence $C_{\tilde{R}}(\tilde{a}) \cong Z_2 \times Z_2 \times Z_2$. If R_0,R_1 denote respectively $C_R(a)$ and the inverse image of $C_{\tilde{R}}(\tilde{a})$ in R, it follows that $|R_1 : R_0| \le 2$. This implies that $\tilde{R}_0$ contains a four group. On the other hand, using the elements

$z_1, z_2, a_1, a_2, b_1, b_2, u$ above as generators of $\tilde{S}$, it follows from the action of $N_{\tilde{M}}(\tilde{R})$ on $\tilde{R}$ that $z_1z_2 \sim b_1z_1z_2 | z_1 \sim b_1 \sim b_1b_2$ in $\tilde{M}$. Moreover, z_1z_2, z_1 are the images of an involution and an element of order 4 of $\bar{R}$ respectively. From this we see that $\tilde{R}$ does not possess a four subgroup whose involutions are each the image of an involution of R. We thus conclude that $C_R(a)$ contains an element of order 4. This means that $z \in \Phi(R_0) \subseteq \Phi(C_S(a))$.

But now as $z \sim a$ in G, there exists an element g in G such that

$$a^g = z \quad \text{and} \quad C_S(a)^g \subseteq S.$$

However, as $z \in \Phi(C_S(a))$, it follows that $z^g \in \Phi(S) \subseteq S_1$. Since (i) is assumed false, z is not conjugate to any involution of $S_1 - R$ and so $z^g \in R$. But under our present assumptions z is also not conjugate to any involution of $R - \langle z\rangle$, as noted before Lemma 6.3. Thus the only possibility is that $z^g = z$, which is clearly a contradiction.

Finally we prove

Lemma 6.5. The case $\bar{M}/\bar{R} \cong S_5$ does not occur.

Proof: Continuing the above analysis, we shall derive a contradiction from the assumption that $\bar{M}/\bar{R} \cong S_5$. We have that S_1 is of type J_2 and consequently S_1 possesses a maximal subgroup S_0 of type $L_3(4)$ by [16, Lemma 5.1 (iii)]. Furthermore, $\bar{M}$ contains an element $\bar{x}$ of order 3 which normalizes $\bar{S}_1$ and hence also normalizes $\bar{S}_0$. Since $\bar{x}$ centralizes $\bar{z}$, $\bar{x}$ centralizes $Z(\bar{S}_0)$ and consequently $\bar{x}$ acts regularly on $\bar{S}_0/\Phi(\bar{S}_0)$. We are therefore in a position to apply Lemma 2.4 (iii). We set $\bar{K} = N_{\bar{M}}(\bar{S}_0)$, so that $\bar{K} = \bar{S}\langle\bar{x}\rangle$. Moreover, $\bar{x}$ centralizes $\bar{S}_1/\bar{S}_0 \cong Z_2$ and $\bar{x}$ is inverted by a 2-element of $\bar{M}$. Hence without loss we can assume that $\bar{S} = \bar{S}_0\bar{V}$, where

$\bar{V} = N_{\bar{S}}(\langle\bar{x}\rangle)$ and $|\bar{V}| = 16$. We note that $Z(\bar{S}_0) = \bar{U}$ and so $\bar{U} \subseteq \bar{V}$. We claim also that $\bar{V}/\bar{U} \cong Z_2 \times Z_2$. Indeed, the elements of $\bar{V} \cap \bar{S}_1$ not contained in $\bar{S}_0$ clearly induce automorphisms of $\bar{S}_0$ which do not stabilize the chain $\bar{S}_0 \supset \Phi(\bar{S}_0) \supset 1$. Since the elements of $\bar{V}$ not contained in $\bar{S}_1$ invert $\bar{x}$, they also induce automorphisms of $\bar{S}_0$ with the same property. Hence $\bar{V}/\bar{U}$ induces a subgroup of $\mathrm{Aut}(\bar{S}_0)$ disjoint from $O_2(\mathrm{Aut}(\bar{S}_0))$ and now [Lemma 7.4 (ii), Part I] implies that $\bar{V}/\bar{U} \cong Z_2 \times Z_2$.

If $\bar{V}$ splits over $\bar{U}$, then Lemma 2.4 (iii) is applicable and yields that $\bar{S}$ and hence S is of type $\hat{A}_{10}$. But then $G \cong L$ by [18, Theorem B]. Therefore $M \cong \hat{A}_{11}$, contrary to the 2-constraint of M. Hence $\bar{V}$ does not split over $\bar{U}$.

We let V be the inverse image of $\bar{V}$ in S. Then $S = S_0V$, $V \supseteq U$, $V/U \cong Z_2 \times Z_2$, and V does not split over U. We have $V = \langle U,t,v\rangle$ with $S_1 = \langle S_0,t\rangle$. Since $Z(S_1) \cong Z_2$, we have that $\langle U,t\rangle \cong D_8$ and consequently $\langle U,t\rangle - U$ contains an involution. Hence without loss we can assume that t is an involution. Either v or tv does not centralize U. Hence interchanging v and tv, if necessary, we can also assume without loss that $\langle U,v\rangle \cong D_8$ and that v is an involution. Since V does not split over U, $(tv)^2 \neq 1$. Since $(tv)^2 \in U$ and is inverted (and hence centralized) by t, the only possibility is that $(tv)^2 = z$.

We know that $\langle S_0,t\rangle = S_1$ is of type J_2. Moreover, $\langle S_0,v\rangle$ is of type J_2, $\hat{A}_8$, or $L_3(4)^{(1)}$ by Lemma 2.4 (iv). But $\langle S_0,v\rangle$ is not of type J_2, otherwise tv would stabilize the chain $S_0 \supset \Phi(S_0) \supset 1$, which is not the case. Furthermore, as $\langle S_0,v\rangle$ has a center of order 2, it is not of type $L_3(4)^{(1)}$. Thus, in fact, $\langle S_0,v\rangle$ is of type $\hat{A}_8$. This means that the action of the element tv on S_0 is uniquely determined. Since $(tv)^2 = z$, we see that the structure of S itself is uniquely determined. It is now easy to verify that S has exactly two elementary abelian subgroups A, B of order 16, namely, the two such subgroups of S_0. We have $S_0 = \langle A, B\rangle$ with t interchanging A and B as $S_1 = \langle S_0, t\rangle$ is of type J_2. Since $z^t = z$, we conclude now by a standard argument (cf. [16, Lemma 6.7]) that any element of $A^{\#}$ which is conjugate to z in G is conjugate to z in $N_G(A)$.

Finally, again as S_1 is of type J_2, we find that every involution of $S_1 - R$ is conjugate in S_1 to an involution of $A - U$. Since $\bar{x}$ normalizes, but does not centralize $\bar{A}/\bar{U}$, it follows that every involution of $S_1 - R$ is conjugate in M to a fixed involution a of $A - U$. But then $z \sim a$ in G by Lemma 6.4 (i). On the other hand, by assumption, z is not conjugate in G to the elements u or uz of U. Hence z is conjugate in G to exactly 13 involutions of A. We conclude therefore from the preceding paragraph that 13 divides $|N_G(A)/C_G(A)|$. However, this factor group is isomorphic to a subgroup of $GL(4,2)$, whose order is not divisible by 13.

This contradiction establishes the lemma and completes the proof of Proposition 6.1.

We next prove

Proposition 6.6. If $\bar{M}$ acts transitively on $\bar{R}^{\#}$, then S is of type $L_3(4)$, $L_3(4)^{(1)}$, J_2, $\hat{A}_8$, or $\hat{A}_{10}$.

Proof: In this case, clearly $\bar{R} \cong E_{16}$ and no nontrivial subgroup of $\bar{R}$ is normal in $\bar{M}$. Since M is a 2-local subgroup of G, it follows at once that $M = N_G(R)$. Furthermore, as $\tilde{M} = \bar{M}/\bar{R}$ acts transitively on $\bar{R}^{\#}$ and $\tilde{M}$ is isomorphic to a subgroup of $\mathrm{Aut}(\bar{R})$, Lemma 2.3 (i) implies that $\tilde{M}$ contains a subgroup $\tilde{L}$ isomorphic to A_5 which acts transitively on

$\overline{R}^{\#}$. By Lemma 2.6, the inverse image of $\tilde{L}$ in $\overline{M}$ has the form $\overline{RL}$, where $\overline{L} \cong A_5$ and $\overline{R} \cap \overline{L} = 1$. Without loss we can assume that $\overline{T} \cap \overline{L}$ is a Sylow 2-subgroup of $\overline{L}$. (Recall that by definition T is a Sylow 2-subgroup M containing R). Since $\overline{L}$ acts transitively on $\overline{R}^{\#}$, $\overline{LR} \cong A_5 \cdot E_{16}^{(2)}$ and consequently $\overline{T}_1 = \overline{R}(\overline{T} \cap \overline{L})$ is of type $L_3(4)$. It also follows that $\overline{L}$ contains a 3-element $\overline{x}$ which acts regularly on $\overline{T}_1$.

We let T_1 be the inverse image of $\overline{T}_1$ in T and set $K = N_G(T_1)$, $C = C_G(T_1)$. Then T_1 is of type $L_3(4)$, and K contains a 3-element x which acts regularly on T_1. Moreover, $C \subseteq C_G(R) \subseteq M$ and so $Z(T_1)$ is a Sylow 2-subgroup of C. Burnside's transfer theorem implies that $T_1C = T_1 \times O(C)$. We can identify $K/Z(T_1)C$ with a subgroup of $\mathrm{Aut}(T_1)$. Moreover no 2-element u of $K - T_1$ stabilizes the chain $T_1 \supset \Phi(T_1) \supset 1$. Indeed, if such a u exists, then u normalizes R, whence $u \in M$ and so $u \in T_1$ as $r(\langle u, T_1 \rangle) \leq 4$, a contradiction.

We can therefore apply Lemma 2.4 (iii, iv) to K to conclude that there exists a Sylow 2-subgroup S_1 of K of the form $S_1 = AT_1$, where A normalizes $O(C)\langle x \rangle$. Since x normalizes T_1, x centralizes $A \cap T_1$ and as x acts regularly on T_1, it follows that $A \cap T_1 = 1$. Now Lemma 2.4 (iii,iv) yields that S_1 is of type $L_3(4)$, $L_3(4)^{(1)}$, J_2, $\hat{A}_8$, or $\hat{A}_{10}$. However, in each case we check that T_1 is characteristic in S_1 (cf. [16], [17], [18]). Hence S_1, is, in fact, a Sylow 2-subgroup of G, and the proposition is proved.

Using the results of the preceding section, we now treat Case III.

Proposition 6.7. If $\overline{M}$ acts intransitively on $\overline{R}$ and if the centralizer of every central involution of G is solvable, then S is of type A_8, A_{10}, or $PSp(4,q)$, $q \equiv 1,7 \pmod 8$.

Proof: By the assumption of this section, $\tilde{M} = \bar{M}/\bar{R}$ contains a subgroup isomorphic to A_5. If $\bar{R} \cong Q_8 * D_8$, then $Z(S) = Z(R) \cong Z_2$. Hence, $C_G(Z(S))$ is nonsolvable, contrary to the hypothesis of the proposition. Thus $\bar{R} \cong E_{16}$. Hence by Lemma 2.3 (ii), $\tilde{M} \cong A_5$ or S_5. It follows therefore from Lemma 2.6 that $\bar{M}$ contains a subgroup $\bar{L} \cong A_5$ or S_5 with $\bar{L} \cap \bar{R} = 1$. In the present case $\bar{L}'$ clearly acts intransitively on $\bar{R}^{\#}$. As in the preceding proposition, we also have that $M = N_G(R)$.

Consider first the case that $\bar{L} \cong A_5$, whence $\overline{LR} \cong A_5 \cdot E_{16}^{(1)}$ and so $\bar{T}$ is of type A_8. Thus T is of type A_8 and consequently R is the unique elementary abelian subgroup of T of order 16. Hence R is characteristic in T and as T is a Sylow 2-subgroup of $M = N_G(R)$, we conclude that T is a Sylow 2-subgroup of G. Thus $S = T$ and so S is of type A_8.

We can therefore assume that $\bar{L} \cong S_5$. By Lemma 2.2 (vi), $\bar{T}$ and hence T, is of type A_{10}. In particular, the proposition holds if $T = S$, so we can suppose that $T \subset S$. Thus $|S| \geq 2^8$.

Setting $\bar{T}_1 = \bar{T} \cap \overline{RL'}$, we have that $\bar{T}_1$ is of type A_8 and so $Z(\bar{T}_1) \cong Z_2$. Hence by Lemma 2.2 (v), $\bar{T}_1$ possesses a subgroup $\bar{Q}$ of index 2 with $\bar{Q} \cong Q_8 * Q_8$ and $\overline{RL}'$ contains an element $\bar{x}$ of order 3 which acts regularly on $\bar{Q}/Z(\bar{Q})$. Thus T possesses a subgroup $Q \cong Q_8 * Q_8$ and $N = N_G(Q)$ contains a 3-element which acts regularly on $Q/Z(Q)$. Since the centralizer of every central involution of G is solvable by hypothesis all the hypotheses of Lemmas 5.3 and 5.4 are satisfied. We therefore conclude that $S = (S_1 * S_2)\langle t\rangle$, where $S_1 \cong S_2 \cong Q_{16}$ or QD_{16} and t is an involution which interchanges S_1 and S_2.

If $S_1 \cong Q_{16}$, then S is of type $PSp(4,q)$ for some $q \equiv 1,7 \pmod 8$ and again the proposition holds. On the other hand, if $S_1 \cong QD_{16}$, G is not fusion-simple by Proposition 5.5, a contradiction. Thus the proposition is proved.

By Propositions 6.1, 6.7, and 6.8, a Sylow 2-subgroup of G is either of type $PSp(4,q)$, $q \equiv 1,7 \pmod 8$, $L_3(4)$, $L_3(4)^{(1)}$, A_8, A_{10}, $\hat{A}_8$, $\hat{A}_{10}$, or J_2. Now Proposition 5.1 shows that Theorems C and D hold for G when M is of type A_5.

7. Theorems C and D, the $GL(3,2)$ case. We now establish Theorems C and D when M is of type $GL(3,2)$. We preserve the notation R, T, S and $\bar{M} = M/O(M)$ of the previous section. We know that $\bar{M}/\bar{R} \cong GL(3,2)$ and $R \cong \bar{R} \cong E_8$, E_{16}, or $Z_4 \times E_8$. As before, M is a 2-constrained 2-local subgroup of G and G is a simple group of sectional 2-rank at most 4.

We first prove

Proposition 7.1. R is not isomorphic to $Z_4 \times E_8$.

Proof: Assume the contrary. It will suffice to prove that $\Omega_1(R) = \Omega_1(T)$. Indeed, if this is the case, then $\Omega_1(R)$ is characteristic in T. Since $R = C_T(\Omega_1(R))$, it follows that R is characteristic in T. Since M is a 2-constrained 2-local subgroup of G which involves $GL(3,2)$, so also is $M_1 = N_G(R)$. Hence applying Theorem B to $M_1/O(M_1)$, we conclude that R is a Sylow 2-subgroup of $O_{2',2}(M_1)$ and that $M_1/O_{2',2}(M_1) \cong GL(3,2)$. The latter condition implies that $M_1 = O_{2',2}(M_1)(M \cap M_1)$ and consequently T is a Sylow 2-subgroup of M_1. Thus $T = S$ and T is a Sylow 2-subgroup of G. We have $\langle z\rangle = \mho^1(R)$ and z is isolated in $M_1 = N_G(R)$. Since $\Omega_1(R) = \Omega_1(T)$ under our present assumption, clearly R is the unique abelian subgroup of its structure in T and so R is weakly closed in T with respect to G. We now conclude that z is isolated in G, contrary to the simplicity of G.

Set $R_1 = \Omega_1(R)$. By Lemma 3.12, we have that $\overline{M}/\overline{R}_1 \cong Z_2 \times GL(3,2)$. Set $\overline{M}_2 = O^2(\overline{M})$. Then $\overline{M}_2$ acts on $\overline{R}_1$ and $\overline{M}_2\overline{R}_1/\overline{R}_1 \cong GL(3,2)$. Since $\mho^1(\overline{R}) \subseteq Z(\overline{M})$ and $\mho^1(\overline{R}) \subseteq \overline{R}_1$, we have that $Z(\overline{M}_2\overline{R}_1) \neq 1$. We can therefore apply Lemma 3.8 to $\overline{M}_2\overline{R}_1$. If $\overline{M}_2$ acts indecomposably on $\overline{R}_1$, we conclude that $\overline{T}_1 = \overline{T} \cap \overline{M}_2\overline{R}_1$ is of type $\hat{A}_8$. But then $\overline{T}_1$ possesses a characteristic subgroup $\overline{T}_0$ of type $L_3(4)$ and so $\overline{T}_0 \lhd \overline{T}$ with $|\overline{T} : \overline{T}_0| = 4$. Since $\overline{R}_1 \lhd \overline{T}$, it follows now from Lemma 2.4 (v) that some element of $\overline{T} - \overline{T}_0$ induces an automorphism of $\overline{T}_0$ which stabilizes the chain $\overline{T}_0 \supset \Phi(\overline{T}_0) \supset 1$. But this implies that $r(\overline{T}) \geq 5$, contrary to the fact that $r(G) \leq 4$. On the other hand, if $\overline{M}_2\overline{R}_1 = Z(\overline{M}_2\overline{R}_1) \times O^2(\overline{M}_2\overline{R}_1)$, $O^2(\overline{M}_2\overline{R}_1)$ is isomorphic to a nontrivial extension of E_8 by $GL(3,2)$ and so $\overline{T} \cap O^2(\overline{M}_2\overline{R}_1)$ is of type A_8 or M_{12} by Lemma 3.4. In particular, $O^2(\overline{M}_2\overline{R}_1)$ has sectional 2-rank 4 and so $\overline{M}_2\overline{R}_1$ has sectional 2-rank 5, again contradicting $r(G) \leq 4$. We thus conclude that $\overline{M}_2$ acts decomposably on $\overline{R}_1$, that $[\overline{R}_1, \overline{M}_2] \cong E_8$, that $\overline{M}_2\overline{R}_1/[\overline{R}_1,\overline{M}_2] \cong SL(2,7)$, and all involutions of $\overline{T} \cap \overline{M}_2\overline{R}_1$ lie in $\overline{R}_1$. In particular, $\overline{R}_1 \subseteq O^2(\overline{M}_2\overline{R}_1)$, whence $\overline{R}_1 \subseteq \overline{M}_2 = O^2(\overline{M})$ and so $\overline{M}_2\overline{R}_1 = \overline{M}_2$.

Finally set $\tilde{M} = \overline{M}/\langle\overline{z}\rangle$. Then $\tilde{R} \cong E_{16}$ and $\tilde{R}_1 \cong E_8$ with $\tilde{R}_1 \lhd \tilde{M}$. If $Z(\tilde{M}) \neq 1$, then $\tilde{M}$ acts decomposably on $\tilde{R}$. But $\tilde{M}/\tilde{R}_1 \cong Z_2 \times GL(3,2)$ and so Lemma 3.8 yields that $\tilde{M} = Z(\tilde{M}) \times O^2(\tilde{M})$, which leads to the same contradiction as in the preceding paragraph. Hence $Z(\tilde{M}) = 1$ and now Lemma 3.9 (iii) yields that all involutions of $\tilde{T} - (\tilde{T} \cap O^2(\tilde{M}))$ are conjugate in $\tilde{T}$. Note also that $\tilde{M}_2 = O^2(\tilde{M})$. Thus if $\overline{t}$ is an element of $\overline{R}$ of order 4, it follows that every involution of $\tilde{T} - (\tilde{T} \cap \tilde{M}_2)$ is conjugate in $\tilde{T}$ to $\tilde{t}$.

Since $|\bar{t}| = 4$, we conclude therefore that $\bar{T} - (\bar{T} \cap \bar{M}_2)$ contains no involutions. However, we have shown in the preceding paragraph that $(\bar{T} \cap \bar{M}_2) - \bar{R}_1$ contains no involutions. Hence $\bar{R}_1 = \Omega_1(\bar{T})$ and we reach the desired conclusion $R_1 = \Omega_1(T)$. This completes the proof.

By the proposition, we have that $R \cong E_8$ or E_{16}. We next prove

Proposition 7.2. If M is the centralizer of some involution of G, then S is of type $\hat{A}_8$.

Proof: Since $R \not\cong Z_4 \times E_8$ by the previous proposition, we have $R \cong E_{16}$ in this case and $M = C_G(z)$ for some involution z of R. We set $N = N_G(R)$. Since M is 2-constrained, $C_M(R) \subseteq O_{2',2}(M)$. Since $z \in R$, we have $C_G(R) \subseteq C_M(R) \subseteq O_{2',2}(M)$. Furthermore, $N_M(R)$ covers $\bar{M} = M/O(M)$ by the Frattini argument and consequently $N/O_{2',2}(N)$ is isomorphic to a subgroup of $GL(4,2)$ containing $GL(3,2)$. By the previous proposition, R is a Sylow 2-subgroup of $O_{2',2}(N)$ and our argument yields that $C_G(R) = C_N(R) = O(N) \times R$. We set $\bar{N} = N/O(N)$.

Suppose first that $|N|>|M|$. Then we must have that $\bar{N}/\bar{R} \cong A_7$ or A_8 and so $\bar{N}$ is transitive on $\bar{R}$. Since N is a 2-constrained 2-local subgroup of G of type A_5 and $r(G) \leq 4$, we can apply Proposition 6.6 (which has been proved under these conditions) with N in the role of M and we conclude that S is of type $L_3(4)$, $L_3(4)^{(1)}$, J_2, $\hat{A}_8$, or $\hat{A}_{10}$. However, as $|S| \geq 2^7$ in the present case, S is not of type $L_3(4)$. Since $R \triangleleft T$ and $|T| = 2^7$, we have that S is not of type J_2 or $L_3(4)^{(1)}$ and $Z(T) = Z(S) = \langle z \rangle \cong Z_2$. Hence $S = T$, as $M = C_G(z)$. Since $R \triangleleft S$, $SCN_4(S)$ is nonempty and so also S is not of type $\hat{A}_{10}$. Hence S is of type $\hat{A}_8$ and so the proposition holds when $|\bar{N}| > |\bar{M}|$.

Suppose then that $|\bar{N}| = |\bar{M}|$, in which case $\bar{z} \in Z(\bar{N})$. We can therefore apply Lemma 3.8 to $\bar{N}$. If $\bar{N}$ acts indecomposably on $\bar{R}$, we conclude that $\bar{T}$, and hence also T, is of type $\hat{A}_8$. Hence $Z(T) = Z(S) = \langle z \rangle \cong Z_2$. This implies that $T = S$ and so S is of type $\hat{A}_8$. Thus the proposition holds in this case. If $\bar{N}$ acts decomposably on $\bar{R}$ and $\bar{N}/[\bar{R},\bar{N}] \cong SL(2,7)$, we conclude that $\bar{T} - \bar{R}$ contains no involutions. Thus $S = T$ and $R = \Omega_1(S)$. Since z is isolated in N, it follows at once that z is isolated in G, contrary to the simplicity of G. Finally if $\bar{N} = Z(\bar{N}) \times O^2(\bar{N})$, we conclude as in the proof of the preceding proposition, with the aid of Lemma 3.4, that $r(\bar{N}) \geq 5$, which is a contradiction. Thus the proposition also holds when $|\bar{N}| = |\bar{M}|$.

Finally we prove

Proposition 7.3. If the centralizer of every central involution of G is solvable, then S is of type M_{12}, A_8, A_{10}, or $\hat{A}_8$.

We shall consider three cases separately. Note that if $R \cong E_8$, then Lemma 3.4, applied to $\bar{M}$, yields that T is of type M_{12} or A_8. Since $R \cong E_8$ or E_{16} we thus have the following three possibilities:

(I) $R \cong E_8$ and T is of type M_{12};

(II) $R \cong E_8$ and T is of type A_8;

(III) $R \cong E_{16}$.

We first prove

Lemma 7.4. If $R \cong E_8$ and T is of type M_{12}, then S is of type M_{12}.

Proof: We must prove that $S = T$ to establish the lemma. Assume false, in which case $S - T$ contains an element x such that $T^x = T$, $x^2 \in T$, and $R^x \neq R$. We can describe T as follows (cf. [19]):

$$T = \langle a,b,t,u \mid a^4 = b^4 = t^2 = u^2 = 1,\ a^t = b,\ a^u = a^{-1},\ b^u = b^{-1},$$
$$[a,b] = [t,u] = 1\rangle.$$

One checks directly from these relations that $Z(T) = \langle z\rangle = \langle a^2b^2\rangle \cong Z_2$ and that T contains exactly two elementary normal subgroups of order 8; namely, $\langle a^2,b^2,u\rangle$ and $\langle a^2,b^2,uab\rangle$. Without loss we may assume that $R = \langle a^2,b^2,u\rangle$, in which case $R^x = \langle a^2,b^2,uab\rangle$. Furthermore, T contains a characteristic subgroup $Q = \langle a^{-1}b,ab,t,u\rangle \cong Q_8 * Q_8$. Note also that as M is a 2-local subgroup, $Z(S) \subseteq T$, whence $Z(S) = Z(T) = Z(Q) \cong Z_2$. Also by the structure of M, $N_M(Q)$ contains a 3-element which acts regularly on $Q/Z(Q)$.

We shall argue first that $|S:T| = 2$. Examining the structure of $\bar{C} = C_{\bar{M}}(\bar{z})$, we find that $\bar{Q} = O_2(\bar{C})$ and that $\bar{C}/\bar{Q} \cong S_3$. One also has that $\bar{R} \cong E_8$ and that $\bar{C}/\bar{R} \cong S_4$ acts faithfully on $\bar{R}$. Thus $\bar{C}/\bar{R}$ can be regarded as the stabilizer of $\bar{z}$ in $\bar{M}/\bar{R} \cong GL(3,2)$. Since $\bar{M}/\bar{R}$ acts doubly transitively on $\bar{R}^\#, \bar{C}$ is thus transitive on $\bar{R} - \langle\bar{z}\rangle$. Hence $\bar{u} \sim \bar{a}^2$ in $\bar{C}$. Since $\bar{Q} = O_2(\bar{C})$, it follows therefore, if we set $N = N_G(Q)$, that $u \sim a^2$ in N. Furthermore, we have $\bar{Q} = \bar{R}\langle\bar{t},\overline{uab}\rangle$ with $\langle\bar{t},\overline{uab}\rangle \cong Z_2 \times Z_2$. Let $\bar{y}$ be a 3-element of $\bar{C}$. Since $\bar{Q} \cong Q_8 * Q_8$ splits over $\bar{R}$, also $\langle\bar{y},\bar{Q}\rangle$ splits over $\bar{R}$. This implies that $\bar{Q}$ contains a $\bar{y}$-invariant subgroup $\bar{R}_1 \cong E_8$ with $\bar{R}_1 \cap \bar{R} = \langle\bar{z}\rangle$. It follows from the description of Q and T given above that the only possibilities are $\bar{R}_1 = \langle\bar{t},\overline{uab},\bar{z}\rangle$ or $\bar{R}_1 = \langle\bar{t},\overline{uab},\bar{z}\rangle^{\bar{a}}$. In either case we obtain immediately that $\bar{t} \sim \overline{tu} \sim \overline{uab}$ in $\bar{N}$ whence $t \sim tu \sim uab$ in N.

Now consider the same situation in M^x. We have $R^x = \langle a^2,b^2,uab\rangle$. Since Q is characteristic in T, $Q^x = Q$ and so $Q^x = R^x\langle t,u\rangle$. This time our argument yields that $uab \sim a^2$ and $t \sim u \sim tu$ in $N^x = N$. We see then that all 18 non-central involutions of Q are conjugate in N.

But all the assumptions of Lemma 5.3 are satisfied with G in the role of X. Hence $\overline{N} = N/O(N)Q$ is isomorphic to a subgroup of a Sylow 3-normalizer in A_8. It follows therefore from the preceding paragraph that $|\overline{N}|$ is divisible by 9. Moreover, if we set $K = C_G(z)$, Lemma 5.3 implies that N covers $K/O(K)$. Since $S \subseteq K$, we see that N contains a Sylow 2-subgroup of G, which without loss we may assume to be S. By the structure of $\overline{N}$, we have that $|\overline{S}| \leq 8$. On the other hand, as $S \supset T$, we have $|\overline{S}| \geq 4$.

Now we can verify that $|S:T| = 2$. Suppose false, in which case $|S:T| = 4$ and $|\overline{S}| = 8$. But as $Q \cong Q_8 * Q_8$, Q contains exactly 6 elementary abelian subgroups of order 8. Since $\overline{N}$ is isomorphic to a Sylow 3-normalizer in A_8, we see that R has 6 conjugates in N. Since $|\overline{S}| = 8$, it follows that $|N_S(R):Q| = 4$. However, this is impossible as $T = N_S(R)$ and $|T:Q| = 2$. This proves our assertion.

Now we can quickly establish the lemma. Since $U = \langle a^2, b^2 \rangle$ is the unique element of $U(T)$, it is an element of $U(S)$. Since $\overline{M}$ acts transitively on $\overline{R}^{\#}$, we have that $a^2 \sim b^2 \sim a^2b^2$ in G. It will therefore suffice to prove that $SCN_3(S)$ is empty, for then we can apply Proposition 5.2 to obtain that S is of type J_2 (as $|S| = 2^7$). However, it is easily checked that a 2-group of type J_2 does not possess a maximal subgroup of type M_{12}. Hence the proposition will follow once this assertion is proved.

Suppose then that S contains an elementary abelian normal subgroup A of order 8. Since every elementary abelian subgroup of order 8 of Q is conjugate in N to R, the assumption $A \subseteq Q$ would imply that R is normal in a Sylow 2-subgroup S^* of N and hence of G. But then $S^* \subseteq M = N_G(R)$, contrary to the fact that T is a Sylow 2-subgroup of M and $T \subset S$. Thus

$A \not\subseteq Q$ and so by the structure of T, $A \not\subseteq T$. Since $|S:T| = 2$, we have $S = AT$. Also $U = A \cap T$ as $A \cap T \triangleleft T$ and $A \cap T \cong Z_2 \times Z_2$. Since the element x can be taken as any element of $N_S(T) - T$, we can assume without loss that $x \in A$, whence x centralizes $U = \langle a^2, b^2 \rangle$. Comparing the generators of R and R^x, it follows now that $u^x = uabv$ for some element v in U. Hence $[u,x] = abv$. However, $[u,x] \in A$ as $x \in A \triangleleft S$, so $[u,x]$ is of order 2. Since $|abv| = 4$ for any v in U, we reach a contradiction. Thus $SCN_3(S)$ is empty and the proposition is proved.

We next prove

Lemma 7.5. If $R \cong E_8$ and T is of type A_8, then S is of type A_8, A_{10}, or $\hat{A}_8$.

Proof: We can assume that $S \supset T$. T is generated by involutions a,b,c,d,e,f satisfying

$$[c,e] = [b,f] = a, \ [d,e] = b, \ [d,f] = c$$

with all other commutators of pairs of generators being trivial (cf. [17]). One checks that T has exactly three self-centralizing normal elementary abelian subgroups of order 8; namely, $\langle a,b,e \rangle$, $\langle a,c,f \rangle$, and $\langle a,bc,ef \rangle$. One of these must be R. We have $Z(T) = \langle a \rangle \cong Z_2$ and again as M is a 2-local subgroup, we have $Z(T) = Z(S)$. In addition, $Q = \langle a,b,c,e,f \rangle$ is the unique subgroup of T isomorphic to $Q_8 * Q_8$ and $A = \langle a,b,c,d \rangle$ the unique subgroup of T isomorphic to E_{16}.

If $\bar{C} = C_{\bar{M}}(\bar{a})$, we again have as in the preceding lemma that $\bar{Q} = O_2(\bar{C})$, that $\bar{C}/\bar{Q} \cong S_3$, and that $N_{\bar{M}}(\bar{Q})$ contains a 3-element which acts regularly on $\bar{Q}/Z(\bar{Q})$. Thus Lemma 5.3 can be applied once again. Setting $N = N_G(Q)$ and $K = C_G(a)$, we have that N covers $K/O(K)$. Since $N \subseteq K$, N thus contains a Sylow 2-subgroup of K, which without loss we can assume to be S.

We claim that $\bar{N} = N/O(N)Q$ does not have order divisible by 9. Indeed, if that were the case, then all 6 elementary subgroups of Q would be conjugate in N as in the preceding lemma. But then $R \sim \langle a,b,c \rangle$ in N. However, this is impossible as, on the one hand, R is a Sylow 2-subgroup of $C_G(R) \subseteq M$, while, on the other, $C_G(\langle a,b,c \rangle) \supseteq A = \langle a,b,c,d \rangle$. This proves our assertion. Since $\bar{N}$ is isomorphic to a subgroup of a Sylow 3-normalizer in A_8, it follows that $\bar{N}$ is isomorphic to a subgroup of $Z_2 \times S_3$. In particular, $|\bar{S}| \leq 4$ and so $|S| \leq 2^7$. But as $S \supset T$ and $|T| = 2^6$, we conclude that $|S| = 2^7$ and that $\bar{N} \cong Z_2 \times S_3$.

Now we turn our attention to $E = N_G(A)$. Since A is characteristic in T, we have $S \subseteq E$. Since T is of type A_8, Aut (T) contains a subgroup isomorphic to S_3 which transitively permutes the three self-centralizing elementary abelian normal subgroups of T of order 8. (This can be seen from the structure of $C_{A_{10}}((12)(34)(56)(78))$. Hence without loss we may assume that $R = \langle a,b,e \rangle$ and that $R^t = \langle a,c,f \rangle$, where $t \in S - T$. In particular, it follows that $\langle A,e \rangle^t = \langle A,f \rangle$, whence $e^t = fk$ for some k in A. Since e and f are involutions, we conclude that $S/A \cong D_8$. By the structure of $\bar{M}$, we also have that $N_{\bar{M}}(\langle \bar{a},\bar{b} \rangle)$ splits over $\bar{R}$ with the factor group isomorphic to S_4 and that $C_{\bar{M}}(\langle a,b \rangle) \cong Z_2 \times Z_2 \int Z_2$. From this we see that $N_{\bar{M}}(\langle \bar{a},\bar{b} \rangle)/\bar{A} \cong Z_2 \times S_3$ and a 3-element of $N_{\bar{M}}(\langle \bar{a},\bar{b} \rangle)$ acts regularly on $\bar{A}$. It follows therefore that E contains a 3-element u which acts regularly on A and that $\tilde{E} = E/C_G(A)$ contains a subgroup isomorphic to $Z_2 \times S_3$. Since $C_T(A) = A$, $B = C_S(A)$ has order at most 2^5 and $B \supseteq A$. Since $r(G) \leq 4$, it follows that either $B = A$ or $B \cong Z_4 \times E_8$. But B is a Sylow 2-subgroup of $C_G(A)$ and so we can suppose without loss that u normalizes B. Hence if $B \supset A$, u has a nontrivial fixed point on

$A = \Omega_1(B)$, which is not the case. Thus $B = A$ and so A is a Sylow 2-subgroup of $C_G(A)$.

Finally $\tilde{E}$ is isomorphic to a subgroup of A_8 with dihedral Sylow 2-subgroups of order 8 and containing a subgroup isomorphic to $Z_2 \times S_3$. If $\tilde{E}$ is solvable, the only possibility is that $\tilde{E}$ is isomorphic to a full Sylow 3-normalizer in A_8 or a subgroup of $N_{A_8}((\langle 123\rangle))$. Since $O(\tilde{E})$ has no nontrivial fixed points on A, in either case it is immediate that S splits over A. We conclude therefore from Lemma 2.2 (vii) or (viii) that S is of type A_{10} or $\hat{A}_8$. on the other hand, if $\tilde{E}$ is nonsolvable, then $\tilde{E}$ is not isomorphic to $GL(3,2)$ as the 3-element $\tilde{u} \in \tilde{E}$ acts regularly on A. Hence $\tilde{E}$ contains a subgroup isomorphic to A_5 and so E is a non-solvable 2-constrained 2-local subgroup of type A_5. We conclude therefore from Propositions 6.6 and 6.7, with K in the role of M, that S is of type J_2, $L_3(4)^{(1)}$ $\hat{A}_8$, or A_{10} (as $|S| = 2^7$). However, the first case is excluded as then the centralizer of a central involution or G is nonsolvable by [16]. The second case is excluded as such a 2-group cannot be a Sylow 2-subgroup of a simple group. Thus S has one of the asserted structures.

Finally we prove

Lemma 7.6. If $R \cong E_{16}$, then $S = T$ and S is of type $\hat{A}_8$.

Proof: We first apply the results of Lemmas 3.8 and 3.9. Consider first the case that $Z(\bar{M}) \neq 1$. Since $r(G) \leq 4$, it follows, as usual, that $\bar{M} \neq Z(\bar{M}) \times O^2(\bar{M})$. If $\bar{T}$ is of type $\hat{A}_8$, then $Z(\bar{T}) \cong Z_2$ and consequently $Z(T) = Z(S)$. But then $Z(\bar{S}) = Z(\bar{M})$ and consequently $C_G(Z(S))$ is nonsolvable, contrary to hypothesis.

Hence Lemma 3.8 (ii) necessarily holds and, in particular, $R = \Omega_1(T)$ and $\bar{M}$ acts decomposably on $\bar{R}$. Setting $M_1 = N_G(R)$, we have that M_1 is a 2-constrained 2-local subgroup of G which involves $GL(3,2)$ since M is such a 2-local subgroup. Set $\tilde{M}_1 = M_1/O(M_1)$. If $O_2(\tilde{M}_1) \cong Z_4 \times E_8$, then $\tilde{M}_1/O_2(\tilde{M}_1) \cong GL(3,2)$ and so Proposition 7.1 applies with M_1 in the role of M to yield a contradiction. We conclude therefore from Theorem B that $\tilde{R} = O_2(\tilde{M}_1)$. We have that M_1 covers $M/O(M)$ and as $R = \Omega_1(T)$, $T \subset M_1$. If T is not a Sylow 2-subgroup of M_1, then as M_1 involves $GL(3,2)$ and 2^8 divides $|M_1|$, the only possibility is that $\tilde{M}_1/\tilde{R} \cong A_8$. Thus $\tilde{M}_1$ acts transitively on $\tilde{R}^{\#}$ and now applying Proposition 6.6 with M_1 in the role of M, we conclude that S is of type $L_3(4)$, $L_3(4)^{(1)}$, J_2, $\hat{A}_8$, or $\hat{A}_{10}$. In particular, $|S| \leq 2^8$. However, as $\tilde{M}_1/\tilde{R} \cong A_8$, we see that $|S| \geq 2^{10}$. This contradiction shows that T is a Sylow 2-subgroup of M_1, which implies that T is a Sylow 2-subgroup of G. Thus $Z(T) = Z(S)$ and so as in the preceding paragraph the centralizer of a central involution of G is non-solvable. We therefore conclude that $Z(\bar{M}) = 1$. Thus Lemma 3.9 is applicable.

We argue next that $S = T$. Assume false, in which case there exists x in $S - T$ such that $T^x = T$, $x^2 \in T$, and $R^x \neq R$. In particular, T contains two elementary abelian subgroups of order 16 and so $\bar{M}$ splits over $\bar{R}$ and $\bar{T}$ is of type $\hat{A}_8$ by Lemma 3.9 (iv) and (v). Furthermore, by Lemmas 3.9 (ii) and 2.2 (v), T contains a subgroup $Q \cong Q_8 * Q_8$ and $N = N_G(Q)$ contains a 3-element which acts regularly on $Q/Z(Q)$. Thus Lemma 5.3 is applicable and if we set $K = C_G(Z(S))$, we conclude, as usual, that N covers $K/O(K)$, that N contains a Sylow 2-subgroup of K which we can assume without loss is S, and that $\tilde{N} = N/O(N)Q$ is isomorphic to a subgroup of a Sylow 3-normalizer in A_8. In particular, $|S| \leq 2^8$. Since $S \supset T$ and $|T| = 2^7$, we have, in fact, $|S| = 2^8$ and $\tilde{N}$ is isomorphic to a full Sylow 3-normalizer in A_8.

By Lemma 2.2 (vii), we have that $S/Z(S)$ is of type A_{10}. This implies that the second center $Z_2(S) \cong Z_2 \times Z_2$ and that $S/Z_2(S)$ is of type A_8. Hence $|Z_3(S)| = 8$ and $Z_3(S)$ is the unique normal subgroup of S of order 8. Furthermore, since $\tilde{N}$ is isomorphic to a full Sylow 3-normalizer in A_8, all six elementary subgroups of Q of order 8 are conjugate in N. This implies that no one of them is normal in S. Hence $Z_3(S) \not\cong E_8$ and consequently $SCN_3(S)$ is empty. Observe next that T has exactly two elementary abelian subgroups of order 16; namely, R and R^x. Moreover, $\langle R, R^x \rangle$ is of type $L_3(4)$. Setting $U = R \cap R^x$, we see that U is the unique element of $U(T)$ and hence $U \in U(S)$. Since $U \subseteq R \cap T'$, $\bar{U} \subseteq \bar{R} \cap O^2(\bar{M})$. However, as $Z(\bar{M}) = 1$ and $\bar{M}$ splits over $\bar{R}$, $\bar{R} \cap O^2(\bar{M}) = [\bar{R}, \bar{M}] \cong E_8$ and so $\bar{M}$ acts transitively on the involutions of $\bar{R} \cap O^2(\bar{M})$. We conclude therefore that the involutions of U are all conjugate in G. Since T is of type $\hat{A}_8$, Proposition 5.2 implies that S is of type $\hat{A}_{10}$. But then the centralizer of every involution of G is nonsolvable by [18, Theorem B]. This contradiction establishes that $S = T$.

If S is of type $\hat{A}_8$, the lemma holds, so we can assume this is not the case, whence $\bar{M}$ does not split over $\bar{R}$ and R is the unique elementary subgroup of S of order 16 by Lemma 3.9 (iv) and (v). Setting $\bar{R}_1 = [\bar{R}, \bar{M}]$, we have that $\bar{M}/\bar{R}_1 \cong Z_2 \times GL(3,2)$ by Lemma 3.9 (i). Hence $O^2(\bar{M})/\bar{R}_1 \cong GL(3,2)$ and $O^2(\bar{M})$ does not split over $\bar{R}_1$, whence $\bar{S}_1 = \bar{S} \cap O^2(\bar{M})$ is of type M_{12} by Lemma 3.4. Let S_1 be the inverse image of $\bar{S}_1$ in S.

Finally let $y \in R - R_1$. By Thompson's fusion lemma, y must be conjugate to some involution v of S_1. But R is weakly closed in S. Furthermore, one concludes, as usual, that M covers $N_G(R)/O(N_G(R))$, whence $R_1 \triangleleft N_G(R)$.

Hence $v \in R_1$. However, as $O^2(\overline{M})$ is a nonsplit extension of $GL(3,2)$ by E_8, one checks easily that $O^2(\overline{M})$ has exactly two conjugacy classes of involutions. Representing S_1 by generators a,b,t,u, as in Lemma 7.4, then $z = a^2b^2$ and t are representatives of these two classes. It follows therefore that $y \sim t$ in G. By the action of $\overline{M}$ on $\overline{R}$, we have that $C_S(y) = R$. On the other hand, the involutions of $S - S_1$ are conjugate in S by Lemma 3.9 (iii) and since $y \not\sim z$ in G, G has exactly two conjugacy classes of involutions represented by z and t.

On the other hand, we check that $C_{S_1}(t) \cong Z_2 \times D_8$. Hence if w is an extremal involution of S that is conjugate to y, then $C_S(w)$ contains subgroups isomorphic to both E_{16} and $Z_2 \times D_8$ and so $|C_S(w)| \geq 2^5$. But as all involutions of $S - S_1$ are conjugate to y and $C_S(y) = R$, this in turn implies that $w \in S_1$, whence w does not centralize R_1. Since R is the unique elementary abelian subgroup of S of order 16, clearly $C_S(w)$ can contain no such subgroup and we reach a contradiction. This completes the proof of the lemma.

Together Lemmas 7.4, 7.5, and 7.6 yield Proposition 7.3. Together Propositions 7.2 and 7.3 yield that a Sylow 2-subgroup of G is of type A_8, A_{10}, $\hat{A}_8$, or M_{12}. Now Proposition 5.1 shows that Theorems C and D also hold for G when M is of type $GL(3,2)$.

PART III

NON 2-CONSTRAINED CENTRALIZERS OF INVOLUTIONS;

SOME SPECIAL CASES

1. Introduction. In Parts I and II of this paper, we have determined all finite simple groups of sectional 2-rank at most 4 in which all 2-local subgroups are 2-constrained. In particular, any such group satisfies the conclusion of our Main Theorem, stated in the Introduction of this paper. In the remaining Parts III, IV, V, and VI we shall study simple groups of sectional 2-rank at most 4 in which some 2-local subgroup is non 2-constrained and shall show ultimately that they, too, satisfy the conclusion of the Main Theorem.

To describe the results to be proved in this part, we first introduce some terminology. For any group X, we define L(X) to be the unique largest normal semisimple subgroup of X. Thus L(X) is either trivial or a central product of quasisimple groups - that is, perfect central extensions of nonabelian simple groups. When O(X) = 1, it is immediate that X is 2-constrained if and only if L(X) = 1. Furthermore, we note that the quasisimple factors of L(X) are uniquely determined by L(X) as they can equivalently be defined to be the collection of all minimal normal perfect subgroups of L(X).

When X is non 2-constrained of sectional 2-rank at most 4 with O(X) = 1, as will be the case for the groups X that we shall be considering, L(X) can be equivalently defined to be the ultimate term $X^{(\infty)}$ of the derived series of X (cf. Proposition 2.1 below). We note also that, in general, when $O(X) = 1$, $L(X)$ is the 2-layer of X as this term is defined in [15], [23], and [24].

Furthermore, for any group G, we shall denote by $\mathcal{L}(G)$ the set of quasisimple factors of

$$L(C_G(x)/O(C_G(x)))$$

as x ranges over the involutions of G.

In view of the preceding remark, $\mathcal{L}(G)$ is empty if and only if the centralizer of every involution of G is 2-constrained. However, by [14, Theorem 4] the latter condition holds if and only if every 2-local subgroup of G is 2-constrained. Hence in the present situation, our concern will be with simple groups G of sectional 2-rank at most 4 in which $\mathcal{L}(G)$ is nonempty.

In proving the Main Theorem, we shall proceed by induction on $|G|$ and so we can restrict ourselves to the case in which the nonsolvable composition factors of the proper subgroups of G are of known type- that is, are isomorphic to groups listed in the Main Theorem. In particular, the nonabelian simple factors of the element of $\mathcal{L}(G)$ are then of known type. By studying the central extensions of the known simple groups, we can determine quite easily the various possibilities for the elements of $\mathcal{C}(G)$. This constitutes our first major result.

Theorem A. If G is a simple group of sectional 2-rank at most 4 in which the nonsolvable composition factors of the proper subgroups of G are of known type, then every element of $\mathcal{C}(G)$ is isomorphic to one of the following groups:

I. Simple groups:

$L_2(q)$, $L_3(q)$, $U_3(q)$, q odd, $Re(q)$,* q an odd power of 3; $L_2(8)$, $Sz(8)$, A_7, M_{11}, or J_1;

II. Nonsimple groups:

$SL(2,q)$, q odd, $SL(4,q)$, $q \equiv 3 \pmod 4$, $SU(4,q)$, $q \equiv 1 \pmod 4$, $Sp(4,q)$, q odd, $\hat{S}z(8)$, $\hat{A}_n$, $7 \leq n \leq 11$, or $\hat{M}_{12}$.

Here $\hat{S}z(8)$, $\hat{A}_n$, and $\hat{M}_{12}$ denote the unique perfect central extension of $Sz(8)$, A_n, M_{12} respectively by Z_2.

For simplicity, for each involution x of G we shall set $L_x = L(C_G(x)/O(C_G(x)))$. Using the condition that G has 2-rank at most 4, one obtains as an immediate corollary of Theorem A,

Corollary A. For each involution x of G, either L_x is quasi-simple or L_x is a central product of two components L_1, L_2 such that

(i) $L_1 \cong SL(2,q_1)$, q_1 odd, or $\hat{A}_7$; and

(ii) $L_2 \cong L_2(q_2)$, $SL(2,q_2)$, q_2 odd, $L_3(q_2)$, $q_2 \equiv -1 \pmod 4$, $U_3(q_2)$, $q_2 \equiv 1 \pmod 4$, A_7, $\hat{A}_7$, or M_{11}.

We note also the following consequence of the fact that G has sectional 2-rank at most 4. If for some involution x of G, L_x has sectional 2-rank 4, then L_x contains its own centralizer in $H_x = C_G(x)/O((C_G(x))$. In other words, H_x/L_x is isomorphic to a group of outer automorphisms of L_x. In particular, this will be the case if $L_x \cong SL(4,q)$, $q \equiv 3 \pmod 4$, $SU(4,q)$, $q \equiv 1 \pmod 4$, $Sp(4,q)$, $\hat{S}z(8)$, $\hat{A}_n$, $8 \leq n \leq 11$, or $\hat{M}_{12}$ or if L_x has two components.

The main object of this part is to treat the case in which, for some involution x of G, L_x has sectional 2-rank 4 and is either quasisimple (and hence nonsimple) or is the product of two nonsimple components. We remark that for technical reasons, it is preferable to postpone to a later part the analysis of the case in which some L_x is a product of two components, one simple and one (necessarily) nonsimple.

We shall **here** prove

Theorem B. Let G be a simple group of sectional 2-rank at most 4 in which the nonsolvable composition factors of the proper subgroups of G are of known type. If for some involution x of G, L_x has sectional 2-rank 4 and each of its components is nonsimple, then G is isomorphic to one of the following groups:

I. $G_2(q)$, $^2D_4(q)$, $PSp(4,q)$, q odd, $q \geq 5$; or $L_4(q)$, $q \not\equiv 1 \pmod 8$, $U_4(q)$, $q \not\equiv 7 \pmod 8$, $q \geq 5$;

II. M^c or L.

In particular, G satisfies the conclusion of the Main Theorem. Theorem B also shows that the only possibilities for L_x, under the given hypotheses, are $L_x \cong SL(2,q) * SL(2,q)$, $SL(2,q) * SL(2,q^3)$, q odd, $q \geq 5$, $SL(4,q)$, $q \equiv -1 \pmod 4$, $SU(4,q)$, $q \equiv 1 \pmod 4$, $\hat{A}_8$, or $\hat{A}_{11}$. As usual, the proof of Theorem B is accomplished by determining the possible structures of a Sylow 2-subgroup of G and then invoking known classification theorems.

2. **Theorem** A. Let G be a simple group with $r(G) \leq 4$ in which the nonsolvable composition factors of its proper subgroups satisfy the conclusion of the Main Theorem. Let x be an involution of G such that $H = C_G(x)$ is not 2-constrained. Setting $\overline{H} = H/O(H)$, we have that $L(\overline{H}) \neq 1$. We let $\overline{L}$ be a component of $L(\overline{H})$. To establish Theorem A, we must determine the possibilities for $\overline{L}$. By our assumption on the subgroup structure of G, we know, to begin with, that $\overline{L}/Z(\overline{L})$ is isomorphic to one of the simple groups listed in the Main Theorem.

First of all, $\overline{L}$ centralizes $\overline{x}$. Hence if $r(\overline{L}) = 4$, we must have $\overline{x} \in Z(\overline{L})$, otherwise $\overline{x} \notin \overline{L}$ and so $r(\overline{L}\langle\overline{x}\rangle) = 5$, contrary to $r(G) \leq 4$. In particular, if $\overline{L}$ is simple, then $r(\overline{L}) \leq 3$. From this observation, we immediately obtain Theorem A(I).

Hence we can suppose that $\overline{L}$ is nonsimple. The Schur multipliers of each of the groups of the Main Theorem are known [8], [25], and the only ones which possess a perfect central extension by a group of even order are

$L_2(q)$, $L_4(q)$, $U_4(q)$, $PSp(4,q)$, q odd;

$L_3(4)$, $Sz(8)$;

A_n, $7 \leq n \leq 11$;

M_{12}, M_{22}, J_2.

Thus $\tilde{L} = \overline{L}/Z(\overline{L})$ is isomorphic to one of these groups. We make a preliminary observation. Suppose $Z(\bar{L}) \cong Z_2$ and $\tilde{L}$ possesses a split extension $\tilde{E}\tilde{K}$ with $\tilde{E} \cong E_{16}$, $\tilde{K} \cong A_5$, and $\tilde{K}$ acting transitively on $\tilde{E}^{\#}$. If $\bar{E}$ denotes the inverse image of $\tilde{E}$ in $\bar{L}$, then $\tilde{K}$ acts faithfully on $\bar{E}$ and so either $\bar{E} \cong Q_8 * D_8$ or E_{32}. However, the transitive action of $\tilde{K}$ on $\tilde{E}$ clearly excludes the first possibility, while the fact that $r(G) \leq 4$ excludes the second. Hence $\tilde{L}$ cannot contain such a split extension.

But each of the groups $L_4(q)$, $q \equiv 1 \pmod 4$, $U_4(q)$, $q \equiv -1 \pmod 4$, $L_3(4)$, and M_{22} possess a split extension of this form [16], [18], [37]. Hence by the argument of the preceding paragraph, $\bar{L}$ is not isomorphic to any one of these groups.

Now apart from the group $Sz(8)$, for the remaining groups listed, the 2-primary component of their Schur multiplier is known to be of order 2 and the resulting extension is known to be uniquely determined [8], [25]. Hence in any of these cases, $Z(\bar{L}) \cong Z_2$ and correspondingly $\bar{L} \cong SL(2,q)$, $Sp(4,q)$, q odd, $SL(4,q)$, $q \equiv 3 \pmod 4$, $SU(4,q)$, $q \equiv 1 \pmod 4$, $\hat{A}_n$, $7 \leq n \leq 11$, $\hat{M}_{12}$, or $\hat{J}_2$, where, as usual, $\hat{J}_2$ denotes the unique perfect central extension of J_2 by a group of order 2.

Except in the case of $\hat{J}_2$, $\bar{L}$ has one of the forms listed in Theorem A(II). We shall show that $\bar{L} \cong \hat{J}_2$ cannot occur, so assume the contrary. By the structure of J_2, $\tilde{L} = \bar{L}/Z(\bar{L})$ then contains a subgroup $\tilde{E} \cong Q_8 * D_8$ such that $N_{\tilde{L}}(\tilde{E}) = \tilde{E}\tilde{K}$ with $\tilde{K} \cong A_5$ with $\tilde{K}$ acting faithfully on $\tilde{E}$. Hence if $\bar{E}$ denotes the inverse image of $\tilde{E}$ in $\bar{L}$, then $N_{\bar{L}}(\bar{E})/\bar{E} \cong A_5$ and $C_{\bar{L}}(\bar{E}) \subseteq \bar{E}$. Since $r\left(N_{\bar{L}}(\bar{E})\right) \leq 4$, Theorem B of Part II is applicable and yields that $\bar{E} \cong E_{16}$ or $Q_8 * D_8$. Since $|\bar{E}| = 64$, this is a contradiction.

Finally suppose $\tilde{L} \cong Sz(8)$. By the results of [4], $Sz(8)$ possesses, up to isomorphism, two perfect central extensions - one by Z_2 and one by $Z_2 \times Z_2$, these extensions being correspondingly of 2-ranks 4 or 5. Since $r(G) \leq 4$, the latter possibility is excluded for $\bar{L}$ and so necessarily $\bar{L} \cong \hat{Sz}(8)$.

We see then that if $\bar{L}$ is nonsimple, then $\bar{L}$ is isomorphic to one of the groups listed in Theorem A (II). Thus Theorem A is proved.

To establish Corollary A, suppose that $L_x = L(\bar{H})$ is not quasisimple. Since any simple component of $L(\bar{H})$ has 2-rank at least 2, this forces $\bar{x} \in L(\bar{H})$, otherwise $r(L(\bar{H})\langle\bar{x}\rangle) \geq 5$. Likewise $L(\bar{H})$ has exactly one other component $\bar{L}_1$ in addition to $\bar{L}$ and $r(\bar{L}) = r(\bar{L}_1) = 2$, otherwise $L(\bar{H})Z(L(\bar{H}))$ would have 2-rank at least 5. Since $\bar{x} \in Z(L(\bar{H}))$, at least one of $\bar{L}$ or $\bar{L}_1$ is nonsimple. But now if use Theorem A to check the possibilities for $\bar{L}$ and $\bar{L}_1$, we immediately obtain the conclusion of Corollary A.

As a further consequence of Theorem A, we have

Proposition 2.1. If x is an involution of G such that $H = C_G(x)$ is not 2-constrained and if we set $\bar{H} = H/O(H)$, then $\bar{H}/L(\bar{H})$ is solvable.

Proof: By assumption, $\bar{L} = L(\bar{H}) \neq 1$. We set $\bar{C} = C_{\bar{H}}(\bar{L})$ and argue first that $\bar{C}$ is solvable. We have that $\bar{L} \cap \bar{C} = Z(\bar{L})$. Since $\bar{C} \triangleleft \bar{H}$, also $L(\bar{C}) \triangleleft \bar{H}$ and so $L(\bar{C}) \subseteq L(\bar{H})$. But then $L(\bar{C}) \subseteq \bar{L} \cap \bar{C} = Z(\bar{L})$, an abelian group, thus forcing $L(\bar{C}) = 1$. Likewise $O(\bar{C}) \triangleleft \bar{H}$ and as $O(\bar{H}) = 1$, it follows that $O(\overline{C}) = 1$. We conclude now that $\overline{C}$ is 2-constrained. Hence if we set $\overline{T} = O_2(\overline{C})$, we have that $C_{\overline{C}}(\overline{T}) \subseteq \overline{T}$.

On the other hand, $r(\overline{L}/Z(\overline{L})) \geq 2$ and as $\overline{T}$ centralizes $\overline{L}$ with $\overline{L} \cap \overline{T} = Z(\overline{L})$, it follows that $\tilde{T} = \overline{T}/Z(\overline{L})$ has sectional 2-rank at most 2. Thus $|\tilde{T}/\Phi(\tilde{T})| = 4$ and consequently any element $\overline{u}$ of $\overline{C}$ of prime order $p \geq 5$ acts trivially on $\tilde{T}$. But by Theorem A and Corollary A, $Z(\overline{L})$ is elementary of order at most 4 and so $\overline{u}$ also centralizes $Z(\overline{L})$. Thus $\overline{u}$ stabilizes the chain $\overline{T} \supset Z(\overline{L}) \supseteq 1$ and so centralizes $\overline{T}$. Since $C_{\overline{C}}(\overline{T}) \subseteq \overline{T}$, our argument yields that $\overline{C}$ must be a $\{2,3\}$-group. Hence $\overline{C}$ is solvable, as asserted.

Now $\overline{H}/\overline{L}\,\overline{C}$ is isomorphic to a group of outer automorphisms of $\overline{L}$. The possible components of $\overline{L}$ are listed in Theorem A and each has a solvable outer automorphism group. Since $\overline{L}$ has at most 2 components by Corollary A, it follows at once that $\overline{H}/\overline{L}\,\overline{C}$ is solvable. Since $\overline{C}$ is solvable, we conclude that $\overline{H}/\overline{L}$ is solvable and the proposition is proved.

Proposition 2.1 shows that in the case of non 2-constrained groups X with $O(X) = 1$ and $r(X) \leq 4$, $L(X)$ can equally well be defined to be $X^{(\infty)}$, the ultimate term of the derived series of X.

3. The $\hat{S}z(8)$ case. Henceforth G will denote a simple group of sectional 2-rank at most 4 in which the nonsolvable composition factors of the proper subgroups of G are of known type and which possesses an involution x such that L_x has sectional 2-rank 4 with each of its components nonsimple.

The possibilities for L_x are determined from Theorem A and Corollary A. We shall separate these into six distinct cases as follows:

(I) $L_x \cong \hat{S}z(8)$;

(II) $L_x \cong \hat{A}_n$, $n = 8,9,10$, or 11;

(III) $L_x \cong \hat{M}_{12}$;

(IV) $L_x \cong Sp(4,q)$, q odd, $SL(4,q)$, $q \equiv 3 \pmod 4$, or $SU(4,q)$, $q \equiv 1 \pmod 4$;

(V) $L_x \cong L_1 \times L_2$, where $L_i \cong SL(2,q_i)$ or $\hat{A}_7$, $i = 1,2$;

(VI) $L_x \cong L_1 * L_2$, where $L_i \cong SL(2,q_i)$ or $\hat{A}_7$, $i = 1,2$.

We set $M = C_G(x)$ and $\bar{M} = M/O(M)$, so that $L_x = L(\bar{M})$. If L_x is quasisimple, we write $\bar{L} = L(\bar{M})$ for brevity. We let T be a Sylow 2-subgroup of M and we let R be the inverse image in T of $\bar{T} \cap L(\bar{M})$. We also let S be a Sylow 2-subgroup of G containing T and fix all this notation for the balance of Part III.

In this short section, we eliminate the $\hat{S}z(8)$ case.

Proposition 3.1. L_x is not isomorphic to $\hat{S}z(8)$.

Proof. Assume false, so that $L(\bar{M}) = \bar{L} \cong \hat{S}z(8)$. Setting $\bar{C} = C_{\bar{H}}(\bar{L})$, we have that $\bar{H}/\bar{C}$ is isomorphic to a subgroup of $\mathrm{Aut}(\bar{L})$. Since $|\mathrm{Aut}(\bar{L})/\mathrm{Inn}(\bar{L})| = 3$, it follows that $\bar{T} \subseteq \bar{L}\bar{C}$. Hence if we set $\bar{R}_1 = \bar{T} \cap \bar{C}$, we have that $\bar{T} = \bar{R}\bar{R}_1$ with $\bar{R}_1$ centralizing $\bar{R}$ and $\bar{R} \cap \bar{R}_1 = \bar{R} \cap Z(\bar{L}) \cong Z_2$. We have that $m(\bar{R}) = 4$. Hence if $\bar{R}_1$ contained an involution $\bar{r}_1$ not in $\bar{R}$, it would follow that $m(\bar{R}\langle\bar{r}_1\rangle) = 5$, contrary to $r(\bar{T}) \leq 4$. Thus $\Omega_1(\bar{R}_1) \subseteq \bar{R} \cap \bar{R}_1$. Since clearly $\bar{x} \in \bar{R}_1$, we have, in fact, $\langle\bar{x}\rangle = \Omega_1(\bar{R}_1) = \bar{R} \cap \bar{R}_1$. Furthermore, as $m(\bar{R}/\langle\bar{x}\rangle) = 3$, it follows likewise that $r(\bar{R}_1/\langle\bar{x}\rangle) \leq 1$, otherwise $r(\bar{T}) \geq 5$. Thus $\bar{R}_1/\langle\bar{x}\rangle$ is cyclic and consequently $\bar{R}_1$ is abelian. Since $\langle\bar{x}\rangle = \Omega_1(\bar{R}_1)$, this forces $\bar{R}_1$ to be cyclic. We see then that $\bar{R}_1 = Z(\bar{T})$ with $\langle\bar{x}\rangle = \Omega_1(Z(\bar{T}))$. We therefore conclude that $\langle x\rangle = \Omega_1(Z(T))$.

Since $\langle x\rangle$ is characteristic in T, we obtain now that $T = S$ and so x is a central involution. Moreover, as $\bar{R}_1$ is cyclic and $O(\bar{C}) \subseteq O(\bar{M}) = 1$, we also have that $\bar{R}_1 = \bar{C}$. Hence M is exceptional in the sense of [24] and now [24, Lemma 7.2] yields a contradiction.

Now that the $\hat{Sz}(8)$ case is eliminated, we can establish the following property of $\bar{M}$.

Lemma 3.2. We have $C_{\bar{M}}(L(\bar{M})) = Z(L(\bar{M}))$.

Proof: Again set $\bar{C} = C_{\bar{M}}(L(\bar{M}))$, so that $\bar{C} \cap L(\bar{M}) = Z(L(\bar{M}))$. In each of cases (II)-(VI), we have that $m(L(\bar{M})/Z(L(\bar{M})) = 4$. Since case (I) has been eliminated by the preceding proposition, it follows now that $|\bar{C}/Z(L(\bar{M}))|$ must be odd, otherwise $r(\bar{M}) \geq 5$. Hence $Z(L(\bar{M}))$ is a Sylow 2-subgroup of $\bar{C}$ and is in the center of $\bar{C}$, whence $\bar{C} = Z(L(\bar{M})) \times O(\bar{C})$ by Burnside's transfer theorem. Since $O(\bar{C}) \subseteq O(\bar{M}) = 1$, the lemma follows.

4. The $\hat{A}_n$ case.

In this section we shall prove

Proposition 4.1. If $L_x \cong \hat{A}_n$, $n = 8,9,10$, or 11, then $G \cong M^c$ or L.

We carry out the proof in a short sequence of lemmas. We have that $\bar{L} = L(\bar{M}) \cong \hat{A}_n$, $n = 8,9,10$, or 11.

Lemma 4.2. If $\bar{M} = \bar{L}$, then $G \cong M^c$ or L.

Proof: In this case $\langle x\rangle = Z(T)$ is characteristic in T and consequently $S = T$. Thus S is of type $\hat{A}_8$ or $\hat{A}_{10}$. Now [18, Theorems A,B] yield the lemma.

We can therefore assume henceforth that $\bar{M} \supset \bar{L}$. Note that by Lemma 3.2, we have that $C_{\bar{M}}(\bar{L}) = \langle\bar{x}\rangle$. This enables us to prove

Lemma 4.3. We have $\bar{M}/\langle\bar{x}\rangle \cong S_8$ or S_9.

Proof: Setting $\tilde{M} = \bar{M}/\langle\bar{x}\rangle$, our conditions imply that $\tilde{L} = A_n, n=8,9,10,$ or 11, $\tilde{M} \supset \tilde{L}$, and $C_{\tilde{M}}(\tilde{L}) = 1$. Hence $\tilde{M} \cong S_n$. However, if $n = 10$ or 11, then $r(\tilde{M}) = 5$, contrary to $r(G) \leq 4$. Hence $n = 8$ or 9 and the lemma holds.

As is well-known, S_n has two nonisomorphic proper coverings by Z_2, in which respectively the involutions (12) and $(12)(34)(56)$ of S_n are represented by involutions in the covering group. We shall denote these covering groups by $\hat{S}_n^{(1)}$ and $\hat{S}_n^{(2)}$ respectively.

We need the following facts:

Lemma 4.4. The following conditions hold:

(i) $\hat{S}_8^{(2)}$ and $\hat{S}_9^{(2)}$ have Sylow 2-subgroups of type $\hat{A}_{10}$;

(ii) $SCN_3(2)$ is empty in $\hat{S}_8^{(1)}$ and $\hat{S}_9^{(1)}$.

Proof: Since $S_8 \subseteq A_{10}$, $\hat{A}_{10}$ must contain $\hat{S}_8^{(1)}$ or $\hat{S}_8^{(2)}$. One checks directly that, in fact, $\hat{A}_{10}$ contains $\hat{S}_8^{(2)}$. Similarly $\hat{S}_9^{(2)} \subseteq \hat{A}_{11}$. This yields (i) at once.

Now set $X = \hat{S}_n^{(1)}$, $n = 8$ or 9, $\bar{X} = X/Z(X)$, let P be a Sylow 2-subgroup of X and set $Q = P \cap X'$. We have that $\bar{P}$ is of type S_8 and so $\bar{P}$ is of type A_{10}. In addition, $X' \cong \hat{A}_n$, so Q is of type

$\hat{A}_8$ and hence $\bar{Q}$ is of type $\hat{A}_8$ and of index 2 in $\bar{P}$.

Suppose $A \in SCN_3(P)$. Then $\bar{A}$ is a normal abelian subgroup of $\bar{P}$. By the structure of $\bar{P}$, we must then have $\bar{A} \subset \bar{Q}$. Thus $A \subset Q \subset X'$. On the other hand, the group X' possesses an extension X^* by a group of order 2 with $X^* \cong S_n^{(2)}$. We identify the groups $X^*/Z(X)$ and $X/Z(X)$, which are isomorphic to S_n and we choose a Sylow 2-subgroup P^* of X^* containing Q such that $P^*/Z(X) = P/Z(X)$. But now by the definition of $S_n^{(1)}$ and $S_n^{(2)}$, we see that corresponding elements of $P - Q$ and $P^* - Q$ have the same action on Q. Since $A \triangleleft P$, we conclude therefore that $A \triangleleft P^*$. However, P^* is of type $\hat{A}_{10}$ and hence of type L and so $SCN_3(P^*)$ is empty. This contradiction establishes (ii).

<u>Lemma 4.5</u>. We have $\bar{M} \cong \hat{S}_8^{(1)}$ or $\hat{S}_9^{(1)}$.

<u>Proof</u>: If false, then $\bar{M} \cong \hat{S}_8^{(2)}$ or $\hat{S}_9^{(2)}$ and T is of type $\hat{A}_{10}$ by the preceding lemma. Again $\langle x \rangle = Z(T)$ and so $S = T$. Now [18, Theorem B] shows that there is no simple group in which the centralizer of a central involution has the form of M.

It remains to eliminate these two cases. The structure of T is, of course, uniquely determined as $\hat{S}_8^{(1)}$ and $\hat{S}_9^{(1)}$ have isomorphic Sylow 2-subgroups. In particular, we find that $\langle x \rangle = Z(T)$, so again we have $S = T$. We set $\tilde{M} = \bar{M}/\langle\bar{x}\rangle$. Since $\bar{M} \cong \hat{S}_8^{(1)}$ or $\hat{S}_9^{(1)}$, $S - R$ contains an involution y such that $\tilde{y}$ corresponds to a transposition in $\tilde{M} \cong S_8$ or S_9.

Let U be the unique element of $U(S)$, so that $U = \langle x,u \rangle$ for some u in U. If $u \sim x$ in G, then all involutions of U are conjugate and now the main results of [29] and [35] apply to yield that S is of type $G_2(q)$, $q \equiv 1$ or $7 \pmod 8$, J_2, or $\hat{A}_{10}$. However, one checks directly that our 2-group S is of none of these forms. Hence u is not conjugate to x in G.

On the other hand, as $\bar{L} \cong \hat{A}_8$ or $\hat{A}_9$, every involution of $R - \langle x \rangle$ is conjugate to u in M. Hence by Thompson's transfer lemma, we must have that $y \sim u$ or $y \sim x$ in G.

Observe next that $C_{\bar{L}}(\bar{y}) \cong \hat{A}_6$ or $\hat{A}_7$ and consequently $C_S(y)$ contains a subgroup $Q \cong Q_{16}$ with $Z(Q) = \langle x \rangle$ and $\bar{Q}$ a Sylow 2-subgroup of $C_{\bar{L}}(\bar{y})$.

If $y \sim x$ in G, then $y^g = x$ for some g in G such that $C_S(y)^g \subset S$. But then $Q^g \subset S$ and consequently $(Q')^g \subset R$, whence $x^g \in R$. However, as x is conjugate to no involution of $R - \langle x \rangle$, this forces $x^g = x$, which is clearly impossible.

Hence $y \sim u$ in G. Since $|S : C_S(u)| = 2$ and u is not conjugate to x in G, u is extremal and so $y^g = u$ for some g such that $C_S(y)^g \subset C_S(u)$. Then, as before, this forces $x^g = x$, whence $g \in M$. However, $\bar{u} \in \bar{L}$, $\bar{y} \in \bar{M} - \bar{L}$, and $\bar{L} \triangleleft \bar{M}$, so clearly y cannot be conjugate to u in M. This contradiction establishes the proposition.

5. <u>The $\hat{M}_{12}$ case</u>. In this section we shall treat the case in which $L_x \cong \hat{M}_{12}$. In contrast to all other cases of Theorem B, the involution x need not be central under this assumption and so we need not have $S = T$.

We first introduce some notation and establish some preliminary results concerning the structure of M which are independent of our assumption that $r(G) \leq 4$. We set $\bar{L} = L(\bar{M})$, so that $\bar{L} = L_x \cong \hat{M}_{12}$. By Lemma 3.2,

$C_{\bar{M}}(\bar{L}) = \langle\bar{x}\rangle$. Hence if we set $\tilde{M} = \bar{M}/\langle\bar{x}\rangle$, $C_{\tilde{M}}(\tilde{L}) = 1$ and so $\tilde{M}$ is isomorphic to a subgroup of $Aut(M_{12})$ containing $Inn(M_{12})$. As is well known (cf. [8]), $|Aut(M_{12}): Inn(M_{12})| = 2$, so either $\tilde{M} = \tilde{L}$ or $|\tilde{M}:\tilde{L}| = 2$ and $\tilde{M} \cong Aut(M_{12})$. In the first case, $\bar{M} = \bar{L}$ is, of course, isomorphic to M_{12}, while in the second case $\bar{M}$ is isomorphic to an extension of M_{12} by Z_2. We make some remarks about the second case.

It is well-known that if X is a perfect group, then X possesses a unique maximal perfect central extension Y by an abelian group A and that Y is uniquely determined up to isomorphism. Moreover, this assertion can be established by representing G as F/R, where F is a free group and using the fact that $R/[R,F] = (R \cap F')/[R,F] \times T/[R,F]$, where T is a torsion free abelian group and then taking Y to be F/T. Furthermore, Thompson has shown, using this representation, that any <u>single</u> automorphism α (but not necessarily group of automorphisms) of X with $\langle\alpha\rangle \cap Inn(X) = 1$ can be lifted to an automorphism α^* of Y.

Applying this result to the case that $X = M_{12}$ and $Y = \hat{M}_{12}$ with $X\langle\alpha\rangle = Aut(M_{12})$ and α of order 2, we thus obtain precisely two extensions of $\hat{M}_{12}$ by Z_2, corresponding to the cases in which $(\alpha^*)^2 = 1$ and $\langle\alpha^{*2}\rangle = Z(\hat{M}_{12})$. We denote these respectively by $\widehat{Aut}(M_{12})^{(1)}$ and $\widehat{Aut}(M_{12})^{(2)}$. This discussion shows that in the case that $\tilde{M} \cong Aut(M_{12})$, $\bar{M}$ is necessarily isomorphic to $Aut(\hat{M}_{12})^{(i)}$, $i = 1$ or 2.

Now returning to our situation, we set $R = T \cap M'$, so that $\bar{R}$ is a Sylow 2-subgroup of $\bar{L}$. Thus R and $\bar{R}$ are of type $\hat{M}_{12}$, while $\tilde{R}$ is of type M_{12}. We set

$$R = \langle x,z,z_1,a,b,t,u\rangle,$$

where $\tilde{z}, \tilde{z}_1, \tilde{a}, \tilde{b}, \tilde{t}, \tilde{u}$ (which generate $\tilde{R}$) satisfy the relations given in [Part II, Lemma 7.4] and $a^2b^2 = z$, $a^2 = z_1$. In addition, we set

$$Q = \langle x, z, z_1, ab, t, u \rangle,$$

so that $\tilde{Q} \cong Q_8 * Q_8$. We also let Q_i be subgroups of Q containing x such that $\tilde{Q} = \tilde{Q}_1\tilde{Q}_2 = \tilde{Q}_1 * \tilde{Q}_2$ with $\tilde{Q}_1 \cong \tilde{Q}_2 \cong \tilde{Q}_8$. Clearly $Q = Q_1Q_2$.

Finally by the structure of M_{12}, $N_M(Q)$ contains a 3-element y which normalizes both Q_1 and Q_2 such that $\tilde{y}$ has order 3, $\tilde{y} \in \tilde{L}$, and $\tilde{y}$ acts regularly on $\tilde{Q}/Z(\tilde{Q})$. Observing the possible structures of a 2-group of order 16, we conclude at once that

$$Q_1 \cong Q_2 \cong Z_2 \times Q_8.$$

We set $V_i = [Q_i, y]$, $i = 1,2$, so that $V_i \cong Q_8$ and $Q_i = \langle x \rangle \times V_i$, $i = 1, 2$.

The following general omnibus lemma will be useful.

Lemma 5.1. The following conditions hold:

(i) $\tilde{L}$ has two conjugacy classes of involutions and the involution fusion pattern is $\tilde{z} \sim \tilde{z}_1 \sim \tilde{u}\tilde{a} \sim \tilde{u}\tilde{a}\tilde{b} | \tilde{t} \sim \tilde{u} \sim \tilde{t}\tilde{u}$;

(ii) z is an involution;

(iii) Q is a special 2-group with $Z(Q) = \langle x, z \rangle$;

(iv) t is of order 4;

(v) $\Omega_1(Q) = \langle x,z,z_1,uab\rangle \cong E_{16}$. In particular, $\Omega_1(Q)$ is the unique elementary abelian subgroup of Q of order 16;

(vi) Replacing z by zx, if necessary, we have

$$Q/\langle x\rangle \cong Q/\langle xz\rangle \cong Q_8 * Q_8 \quad\text{and}\quad Q/\langle z\rangle \cong Z_2 \times Z_2 \times Q_8.$$

(vii) $N_{\bar{M}}(\bar{Q})/\bar{Q} \cong S_3$ or $Z_2 \times S_3$ according as $\tilde{M} \cong M_{12}$ or $\mathrm{Aut}(M_{12})$.

(viii) $C_{\bar{M}}(\bar{Q}) \subseteq \bar{Q}$ and an element of $N_{\bar{M}}(\bar{Q})$ of order 3 acts trivially on $Z(\bar{Q})$ and regularly on $\bar{Q}/Z(\bar{Q})$;

(ix) $C_{\tilde{M}}(\tilde{v}) \subseteq \tilde{L}$ for any element $\tilde{v}$ of $\tilde{R}$ of order 4.

Proof: The involution fusion pattern of M_{12} is well known (cf. [19]). Since $\tilde{L} \cong M_{12}$, we immediately obtain (i). Since $Z(\tilde{Q}_i) = \langle\tilde{z}\rangle$ and $Q_i \cong Z_2 \times Q_8$, clearly z must be an involution **and** $\langle x,y\rangle = Z(Q_i)$, $i = 1,2$. In particular, (ii) also holds.

Set $Z = \langle x,z\rangle$. Since $Q = Q_1Q_2$ and $Z = Z(Q_i)$, $i = 1,2$, $Z \subseteq Z(Q)$. On the other hand, as $\tilde{Q} \cong Q/\langle x\rangle \cong Q_8 * Q_8$, we also have that $Z(Q) \subseteq Z$, so $Z = Z(Q)$. If $x \in Q'$, then by the structure of Q_1, certainly $Z \subseteq Q'$. Since $Q/Z \cong E_{16}$, this implies that $Z = Q' = \Phi(Q)$ and hence Q is special. Thus (iii) holds in this case.

Suppose, on the other hand, that $x \notin Q'$, in which case $Q = \langle x\rangle \times Q_0$, where $Q_0 = [Q,y] \cong Q_8 * Q_8$. By the structure of M_{12}, $\tilde{y}$ is inverted by an involution $\tilde{v}$ of $\tilde{L}$, which without loss we can suppose lies in $\tilde{S}$. But then if v is a representative of $\tilde{v}$ in S, we see that v leaves $Q_0 = [Q,y]$ invariant. On the other hand, as $\tilde{Q} = \langle\tilde{z},\tilde{z}_1,\tilde{a}\tilde{b},\tilde{t},\tilde{u}\rangle$ and $\tilde{v} \notin \tilde{Q}$, $\tilde{v} \sim \tilde{z}$ in $\tilde{L}$ by (i). Since z is an involution, it follows that v must be as well. Hence $Q_0\langle v\rangle$ is a complement to $\langle x\rangle$ in R and so

R splits over $\langle x\rangle$. But now Gaschütz' theorem yields that $\bar{L}$ splits over $\langle\bar{x}\rangle$, contrary to the fact that $\bar{L} \cong \hat{M}_{12}$ is perfect.

Observe next that two involutions of $\tilde{Q}$ that are conjugate in $\tilde{L}$ are either both images of involutions of Q or elements of Q of order 4 with square x. Since z is an involution, all conjugates of $\tilde{z}$ in $\tilde{L}$ which lie in $\tilde{Q}$ are thus represented by involutions of Q. If t is also an involution, then the same assertion holds for t in place of z. Since $\tilde{L}$ has only two conjugacy classes of involutions, it follows in this case that every involution of $\tilde{Q}$ is represented by an involution of Q.

To derive a contradiction in this case, we prove the preliminary assertion that $V_1' = V_2'$, which is independent of whether or not t is an involution. Assume false. Since Q is special with center $Z = \langle x,z\rangle$, it follows that $Z = \langle V_1', V_2'\rangle$. Since $Q = Q_1Q_2 = \langle x\rangle V_1V_2$, we obtain that $Q = V_1V_2$. Hence Q satisfies all the assumptions of [28, Lemmas 9 and 10] with Q in the role of S. We thus conclude that either $Q \cong Q_8 \times Q_8$ or Q is of type $L_3(4)$. In the first case, $\Omega_1(Q) = Z$ and by the discussion of the preceding paragraph, no involution of $\tilde{Q} - \langle\tilde{z}\rangle$ is conjugate to $\tilde{z}$ in $\tilde{L}$, which is false. Hence Q is of type $L_3(4)$. By the structure of Q, x is a square in Q. Since $\tilde{L}$ has only two conjugacy classes of involutions, it follows that every involution of Q maps in $\tilde{Q}$ to a conjugate of $\tilde{z}$ (and every element of Q with square x maps in $\tilde{Q}$ to a conjugate of $\tilde{t}$). Computing the number of involutions in $Q - \langle x\rangle$, we thus find that $\tilde{z}$ has 13 conjugates in $\tilde{L}$ which lie in $\tilde{Q}$. However, by our fusion pattern, we easily check that the actual number of such conjugates of $\tilde{z}$ is 7. This contradiction establishes our assertion.

Now assume t is an involution, in which case every involution of $\tilde{Q}$ is represented by an involution of Q. We have $Q = \langle x\rangle V_1V_2$ and we argue that V_1 centralizes V_2. Setting $V_i = \langle v_i, w_i\rangle$, $i = 1,2$, we need only show by symmetry that v_1 centralizes v_2. Since $\tilde{v}_1\tilde{v}_2$ is an involution, v_1v_2 or xv_1v_2 must be an involution, so, in fact, both are involutions. Since $Q' \subseteq Z(Q)$, we have $1 = (v_1v_2)^2 = v_1^2v_2^2[v_1,v_2]$. But $\langle v_i^2\rangle = V_i'$ as $V_i \cong Q_8$, $i = 1,2$, and as $V_1' = V_2'$, it follows that $v_1^2 = v_2^2$, whence $v_1^2v_2^2 = 1$. Thus $1 = [v_1,v_2]$ and v_1 centralizes v_2. Hence V_1 centralizes V_2, as asserted. Since x also centralizes V_1 and V_2, we compute now that $Q' = \langle V_1', V_2'\rangle = V_1'$. Since $x \notin V_1'$, we conclude that $x \notin Q'$, contrary to what we have proved above. Thus t must be of order 4 and (iv) holds.

Setting $A = \langle x, z, z_1, uab\rangle$, we check from our fusion pattern that the 7 involutions of $\tilde{A}$ include precisely all the conjugates of $\tilde{z}$ in $\tilde{L}$ which lie in $\tilde{Q}$. Since z is an involution, while t is of order 4, we conclude at once that $A \cong E_{16}$ and that A contains every involution of Q. Thus (v) also holds.

Replacing z by xz, if necessary, we can assume without loss that $V_i' = \langle xz\rangle$, $i = 1,2$. We now prove (vi). Once again let y be the element of $N_M(Q)$ such that $V_i = [Q_i, y]$, $i = 1,2$. We have that $Q = \langle x\rangle V_1V_2 = V_1A$ with y normalizing V_1 and A. Clearly y^3 centralizes Q. We set $K = Q\langle y\rangle$ and $\bar{K} = K/\langle xz, y^3\rangle$, so that $\bar{Q} \cong Q/\langle xz\rangle$ and $|\bar{y}| = 3$. Moreover, $\bar{A} \cong E_8$, $\bar{V}_1 \cong Z_2 \quad Z_2$, $\bar{A} \cap \bar{V}_1 = 1$, $\bar{Z} = \langle\bar{x}\rangle \subseteq Z(\bar{K})$, $\bar{V}_1\langle\bar{y}\rangle \cong A_4$, and $\bar{A} = \bar{Z} \times [\bar{A},\bar{y}]$. In particular, $\bar{K}$ is a split extension of E_8 by A_4. But now [Part II, Lemma 3.2] yields that $Q/\langle xz\rangle \cong \bar{Q} \cong Q_8 * Q_8$.

Observe that Q is of exponent 4 with 15 involutions, so Q has $48 = 64 - 16$ elements of order 4. Moreover, by our fusion pattern, $\tilde{t}$ has 12 conjugates in $\tilde{L}$ which lie in $\tilde{Q}$, so Q has 24 elements of order 4 whose square is x. We claim that Q also has 24 elements of order 4 whose square is xz. Indeed, $Q/\langle xz\rangle \cong Q_8 * Q_8$ contains exactly 19 involutions. On the other hand, $A/\langle xz\rangle \cong E_8$ contains exactly 7 involutions. Since $A = \Omega_1(Q)$, it follows that 12 of the 19 involutions of $Q/\langle xz\rangle$ must lift to elements of order 4 in Q and consequently Q must have 24 elements whose square is xz.

Our argument yields that z is not a square in Q. Hence if we set $\bar{Q} = Q/\langle z\rangle$, we have that $\bar{A} = \Omega_1(\bar{Q}) \cong E_8$. In addition, we have that $\bar{Q} = \bar{V}_1\bar{A}$ with $\bar{V}_1 \triangleleft \bar{Q}$, $\bar{V}_1 \cong Q_8$, and $\bar{Q}' = \langle\bar{x}\rangle$. If $\bar{V}_1$ does not centralize $\bar{A}$, then $[\bar{v}_1,\bar{a}] = \bar{x}$ for some $\bar{v}_1$ in $\bar{V}_1 - \langle\bar{x}\rangle$ and $\bar{a}$ in $\bar{A} - \langle\bar{x}\rangle$. But then as $\bar{v}_1^2 = \bar{x}$ and $\bar{a}^2 = 1$, it follows that $\bar{v}_1\bar{a}$ is an involution, contrary to the fact that $\bar{A} = \Omega_1(\bar{Q})$. Hence $\bar{V}_1$ centralizes $\bar{A}$ and so $\bar{Q} = \bar{V}_1 \times \bar{A}_1 \cong Q_8 \times Z_2 \times Z_2$ for some four subgroup $\bar{A}_1$ of $\bar{A}$. This establishes (vi).

Next we prove (vii) and (viii). We know from the structure of M_{12} that $C_{\tilde{L}}(\tilde{z})$ is a split extension of $\tilde{Q} \cong Q_8 * Q_8$ by S_3 and that $\tilde{R}$ is a Sylow 2-subgroup of $C_{\tilde{L}}(\tilde{z})$. In particular, $C_{\tilde{L}}(\tilde{z}) \subseteq N_{\tilde{L}}(\tilde{Q})$. On the hand, as $\langle\tilde{z}\rangle = Z(\tilde{Q})$, the reverse inclusion also holds and so $N_{\tilde{L}}(\tilde{Q}) = C_{\tilde{L}}(\tilde{z}) = \tilde{R}\langle\tilde{y}\rangle$. Thus $N_{\bar{L}}(\bar{Q}) = \bar{R}\langle\bar{y}\rangle$ and $N_{\bar{L}}(\bar{Q})/\bar{Q} \cong S_3$. In addition, $\bar{y}$ acts regularly on $\bar{Q}/\bar{Z}$. Since $\bar{y}$ obviously centralizes $\bar{x}$, $\bar{y}$ acts trivially on $\bar{Z} = \langle\bar{x},\bar{z}\rangle$. Furthermore, clearly $C_{\bar{L}}(\bar{Q}) \subseteq \bar{Z}$. Hence if $\bar{M} = \bar{L}$, we conclude that (vii) and (viii) hold.

Suppose then that $\bar{M} \supset \bar{L}$, whence $|\bar{M}:\bar{L}| = 2$ and consequently $C_{\tilde{M}}(\tilde{z}) = \tilde{T}\langle\tilde{y}\rangle$ and $\tilde{T}\langle\tilde{y}\rangle/\tilde{Q}$ is of order 12 and contains $\tilde{R}\langle\tilde{y}\rangle/\tilde{Q} \cong S_3$ as a subgroup of index 2. The only possibility is that $\tilde{T}\langle\tilde{y}\rangle/\tilde{Q} \cong S_3 \times Z_2$. Hence by the action of $\tilde{y}$ on $\tilde{Q}$, we have that $\tilde{Q} = [O_2(C_{\tilde{M}}(\tilde{z})), \tilde{y}]$ and so $\tilde{Q} \lhd C_{\tilde{M}}(\tilde{z})$. As above, this implies that $N_{\tilde{M}}(\tilde{Q}) = C_{\tilde{M}}(\tilde{z})$, whence $N_{\bar{M}}(\bar{Q})/\bar{Q} \cong S_3 \times Z_2$, so (vii) and the second assertion of (viii) hold in this case as well.

Suppose (ix) holds. Since $\bar{Q}$ contains an element whose image in $\tilde{Q}$ is of order 4, it then follows that $C_{\bar{M}}(\bar{Q}) \subseteq \bar{L}$. Since $C_{\tilde{L}}(\tilde{Q}) \subseteq \tilde{Q}$ by the structure of M_{12}, this yields that $C_{\bar{M}}(\bar{Q}) \subseteq \bar{Q}$, giving the first assertion of (viii).

Thus it remains to establish (ix). By [10], it is known that M_{12} has exactly two conjugacy classes of elements of order 4 and any element of $\mathrm{Aut}(M_{12}) - M_{12}$ conjugates one class into the other. Clearly then $C_{\mathrm{Aut}(M_{12})}(v) \subseteq M_{12}$ for any element v of order 4 of M_{12}. Since $\tilde{M} \cong M_{12}$ or $\mathrm{Aut}(M_{12})$ and $\tilde{L} \cong M_{12}$, this yields (ix) at once.

We next prove

Lemma 5.2. The following conditions hold:

(i) $Z(T) = \langle x,z\rangle$;

(ii) Q is characteristic in T.

Proof: We have $|R:Q| = 2$ and $|T:R| \leq 2$. Clearly $x \in Z(T)$. Since $Z = \langle x,z\rangle = Z(Q)$ by the preceding lemma, either $Z(Q) = Z(R)$ or $R - Q$ contains an element v such that $z^v = zx$. Since v normalizes Q, it follows then that $Q/\langle z\rangle \cong Q/\langle zx\rangle$. However, this contradicts Lemma 5.1 (vi). Hence $Z = Z(R)$.

Furthermore, by the analysis of the preceding lemma, $N_{\tilde{L}}(\tilde{Q})/\tilde{Z}$ is a split extension of $\tilde{Q}/\tilde{Z} \cong E_{16}$ by S_3 and an element $\tilde{y}$ of $N_{\tilde{L}}(\tilde{Q})$ of order 3 acts regularly on $\tilde{Q}/\tilde{Z}$. Since $\tilde{R} \subseteq N_{\tilde{L}}(\tilde{Q})$, this implies that $\tilde{R}/\tilde{Z} \cong Z_2 \times Z_2 \int Z_2$, whence $\tilde{Q}/\tilde{Z}$ is characteristic in $\tilde{R}/\tilde{Z}$. Thus Q/Z is characteristic in R/Z and as $Z = Z(Q)$, it follows that Q is characteristic in R.

If $T = R$, we see that both parts of the lemma hold; hence we can suppose that $T \supset R$. Since Q is characteristic in R, $Q < T$. Hence if $z^v = zx$ for some v in $T - Q$, we reach the same contradiction as in the first paragraph. We conclude therefore that $Z \subset Z(T)$. On the other hand, Lemma 5.1(viii) implies that $Z(T) \subseteq Q$, whence $Z = Z(T)$ and (i) is proved.

Finally setting $\tilde{K} = N_{\tilde{M}}(\tilde{Q})$ and $\bar{K} = \tilde{K}/\tilde{Z}$, we know that $\bar{K}$ is a split extension of $\bar{Q} \cong E_{16}$ by $N_{\bar{K}}(\langle\bar{y}\rangle) \cong S_3 \times Z_2$ with $\bar{y}$ acting regularly on $\bar{Q}$. We conclude therefore from [Part II, Lemma 2.2 (viii)] that the Sylow 2-subgroup $\bar{T}$ of $\bar{K}$ is of type A_8. Hence $\bar{Q}$ is characteristic in $\bar{T}$ and so Q is characteristic in T, proving (ii).

We need one further fact.

<u>Lemma 5.3</u>. Suppose Q is normal in a group H and the following conditions hold:

(a) $C_H(Q) = Z(Q)$;

(b) $H/Q \cong S_3 \times Z_2$;

(c) An element of H of order 3 centralizes $Z(Q)$ and acts regularly on $Q/Z(Q)$.

Under these conditions, $Z(Q) \subseteq Z(O_2(H))$.

Proof: Our conditions imply that $H = KQ$, where $K \cap Q = Z(Q)$ and $K/Z(Q) \cong S_3 \times Z_2$. Set $Z = Z(Q)$, $\bar{H} = H/Z$ and let v, w be respectively a 2-element and 3-element of K such that $\langle\bar{v}\rangle = O_2(\bar{K})$ and $\langle\bar{w}\rangle = O_3(\bar{K})$. Since clearly $O_2(H) = QO_2(K) = Q\langle v\rangle$, we must show that v centralizes $Z(Q)$.

If $\bar{v}$ centralizes $\bar{Q}$, then $[v,Q] \subseteq Z$, whence $[v,Q,w] = 1$. Since w centralizes Z and $\bar{w}$ centralizes $\bar{v}$, w centralizes v, so also $[w,v,Q] = 1$. Hence $[Q,w,v] = 1$ by the three subgroup lemma and so v centralizes $[Q,w]$. But $[Q,w] = Q$ by the action of w on Q, so the lemma holds in this case. We can therefore assume that $\bar{v}$ does not centralize $\bar{Q}$.

Since $\bar{Q} \cong E_{16}$, $\bar{K}$ can therefore be identified with a subgroup of A_8. Since $\bar{w}$ acts regularly on $\bar{Q}$, $\bar{w}$ is necessarily generated by a 3-cycle and therefore $N_{A_8}(\langle\bar{w}\rangle)/\langle\bar{w}\rangle \cong S_5$. Thus a Sylow 2-subgroup $\bar{D}$ of $N_{A_8}(\langle\bar{w}\rangle)$ is dihedral of order 8. If we choose $\bar{D}$ suitably, then we can assume that $\langle\bar{v}\rangle = Z(\bar{D})$. Since A_8 has only one conjugacy class of 3-cycles, we conclude that the group $\langle\bar{w}\rangle \times \langle\bar{v}\rangle$ is uniquely determined up to conjugacy in A_8. This means that the action of v on Q is uniquely determined up to conjugacy modulo the subgroup of $\mathrm{Aut}(Q)$ which stabilizes the chain $Q \supset Z \supset 1$.

On the other hand, if we set $M^* = \widehat{\mathrm{Aut}}(M_{12})^{(1)}$, we can identify Q with a subgroup of $(M^*)' \cong \hat{M}_{12}$. Moreover, if we let T^* be a Sylow 2-subgroup of M^* containing Q, the proof of Lemma 5.2 shows that $Q \lhd T^*$ and that $Z = Z(T^*)$. By the structure of M^*, $C_{M^*}(Q)$ has a normal

2-complement. Hence if we set $H^* = N_{M^*}(Q)$ and $\bar{H}^* = H^*/O(H^*)$, $\bar{T}^*$ is a Sylow 2-subgroup of $\bar{H}^*$ and $C_{\bar{H}^*}(\bar{Q}) = \bar{Z}$. Furthermore, the preceding analysis yields that $\bar{H}^*/\bar{Q} \cong S_3 \times Z_2$ and an element of $\bar{H}^*$ of order 3 acts trivially on $\bar{Z}$ and regularly on $\bar{Q}/\bar{Z}$. Thus $\bar{H}^*, \bar{Q}$ satisfy the same hypotheses as H, Q (since $Q \cong \bar{Q}$). But as $Z = Z(T^*)$, $O_2(\bar{H}^*)$ centralizes $\bar{Z}$. We see then that the lemma holds for at least one choice of H. However, as the action of v on Q is uniquely determined up to conjugacy modulo the subgroup of $\mathrm{Aut}(Q)$ which stabilizes the chain $Q \supset Z \supset 1$, it follows now that v must centralize Z in the given group H and the lemma is proved.

With these preliminaries we can now eliminate the noncentral involution case.

<u>Proposition 5.4</u>. The involution x is central.

<u>Proof</u>: Suppose false, in which case $S \supset T$. Setting $W = N_S(T)$, we have that $W \supset T$. Moreover, as Q is characteristic in T, W normalizes Q. Thus W acts on $Z(Q) = \langle x,z \rangle$ and as T is a Sylow 2-subgroup of $M = C_G(x)$, it follows that $|W:T| = 2$.

We now set $N = N_G(Q)$ and $\bar{N} = N/O(N)$. If W_1 is a Sylow 2-subgroup of N containing W, then W_1 acts on $Z(Q)$ and the same reasoning shows that $|W_1:T| = 2$, so $W_1 = W$. Furthermore, as $C_G(Q) \subseteq M$, $C_G(Q)$ has a normal 2-complement by the structure of M. Hence $C_{\bar{N}}(\bar{Q}) = Z(\bar{Q})$. In addition, if we put $K = M \cap N = N_M(Q)$, we know that $\bar{K}/\bar{Q} \cong S_3$ or $S_3 \times Z_2$, that $\bar{T}$ is a Sylow 2-subgroup of $\bar{K}$, and that an element of $\bar{K}$ of order 3 acts trivially on $Z(\bar{Q})$ and regularly on $\bar{Q}/Z(\bar{Q})$. Since $Q/\langle x \rangle \not\cong Q/\langle z \rangle$ by Lemma 5.1 (vi), N does not contain a 3-element which acts nontrivially on $Z(Q)$. Since N acts on $Z(Q)$, this implies that $|N:K| \leq 2$, whence $N = KW$ and so also $\bar{N} = \bar{K}\bar{W}$ with $\bar{K}$ of index 2 in $\bar{N}$.

Consider first the case that $\tilde{K} = \bar{K}/\bar{Q} \cong S_3$. Clearly then the only possibility is that $\tilde{N} = \bar{N}/\bar{Q} \cong S_3 \times Z_2$. But now we see that all the assumptions of the preceding lemma are satisfied with $\bar{N}$ as H (inasmuch as $\bar{Q} \cong Q$). We therefore conclude that $Z(\bar{Q}) \subseteq Z(O_2(\bar{N}))$. However, $\bar{W} = \bar{T}O_2(\bar{N})$ and consequently $Z(\bar{Q}) \subseteq Z(\bar{W})$. Thus W centralizes $Z(Q)$ and, in particular, centralizes x, which implies that $W = T$, a contradiction. We therefore conclude that $\tilde{K} \cong S_3 \times Z_2$.

If $\tilde{N} \cong S_3 \times Z_2 \times Z_2$, then clearly $\bar{N}$ possesses a subgroup $\bar{N}_1$ of index 2 containing $\bar{Q}$ such that $\tilde{N}_1 \cong S_3 \times Z_2$ and $\bar{N} = \bar{N}_1\bar{T}$. In this case the preceding lemma applies to $\bar{N}_1$ and so $Z(\bar{Q}) \subseteq Z(O_2(\bar{N}_1))$. But $\bar{W} = O_2(\bar{N}_1)\bar{T}$, so $Z(\bar{Q}) \subseteq Z(\bar{W})$, which leads to the same contradiction as in the preceding case. Hence $\tilde{N} \not\cong S_3 \times Z_2 \times Z_2$.

We use this to prove that $\tilde{N}$ acts faithfully on $\bar{Q}/Z(\bar{Q})$. First of all, suppose $\langle\tilde{v}\rangle = O_2(\tilde{K})$ acts trivially on $\bar{Q}/Z(\bar{Q})$, whence $\tilde{v}$ stabilizes the chain $\bar{Q} \supset Z(\bar{Q}) \supset 1$. But clearly $O_3(\tilde{K}) = \langle\tilde{y}\rangle$, where y is the 3-element of $N_M(Q)$ defined at the beginning of the section. Since $[\bar{Q},\tilde{y}] = \bar{Q}$ and $\tilde{y}$ centralizes $\tilde{v}$, we conclude now by the three subgroup lemma as we did in Lemma 5.3 that $\tilde{v}$ acts trivially on $\bar{Q}$. But then $C_{\bar{M}}(\bar{Q}) \not\subseteq \bar{Q}$, contrary to Lemma 5.1 (viii). We conclude that $\tilde{K}$ acts faithfully on $\bar{Q}/Z(\bar{Q})$. Hence if $\tilde{N}$ does not act faithfully on $\bar{Q}/Z(\bar{Q})$, it follows that $\tilde{N} = \tilde{K} \times \tilde{W}_1$, where $\tilde{W}_1$ is of order 2 and acts trivially on $\bar{Q}/Z(\bar{Q})$. But then $\tilde{N} \cong S_3 \times Z_2 \times Z_2$, which is not the case.

We conclude now that $\tilde{N}$ is isomorphic to a subgroup of $A_8 \cong \mathrm{Aut}(\bar{Q}/Z(\bar{Q}))$ Since $\tilde{y}$ acts regularly on $\bar{Q}/Z(\bar{Q})$, $\tilde{N}$ is, in fact, isomorphic to a subgroup of $N_{A_8}(\langle(123)\rangle)$. Thus $\tilde{W} \cong D_8$ and $O_2(\tilde{N}) \cong Z_2 \times Z_2$. Since

$C_{\bar{Q}}(\bar{y}) = Z(\bar{Q})$ and $C_{\tilde{N}}(\tilde{y}) = \langle\tilde{y}\rangle \times O_2(\tilde{N})$, we see that $\bar{W}_0 = C_{\bar{W}}(\bar{y})$ is of order 16, contains $Z(\bar{Q})$, and $\tilde{W}_0 = O_2(\tilde{N}) \cong Z_2 \times Z_2$.

Setting $\bar{T}_0 = \bar{T} \cap \bar{W}_0$, we have that $\bar{T}_0$ is of order 8 and contains $Z(\bar{Q})$. Since $\tilde{W}_0 \cong Z_2 \times Z_2$, we can therefore write $\bar{W}_0 = \bar{T}_0\langle\bar{w}\rangle$, where $\bar{w}^2 \in Z(\bar{Q})$. Then $\bar{W} = \bar{T}\langle\bar{w}\rangle$ and so if $\bar{w}$ centralized $Z(\bar{Q})$, it would follow that W centralized x, which we know is not the case. Hence $Z(\bar{Q})\langle\bar{w}\rangle \cong D_8$ and so we can choose $\bar{w}$ to be an involution. Recall that $\bar{Q}/\langle\bar{x}\rangle \cong \bar{Q}/\langle\overline{xz}\rangle \not\cong \bar{Q}/\langle\bar{z}\rangle$. Since $\bar{w}$ normalizes $\bar{Q}$ and does not centralize $\bar{x}$, it follows that $\bar{w}$ centralizes $\bar{z}$, whence $\langle\bar{z}\rangle = Z(Z(\bar{Q})\langle\bar{w}\rangle)$. On the other hand, $\bar{T}_0$ is abelian as $Z(\bar{Q}) \subseteq Z(\bar{T})$ and so $\bar{T}_0 \cong Z_4 \times Z_2$ or E_8.

We claim next that $\bar{W}_0/\langle\bar{z}\rangle \cong E_8$. If $\bar{T}_0 \cong E_8$, then $[\bar{T}_0,\bar{w}]$, being of order at most 2, must equal $\langle\bar{z}\rangle$ and so our assertion is clear. On the other hand, if $\bar{T}_0 \cong Z_4 \times Z_2$, then $\mho^1(\bar{T}_0)$ is of order 2 and normal in $\bar{W}_0$, so $\bar{w}$ centralizes $\mho^1(\bar{T}_0)$. Since $\mho^1(\bar{T}_0) \subseteq Z(\bar{Q})$, it follows that $\mho^1(\bar{T}_0) = \langle\bar{z}\rangle$. Hence $\bar{T}_0/\langle\bar{z}\rangle \cong Z_2 \times Z_2$ and so either our assertion holds or $\bar{W}_0/\langle\bar{z}\rangle \cong D_8$. But then $\bar{W}_0 \cong D_{16}$ or QD_{16}, contrary to the fact that neither group possesses a normal four subgroup.

Finally we consider $\bar{A} = \Omega_1(\bar{Q}) = \langle\bar{x},\bar{z},\bar{z}_1, \overline{uab}\rangle \cong E_{16}$ and we set $\widetilde{AW}_0 = \overline{AW}_0/\langle\bar{z}\rangle$. Then $\tilde{A} \cong \tilde{W}_0 \cong E_8$ and $\tilde{A} \cap \tilde{W}_0 = \langle\tilde{x}\rangle$. Setting $\tilde{A}_0 = [\tilde{A},\tilde{y}]$, we have that $\tilde{A}_0 \cong Z_2 \times Z_2$ and that $\tilde{A} = \langle\tilde{x}\rangle \times \tilde{A}_0$. But as $\bar{W}_0$ centralizes $\bar{y}$, it follows that $\tilde{W}_0$ centralizes $\tilde{A}_0$. Hence $\tilde{A}_0\tilde{W}_0 \cong E_{32}$, contrary to $r(G) \le 4$. This completes the proof of the proposition.

By Proposition 5.4, we have $S = T$. To treat this case it will be helpful to have a more exact description of $\tilde{S}$ in the case that $\tilde{M} \cong \mathrm{Aut}(M_{12})$. As we shall see, this group has already been described in Lemma 2.1 of Part II, which dealt with the structure of $B = \mathrm{Aut}(Z_4 \times Z_4)$. In particular we described the Sylow 2-subgroup $\langle\alpha,\beta,\delta,\epsilon\rangle$ of the semi-direct product of $Z_4 \times Z_4$ by $N_B(E)$, where E is a Sylow 3-subgroup of B. Here $\langle\alpha,\beta\rangle \cong Z_4 \times Z_4$, $\langle\delta,\epsilon\rangle \cong D_8$ with $\delta^2 = \epsilon^2 = 1$. Moreover, if $\gamma = (\delta\epsilon)^2$, then $\langle\alpha,\beta,\gamma,\epsilon\rangle$ is of type M_{12}. Hence we can identify $\tilde{R}$ with $\langle\alpha,\beta,\gamma,\epsilon\rangle$ under the correspondence $\tilde{a} = \alpha$, $\tilde{b} = \beta$, $\tilde{u} = \gamma$, $\tilde{t} = \epsilon$.

We shall prove

<u>Lemma 5.5</u>. If $S \supset R$, then we have

(i) $\tilde{S} \cong \langle\alpha,\beta,\delta,\epsilon\rangle$;

(ii) Every involution of $\langle\alpha,\beta,\delta,\epsilon\rangle - \langle\alpha,\beta,\gamma,\epsilon\rangle$ is conjugate in $\langle\alpha,\beta,\delta,\epsilon\rangle$ to δ;

(iii) $C_{\langle\alpha,\beta,\delta,\epsilon\rangle}(\delta) = \langle\alpha^2,\beta^2,\gamma,\delta\rangle \cong E_{16}$.

<u>Proof</u>: First of all, (ii) and (iii) can be verified by direct computation, so we need only prove (i).

We have that $\tilde{R} = \langle\tilde{a},\tilde{b},\tilde{t},\tilde{u}\rangle$ is of type M_{12} with $\tilde{A} = \langle\tilde{a},\tilde{b}\rangle = \langle\tilde{a}\rangle \times \langle\tilde{b}\rangle \cong Z_4 \times Z_4$ and characteristic in $\tilde{R}$. Furthermore, by the structure of M_{12}, we have that $C_{\tilde{L}}(\tilde{A}) = \tilde{A}$ and $N_{\tilde{L}}(\tilde{A})$ is a split extension of $Z_4 \times Z_4$ by $S_3 \times Z_2$. Now set $\tilde{N} = N_{\tilde{M}}(\tilde{A})$. Since $\tilde{A}$ is characteristic in $\tilde{R}$, $\tilde{S} \subseteq \tilde{N}$ and so $\tilde{N} = \tilde{S}(\tilde{N} \cap \tilde{L})$ with $|\tilde{N}:\tilde{N} \cap \tilde{L}| = 2$. Moreover, by Lemma 5.1 (ix), no element of $\tilde{N} - \tilde{L}$ centralizes $\tilde{A}$. Hence $\tilde{N}/\tilde{A}$ is isomorphic to a subgroup of $\tilde{B} = \mathrm{Aut}(\tilde{A})$ and we conclude at once that $\tilde{N} = \tilde{A}N_{\tilde{B}}(\tilde{E})$ where $\tilde{E}$ is a Sylow 3-subgroup of $\tilde{B}$. Now the discussion preceding the lemma shows that (i) holds.

We next prove

Lemma 5.6. The following conditions hold:

(i) x is not conjugate in G to an involution of $R - \langle x\rangle$;

(ii) $S = R\langle y\rangle$ for some involution y conjugate to x in G.

Proof: Suppose $x \sim y$ in G for some y in $R-\langle x\rangle$. We know that $\tilde{L}$ has two conjugacy classes of involutions represented by $\tilde{z}$ and $\tilde{t}$ and also that t is of order 4. Hence we can assume without loss that $y = z$ or zx. By Lemma 5.2 (i), $Z(S) = \langle x,z\rangle$. Hence by Burnside's lemma, $x \sim z$ in $N = N_G(S)$ and consequently N contains a 3-element v which cyclically permutes x, z, and xz. But Q is characteristic in S by Lemma 5.2 (ii) and so v permutes the groups $Q/\langle x\rangle$, $Q/\langle z\rangle$, and $Q/\langle xz\rangle$. However, this is impossible as $Q/\langle x\rangle \not\cong Q/\langle z\rangle$ by lemma 5.1 (vi). Thus (i) holds.

Now Glauberman's Z^*-theorem implies that $x \sim y$ in G for some y in $S-R$. Since $|S:R| \leq 2$, this forces $S = R\langle y\rangle$ and so (ii) also holds.

We can now derive a final contradiction:

Proposition 5.7. The involution x is not central.

Proof: We continue the above analysis. We have $S = R\langle y\rangle$ with $y \sim x$ in G and $y \in S-R$. Thus $S \supset R$ and so we can identify $\tilde{S}$ with $\langle \alpha,\beta,\delta,\epsilon\rangle$. In view of Lemma 5.5 (ii), every involution of $S-R$ is conjugate in S to δ or δx and so without loss we can assume that $y = \delta$ or δx. Thus every involution of $S-R$ is conjugate in S to y or yx. By Lemma 5.5 (iii), $C_{\tilde{S}}(\tilde{y}) = C_{\tilde{S}}(\delta) = \langle \delta,\gamma,\alpha^2,\beta^2\rangle \cong E_{16}$. Reverting to our original notation, we have $C_{\tilde{S}}(\tilde{y}) = \langle \tilde{a}^2,\tilde{b}^2,\tilde{u},\tilde{y}\rangle$.

We have $y^g = x$ with $C_S(y)^g \subseteq S$ for some g in G. If $x \in \Phi(C_S(y))$, then $x^g \in \Phi(S) \subseteq R$, whence $x^g = x$ by Lemma 5.6 (i), contrary to the fact that $y^g = x$. Hence $x \notin \mho(C_S(y))$. Since $C_{\tilde{S}}(\tilde{y}) \cong E_{16}$ and $\tilde{S} \cong S/\langle x\rangle$, it follows now that $C_S(y)$ is elementary abelian. But clearly $C_S(y)$ contains a maximal subgroup of the inverse image $\langle a^2,b^2,u,y,x\rangle$ of $C_{\tilde{S}}(\tilde{y})$. However $|u| = 4$ as $|t| = 4$ and $\tilde{u} \sim \tilde{t}$ in $\tilde{M}$ and so $u \notin C_S(y)$. We therefore conclude that $C_S(y) \cong E_{16}$. Furthermore, our argument yields that $[u,y] = x$, whence $y \sim yx$ in S. In particular, we see that every involution of $S-R$ is conjugate in S to y.

Finally consider $C_R(y) = C_S(y) \cap R$, which is clearly isomorphic to E_8 and contains 6 involutions v_i, $1 \leq i \leq 6$, which lie in $R-\langle x\rangle$. By Lemma 5.6 (i), no $v_i \sim x$ in G. Since $x \sim y$ in G, it follows therefore from the preceding paragraph that $v_i^g \notin S-R$, $1 \leq i \leq 6$. Hence $v_i^g \in R$, $1 \leq i \leq 6$. But clearly $C_S(y)$ is generated by the involutions v_i and y, so $C_S(y)^g \subseteq R$. Thus $x^g \in R$, whence $x^g = x$ by Lemma 5.6 (i), contrary to the fact that $y^g = x$.

Together Propositions 5.4 and 5.7 show that L_x is not isomorphic to $\hat{M}_{12}$.

<u>Remark</u>. The portion of the preceding argument which deals with the case that x is a central involution (including Lemmas 5.1 and 5.2) does not use the assumption that $r(G) \leq 4$. Hence our proof yields the following general result.

<u>Theorem C</u>. There exists no simple group which possesses a central involution x such that

$$C_G(x)/O(C_G(x)) \cong \hat{M}_{12} \text{ or } \widehat{\mathrm{Aut}}(M_{12})^{(i)}, \quad i = 1,2.$$

6. Some lemmas. For the analysis of the next three cases, we shall need a number of preliminary results. It will be preferable to collect them all in the present section.

If $L = SL(2,q)$, q an odd square, then L possesses an automorphism y of period 2 induced from the Galois group of $GF(q)$. Any involution y' of $L\langle y\rangle - L$ is conjugate in $L\langle y\rangle$ to y. Thus the action of y' on L is essentially the same as that of y. Similarly if $L = L_2(q)$, q an odd square, L possesses an automorphism of period 2 induced from the Galois group of $GF(q)$ and any involution of $L\langle y\rangle - L$ is conjugate in $Aut(L)$ to y. It will be convenient in either case to say that any involution of $L\langle y\rangle - L$ induces a "field automorphism" of L. In addition, we shall also say that any element which centralizes L induces a field automorphism of L. Of course, if q is not a square, L does not possess a field automorphism of period 2.

In our first lemma, we consider a group H which possesses a normal subgroup $L \cong SL(2,q)$, q odd, or $\hat{A}_7$ with $C_H(L) = Z(L)$.

Lemma 6.1. The following conditions hold:

(i) If $L \cong SL(2,q)$ and y is an element of $H-L$ with $y^2 \in Z(L)$ which induces a field automorphism of L, then $C_L(y) \cong SL(2,q_0)$ with $q_0^2 = q$. In particular, $C_L(y)$ contains a quaternion group;

(ii) If $L \cong \hat{A}_7$ and y is an element of $H - L$ with $y^2 \in Z(L)$, then $C_L(y)$ contains a quaternion group;

(iii) If T is a Sylow 2-subgroup of H, then $Z(T) = Z(L)$.

Proof: First, (i) is immediate from the action of y on L. To prove (ii), set $\bar{H} = H/Z(L)$. Then $\bar{H} = \bar{L}\langle\bar{y}\rangle \cong S_7$ and so $C_{\bar{L}}(\bar{y}) \cong S_5$ or S_4

according as $\bar{y}$ corresponds to a transposition or a product or 3 transpositions in S_7. Correspondingly we see that $C_L(y)$ contains a subgroup isomorphic to $\hat{A}_5 \cong SL(2,5)$ or $\hat{A}_4 \cong SL(2,3)$. In either case, we obtain (ii).

Suppose (iii) is false, in which case $Z(T)$ contains an element y with $y^2 \in Z(L)$. Since $T \cap L$ is generalized quaternion, $y \notin T \cap L$. Again set $\bar{H} = H/Z(L)$. Suppose first that $L \cong SL(2,q)$, in which case $\bar{L} \cong L_2(q)$, $C_{\bar{H}}(\bar{L}) = 1$, and $\bar{y}$ is an involution which centralizes the Sylow 2-subgroup $\bar{T} \cap \bar{L}$ of $\bar{L}$. Since $Aut(\bar{L}) \cong P\Gamma L(2,q)$, it is immediate from the structure of $P\Gamma L(2,q)$ that $\bar{y}$ must induce a field automorphism of $\bar{L}$ and so y induces a field automorphism of L. Hence $C_L(y) \cong SL(2,q_0)$ with $q_0^2 = q$ by (i). But a Sylow 2-subgroup of $SL(2,q_0)$ has lower order than that of $SL(2,q)$ and so y does not centralize $T \cap L$, contrary to the fact that $y \in Z(T)$. Thus (iii) holds in this case.

Finally if $L \cong \hat{A}_7$, then $\bar{H} = \bar{L}\langle\bar{y}\rangle \cong S_7$. Let $\bar{H}_1$ be a subgroup of $\bar{H}$ containing $\bar{T}$ and isomorphic to S_6. Then $\bar{H}_1 \cap \bar{L} \cong A_6 \cong L_2(9)$. Hence if H_1 denotes the inverse image of $\bar{H}_1$ in H, then T is a Sylow 2-subgroup of H_1, H_1 contains a normal subgroup $L_1 \cong SL(2,9)$, and $C_{H_1}(L_1) = Z(L_1) = Z(L)$. Thus $Z(T) = Z(L_1) = Z(L)$ by the preceding paragraph, applied to H_1, so (iii) holds in this case as well.

Again consider $L = L_2(q)$, q odd, and set $H = Aut(L) \cong P\Gamma L(2,q)$, identifying L with $Inn(L)$. As is well-known, if q is not a square, H possesses a unique subgroup K_1 containing L as a subgroup of index 2 and $K_1 \cong PGL(2,q)$; while if q is a square, H possesses three such subgroups K_1, K_2, K_3 with $K_1 \cong PGL(2,q)$, $K_2 = L\langle y\rangle$, where y is an involution which induces a field automorphism of L of period 2, and $K_3 = L\langle t\rangle$, where $t = xy$ with x an involution of $K_1 - L$. The group K_3 is denoted by $PGL^*(2,q)$ and has quasi-dihedral Sylow 2-subgroups. Thus every involution of H lies in K_1 or K_2.

In particular, we have

<u>Lemma 6.2</u>. The following conditions hold:

(i) If y is an involution of $H-L$, either y induces a field automorphism of L or $L\langle y\rangle \cong PGL(2,q)$;

(ii) If L is of index 2 in a group K, either $K-L$ contains an involution or $K \cong PGL^*(2,q)$;

(iii) If y is an involution of $H-L$ which induces a field automorphism of L, all involutions of $L\langle y\rangle - L$ are conjugate to y in H and to y or yz in $L\langle y\rangle$, where $\langle z\rangle = Z(T)$ and T is a Sylow 2-subgroup of L centralized by y. Moreover, y centralizes T;

(iv) H does not contain a four subgroup disjoint from L.

If $L = L_3(q)$, q odd, then as is also well-known, $H = \mathrm{Aut}(L) \cong P\Gamma L(3,q)\langle y\rangle$, where y is an involution induced from the transpose-inverse map of $SL(3,q)$ and $C_L(y) \cong PGL(2,q)$. In addition, if $q \equiv -1 \pmod 4$, then $|P\Gamma L(3,q)/L_3(q)|$ is odd and so $L\langle y\rangle$ contains a Sylow 2-subgroup of H in this case (assuming L is identified with $\mathrm{Inn}(L)$). Similarly if $L = U_3(q)$, q odd, then $H = \mathrm{Aut}(L) \cong P\Gamma U(3,q)$, $H-L$ contains an involution y which is induced from the field automorphism of period 2 of

$GF(q^2)$ over $GF(q)$, and $C_L(y) \cong PGL(2,q)$. Furthermore, it can be checked that every involution of $L\langle y\rangle - L$ is conjugate in $L\langle y\rangle$ to y. We shall give a group-theoretic proof of this assertion in Lemma 2.3(iv) of Part IV when L has quasi-dihedral Sylow 2-subgroups, which is the only case in which we need the result. Here we use only the following consequence of this assertion: If y' is any involution of $L\langle y\rangle - L$, then $C_L(y') \cong PGL(2,q)$. Finally we note that $Aut(M_{11}) \cong M_{11}$.

Together these facts yield the following lemma.

Lemma 6.3. If $L \lhd H$ with $L \cong L_3(q)$, $q = -1 (\mathrm{mod}\ 4)$, $U_3(q)$, $q = 1 (\mathrm{mod}\ 4)$ or M_{11} and $C_H(L) = 1$; then we have

(i) If $|H/L|$ is even, then $H-L$ contains an involution and $L \not\cong M_{11}$;

(ii) If y is an involution of $H-L$, then $C_L(y) \cong PGL(2,q)$. In particular, no 2-element of $H-L$ centralizes a Sylow 2-subgroup of L.

In the next lemma, we consider a group H with $r(H) \leq 4$ which possesses a normal subgroup $L = L_1 \times L_2$ with $L_i \cong L_2(q)$, q odd, $q_i \geq 5$ or A_7, $i = 1,2$ and with $C_H(L) = 1$. We let T be a Sylow 2-subgroup of H.

Lemma 6.4. The following conditions hold:

(i) $T/T \cap L$ is isomorphic to a subgroup of D_8;

(ii) $N_T(L_i)/T \cap L$ is isomorphic to a subgroup of $Z_2 \times Z_2$, $i = 1,2$;

(iii) If $L_i \cong A_7$, $i = 1,2$, then $|T/T \cap L| \leq 2$ and every element of $T-L$ interchanges L_1 and L_2;

(iv) If $L_i \cong L_2(q_i)$, $i = 1,2$, then no involution of $T-L$ induces a field automorphism or inner automorphism of both L_1 and L_2;

(v) If y is an involution of T such that $L_i\langle y\rangle \cong PGL(2,q_i)$, $i = 1,2$, then $(T \cap L)\langle y\rangle$ is normal in T;

(vi) If y, y' are involutions of T which leave L_i invariant such that $L_i\langle y\rangle \cong L_i\langle y'\rangle$, $i = 1,2$, then $(T \cap L)\langle y\rangle = (T \cap L)\langle y'\rangle$

(vii) $|C_T(L_i): T \cap L_j| \leq 2$, $i \neq j$, $1 \leq i, j \leq 2$;

Proof: Let Y be the subgroup of T leaving L_1 and L_2 invariant, so that $|T:Y| \leq 2$ and $Y \supseteq T \cap L$. All parts of the lemma hold trivially if $Y = T \cap L$, so we can assume $Y \supset T \cap L$.

We claim first that $Y/T \cap L$ is elementary. Indeed, if not, there exists y in Y with $y^2 \notin T \cap L$ and $y^4 \in T \cap L$. Then for $i = 1$ and 2, either y^2 induces an inner automorphism of L_i or $L_i \cong L_2(q_i)$ and y^2 induces a field automorphism of L_i of order 2. Lemma 6.2 now yields that $(T \cap L)\langle y^2\rangle = (T \cap L) \times \langle y'\rangle$ for some involution y'. Since $m(Y \cap T) = 4$, this yields $m(Y) \geq 5$, contrary to $r(H) \leq 4$. Furthermore, the same argument shows that if $L_i \cong L_2(q_i)$, $i = 1,2$, then no element of $T - L$ induces a field automorphism or an inner automorphism of L_i for both $i = 1$ and 2, so (iv) holds.

We know that $T \cap L_i \cong D_{2^{n_i}}$, $n_i \geq 2$, $i = 1,2$. If $n_i \geq 3$, we set $\langle z_i\rangle = Z(T \cap L_i)$, $i = 1,2$. Consider a coset X of $T \cap L$ in Y with $X \neq T \cap L$. Since $Y/T \cap L$ is elementary, we know the possible structures of $L\langle X\rangle/L_i$, $i = 1$ and 2. Examining these in succession, we easily verify that there exists a representative x of X such that one of the following holds:

(a) $\langle x\rangle \cap L = 1$;

(b) $x^2 = z_i$ and $L_i\langle x\rangle \cong PGL^*(2,q_i)$, $i = 1$ or 2;

(c) $x^2 = z_1z_2$ and $L_i\langle x,z_j\rangle \cong PGL^*(2,q_i) \times Z_2$, $i \neq j$, $i,j = 1,2$.

If $L_i \cong A_7$ for both $i = 1$ and 2, then (a) must hold. But then as $Aut(A_7) \cong S_7$, x centralizes a four subgroup of $T \cap L_i$ for both $i=1$ and 2 and consequently $r(T) \geq 5$, contrary to $r(H) \leq 4$. Hence $L_i \not\cong A_7$ for some i, say $i = 1$. In particular, (iii) holds.

Again by the structure of $P\Gamma L(2,q_1)$, Y contains a subgroup Y_1 of index at most 2 such that each coset of $T \cap L_1$ in Y_1 contains an element which induces a field automorphism or an inner automorphism of L_1. Hence for each coset X in Y_1 with $X \neq T \cap L$, our representative x will centralize a four subgroup of $T \cap L_1$. If $L_2 \cong A_7$, then x will centralize a four subgroup of $T \cap L_2$ and again we obtain the contradiction $r(T) \geq 5$. Hence in this case, we must have $Y_1 = T \cap L$, whence $Y/T \cap L \cong Z_2$ and (ii) holds.

On the other hand, if $L_2 \cong L_2(q_2)$ and Y_2 is defined for L_2 as Y_1 was defined for L_1, then we must have $Y_1 \cap Y_2 = T \cap L$, otherwise we could choose our coset X in $Y_1 \cap Y_2$ and x would centralize a four subgroup of both $T \cap L_1$ and $T \cap L_2$, giving the same contradiction. We conclude at once from this that $Y/T \cap L$ is isomorphic to a subgroup of $Z_2 \times Z_2$, so (ii) holds in this case as well.

Clearly $C_T(L_i) \subseteq Y$, $i = 1$ or 2. Suppose $|C_T(L_i) : T \cap L_j| > 2$ for some i, $i \neq j$, in which case $Y/T \cap L \cong Z_2 \times Z_2$. By the preceding paragraph $L_k \cong L_2(q_k)$ for both $k = 1$ and 2 and so Y_k is defined for both $k = 1$ and 2. Our condition on L_i implies, in fact, that $Y_i = Y$. But then we see that $Y_i \cap Y_j = Y_1 \cap Y_2 \supset T \cap L$, contrary to what we have shown above. Hence (vii) also holds.

Since $|T:Y| \leq 2$, clearly either (i) holds or $Y/T \cap L \cong Z_2 \times Z_2$ and $T/T \cap L \cong Z_4 \times Z_2$ or E_8. In the latter case, $T \supset Y$ and there exists $t \in T$ interchanging L_1 and L_2, so $L_1 \cong L_2$ and hence $L_2 \cong L_2(q_2)$. Furthermore, we see that $Y_1^t = Y_2$. However, as $T/T \cap L$ is abelian, $Y_1^t = Y_1$ and consequently $Y_1 = Y_1 \cap Y_2$. But $Y_1 \supset T \cap L$ as $Y/T \cap L \cong Z_2 \times Z_2$, so $Y_1 \cap Y_2 \supset T \cap L$, contrary to what we have shown above. Thus (i) holds.

Suppose next that (v) is false, in which case clearly $T/T \cap L \cong D_8$. If Z denotes the subgroup of T containing $T \cap L$ whose image is $Z(T/T \cap L)$, then $Z \lhd T$ and $|Z/T \cap L| = 2$. Moreover, $Y/T \cap L \cong Z_2 \times Z_2$ in the present case and so it is normal in $T/T \cap L$, which implies that $Z \subseteq Y$. Choosing t in $T-Y$, we have $Z^t = Z$ and $Y_1^t = Y_2$. Hence $Z \not\subseteq Y_1$ or Y_2, otherwise $Y_1 \cap Y_2 \supset T \cap L$, which is not the case. Thus no element of $Z-L$ induces a field automorphism of L_1 or L_2. Choosing our coset X now to lie in Z and using the fact that $Z^t = Z$, we see that our element x can be chosen so that either $x^2 = 1$ with $L_i\langle x\rangle \cong PGL(2,q_i)$, $i = 1,2$ or (c) holds.

On the other hand, $y \notin Z$ as (v) is assumed to be false, so $yx \in Y-L$. Since $L_i\langle y\rangle \cong PGL(2,q_i)$, $i = 1,2$, it follows in either case that the coset $U = (T \cap L)yx$ contains an involution u which induces a field automorphism of both L_1 and L_2 (possibly the trivial automorphism), which yields the usual contradiction. Therefore (v) also holds.

Similarly if y,y' are as in (vi) and $(T \cap L)\langle y\rangle \neq (T \cap L)\langle y'\rangle$, then $yy' \notin T \cap L$ and the coset $U = (T \cap L)yy'$ necessarily contains an involution u which centralizes both L_1 and L_2, so again the assumption $r(H) \leq 4$ is contradicted. Hence (vi) holds as well and the lemma is proved.

Next assume that H is a group with $r(H) \leq 4$ which possesses a normal subgroup $L = L_1 \times L_2$ with $L_i \cong SL(2,q_i)$, q_i odd, $q_i \geq 5$, or $\hat{A}_7$, $i=1,2$, and with $C_H(L) = Z(L)$. Again let T be a Sylow 2-subgroup of H.

Lemma 6.5. The following conditions hold:

(i) $Z(T) \cong Z_2$ or $Z_2 \times Z_2$ according as T does or does not contain an element which interchanges L_1 and L_2;

(ii) $T/T \cap L$ is isomorphic to a subgroup of D_8;

(iii) $C_T(Z(L_i))/T \cap L$ is isomorphic to a subgroup of $Z_2 \times Z_2$, $i = 1,2$;

(iv) If $Z(T) \cong Z_2 \times Z_2$, then $\Omega_1(T') = Z(T) = Z(L)$;

Proof: Set $\overline{H} = H/Z(L)$. Since $T \cap L \cong Q_{2^n} \times Q_{2^m}$, either $Z(T) \subseteq L$ or $Z(T)-L$ contains an element z such that $z^2 \in Z(L)$. But then $\bar{z}$ is an involution which centralizes $\overline{T} \cap \overline{L}$ and consequently $r(\overline{T}) \geq 5$, a contradiction. Thus $Z(T) \subseteq Z(L)$. If each $L_i \lhd H$, then $Z(L_i) \subseteq Z(H)$, $i = 1,2$, and in this case $Z(T) = Z(L_1) \times Z(L_2) = Z(L) \cong Z_2 \times Z_2$. In the contrary case, T must contain an element t which interchanges L_1 and L_2. Then t interchanges $Z(L_1)$ and $Z(L_2)$ and so $Z(T) \cong Z_2$ in this case. Thus (i) holds.

Since $T/T \cap L \cong \overline{T}/\overline{T} \cap \overline{L}$, (ii) follows from Lemma 6.4 (i). Furthermore, $C_T(Z(L_i))$ is precisely the subgroup of T which leaves both L_1 and L_2 invariant and now (iii) follows similarly from Lemma 6.4 (ii).

Finally assume $Z(T) \cong Z_2 \times Z_2$, in which case $L_i \lhd H$, $i = 1,2$. Then $\overline{H}$ is isomorphic to a subgroup of $\mathrm{Aut}(\overline{L}_1) \times \mathrm{Aut}(\overline{L}_2)$. But $\mathrm{Aut}(\overline{L}_i) \cong P\Gamma L(2,q_i)$ or S_7, $i = 1,2$. By the structure of $P\Gamma L(2,q_i)$ or S_7, the derived group of a Sylow 2-subgroup of $\mathrm{Aut}(\bar{L}_i)$ lies in $\bar{L}_i$. It follows therefore that $\bar{T}' \subseteq \bar{L}$. Hence $T' \subseteq L$ and as $T \cap L \cong Q_{2^n} \times Q_{2^m}$, we conclude at once that $\Omega_1(T') = Z(L)$. Thus (iv) holds and the lemma is proved.

Now assume that H is a group with $r(H) \leq 4$ which possesses a normal subgroup $L = L_1 * L_2$ with $L_i \cong SL(2,q_i)$, $q_i \geq 5$, or $\hat{A}_7$, $i = 1,2$, and with $C_H(L) = Z(L)$. Again let T be a Sylow 2-subgroup of H.

<u>Lemma 6.6</u>. The following conditions hold:

(i) $Z(T) = Z(L)$;

(ii) Every elementary abelian normal subgroup of T of order at most 8 lies in L.

<u>Proof</u>: We have $T \cap L \cong Q_{2^n} * Q_{2^m}$ for some m and n, so $Z(T \cap L) = Z(L)$. Hence if $Z(T) \supset Z(L)$, $T-L$ contains an element y such that $y^2 \in Z(L)$. Set $\bar{H} = H/Z(L)$, so that $\bar{y}$ is an involution and $\bar{y}$ centralizes $\bar{T} \cap \bar{L}$. In particular, $\bar{y}$ leaves $\bar{L}_i$ invariant and $\bar{L}_i\langle\bar{y}\rangle \not\cong PGL(2,q_i)$, $i = 1$ or 2. Hence either $\bar{L}_i \cong L_2(q_i)$ and $\bar{y}$ induces a field (or inner) automorphism of $\bar{L}_i$ or $\bar{L}_i \cong A_7$, $i = 1,2$. Correspondingly Lemma 6.1 (i) or (ii), applied to $L_i\langle y\rangle$, implies that $C_{\bar{L}_i}(\bar{y})$ contains a four group, $i = 1,2$. Thus $r(\bar{H}) \geq 5$, a contradiction, and so (i) holds.

Next let $A \lhd T$ with A elementary of order at most 8 and suppose $A \not\subseteq L$. Then $A \supset Z(T) = Z(L)$ and $|\bar{A}| = 2$ or 4. Since $\bar{A} \lhd \bar{T}$, it follows that $|[\bar{A},\bar{T}]| \leq 2$. Let $a \in A-L$ and suppose first that a interchanges $\bar{L}_1$ and $\bar{L}_2$, whence $\bar{a}$ interchanges $\bar{T} \cap \bar{L}_1$ and $\bar{T} \cap \bar{L}_2$. But then $|[\bar{a}, \bar{T} \cap \bar{L}_1]| \geq 4$, contrary to $|[\bar{A},\bar{T}]| \leq 2$. Thus $\bar{a}$ leaves $\bar{L}_1$

and $\bar{L}_2$ invariant. Suppose next $\bar{L}_i\langle\bar{a}\rangle \cong PGL(2,q_i)$, $i = 1$ or 2. Then $(\bar{T} \cap \bar{L}_i)\langle\bar{a}\rangle \cong D_{2^n}$ for some n and consequently $(T \cap L_i)\langle a\rangle \cong QD_{2^{n+1}}$. We see then that $[a, T \cap L_i]$ contains an element of order 4. However, this is impossible as $[a,T] \subseteq A$ and A is elementary. We therefore conclude that, for both $i = 1$ and 2, either $\bar{L}_i \cong A_7$ or $\bar{a}$ induces a field automorphism of $\bar{L}_i$. But now it follows as in the preceding paragraph that $r(\bar{H}) \geq 5$, a contradiction. Thus (ii) also holds.

Finally we consider a group H with $r(H) \leq 4$ which possesses a normal subgroup $K \cong SL(4,q)$, $q \equiv 3 (\mathrm{mod}\ 4)$, $SU(4,q)$, $q \equiv 1 (\mathrm{mod}\ 4)$, or $Sp(4,q)$, q odd, with $C_H(K) = Z(K)$ and we let T be a Sylow 2-subgroup of H.

<u>Lemma</u> 6.7. The following conditions hold:

(i) K possesses a T-invariant subgroup $L = L_1 \times L_2$ with $L_i \cong SL(2,q)$, $i = 1,2$;

(ii) $T \cap K$ contains an involution which interchanges L_1 and L_2;

(iii) $T \cap K/T \cap L \cong Z_2 \times Z_2$, $Z_2 \times Z_2$, or Z_2 according as $K \cong SL(4,q)$, $SU(4,q)$, or $Sp(4,q)$;

(iv) $|T/T \cap K| \leq 2$ and $T/T \cap L$ is isomorphic to a subgroup of D_8;

(v) Every element of T either interchanges L_1 and L_2 or leaves them invariant;

(vi) $Z(T) = Z(K) \cong Z_2$;

(vii) $Z(L) \cong Z_2 \times Z_2$ and L is characteristic in $C_H(Z(L))$.

(viii) If $q = 3$, then $C_H(Z(L))$ is solvable;

Proof: Set $\bar{H} = H/Z(H)$. If $\bar{z}$ is a central involution of $\bar{T}$ with $\bar{z} \in \bar{K}$, the inverse image Z of $\langle \bar{z} \rangle$ in K is a normal four subgroup of T. Using the known structure of the centralizer of an involution in $L_4(q)$, $U_4(q)$, and $PSp(4,q)$ or the natural matrix representations of $SL(4,q)$, $SU(4,q)$, and $Sp(4,q)$, one easily concludes that $C_K(Z)$ has the form $L(T \cap L)$ with $L \lhd C_K(Z)$ and $L = L_1 \times L_2$ with $L_i \cong SL(2,q)$, $i = 1,2$, that L satisfies conditions (ii) and (iii), and that $L = C_K(Z)'$ and so is T-invariant. Thus (i) also holds.

If $q > 3$, then Lemma 6.5 applies to the group LT and yields that $T/T \cap L$ is isomorphic to a subgroup of D_8 and that $Z(T) \cong Z_2$. In particular, $|T/T \cap K| \leq 2$ if $K \cong SL(4,q)$ or $SU(4,q)$. We claim the same holds if $K \cong Sp(4,q)$. Indeed, if y is an involution of $T \cap K$ which interchanges L_1 and L_2, all involutions of $(T \cap L)\langle y \rangle = T \cap K$ not in $T \cap L$ are conjugate in $T \cap K$ to y inasmuch as $T \cap K \cong Q_{2^n} \int Z_2$ for some n. Since $T \cap L$ and $T \cap K$ are normal in T, it follows that $T = (T \cap L)C_T(y)$. If our assertion is false, then $|C_T(y)/C_{T \cap K}(y)| = 4$ and consequently some element t of $C_T(y)$ not in K induces a field automorphism of L_1. Since t centralizes y, t also induces a field automorphism of L_2. As usual, this implies that $r(T) \geq 5$, a contradiction, so our assertion holds. Hence (iv) and (vi) hold when $q > 3$. Likewise (v) holds in this case as then L_1 and L_2 are the uniquely determined components of L. Furthermore, the structure of $Aut(K)$ implies that H/K is solvable, so if $q > 3$, $L = L(C_H(Z))$ and consequently (vii) also holds.

On the other hand, if $q = 3$, $K \cong SL(4,3)$ or $Sp(4,3)$ and again using the structure of $Aut(K)$, we conclude that $|H:K| \leq 2$ and that (iv), (v), (vi), and (vii) hold. Moreover, $N_H(Z) = LT$ and so (viii) also holds.

The following two results about 2-groups will be very useful.

Lemma 6.8. If T is a 2-group of the form $(T_1 \times T_2)\langle t\rangle$, where t interchanges T_1, T_2 under conjugation and $t^2 \in T_1T_2$, then

$$T = (T_1 \times T_2)\langle x\rangle$$

for some involution x. In particular, $T \cong T_1 \int Z_2$ and $T-(T_1 \times T_2)$ contains an involution.

Proof: We proceed by induction on $|T_1|$, the lemma being obvious if $T_1 = 1$. Let z_1 be an involution of $Z(T_1)$ and set $Z = \langle z_1, z_1^t\rangle$. Note that as $T_1^t = T_2$, $z_1^t \in Z(T_2)$, whence $Z \subseteq Z(T_1T_2)$. Moreover, as $t^2 \in T_1T_2$, clearly t normalizes Z. Thus $Z \triangleleft T$. By induction, the lemma holds in $\bar{T} = T/Z$, and so $T = T_1T_2$ contains an element u such that $\bar{u}^2 = 1$ and $\bar{T}_1^{\bar{u}} = \bar{T}_2$. Then $u^2 \in Z$ and, as $Z \cong Z_2 \times Z_2$, $Z\langle u\rangle$ is of order 8. But as $u \notin T_1T_2$, u does not centralizes Z and so $Z\langle u\rangle \cong D_8$. Hence $Z\langle u\rangle - Z$ contains an involution x. Since $u \notin T_1T_2$, also $x \notin T_1T_2$ and so $T = T_1T_2\langle x\rangle$. The lemma follows.

Lemma 6.9. If T is a 2-group with $r(T) \leq 2$, then either T is cyclic, T is of maximal class, or $\Omega_1(T) \cong Z_2 \times Z_2$.(*)

Proof: Assume T is neither cyclic nor of maximal class, in which case $SCN_2(T)$ is nonempty. Let A be an element of $SCN_2(T)$. If $\Omega_1(T) = \Omega_1(A)$, the lemma holds, so we can also assume that $T - A$ contains an involution. If $|A| = 4$, then $|T| = 8$ and $T \cong D_8$. Again the

(*) Thompson has shown that, in fact, T is either cyclic or metacyclic.

lemma holds. Thus we can also assume $|A| > 4$.

Setting $B = \Omega_2(A)$, we have that $B \cong Z_4 \times Z_4$ or $Z_4 \times Z_2$. Consider the first case and let t be an involution of $T - A$. If t interchanges some basis of B, then $B\langle t\rangle$ is wreathed of order 32. But then $r(B\langle t\rangle) = 3$, contrary to $r(T) \leq 2$. Hence t leaves the elements of a basis of B invariant, whence t centralizes $\Omega_1(B)\langle t\rangle \cong E_8$, giving the same contradiction. We conclude that $B \cong Z_4 \times Z_2$ and hence that $A \cong Z_{2^n} \times Z_2$ for some $n \geq 2$.

Suppose next that $n > 2$. With t as above, we have that t normalizes $C = \mho^{n-2}(A) \cong Z_4$ and $D = \Omega_1(A) \cong Z_2 \times Z_2$. Then $C \cap D \cong Z_2$ and t centralizes $CD/C \cap D$, whence $CD\langle t\rangle/C \cap D \cong E_8$, again contradicting $r(T) \leq 2$.

Assume finally that $n = 2$, whence $A \cong Z_4 \times Z_2$. Let $A = \langle a,b\rangle$ with $|a| = 4$ and $|b| = 2$. If t above normalizes $\langle a\rangle$, then $A\langle t\rangle/\langle a^2\rangle \cong E_8$, giving the same contradiction. Thus $a^t = au$ for some u in $\Omega_1(A) - \langle a^2\rangle$ and so without loss we can assume that $u = b$. Setting $R = A\langle t\rangle$, we have that $[a,t] = b \in R'$. But also t does not centralize $\Omega_1(A)$, otherwise $\Omega_1(A)\langle t\rangle \cong E_8$. Hence $[b,t] = a^2 \in R'$. Hence $a^{t^2} = (ab)^t = (ab)(ba^2) = a^3 \neq a$, contradicting $t^2 = 1$.

We also have

<u>Lemma 6.10</u>. If T is a 2-group with $r(T) \leq 2$ which admits a nontrivial automorphism of odd order, then $T \cong Q_8$ or $Z_{2^n} \times Z_{2^n}$ for some n.

<u>Proof</u>: Let α be such an automorphism. Since $r(T) \leq 2$, $T/\Phi(T) \cong Z_2 \times Z_2$ and so α^3 acts trivially on $T/\Phi(T)$, whence $\alpha^3 = 1$.

Thus $|\alpha| = 3$. If T is abelian, clearly $T \cong Z_{2^n} \times Z_{2^n}$, so we can assume that T is not abelian.

Let Z be a minimal α-invariant subgroup of $Z(T) \cap T'$. Since $r(T) \leq 2$, $Z \cong Z_2$ or $Z_2 \times Z_2$ and $\bar{T} = T/Z \cong Q_8$ or $Z_{2^n} \times Z_{2^n}$ by induction. Suppose $\bar{T} \cong Q_8$. If $Z \cong Z_2$, then $T \cong Q_8 \times Z_2$ as the Schur multiplier of Q_8 is trivial, contrary to $r(T) \leq 2$. Hence $Z \cong Z_2 \times Z_2$ and so α acts regularly on Z. But $C_T(\alpha) \neq 1$ as $C_{\bar{T}}(\alpha) \neq 1$. Thompson's $A \times B$ - lemma shows that $C_T(\alpha)$ centralizes Z. Since $Z \cap C_T(\alpha) = 1$, it follows that $r(T) \geq 3$, which is not the case. Thus $\bar{T} \cong Z_{2^n} \times Z_{2^n}$ for some n.

Since α acts regularly on $T/\Phi(T)$ and $Z \subseteq \Phi(T)$, α acts regularly on $\bar{T}$. If α acts regularly on Z, then α acts regularly on T and as $|\alpha| = 3$, it follows that T has class 2. But T is generated by two elements and consequently T' is cyclic. Hence α centralizes T' and so $T' = 1$, contrary to the fact that T is nonabelian. Thus $Z \cong Z_2$. If Q denotes the inverse image of $\Omega_1(\bar{T})$ in T, then $|Q| = 8$ and α acts nontrivially on Q, so we must have $Q \cong Q_8$. But as $\bar{T}$ acts trivially on $\bar{Q}$ and $Q \lhd T$, we have that $T = QC_T(Q)$. Since $r(T) \leq 2$, this forces $T = Q$ and the lemma holds in all cases.

Using the preceding two results, we can give a fairly complete description of all groups of sectional 2-rank at most 2. It is well known (by Schur) that $PGL(2,q)$ has two nonisomorphic nonsplit central extensions by Z_2. One has quasi-dihedral Sylow 2-subgroups and the other generalized quaternion. We denote by $SL^{(1)}(2,q)$ or $SL^{(2)}(2,q)$, respectively, those central extentions. One checks easily that $SL^{(1)}(2,q) \cong SL^{\pm}(2,q)$, $q \equiv -1 (\mathrm{mod}\ 4)$, or $SU^{\pm}(2,q)$, $q \equiv 1 (\mathrm{mod}\ 4)$, where $SL^{\pm}(2,q)$ and $SU^{\pm}(2,q)$ denote the subgroups of $GL(2,q)$ and $GU(2,q)$, respectively, consisting of the elements of determinant ± 1. Likewise $PSL_m(2,3)$ will denote the semi-direct product of $Z_{2^n} \times Z_{2^n}$ by Z_3 in which a 3-element acts fixed-point-free on $Z_{2^n} \times Z_{2^n}$.

<u>Proposition 6.11</u>. Let H be a group with $r(H) \leq 2$ and $O(H) = 1$. If T denotes a Sylow 2-subgroup of H, then one of the following holds:

(i) $H = T$;

(ii) $T \cong Z_{2^n} \times Z_{2^n}$ for some n and $H \cong PSL_n(2,3)$;

(iii) $T \cong Q_{2^n}$ for some n and $O^{2'}(H) \cong SL(2,q)$, q odd, $\hat{A}_7$, or $SL^{(2)}(2,q)$;

(iv) $T \cong D_{2^n}$ for some n and $O^{2'}(H) \cong L_2(q)$, $PGL(2,q)$, q odd, or A_7;

(v) $T \cong QD_{2^n}$ for some n and $O^{2'}(H) \cong PGL^*(2,q)$, q odd, $L_3(q)$, $q \equiv -1 \pmod 4$, $U_3(q)$, $q \equiv 1 \pmod 4$, $SL^{(1)}(2,q)$, or M_{11}.

<u>Proof</u>: If T is of type (iii), (iv) or (v), then our assertions follow from the known structure* of groups with dihedral or quasi-dihedral Sylow 2-subgroups [2] and [22] together with the Brauer-Suzuki theorem [7]. Likewise (i) holds if T is cyclic, so we can assume that T is of none of these forms. Hence by Lemma 6.10, $\Omega_1(T) \cong Z_2 \times Z_2$ and $|T| > 4$.

We shall argue that $L(H) = 1$. Suppose false. Since $r(H) \leq 2$, clearly $L = L(H)$ is quasisimple. Since $\bar{L} = L/Z(L)$ simple of sectional

2-rank 2, the main result of [3] implies that $\bar{L}$ has dihedral or quasi-dihedral Sylow 2-subgroups. In the latter case, $\bar{L} \cong L_3(q)$, $q \equiv -1 \pmod 4$, $U_3(q)$, $q \equiv 1 \pmod 4$, or M_{11}. Since $Z(L)$ is a 2-group, this in turn implies that $Z(L) = 1$, whence $T \cap L$ is quasi-dihedral. However, this is impossible as $\Omega_1(T) \cong Z_2 \times Z_2$. We conclude that $\overline{T \cap L}$ is dihedral.

If $Z(L) = 1$, then again as $\Omega_1(T) \cong Z_2 \times Z_2$, we must have $T \cap L \cong Z_2 \times Z_2$ and $L \cong L_2(q)$, $q \equiv 3, 5 \pmod 8$. Likewise no involution of H can centralize T and as $O(H) = 1$, it follows that $C_H(L) = 1$. Thus H is isomorphic to a subgroup of $P\Gamma L(2,q)$. Since $q \equiv 3, 5 \pmod 8$, this implies that T is dihedral of order 4 or 8, contrary to assumption. Thus $Z(L) \neq 1$.

We therefore conclude that $L \cong SL(2,q)$, q odd, or $\hat{A}_7$. Since T is not generalized quaterinon, $T \not\subseteq L$. Since $r(T) \leq 2$, also $Z(L)$ must be a Sylow 2-subgroup of $C_H(L)$. Again as $O(H) = 1$, this forces $C_H(L) = Z(L)$. If $L \cong \hat{A}_7$, it follows now that $\bar{H} = H/Z(L) \cong S_7$, contrary to the fact that $r(S_7) = 3$. Thus $L \cong SL(2,q)$ and $\bar{H}$ is isomorphic to a subgroup of $P\Gamma L(2,q)$. If some element of $\bar{T}$ induced a field automorphism of $\bar{L}$ of period 2, then $r(\bar{T}) \geq 3$, again a contradiction. Hence $O^{2'}(\bar{H}) \cong PGL(2,q)$ and consequently $\bar{T}$ is dihedral. Since T does not split over $Z(L)$, this in turn implies that T is generalized quaternion or quasi-dihedral, neither of which is the case.

We have thus proved that $L(H) = 1$ and hence that H is 2-constrained. Setting $Q = O_2(H)$, we have that $C_H(Q) \subseteq Q$. If $H = Q$, then (i) holds; so we can suppose that $H \supset Q$. Thus H contains an element $x \neq 1$ of odd order and x induces a nontrivial automorphism of Q. Since $r(T) \leq 2$, also $r(Q) \leq 2$ and so by the preceding lemma, $Q \cong Q_8$ or $Z_{2^n} \times Z_{2^n}$ for some n. In either case we see that $|x| = 3$. If $Q \cong Q_8$, it is immediate that $T \cong Q_{16}$ or QD_{16}, neither of which is the case. Hence $Q \cong Z_{2^n} \times Z_{2^n}$.

Finally $H/Q \cong Z_3$ or S_3. In the first case, $H \cong PSL_n(2,3)$ and so (ii) holds. In the second case, $N_H(\langle x \rangle) \cong S_3$ by the Frattini argument and so $T - Q$ contains an involution t. But then either $T = Q\langle t \rangle$ is wreathed or t centralizes $\Omega_1(Q)$. In either case, it follows that $r(T) \geq 3$, a contradiction. This establishes the proposition.

7. <u>The</u> $SL(4,q)$, $SU(4,q)$, $Sp(4,q)$ <u>cases</u>. In this section we shall prove

<u>Proposition 7.1</u>. If L_x is isomorphic to $SL(4,q)$, $q \equiv 3 \pmod 4$, $SU(4,q)$, $q \equiv 1 \pmod 4$, or $Sp(4,q)$, q odd, then $G \cong L_5(q)$, $q \equiv -1 \pmod 4$ or $U_5(q)$, $q \equiv 1 \pmod 4$.

We shall argue in a sequence of lemmas that $S \cong QD_{2^m} \int Z_2$ for some m and shall then invoke [38*] and [46]. First we introduce some notation. By Lemma 6.6(i), $L_x = L(\bar{M})$ possesses a T-invariant subgroup of the form $\bar{L}_1 \times \bar{L}_2$, where $\bar{L}_i \cong SL(2,q)$, $i = 1,2$. Moreover, $\bar{T} \cap L(\bar{M})$ contains an involution which interchanges $\bar{L}_1$ and $\bar{L}_2$. In particular, Lemma 6.7(vi) implies now that $Z(\bar{T}) = \langle \bar{x} \rangle$, whence $Z(T) = \langle x \rangle$. Since $Z(S) \subseteq Z(T)$, we thus conclude that x is a central involution and $S = T$.

Next let Q,R be subgroups of S whose images in $\bar{M}$ are Sylow 2-subgroups of $\bar{L}_1 \times \bar{L}_2$ and $L(\bar{M})$, respectively. Then $Q \subset R$ and $Q = Q_1 \times Q_2$, where $Q_i \cong Q_{2^n}$ for some n and $\bar{Q}_i$ is a Sylow 2-

subgroup of $\bar{L}_i$, $i = 1,2$. We also set $\langle z_i \rangle = Z(Q_i)$, $i = 1,2$, and $Z = \langle z_1, z_2 \rangle$. Then R contains an element which interchanges z_1 and z_2 and, moreover, $Z \lhd S$. Hence $x = z_1 z_2$ and $Z \in U(S)$. In fact, it is easily checked, using Lemma 6.4, that Z is the unique element of $U(S)$. Finally we let L be a minimal S-invariant subgroup of $C_M(Z)$ which covers $\bar{L}_1 \bar{L}_2$. Then Q is a Sylow 2-subgroup of L and $L/O(L) \cong \bar{L}_1 \bar{L}_2$ with L perfect if $q > 3$ and L solvable with $Q \lhd L$ if $q = 3$. Furthermore, we can write $L = L_1 L_2$, where $Q_i \subseteq L_i$ and L_i maps on $\bar{L}_i$, $i = 1,2$.

First of all, we have

<u>Lemma 7.2</u>. The following conditions hold:

(i) $R/Q \cong Z_2 \times Z_2$, $Z_2 \times Z_2$, or Z_2 according as $L_x \cong SL(4,q)$, $SU(4,q)$, or $Sp(4,q)$;

(ii) $|S/R| \leq 2$;

(iii) S/Q is isomorphic to a subgroup of D_8 ;

(iv) Every involution of $R - \langle x \rangle$ is conjugate in M to z_1.

<u>Proof</u>: First of all, (i), (ii), and (iii) follow from Lemmas 6.7(iii), and (iv) respectively. As for (iv), we have noted in [17] and [21] that the groups $L_4(q)$, $q \equiv 3 \pmod 4$, $U_4(q)$, $q \equiv 1 \pmod 4$, and $PSp(4,q)$, q odd, possess exactly two conjugacy classes of involutions. Clearly in the corresponding covering groups $SL(4,q)$, $SU(4,q)$, or $Sp(4,q)$, as the case may be, one of these classes is represented by involutions and one by elements of order 4. Since z_1 is an involution and $z_1 \sim z_1^x = z_2$ in M, (iv) follows at once.

We next prove

Lemma 7.3. The following conditions hold:

(i) x is not conjugate in G to an involution of $R - \langle x \rangle$;

(ii) x is conjugate in G to an involution y of $S - R$;

(iii) $|S/R| = 2$.

Proof: Suppose (i) is false, in which case $x \sim z_1$ in G by Lemma 7.2(iv). Since Z is the unique element of $U(S)$, a standard argument now yields that $N_G(Z)$ contains a 3-element u which does not centralize Z. We set $H = C_G(Z)$, so that $L = L_1L_2 \subseteq H \subseteq M$ and u normalizes H. Setting $\tilde{H} = H/O(H)$, Lemma 6.7(vii) implies that $\tilde{L}$ is characteristic in $\tilde{H}$ and so u acts on $\tilde{L} = \tilde{L}_1 \times \tilde{L}_2 \cong SL(2,q) \times SL(2,q)$. Since u is a 3-element, we conclude at once, whether $\tilde{L}$ is solvable or nonsolvable, that u centralizes $\tilde{Z} = Z(\tilde{L})$. Hence u centralizes Z, which is not the case.

By Glauberman's Z^*-theorem, $x \sim y$ in G for some y in $S - \langle x \rangle$ and so $y \in S - R$ by (i). Thus (ii) holds and now (iii) follows from Lemma 7.2(ii).

With y fixed as in Lemma 7.3(ii), we now prove

Lemma 7.4. The following conditions hold:

(i) $x \notin \Phi(C_S(y))$;

(ii) y normalizes Q_i, $i = 1,2$;

(iii) $Q_i\langle y \rangle \cong QD_{2^{n+1}}$ and $Q_j\langle y \rangle \ncong QD_{2^{n+1}}$, $i \neq j$, $i,j = 1,2$;

(iv) $S/Q \cong D_8$ and $yQ \notin Z(S/Q)$.

Proof: We have $y^g = x$ for some g in G with $C_S(y)^g \subseteq S$. But then if (i) were false, it would follow that $x^g \in \Phi(S)$. But $\Phi(S) \subseteq R$ by Lemma 7.3(iii). Lemma 7.3(i) would then force $x^g = x$, which is not the case. Hence (i) holds.

If y interchanged Q_1 and Q_2, then y centralizes the "diagonal" of $Q_1 \times Q_2$ and so $x \in \Phi(C_Q(y))$, contrary to (i). Hence $\bar{y}$ does not interchange $\bar{L}_1$ and $\bar{L}_2$ and so $\bar{y}$ leaves $\bar{L}_1$ and $\bar{L}_2$ invariant by Lemma 6.7. Thus y normalizes both Q_1 and Q_2 and so (ii) also holds.

Since z_i is the unique involution of Q_i, either $C_{Q_i}(y) = \langle z_i \rangle$ or $z_i \in \Phi(C_{Q_i}(y))$. But then if the latter conclusion held for both $i = 1$ and 2, it would follow that $x = z_1 z_2 \in \Phi(C_Q(y))$, again contradicting (i). Thus $C_{Q_i}(y) = \langle z_i \rangle$ for at least one value of i, say $i = 1$. Since $Q_1 \cong Q_{2^n}$, this forces $Q_1\langle y \rangle \cong QD_{2^{n+1}}$.

Hence to complete the proof of (iii), it remains only to show that $Q_2\langle y \rangle \not\cong QD_{2^{n+1}}$; so assume the contrary. We let t be an element of $R - Q$ which interchanges Q_1 and Q_2 and we let S_0 be the subgroup of S which normalizes Q_1 and Q_2. Then if we set $\bar{S} = S/Q$, we have that $\bar{S} = \bar{S}_0\langle \bar{t} \rangle$ with $|\bar{S}_0| \leq 4$, $\bar{y} \in \bar{S}_0$, and $\bar{S}$ isomorphic to a subgroup of D_8. Since $Q_1\langle y \rangle \cong QD_{2^{n+1}}$, so also $Q_2\langle y^t \rangle \cong QD_{2^{n+1}}$. But $Q_2\langle y \rangle \cong QD_{2^{n+1}}$ and consequently either $\bar{y} = \bar{y}^{\bar{t}}$ or $y^{-1}y^t$ would induce an inner automorphism of both Q_1 and Q_2 and would lie in $S_0 - Q$. However, in the latter case, every element of $\langle y^{-1}y^t, Q \rangle$ would induce an inner automorphism of both Q_1 and Q_2. Hence $\langle y^{-1}y^t, Q \rangle = QV$, where $[V,Q] = 1$ and $Q \cap V \subseteq Z(Q)$. Thus $r(QV) > 4$, a contradiction. Thus $\bar{y}$ centralizes $\bar{t}$. Since $|\bar{S}_0| \leq 4$, $\bar{y}$ also centralizes $\bar{S}_0$ and so $\langle \bar{y} \rangle \subseteq Z(\bar{S})$. We therefore conclude that $Q\langle y \rangle = Q_1Q_2\langle y \rangle \triangleleft S$.

On the other hand, it is immediate from the structure of $Q\langle y \rangle$ that all involutions of $Q\langle y \rangle - Z$ are conjugate to y in $Q\langle y \rangle$.

Hence by a Frattini-type argument, $S = QC_S(y)$. Therefore without loss we can assume that $t \in C_S(y)$. But $z_i \in C_S(y)$, $i = 1,2$, and $z_1^t = z_2$, whence $x = z_1z_2 \in \Phi(C_S(y))$, once again contradicting (i). Thus (iii) also holds.

Finally it follows from (iii) that $\bar{y} \notin Z(\bar{S})$; otherwise $\bar{y}$ would centralize $\bar{t}$ and this would imply that y induced the same action on Q_1 and Q_2, whence $Q_1\langle y\rangle \cong Q_2\langle y\rangle$, which is not the case. This forces $\bar{S} \cong D_8$ and so (iv) also holds.

As an immediate corollary, we obtain

Lemma 7.5. We have $\bar{L} \cong SL(4,q)$ or $SU(4,q)$.

Proof: Indeed, otherwise $\bar{L} \cong Sp(4,q)$ and Lemma 7.2 would imply that $|S/Q| \le 4$, contrary to the fact that $S/Q \cong D_8$.

For definiteness, assume $Q_1\langle y\rangle \cong QD_{2^{n+1}}$. We set $H = C_G(z_2)$ and $\bar{H} = H/O(H)$. In addition, we put $S_0 = C_S(Z)$, so that $|S : S_0| = 2$. We next prove

Lemma 7.6. The following conditions hold:

(i) $x \sim y$ in H;

(ii) S_0 is a Sylow 2-subgroup of H;

(iii) $L(\bar{H})$ possesses a component $\bar{J}_1 \cong L_3(q)$, $U_3(q)$, or M_{11} (and $q = 3$) which centralizes $\bar{Q}_2$ and has Sylow 2-subgroup $\bar{Q}_1\langle \bar{y}\bar{z}_2\rangle$.

Proof: Since $Q_2\langle y\rangle \not\cong QD_{2^{n+1}}$, $C_{Q_2}(y)$ contains an element v of order 4. Again with $y^g = z$ and $C_S(y)^g \subseteq S$, it follows that $(v^2)^g \in R$ (as $|S : R| = 2$). Since $v^2 = z_2$ and $z_2 \not\sim x$ in G, Lemma 7.2(iv) implies that $(v^2)^{gm} = z_2$ for some $m \in M$ and so $gm \in H$. Since $y^{gm} = z$, we thus obtain (i).

Since $z_2 \not\sim x$ in G, z_2 is a noncentral involution. Since $S \subseteq H$ and $|S : S_0| = 2$, clearly (ii) also holds.

Observe next that $\bar{x} \notin O_2(\bar{H})$. Indeed, in the contrary case, also $\bar{y} \in O_2(\bar{H})$. But then $[\bar{y}, \bar{Q}_1] \subseteq O_2(\bar{H})$. Since $Q_1\langle y \rangle \cong QD_{2^{n+1}}$, $\bar{y}$ does not centralize $[\bar{y}, \bar{Q}_1]$ and so $\bar{y} \notin Z(O_2(\bar{H}))$. However, as S_0 is a Sylow 2-subgroup of H and $x \in Z(S_0)$, clearly $\bar{x} \in Z(O_2(\bar{H}))$. Hence $\bar{x}$ cannot be conjugate to $\bar{y}$ in $\bar{H}$, contrary to (i). This proves our assertion. We shall use this fact in proving (iii).

Consider first the case $q > 3$. Then L is perfect with $L/O(L) \cong SL(2,q) \times SL(2,q)$ and $L \subseteq H$. By Proposition 2.1, it follows that $\bar{L} \subseteq L(\bar{H})$. By Theorem A and Corollary A we know the possible structures of $L(\bar{H})$. If $L(\bar{H})$ is quasisimple, then considering the structure of $\bar{L}$, we check directly that $L(\bar{H})$ must be isomorphic to $SL(4,r)$, $r \equiv 3 \pmod 4$, $SU(4,r)$, $r \equiv 1 \pmod 4$, or $Sp(4,r)$, r odd. But then we can use z_2 in the role of x. However, we have seen already that x is a central involution, while z_2 is not. Hence $L(\bar{H})$ is not quasisimple and so $L(\bar{H})$ has two components $\bar{J}_1$, $\bar{J}_2$.

If $\bar{J}_i \cong SL(2,r_i)$, r_i odd, or $\hat{A}_7$ for both $i = 1$ and 2, then $\bar{x} \in Z(L(\bar{H})) \subseteq O_2(\bar{H})$ as $\bar{x} \in \bar{L} \subseteq L(\bar{H})$ and $\bar{x}$ is in the center of the Sylow 2-subgroup $\bar{S}_0 \cap L(\bar{H})$ of $L(\bar{H})$. But $\bar{x} \notin O_2(\bar{H})$, as we have shown above. Hence, say, $\bar{J}_1$ is simple and $\bar{J}_2 \cong SL(2,r_2)$ or $\hat{A}_7$. Since $\bar{x}$ centralizes $\bar{S}_0 \cap \bar{J}_2$, which is a Sylow 2-subgroup of $\bar{J}_2$, $\bar{x}$ centralizes $\bar{J}_2$. Thus $C_M(Z)$ covers $\bar{J}_2$ and we conclude at once from the structure of $C_M(Z)$ that $\bar{J}_2 \subseteq \bar{L}$. Then $\bar{J}_2 \triangleleft \bar{L}$ and so $\bar{J}_2 = \bar{L}_1$ or $\bar{L}_2$. However, as $\bar{x} \notin O_2(\bar{H})$, the only possibility is that $\bar{J}_2 = \bar{L}_2$. But $\bar{J}_1 = C_{L(\bar{H})}(\bar{J}_2)$ and as $\bar{L}_1$ centralizes $\bar{L}_2$, it follows that $\bar{L}_1 \subseteq \bar{J}_1$. Since $\bar{Q}_1$ is generalized quaternion, we conclude now from Corollary A that $\bar{J}_1$ has quasi-dihedral Sylow 2-subgroups, whence $\bar{J}_1 \cong L_3(r_1)$, $r_1 \equiv -1 \pmod 4$, or $U_3(r_1)$, $r_1 \equiv 1 \pmod 4$. However, as $C_M(Z)$ covers $C_{\bar{J}_1}(\bar{z}_1)$, we see at once that $\bar{L}_1 = L(C_{\bar{J}_1}(\bar{z}_1))$, which implies that $r_1 = q$.

We know that $\bar{Q}_2 \subseteq \bar{L}_2$ centralizes $\bar{J}_1$. Hence to complete the proof of (iii) in this case, it remains to show that $\bar{Q}_1\langle\bar{y}\bar{z}_2\rangle$ is a Sylow 2-subgroup of $\bar{J}_1$. We have that $\bar{x} \in \bar{J}_1\langle\bar{z}_2\rangle \triangleleft \bar{H}$ and as $\bar{x} \sim \bar{y}$ in $\bar{H}$, this forces $\bar{y} \in \bar{J}_1 \times \langle\bar{z}_2\rangle$. By the preceding paragraph, the Sylow 2-subgroup $\bar{S}_0 \cap \bar{J}_1$ of $\bar{J}_1$ is isomorphic to $QD_{2^{n+1}}$ and so either $\bar{S}_0 \cap \bar{J}_1 = \bar{Q}_1\langle\bar{y}\rangle$ or $\bar{Q}_1\langle\bar{y}\bar{z}_2\rangle$. However, in the first case $\bar{y} \sim \bar{z}_1$ in $\bar{J}_1$, whence $x \sim y \sim z_1$ in G, which is a contradiction. Thus the desired conclusion holds and (iii) is proved when $q > 3$.

Suppose then that $q = 3$. Since $\bar{x} \in Z(\bar{S}_0)$ and $\bar{S}_0$ is a Sylow 2-subgroup of $\bar{H}$, $\bar{x}$ centralizes $O_2(\bar{H})$. But then if $\bar{x}$ centralizes $L(\bar{H})$, $\bar{x} \in C_{\bar{H}}(L(\bar{H})O_2(\bar{H})) \subseteq O_2(\bar{H})$ by [23, Theorem 2]. However, we have shown above that $\bar{x} \notin O_2(\bar{H})$. Hence $\bar{x}$ does not centralize $L(\bar{H})$; in particular, $L(\bar{H}) \neq 1$. Furthermore, it follows that not every component of $L(\bar{H})$ is isomorphic to $SL(2,r)$, r odd, or $\hat{A}_7$. On the other hand, if $L(\bar{H})$ is quasisimple with $r(L(\bar{H})) = 4$, then the discussion of Sections 3 - 7 together with Theorem A shows that $L(\bar{H}) \cong \hat{M}_{12}$ inasmuch as z_2 is a noncentral involution. But by Proposition 5.4, this case is also excluded. We conclude therefore from Theorem A and Corollary A that $L(\bar{H})$ possesses a simple component $\bar{J}_1$ and $\bar{J}_1 \triangleleft \bar{H}$.

We know the possibilities for $\bar{J}_1$ and in each case $|\mathrm{Aut}(\bar{J}_1)/\mathrm{Inn}(\bar{J}_1)|$ is 2'-closed. Since $L \subseteq H$ and $Q \subseteq L'$, it follows that $\bar{Q} \subseteq \bar{J}_1\bar{D} = \bar{J}_1 \times \bar{D}$, where $\bar{D} = C_{\bar{H}}(\bar{J}_1)$. Since $r(\bar{J}_1) \geq 2$, we have $r(\bar{D}) \leq 2$. Hence $r(\bar{Q}/\bar{Q} \cap \bar{J}_1) \leq 2$, and so $\bar{Q} \cap \bar{J}_1 \not\subseteq \bar{Z}$. But $\bar{Q} \cap \bar{J}_1 \triangleleft \bar{L}$ as $Q \triangleleft L$. It follows at once now from the structure of $\bar{L}$ that $\bar{Q} \cap \bar{J}_1 \supseteq \bar{Q}_1$ or $\bar{Q}_2$. However, as $\bar{z}_2 \notin \bar{J}_1$, $\bar{Q}_2 \not\subseteq \bar{J}_1$ and so $\bar{Q}_1 \subseteq \bar{J}_1$. Thus $\bar{J}_1$ contains a quaternion group and now, considering the various possibilities for $\bar{J}_1$ in Theorem A, we see that necessarily $\bar{J}_1 \cong L_3(r)$, $U_3(r)$, r odd, or M_{11}. In particular, $r(\bar{J}_1) = 2$ and consequently $r(\bar{Q}/\bar{Q} \cap \bar{D}) \leq 2$. Hence $\bar{Q} \cap \bar{D} \not\subseteq \bar{Z}$ and as $\bar{Q} \cap \bar{D} \triangleleft \bar{L}$, we conclude from the structure of $\bar{L}$ that $\bar{Q}_2 \subseteq \bar{Q} \cap \bar{D}$. Thus $\bar{Q}_2$ centralizes $\bar{J}_1$.

Observe next that $\bar{x} \in \bar{J}_1\bar{D} - \bar{D}$ and so $C_{\bar{J}_1}(\bar{x}) = C_{\bar{J}_1}(\bar{Z})$ is isomorphic to the centralizer of an involution of $\bar{J}_1$. But $C_G(Z) = C_M(Z)$ is solvable by Lemma 6.7(viii) as $q = 3$ and consequently we must have $r = 3$. Hence $\bar{J}_1 \cong L_3(3)$, $U_3(3)$, or M_{11}. Furthermore, it follows exactly as in the case $q > 3$ that $\bar{y} \in \bar{J}_1 \times \langle \bar{z}_2 \rangle$ and that $\bar{Q}_1\langle \bar{y}\bar{z}_2 \rangle$ is a Sylow 2-subgroup of $\bar{J}_1$. Hence (iii) holds in this case as well.

Now set $R_1 = Q_1\langle yz_2 \rangle$. Since $\bar{R}_1$ is a Sylow 2-subgroup of $\bar{J}_1$ by the preceding lemma, $R_1 \triangleleft S_0$ and $R_1 \cong QD_{2^{n+1}}$. Since $S = S_0\langle t \rangle$, also $R_2 = R_1^t \triangleleft S_0$. But as $R_1 \cap R_2$ is t-invariant, it follows that $R_1 \cap R_2 \triangleleft S$. Clearly $x \notin R_1 \cap R_2$, and as $Z(S) = \langle x \rangle$, this forces $R_1 \cap R_2 = 1$. However, $|S_0| = 4|Q| = (2|Q_1|)^2 = |R_1|^2$. We therefore conclude that $S_0 = R_1R_2 = R_1 \times R_2$. Now Lemma 6.8 shows that we can take t to be an involution. We have thus proved

<u>Lemma 7.7</u>. We have $S \cong QD_{2^{n+1}} \int Z_2$.

We have therefore shown that a Sylow 2-subgroup of G is quasi-dihedral wreath Z_2. Hence by [38*, Theorem 1.6], G is necessarily a group of type $L_5(q)$ for some $q \equiv -1 \pmod 4$ or of type $U_5(q)$ for some $q \equiv 1 \pmod 4$ as these terms are defined in [38*]. In effect, this means that G has the same involution fusion pattern as $L_5(q)$, $q \equiv -1 \pmod 4$ or $U_5(q)$, $q \equiv 1 \pmod 4$, and the centralizers of involutions in G possess normal subgroups of odd index having "essentially" the same structure as the centralizers of the corresponding involutions in $L_5(q)$, $q \equiv -1 \pmod 4$ or $U_5(q)$, $q \equiv 1 \pmod 4$.

However, at this point we can invoke a recent theorem of Collins and Solomon [46] which asserts that a simple group G of type $L_5(q)$, $q \equiv -1 \pmod 4$ or of type $U_5(q)$, $q \equiv 1 \pmod 4$ is necessarily isomorphic to $L_5(q)$ or $U_5(q)$ itself.

8. The direct product case. In this section we shall prove

Proposition 8.1. L_x is not the direct product of two nonsimple components.

We assume false and, if possible, we choose x to be a central involution. We shall derive a contradiction in a sequence of lemmas.

Our conditions imply that $L_x = L(\overline{M})$ consists of two components $\overline{L}_i \cong SL(2,q_i)$, q_i odd, $q_i \geq 5$, or $\hat{A}_7$, $i = 1,2$. We let Q be a subgroup of T whose image in $\overline{M}$ is a Sylow 2-subgroup of $\overline{L}_1 \times \overline{L}_2$. Then $Q = Q_1 \times Q_2$, where Q_i is generalized quaternion and $\overline{Q}_i$ is a Sylow 2-subgroup of $\overline{L}_i$, $i = 1,2$. We also set $\langle z_i \rangle = Z(Q_i)$, $i = 1,2$, $z = z_1z_2$ and $Z = \langle z_1,z_2 \rangle$. By Lemma 6.5, $Z(T) = \langle z \rangle$ or Z according as T does or does not contain an element which interchanges $\overline{L}_1$ and $\overline{L}_2$. Since clearly $x \in Z(T)$, it follows in the first case that $\langle x \rangle = \langle z \rangle = Z(S)$ and hence that $S = T$.

Since $\bar{Z} = Z(\bar{L}_1\bar{L}_2)$, $C_M(Z)$ covers $\bar{L}_1\bar{L}_2$. We let L be a minimal T-invariant subgroup of $C_M(Z)$ which covers $\bar{L}_1\bar{L}_2$. Then Q is a Sylow 2-subgroup of L, L is perfect, and $L/O(L) \cong \bar{L}_1\bar{L}_2$. We can write $L = L_1L_2$, where $Q_i \subseteq L_i$ and L_i maps on $\bar{L}_i$, $i = 1,2$. We fix all this notation for the proof. We note also that $C_{\bar{M}}(L(\bar{M})) = Z(L(\bar{M})) = \bar{Z}$ by Lemma 3.2.

We first prove

Lemma 8.2. The involution x is central.

Proof: We have seen that this is the case if $Z(T) = \langle z\rangle$, so we may assume that $Z(T) = Z$ and also that $x \notin Z(S)$. Since $Z(S) \subseteq Z(T)$, it follows that $Z = \langle x,u\rangle$ with $\langle u\rangle = Z(S)$.

We set $H = C_G(u)$ and $\tilde{H} = H/O(H)$. Then $S \subseteq H$ and as $u \in Z$, also $L \subseteq H$. In particular, H is nonsolvable. If H were 2-constrained, Theorem D of Part II would imply that $G \cong J_2, J_3$, or M_{23}. However, in none of these cases would the centralizer of an involution of G have the form M. Hence H is not 2-constrained. Since L is perfect, Proposition 2.1 now yields that $\tilde{L} \subseteq L(\tilde{H})$.

Finally $r(\tilde{L}) = 4$ as $\tilde{L}/O(\tilde{L}) \cong \bar{L}_1\bar{L}_2$ and consequently $r(L(\tilde{H})) = 4$. If $L(\tilde{H})$ is quasisimple, it follows from Theorem A, using the fact that $\tilde{L} \subseteq L(\tilde{H})$, that $L(\tilde{H}) \cong Sp(4,r)$, r odd, $SL(4,r)$, $r \equiv 3 \pmod 4$, or $SU(4,r)$, $r \equiv 1 \pmod 4$. However, this contradicts Proposition 7.1. Hence $L(\tilde{H})$ consists of two components. If $L(\tilde{H})$ has a simple component $\tilde{J}$ (which is unique by Corollary A), then $\tilde{J} \lhd \tilde{H}$ and so $\tilde{J} \cap Z(\tilde{S}) \neq 1$. Since $u \notin J$, it follows then that $Z(S)$ is noncyclic, contrary to the fact that $Z(S) = \langle u\rangle$. Thus each component $\tilde{J}_i$ of $L(\tilde{H})$ is isomorphic to $SL(2,r_i)$, r odd, or $\hat{A}_7$, $i = 1,2$. If $\tilde{J}_1\tilde{J}_2 = \tilde{J}_1 * \tilde{J}_2$, then clearly $L(\tilde{H}) = \tilde{J}_1\tilde{J}_2$ would not contain a subgroup of the form of $\tilde{L}$ and consequently $L(\tilde{H}) = \tilde{J}_1 \times \tilde{J}_2$. Thus $L_u = L(\tilde{H})$ has the same form as L_x

and this contradicts our choice of x. We conclude therefore that x is central.

Thus $S = T$. If $x \neq z$, we can repeat the same reasoning with $H = C_G(z)$ to conclude that L_z has the same form as L_x. Hence without loss we can assume that $x = z$.

We next prove

Lemma 8.3. The following conditions hold:

(i) Z is the unique element of $U(S)$;

(ii) $x \not\sim z_i$ in G, $i = 1,2$;

(iii) $z_1 \sim z_2$ in G if and only if $z_1 \sim z_2$ in S.

Proof: If $Z = Z(S)$, then (i) holds by the definition of $U(S)$. Hence if $U \in U(S)$ with $U \neq Z$, then $Z(S) = \langle x \rangle$. Thus $[S,U] = \langle x \rangle$ and $U \cap Z = \langle x \rangle$. Hence $\tilde{U} \subseteq Z(\tilde{S})$, $\tilde{U} \cong Z_2$, and $\tilde{U} \neq \tilde{Z}$. But $\tilde{Q} = \tilde{Q}_1 * \tilde{Q}_2$ and $Z(\tilde{Q}) = \tilde{Z}$, whence $\tilde{Q}\tilde{U} = \tilde{Q} \times \tilde{U}$ has sectional 2-rank 5, contrary to $r(G) \leq 4$.

By symmetry, we need only prove that $x \not\sim z_1$ in G to establish (ii). Suppose false. If $Z = Z(S)$, then by Burnside's lemma, $z \sim z_1$ in $N_G(S)$ and hence in $N_G(Z)$. Thus $N_G(Z)$ contains a 3-element u which does not centralize Z. On the other hand, if $Z(S) = \langle z \rangle$, we reach the same conclusion by a standard argument, as (i) holds.

Now set $H = C_G(Z)$, so that $L = L_1L_2 \subseteq H \subseteq M$. Setting $\tilde{H} = H/O(H)$, we see that $\tilde{L} = L(\tilde{H}) = \tilde{L}_1 \times \tilde{L}_2$. But u normalizes H and so u acts on $\tilde{L}_1 \times \tilde{L}_2$. Since u is a 3-element, u must leave both $\tilde{L}_1$ and $\tilde{L}_2$ invariant, whence u centralizes $\tilde{z}_i$, $i = 1, 2$. Thus u centralizes $Z = \langle z_1, z_2 \rangle$, which is not the case. Hence (ii) holds.

Finally assume $z_1 \sim z_2$ in G, but $z_1 \not\sim z_2$ in S. Thus $Z(S) = Z$ and so $z_1 \sim z_2$ in $N_G(Z)$, again by Burnside's lemma. Hence $N_G(Z)$ contains a 3-element which normalizes, but does not centralize Z. But then $x \sim z_1$ in G, contrary to (ii), so (iii) also holds.

Lemma 8.4. The following conditions hold:

(i) S/Q is isomorphic to a subgroup of D_8;

(ii) $C_S(z_1)/Q$ is isomorphic to a subgroup of $Z_2 \times Z_2$;

Proof: These results are a direct consequence of Lemma 6.5.

At this point, we shall divide the analysis into two cases as follows:

Case 1. x is not isolated in $C_G(z_1)$ or $C_G(z_2)$;

Case 2. x is isolated in $C_G(z_1)$ and $C_G(z_2)$.

We first consider Case 1 and, for definiteness, we assume that x is not isolated in $C_G(z_2)$. Lemmas 8.5-8.7 are proved under this assumption. We set $H = C_G(z_2)$ and $\bar{H} = H/O(H)$. We also set $S_0 = C_S(Z)$ so that $|S : S_0| \leq 2$ and $S_0 = C_S(z_2)$.

We first prove

Lemma 8.5. The following conditions hold:

(i) S_0 is a Sylow 2-subgroup of H;

(ii) $L(\bar{H})$ possesses a component $\bar{J}_1 \cong L_3(q)$ or $U_3(q)$, q odd, which contains $\bar{Q}_1$ and centralizes $\bar{Q}_2$.

Proof: Suppose (i) is false, in which case $z_2 \notin Z(S)$ and $S_0 \subset S^*$ for some Sylow 2-subgroup S^* of H. The first condition implies that $Z(S) = \langle x \rangle$ and the second that S^* is a sylow 2-subgroup of G and $\langle z_2 \rangle = Z(S^*)$. But then $x \sim z_2$ in G, contrary to Lemma 8.3(ii). Thus (i) holds.

The proof of (ii) is essentially the same as that of Lemma 7.6(iii) for the case $q > 3$. We have $L \subseteq H$ with L perfect, so $\bar{L} \subseteq L(\bar{H})$. Using Theorem A and Corollary A together with Proposition 7.1, we now conclude that $L(\bar{H})$ is not quasisimple and so consists of two components $\bar{J}_1$, $\bar{J}_2$. If both of these are nonsimple, then $\bar{x} \in Z(L(\bar{H}))$ and as $\bar{S}_0$ is a Sylow 2-subgroup of $\bar{H}$, it follows that $\bar{x} \in Z(\bar{H})$, so x is isolated in H, contrary to assumption. Thus we can assume that, say, $\bar{J}_1$ is simple and $\bar{J}_2$ is nonsimple. But now we can argue as in Lemma 7.6(iii) that $\bar{J}_2 = \bar{L}_2$ and that $\bar{L}_1 \subseteq \bar{J}_1$ with $\bar{J}_1 \cong L_3(q)$ or $U_3(q)$; so (ii) also holds.

If x is not isolated in $C_G(z_1)$, then an analogous statement holds for $C_G(z_1)/O(C_G(z_1))$. In particular, in this case our conditions are symmetric in z_1 and z_2.

We next prove

Lemma 8.6. The involution z_2 is noncentral.

Proof: Assume the contrary, in which case $S_0 = S$, $Z(S) = Z$, S does not contain an element which interchanges Q_1 and Q_2 and S/Q is isomorphic to a subgroup of $Z_2 \times Z_2$.

Let R_1, R_2 be subgroups of S such that $\bar{R}_1$ is a Sylow 2-subgroup of $\bar{J}_1$ and $\bar{R}_2$ is a Sylow 2-subgroup of $\bar{D} = C_{\bar{H}}(\bar{J}_1)$. Then $R_1R_2 = R_1 \times R_2$, $|R_1 : Q_1| = 2$ with R_1 quasi-dihedral, and $R_2 \supseteq Q_2$. Since $r(\bar{J}_1) = 2$, $r(\bar{D}) \leq 2$ and as $\bar{J}_2 = \bar{L}_2 \triangleleft \bar{D}$, Lemma 6.9 implies that R_2 is either quasi-dihedral or generalized quaternion.

We argue that $R_2 = Q_2$ and $S \supset R_1R_2$. In the contrary case, as $|S/Q| \leq 4$ and $R_1 \supset Q_1$, we see that $S = R_1 \times R_2$. If R_1 and R_2

are both quasi-dihedral, then the main result of [41] implies that G is not simple. On the other hand, suppose R_2 is generalized quaternion. By the Z^*-theorem, $z_2 \sim y$ in G for some $y \neq z_2$ in S. But $\Omega_1(S) = \Omega_1(R_1) \times \langle z_2 \rangle$ and by the structure of $\bar{J}_1$ all involutions of $\Omega_1(R_1)$ are conjugate in H to z_1, whence also all involutions of $\Omega_1(R_1)z_2$ are conjugate in H to $z_1z_2 = x$. Thus $y \sim z_1$ or x in H and so $z_2 \sim z_1$ or x in G. However, this contradicts Lemma 8.3(ii) and (iii). This proves our assertions.

Since $|S/Q| \leq 4$, our argument yields that equality holds and that $|S : R_1R_2| = 2$. By Lemma 6.3, $S - R_1R_2$ contains an element v such that $v^2 \in R_2$ and $C_{R_1}(v)$ is dihedral of index 2 in R_1. Since $R_2 = Q_2$ is generalized quaternion, $\langle v, R_2 \rangle$ is again generalized quaternion or $\langle v, R_2 \rangle - R_2$ contains an involution. Hence we can assume $v^2 \in \langle z_2 \rangle$.

In addition, Lemma 6.3 implies that $C_S(Q_1) = Q_2$. This means that x is necessarily isolated in $C_G(z_1)$. Indeed, otherwise if we set $H_1 = C_G(z_1)$ and $\bar{H}_1 = H_1/O(H_1)$, Lemma 8.4(ii) would hold for $\bar{H}_1$ and so $L(\bar{H}_1)$ would possess a component $\bar{J}_2 \cong L_3(r)$ or $U_3(r)$, r odd, with $\bar{J}_2 \supseteq \bar{Q}_2$ and $[\bar{J}_2, \bar{Q}_1] = 1$. But then $\bar{Q}_1$ would centralize a Sylow 2-subgroup of $\bar{J}_2$ and so $C_S(Q_1)$ would contain a quasi-dihedral subgroup, which is not the case.

Suppose now that v can be taken to be an involution. We have noted above that every involution of $R_1R_2 = R_1Q_2$ is conjugate in H to an element of Z. Hence by Thompson's fusion lemma, $v \sim u$ in G for some u in Z. Thus $v^g = u$ and $(C_S(v))^g \subseteq S$ for some $g \in G$. We have $C_{R_1}(v) \cong D_{2^n}$ for some $n \geq 3$ with $\langle z_1 \rangle = \Omega_1(C_{R_1}(v)')$. On the other hand, as $Z(S) = Z(T) \cong Z_2 \times Z_2$, Lemma 6.5(iv) implies that $\Omega_1(S') = Z$. It follows therefore that $z_1^g \in Z$. But now Lemma 8.3(ii) and (iii) force $z_1^g = z_1$, whence $v \sim u$ in $C_G(z_1)$. Since x is isolated in $C_G(z_1)$, we conclude now that $v \sim u$ in M. However, clearly this is impossible by the structure of M.

Hence v cannot be taken as an involution and so $v^2 = z_2$. By [27, Lemma 16], $v^g = u$ for some $u \in R_1R_2$ of order 4 and $g \in G$. But as R_1 is quasi-dihedral and R_2 is generalized quaternion, $u^2 \in Z$. Thus $z_2^g = (v^2)^g = u^2 \in Z$ and so $z_2^g = z_2$, again by Lemma 8.3(ii) and (iii). Therefore $v \sim u$ in H. But $\bar{u} \in \bar{J}_1C_{\bar{H}}(\bar{J}_1)$, while $\bar{v} \notin \bar{J}_1C_{\bar{H}}(\bar{J}_1)$, so this is clearly impossible and the lemma is proved.

By the lemma, z_2 is noncentral and so S contains an element t which interchanges z_1 and z_2. Thus we have $S = S_0\langle t\rangle$ and $Z(S) = \langle z\rangle$. Again let R_1 denote the subgroup of S such that $\bar{R}_1$ is a Sylow 2-subgroup of $\bar{J}_1$ and set $R_2 = R_1^t$.

Finally we prove

<u>Lemma 8.7</u>. We have $S \cong QD_{2^{n+1}} \int Z_2$.

<u>Proof</u>: We have that R_1 is normal in R_1R_2 and if $Q_1 \cong Q_{2^n}$, that $R_1 \cong QD_{2^{n+1}}$. We can now repeat the proof of Lemma 7.7 to obtain that $S_0 = R_1 \times R_2$ and that $S = (R_1 \times R_2)\langle t\rangle \cong QD_{2^{n+1}} \int Z_2$, as asserted.

Now invoking [38*] and [46] once again, it follows that $G \cong L_5(q)$, $q \equiv -1 \pmod 4$ or $U_5(q)$, $q \equiv 1 \pmod 4$. However this is impossible as a central involution in these groups does not have a centralizer of the form of M.

It thus remains to treat Case 2. Since x is isolated in $C_G(z_i)$, $i = 1,2$, we obtain immediately:

Lemma 8.8. Any two elements of S that are conjugate in $C_G(z_i)$, $i = 1,2$, are conjugate in M.

By Glauberman's Z^*-theorem, $x \sim y$ in G for some $y \neq x$ in S. We choose g in G such that $y^g = x$ and $C_S(y)^g \subseteq S$. We also set $\tilde{M} = \bar{M}/\bar{Z}$ and fix this notation.

Lemma 8.9. The following conditions hold:

(i) $Q_i\langle y\rangle$ is quasi-dihedral, $i = 1,2$;

(ii) $\tilde{L}_i\langle\tilde{y}\rangle \cong PGL(2,q_i)$, $i = 1,2$.

Proof: Suppose first that $\bar{y}$ interchanges $\bar{L}_1$ and $\bar{L}_2$, in which case $C_Q(y) \cong Q_1$ and $Z(C_Q(y)) = \langle x\rangle$. We have $C_Q(y)^g \subseteq S$. Since S/Q is isomorphic to a subgroup of D_8 and Q_1 is generalized quaternion, it follows therfore that $x^g \in Q$, whence $x^g \in Z = \Omega_1(Q)$. Now Lemma 8.3(ii) yields $x^g = x$, which is not the case. Hence $\bar{y}$ leaves both $\bar{L}_1$ and $\bar{L}_2$ invariant.

Suppose next that $\bar{y}$ induces a field automorphism of order 2 on $\bar{L}_1$ or $\bar{L}_2$, say $\bar{L}_1$. Replacing y by a suitable conjugate in M, we can assume without loss that $C_{\bar{Q}_1}(\bar{y})$ is a Sylow 2-subgroup of $C_{\bar{L}_1}(\bar{y})$. Then by Lemma 6.1, $C_{Q_1}(y)$ contains a quaternion subgroup Q_0 and we have $\langle z_1\rangle = Z(Q_0)$. Since $Q_0^g \subseteq S$, it follows as in the preceding paragraph that $z_1^g \in Q$, whence $z_1^g \in Z$. By Lemma 8.3(ii) and (iii), this implies that $z_1^g = z_1$ or z_2 with the latter case occuring only if S contains an element t which interchanges z_1 and z_2. Replacing g by gt, if necessary, we can assume without loss that $z_1^g = z_1$. But now the preceding lemma yields that $y \sim x$ in M, which is clearly impossible. Hence $\bar{y}$ does not induce such an automorphism of $\bar{L}_1$ or $\bar{L}_2$.

On the other hand, if $\bar{y}$ induces an inner automorphism of $\bar{L}_1$ or $\bar{L}_2$, say $\bar{L}_1$, then $\bar{L}_1\langle\bar{y}\rangle \cong SL(2,q_1) \times Z_2$ or $SL(2,q_1) * Z_4$. Again we can assume without loss that $C_{\bar{Q}_1}(\bar{y})$ is a Sylow 2-subgroup of $C_{\bar{L}_1}(\bar{y})$. Hence in either case, it follows that $C_{Q_1\langle y\rangle}(y)$ contains a subgroup $V_1 \cong Z_4 \times Z_2$ with $\mho^1(V_1) = \langle z_1\rangle$. Furthermore, y centralizes z_2 and so $C_{Q\langle y\rangle}(y)$ contains a subgroup $V \cong Z_4 \times Z_2 \times Z_2$. But $V^g \subseteq C_S(y)^g \subseteq S$ and S/Q is isomorphic to a subgroup D_8. Clearly these conditions force z_1^g, z_2^g, or $x^g = (z_1z_2)^g$ to lie in Q. However, as $y^g = x$, $x^g \neq x$. Hence if $x^g \in Q$, it would follow that $x^g = z_1$ or z_2, contrary to Lemma 8.3(ii). We conclude that z_1^g or z_2^g lies in Q. As in the preceding case, we can assume for a suitable choice of g that $y \sim x$ in either $C_G(z_1)$ or $C_G(z_2)$. Again the preceding lemma yields a contradiction and so also $\bar{y}$ does not induce such an automorphism of $\bar{L}_1$ or $\bar{L}_2$.

Furthermore, if some $\bar{L}_i \cong \hat{A}_7$, say for $i = 1$, then we conclude, as above, with the aid of Lemma 6.1(ii) that for a suitable choice y, either $C_{Q_1}(y)$ contains a quaternion subgroup or $C_{Q_1\langle y\rangle}(y)$ contains a subgroup isomorphic to $Z_4 \times Z_2$. In either case we reach a contradiction as before. Hence by Lemma 6.2(i) the only possibility is that $\bar{L}_i \cong SL(2,q_i)$, q_i odd, and $\tilde{L}_i\langle\tilde{y}\rangle \cong PGL(2,q_i)$, $i = 1, 2$. But then $\tilde{Q}_i\langle\tilde{y}\rangle$ is dihedral and it follows at once that $Q_i\langle y\rangle \cong \bar{Q}_i\langle\bar{y}\rangle$ is quasi-dihedral, $i = 1, 2$. Thus both parts of the lemma hold.

As an immediate consequence we have

Lemma 8.10. The following conditions hold:

(i) All involutions of $Q\langle y\rangle - Q$ are conjugate in $Q\langle y\rangle$ to y;

(ii) $C_Q(y) = Z$ and $S = QC_S(y)$.

Proof: First, (i) can be checked directly from the fact that $Q = Q_1 \times Q_2$ and $Q_i\langle y\rangle$ is quasi-dihedral, $i = 1, 2$. Since $C_{Q_i}(y) = \langle z_i\rangle$, $i = 1, 2$, also $C_Q(y) = \langle z_1, z_2\rangle = Z$. Furthermore, by Lemma 6.4(v), $Q\langle y\rangle \triangleleft S$. Since also $Q \triangleleft S$, $y^u \in Q\langle y\rangle - Q$ for any u in S, whence $y^{uv} = y$ for some v in Q. Thus $u = (uv)v^{-1} \in C_S(y)Q$ and we conclude at once that $S = C_S(y)Q = QC_S(y)$.

We have $Z = C_Q(y)$ and so $Z^g \subseteq S$. We set $V = Z^g$ and next prove

Lemma 8.11. The following conditions hold:

(i) $V \cap Q = 1$;

(ii) $\bar{V}$ leaves $\bar{L}_1$ and $\bar{L}_2$ invariant;

(iii) $x^{gu} = y$ for some u in Q.

Proof: Suppose (i) is false, in which case $Z^g \cap Q \neq 1$, whence $Z^g \cap Z \neq 1$. If $x^g \in Z$, then $x^g = x$ by Lemma 8.3(ii), contrary to $y^g = x$. Hence $z_i^g \in Z$, $i = 1$ or 2, say $i = 1$, and for the same reason $z_1^g = z_1$ or z_2. Furthermore, as in Lemma 8.9, the latter case occurs only if $z_1 \sim z_2$ in S and so without loss we can assume that $z_1^g = z_1$. But then $y \sim x$ in M by Lemma 8.8, which is a contradiction. Thus (i) holds.

Set $v = x^g$ and $v_i = z_i^g$, $i = 1, 2$. Then $v \sim x$ in G with $v \neq x$ and so v must satisfy the same conditions as y. Lemma 8.9

shows then that $\bar{v}$ leaves $\bar{L}_1$ and $\bar{L}_2$ invariant. Suppose now that $\bar{v}_1$ or $\bar{v}_2$, say $\bar{v}_1$, does not leave $\bar{L}_1$ and $\bar{L}_2$ invariant. Then $\bar{v}_1$ interchanges $\bar{L}_1$ and $\bar{L}_2$ and consequently $C_Q(v_1) \cong Q_1$ and $\langle x \rangle = Z(C_Q(v_1))$. Since $z_1 \not\sim x$ in G, z_1 is extremal in S and so there exists g_1 in G such that $C_S(v_1)^{g_1} \subseteq C_S(z_1)$. As usual, this successively forces $x^{g_1} \in Q$, $x^{g_1} \in Z$, and $x^{g_1} = x$. Thus $v_1 \sim z_1$ in M. But as $v_1 \in S - Q$, this is clearly impossible by the structure of M. Thus (ii) also holds.

We have already noted that v satisfies the same conditions as y, so $\tilde{L}_i\langle\tilde{v}\rangle \cong \tilde{L}_i\langle\tilde{y}\rangle \cong PGL(2,q_i)$, $i = 1, 2$. Hence by Lemma 6.4(vi), $\tilde{Q}\langle\tilde{v}\rangle = \tilde{Q}\langle\tilde{y}\rangle$ and so $Q\langle x^g\rangle = Q\langle v\rangle = Q\langle y\rangle$. But now Lemma 8.10(i) yields (iii).

Taking u as in Lemma 8.11(iii), we have $x^{gu} = y$ and $y^{gu} = x^u = x$. Hence replacing g by gu, if necessary, we can assume without loss that $y^g = x$ and $x^g = y$. Thus $y \in V$.

Furthermore, as $\bar{V}$ leaves $\bar{L}_1$ and $\bar{L}_2$ invariant, V centralizes z_1 and z_2 and so V centralizes Z. But $S_0 = C_S(Z)$ and $|S_0/Q| \leq 4$. Since $V \cap Q = 1$, it follows that $S_0 = QV$. Note also that as $y \in V \cong Z_2 \times Z_2$ and $C_Q(y) = Z$, this in turn implies that $C_{S_0}(y) = ZV \cong E_{16}$. We have therefore proved.

Lemma 8.12. The following conditions hold:

(i) $S_0 = QV$;

(ii) $C_{S_0}(y) = ZV \cong E_{16}$.

We now set $A = ZV$ and next prove

Lemma 8.13. The following conditions hold:

(i) A is the unique elemetary abelian subgroup of $C_S(y)$ of order 16 ;

(ii) $y \sim yx$, yz_1, and yz_2 in $N_S(A)$;

(iii) $y \sim x$ in $N_G(A)$.

Proof: If $S = S_0$, then $C_S(y) = A$ and (i) is obvious. In the contrary case, $S = S_0\langle t\rangle$, where t interchanges Q_1 and Q_2. Furthermore, by Lemma 8.10(ii), we can take t to centralize y and $C_S(y) = A\langle t\rangle$. Since t interchanges z_1 and z_2, we have $C_Z(t) = \langle x\rangle$. But in this case, $\bar{A}\langle\bar{t}\rangle = A\langle t\rangle/Z \cong S/Q \cong D_8$ and so $C_{\bar{A}}(\bar{t}) \cong Z_2$. It follows therefore that $C_S(y) \cong Z_2 \times Z_2 \int Z_2$ and so (ii) holds in this case as well.

Now put $a = yx$, yz_1, or yz_2. Then $a \in A$ and $a \in Q\langle y\rangle - Q$, so $a^v = y$ for some v in Q by Lemma 8.10(i). But then $A^v \subseteq C_S(a)^v = C_S(y)$ and so $A^v = A$ by (i). Thus (ii) holds.

Finally we have $A^g \subseteq C_S(y)^g \subseteq S$ and $y = x^g \in A^g$, so $A^g \subseteq C_S(y)$. Again (i) forces $A^g = A$ and so (iii) also holds.

Now we can easily eliminate Case 2. We have already noted in the proof of Lemma 8.11(iii) that if an element v of $S - \langle x\rangle$ is conjugate to x in G, then $Q\langle v\rangle = Q\langle y\rangle$. In particular, if $v \in A$, then necessarily $v = y$, yx, yz_1, or yz_2 as these are the only elements of A which lie in $Q\langle y\rangle - \langle x\rangle$. We see then that x has exactly 5 conjugates in G which lie in A. But now by the preceding lemma, it follows that x has exactly 5 conjugates in $N = N_G(A)$. Hence if we set $\bar{N} = N/C_G(A)$, we conclude that $|\bar{N}|$ is divisible by 5.

On the other hand, let u_i be of order 4 in Q_i with $\langle u_i\rangle$ normal in S_0, $i = 1, 2$. Such a u_i exists as $Q_i \lhd S_0$ and Q_i is generalized quaternion. Since $Q_i\langle y\rangle$ is quasi-dihedral, y inverts u_i and as $u_i^2 = z_i \in A$, we see that u_i normalizes, but does not centralize A, $i = 1,2$. Since $\langle u_1, u_2\rangle = \langle u_1\rangle \times \langle u_2\rangle$ with $\langle u_1, u_2\rangle \cap A = \langle z_1, z_2\rangle = Z$, it follows now that $N_S(A)/C_S(A)$ contains a four group. Thus $\bar{N}$ contains a four group.

Since $|\bar{N}|$ is also divisible by 5, we conclude now that $\bar{N}$ contains a subgroup isomorphic to A_5. Since x has exactly 5 conjugates in $\bar{N}$, $|N : C_N(x)| = 5$ and it follows therefore from the structure of N that $C_N(x)$ contains a 3-element w which does not centralize A. Thus $w \in N_M(A)$ and so $\bar{w} \in N_{\bar{M}}(\bar{A}) - C_{\bar{M}}(\bar{A})$. But clearly $\bar{w}$ centralizes $\bar{Z} = Z(\bar{L}_1\bar{L}_2)$ and it is immediate from the structure of $\bar{M}/\bar{L}$ that $\bar{w}$ must also centralize $\bar{A}/\bar{Z}$. Hence $\bar{w}$ centralizes $\bar{A}$ and we reach a final contradiction. Therefore also Case 2 cannot occur and Proposition 8.1 is proved.

9. The central product case. Finally we treat the case in which L_x is the central product of two components isomorphic to $SL(2,q_i)$, q_i odd, or $\hat{A}_7$, $i = 1, 2$.

We let $\bar{L}_1$, $\bar{L}_2$ be the components of $L_x = L(\bar{M})$, so that $L(\bar{M}) = \bar{L}_1 * \bar{L}_2$ with $\bar{L}_i \cong SL(2,q_i)$, q_i odd, or $\hat{A}_7$, $i = 1, 2$. Since $C_{\bar{M}}(L(\bar{M})) = Z(L(\bar{M}))$, we have that $\langle\bar{x}\rangle = Z(L(\bar{M}))$. By Lemma 6.6(i), $\langle\bar{x}\rangle = Z(\bar{T})$ and so $\langle x\rangle = Z(T)$, which implies that $S = T$. Again we let Q be the subgroup of S whose image in $\bar{M}$ is a Sylow 2-subgroup of $L(\bar{M})$. Then $Q = Q_1 * Q_2$ with Q_i generalized quaternion and $\bar{Q}_i$

a Sylow 2-subgroup of $\bar{L}_i$, $i = 1, 2$. We let L be the unique minimal normal subgroup of M which covers $L(\bar{M})$. Then L is perfect, Q is a Sylow 2-subgroup of L, $\bar{L} = L(\bar{M})$, and $L = L_1L_2$ with $Q_i \subseteq L_i$ and L_i mapping on $\bar{L}_i$, $i = 1, 2$. We also set $\tilde{M} = \bar{M}/\langle \bar{x} \rangle$, so $L(\tilde{M}) = \tilde{L}_1 \times \tilde{L}_2$, where $\tilde{L}_i \cong L_2(q_i)$ or A_7, $i = 1, 2$. In particular, Lemma 6.4 applies to $\tilde{M}$. We fix all this notation.

We first treat the case that $SCN_3(S)$ is nonempty.

Proposition 9.1. If $SCN_3(S)$ is nonempty, then S is of type M_{12}, A_8, A_{10}, or $\hat{A}_8$.

We carry out the proof in a short sequence of lemmas. We fix an elementary abelian normal subgroup A of S of order 8. Then by Lemma 6.6(ii), $\bar{A} \subseteq \bar{Q}$ and so $A \subseteq Q$. As an immediate consequence we have

Lemma 9.2. We have $Q \cong Q_8 * Q_8$.

Proof: Since $A \subseteq Q$, $SCN_3(Q)$ is nonempty. However, one checks directly that $Q_{2^m} * Q_{2^n}$ does not possess an elementary abelian normal subgroup of order 8 if $m > 3$ or $n > 3$.

The lemma implies that $Q_i \cong Q_8$ and hence that $\bar{L}_i \cong SL(2, q_i)$, $q_i \equiv 3,5 (\mathrm{mod}\ 8)$, $i = 1,2$.

We next prove

Lemma 9.3. The following conditions hold:

(i) $|S| = 2^6$ or 2^7;

(ii) $\tilde{S}$ is isomorphic to a subgroup of a 2-group of type A_8.

Proof: If $Q = S$, it is easily verified that x is isolated in S, contrary to the fact that G is simple. Thus $S \supset Q$. By the structure of $\bar{L}$, $N_L(Q)/C_L(Q)Q \cong Z_3 \times Z_3$. Hence if $|S| = 2^8$, it would follow that $N_M(Q)/C_M(Q) \cong \mathrm{Aut}(Q_8 * Q_8)$ which, as noted in Lemma 5.3 of Part II, is isomorphic to a Sylow 3-normalizer in A_8. It would then follow that the six elementary subgroups of Q of order 8 are all conjugate in $N_G(Q)$. On the other hand, as $A \triangleleft S$, the number of conjugates of A in $N_G(Q)$ is odd. This contradiction shows that $|S| \neq 2^8$. Since S/Q is isomorphic to a group of outer automorphisms of Q, our argument yields that $|S/Q| = 2$ or 4 and so $|S| = 2^6$ or 2^7.

Setting $K = N_M(Q)$ and $\bar{K} = K/O(K)\langle x \rangle$, Lemma 5.3 of Part II implies that $\bar{K}$ is isomorphic to a split extension of $\bar{Q} \cong E_{16}$ by a subgroup of a Sylow 3-normalizer in A_8 and that $\bar{S}$ is isomorphic to a subgroup of a 2-group of type A_8 (inasmuch as $SCN_3(S) \neq \emptyset$). Since $\bar{S} \cong S/\langle x \rangle \cong \tilde{S}$, we conclude that both parts of the lemma hold.

<u>Lemma 9.4</u>. If $|S| = 2^6$, then S is of type M_{12} or A_8.

<u>Proof</u>: By the preceding lemma, $\tilde{S} \cong Z_2 \times Z_2 \int Z_2$. Hence $S - Q$ contains an element y with $y^2 \in \langle x \rangle$ with y either normalizing both Q_1 and Q_2 or interchanging them. Suppose y can be taken to be an involution. If y interchanges Q_1 and Q_2, it is immediate that S is isomorphic to a 2-group of type A_8. If y normalizes Q_1 and Q_2, then $\bar{y}$ inverts a Sylow 3-subgroup of $N_{\bar{L}_i}(\bar{Q}_i)$ and it follows that $Q_i\langle y \rangle \cong QD_{16}$, $i = 1, 2$. But now we see that S is isomorphic to a Sylow 2-subgroup of M_{12}.

Thus the lemma holds if y can be taken to be an involution. In the contrary case, $\Omega_1(S) = Q$. Applying [27, Lemma 16], it follows that $y \sim u$ in G for some u in Q. Since $y^2 = u^2 = x$, this implies that $y \sim u$ in M. Clearly this is impossible by the structure of M.

Finally we have

Lemma 9.5. If $|S| = 2^7$, then S is of type A_{10} or $\hat{A}_8$.

Proof: Set $K = N_M(Q)$ and $\bar{K} = K/O(K)$. If $\bar{P}$ denotes a Sylow 3-subgroup of $\bar{K}$, we have that $\bar{P} \cong Z_3 \times Z_3$ and that $C_{\bar{Q}}(\bar{P}) = \langle \bar{x} \rangle$. By the Frattini argument, $\bar{K} = \bar{Q}N_{\bar{K}}(\bar{P})$ and consequently a Sylow 2-subgroup $\bar{V}$ of $N_{\bar{K}}(\bar{P})$ has order 8, $\bar{x} \in \bar{V}$, and $\bar{V}/\langle \bar{x} \rangle \cong Z_2 \times Z_2$. Then $\bar{V} \cong E_8$, $Z_2 \times Z_4$, D_8 or Q_8. Without loss we can assume that $\bar{V} \subseteq \bar{S}$. Hence if V denotes the inverse image of $\bar{V}$ in S, we have that $S = QV$, where $Q \cap V = \langle x \rangle$ and $V \cong \bar{V}$.

The structure of such a 2-group S and its occurrence as a Sylow 2-subgroup of a simple group have been studied in [28] under the assumption that S normalizes no nontrivial subgroup of M of odd order. However, the argument of [28] depends only upon 2-group considerations and does not actually require this assumption. Hence the analysis given there yields this lemma.

Together Lemmas 9.3(i), 9.4, and 9.5 establish Proposition 9.1.

We turn now to the $SCN_3(S)$ empty case. We shall prove

Proposition 9.6. If $SCN_3(S)$ is empty, then S is of type $L_4(q)$, $q \equiv 7 \pmod 8$, or $PSp(4,q)$ or $G_2(q)$, $q \equiv 1, 7 \pmod 8$.

Again we proceed in a sequence of lemmas. Since $m(S) \geq 3$, S is not of maximal class and so $U(S)$ possesses a unique element U. Then $x \in U$. Since $\tilde{U}$ is an abelian normal subgroup of $\tilde{S}$, Lemma 6.6(ii) implies, as with A above, that $U \subseteq Q$. We set $U = \langle x,u \rangle$. Since

$\langle x \rangle = Z(S)$, $u \sim xu$ in S. Hence if $u \sim x$ in G, then all involutions of U are conjugate in G and so the main results of [29] and [35] are applicable. They imply that S is of type $G_2(q)$, $q \equiv 1,7 \pmod 8$ or else S is of type J_2 or $\hat{A}_{10}$. However, in the latter two cases $G \cong J_2$, J_3, or L by [16, Theorem A] and [18, Theorem B]. But this is impossible as none of these groups possesses an involution whose centralizer has the form of M. Thus the proposition holds in this case.

<u>Therefore we may assume for the proof that</u> u <u>is not conjugate to</u> x <u>in</u> G.

By the structure of $\bar{L} = \bar{L}_1 * \bar{L}_2$, we see that every involution of $\bar{Q} - \langle \bar{x} \rangle$ is conjugate in $\bar{L}$ to $\bar{u}$. Hence our assumption implies:

<u>Lemma 9.7</u>. No involution of $Q - \langle x \rangle$ is conjugate to x in G.

Since G is simple, Glauberman's Z^*-theorem yields that $x \sim y$ in G for some involution y of $S - Q$. We fix such an involution y.

We shall consider two cases separately:

(A). No element of $\bar{S}$ interchanges $\bar{L}_1$ and $\bar{L}_2$;

(B). Some element of $\bar{S}$ interchanges $\bar{L}_1$ and $\bar{L}_2$ (and so $\bar{L}_1 \cong \bar{L}_2$).

We treat case (A) in the next three lemmas.

<u>Lemma 9.8</u>. The following conditions hold:

(i) $S/Q \cong Z_2$ or $Z_2 \times Z_2$;

(ii) $\tilde{L}_i\langle \tilde{y} \rangle \cong PGL(2, q_i)$, $i = 1, 2$;

(iii) $C_Q(y) = U$ and $x \notin \Phi(C_S(y))$;

(iv) Every involution of $Q\langle y \rangle - Q$ is conjugate to y in S.

<u>Proof</u>: First, (i) follows from Lemma 6.4(ii), applied to $\tilde{M}$. Suppose that $x \in \Phi(C_S(y))$. We know there exists g in G such that

$y^g = x$ and $C_S(y)^g \subseteq S$. Moreover, we can clearly assume without loss that $C_{\bar{Q}}(\bar{y})$ is a Sylow 2-subgroup of $C_{L(\bar{M})}(\bar{y})$. By (i), $\Phi(S) \subseteq Q$ and so $x^g \in Q$. Now Lemma 9.7 forces $x^g = x$, contrary to $y^g = x$. Thus $x \notin \Phi(C_S(y))$. But now our choice of y implies that $\tilde{y}$ does not induce an inner automorphism or field automorphism of $\tilde{L}_i$, $i = 1$ or 2 and now (ii) follows from Lemma 6.2(i).

By (ii), $\tilde{Q}_i\langle\tilde{y}\rangle$ is dihedral and consequently $Q_i\langle y\rangle$ is quasi-dihedral, $i = 1, 2$. A direct computation now yields (iv) and the first assertion of (iii); so all parts of the lemma hold.

Lemma 9.9. We have $S/Q \cong Z_2 \times Z_2$.

Proof: Suppose false, in which case $S/Q \cong Z_2$ and $S = Q\langle y\rangle$. By the preceding lemma, $B = C_S(y) \cong E_8$. Moreover, if E is any elementary abelian subgroup of S of order 8 with $E \not\subseteq Q$, then $y' \in E$ for some involution y' of $S - Q$ and so $y' \sim y$ in S. Thus $E = C_S(y')$ and so $E \sim B$ in S. On the other hand, B is not conjugate in G to a subgroup of Q as such a subgroup of Q contains six involutions conjugate to u, while B contains only two involutions conjugate to u.

Now let $g \in G$ be such that $y^g = x$ and $B^g = C_S(y)^g \subseteq S$. By the preceding paragraph $B^g \sim B$ in S and so replacing g by gt for some t in S, we can assume without loss that $g \in N = N_G(B)$. But by the structure of $N_S(B)$, y has four conjugates in N under the action of $N_S(B)$. Clearly none of these is x, so y has at least five conjugates under the action of N. Since B contains two involutions conjugate to u, it follows that y has exactly five conjugates in N.

Thus $|N/C_G(B)|$ is divisible by 5, contrary to the fact that $N/C_G(B)$ is isomorphic to a subgroup of $\mathrm{Aut}(B) \cong GL(3,2)$ and $|GL(3,2)|$ is not divisible by 5.

Finally we have

<u>Lemma 9.10</u>. The case $S/Q \cong Z_2 \times Z_2$ does not occur.

<u>Proof</u>: Assume false. Consider first the case that $C_{\tilde{S}}(\tilde{L}_1) = \tilde{Q}_2$, in which case $\tilde{L}_1\tilde{S}/\tilde{Q}_2$ is isomorphic to a subgroup of $P\Gamma L(2,q_1)$. There thus exists an element v in $S - Q_1Q_2$ such that $\tilde{v}^2 \in \tilde{Q}_2$ and $\tilde{v}$ induces a field automorphism of $\tilde{L}_1$ of order 2. Then $Q_2\langle v\rangle$ is a group and as Q_2 is generalized quaternion, either $Q_2\langle v\rangle - Q_2$ contains an involution or $Q_2\langle v\rangle$ is also generalized quaternion. Thus $Q_2\langle v\rangle$ contains an element whose square lies in $\langle x\rangle$. Hence without loss we can assume that $v^2 \in \langle x\rangle$, in which case $\tilde{v}$ is an involution.

Since $\tilde{v}$ induces a field automorphism of $\tilde{L}_1$, $C_{\tilde{L}_1}(\tilde{v})$ contains a four group and as $r(G) \leq 4$, it follows that $C_{\tilde{L}_2}(\tilde{v})$ does not contain a four group, which implies that $\tilde{L}_2\langle\tilde{v}\rangle \cong PGL(2,q_2)$. But also $\tilde{L}_2\langle\tilde{y}\rangle \cong PGL(2,q_2)$, so $S - Q$ contains an element v_1 which induces an inner automorphism of $\tilde{L}_2$. Hence $C_{\tilde{S}}(\tilde{L}_2) \supset \tilde{Q}_1$. Without loss, we can assume that $\tilde{v}_1$ centralizes $\tilde{L}_2$, whence $C_{\tilde{S}}(\tilde{L}_2) = \tilde{Q}_1\langle\tilde{v}_1\rangle$. As with v, we can also suppose that $v_1^2 \in \langle x\rangle$, in which case $\tilde{v}_1$ is an involution and $\tilde{L}_1\langle\tilde{v}_1\rangle \cong PGL(2,q_1)$. But as $\tilde{L}_1\langle\tilde{y}\rangle \cong PGL(2,q_1)$, the same argument yields that $C_{\tilde{S}}(\tilde{L}_1) \supset \tilde{Q}_1$, a contradiction.

We therefore conclude that, in fact, $C_{\tilde{S}}(\tilde{L}_1) \supset \tilde{Q}_1$. Repeating the same argument on $\tilde{L}_2$, we obtain likewise that $C_{\tilde{S}}(\tilde{L}_2) \supset \tilde{Q}_2$. We see then that we can write $S = \langle Q_1, v_1, v_2\rangle$, where $C_{\tilde{S}}(\tilde{L}_1) = \tilde{Q}_2\langle\tilde{v}_2\rangle$ and

$C_S(\tilde{L}_2) = \tilde{Q}_1\langle\tilde{v}_1\rangle$. Moreover, as above, we can assume that $v_i^2 \in \langle x\rangle$, $i = 1,2$. Clearly $y \in v_1v_2Q - Q$.

Suppose now that v_1 is an involution, whence v_1 is conjugate in G to some element of $Q\langle y\rangle$ by Thompson's fusion lemma. Since $\tilde{L}_2\langle\tilde{v}_1\rangle \not\cong PGL(2,q)$, v_1 is not conjugate to x in G by Lemma 9.8(ii). But then as $x \sim y$ in G, Lemma 9.8(iv) implies that v_1 must be conjugate to an element of $Q - \langle x\rangle$, whence $v_1 \sim u$ as all involutions of $Q - \langle x\rangle$ are conjugate in M. Since $u \not\sim x$ in G, u is extremal and so for some g in G, we have $v_1^g = u$ and $C_S(v_1)^g \subseteq C_S(u)$. But as $\tilde{v}_1$ centralizes $\tilde{L}_2$, $Q_2 \subseteq C_S(v_1)$ and so $x \in \Phi(C_S(v_1))$, whence $x^g \in \Phi(C_S(u)) \subseteq Q$. This forces $x^g = x$, so $v_1 \sim u$ in M, which is clearly impossible.

Therefore $v_1^2 = x$. This time [27, Lemma 16] yields that either x is fused to an involution v of $S - Q\langle y\rangle$ or v_1 is fused to an element of $Q\langle y\rangle$. However, in the first case, $v \in Qv_1$ or Qv_2 as $y \in Qv_1v_2$ and so $\tilde{L}_i\langle\tilde{v}\rangle \not\cong PGL(2,q_i)$ for some $i = 1$ or 2, contradicting Lemma 9.8(ii). Thus $v_1 \sim t$ for some t in $Q\langle y\rangle$. But then $x = v_1^2 \sim t^2 \in Q$ and so $t^2 = x$ by Lemma 9.7. We conclude at once that $v_1 \sim t$ in M, which is again impossible.

Together Lemmas 9.9 and 9.10 imply that case (A) does not occur. Hence we must have case (B). Let y be as above. Again we assume without loss that $C_{\bar{Q}}(\bar{y})$ is a Sylow 2-subgroup of $C_{L(\bar{M})}(\bar{y})$.

Lemma 9.11. If $S/Q \cong Z_2$, then S is of type $PSp(4,q)$ for some $q \equiv 1,7 \pmod 8$.

Proof: In this case, $S = Q\langle y\rangle$ and y must interchange Q_1 and

Q_2 as we have case (B). Since $SCN_3(S)$ is empty, this forces $|Q_i| \geq 16$, $i = 1,2$, and we see that the lemma holds.

By the lemma, Proposition 9.6 holds if $S/Q \cong Z_2$. We can therefore assume henceforth that $|S/Q| \geq 4$.

Lemma 9.12. The following conditions hold:

(i) $C_S(y)$ does not contain a quaternion subgroup of order 8 with center x;

(ii) Either $\tilde{L}_i\langle\tilde{y}\rangle \cong PGL(2,q)$, q odd, $i = 1, 2$, or $\tilde{y}$ interchanges $\tilde{L}_1$ and $\tilde{L}_2$.

Proof: Suppose $C_S(y) \supseteq V \cong Q_8$ with $\langle x\rangle = Z(V)$. There exists g in G such that $y^g = x$ and $C_S(y)^g \subseteq S$. In particular, $V^g \subseteq S$. But S/Q is isomorphic to a subgroup of D_8 by Lemma 6.4 and so V^gQ/Q is a proper homomorphic image of V^g. Thus $x^g \in Q$ and our fusion pattern now forces $x^g = x$, which is a contradiction. Hence (i) holds. Suppose $\tilde{y}$ leaves $\tilde{L}_i$ invariant, $i = 1, 2$. Considering the possible actions of $\bar{y}$ on $\bar{L}_i$ and using the fact that $C_{\bar{Q}}(\bar{y})$ does not contain a quaternion subgroup by (i), we conclude that $\bar{y}$ does not centralize $\bar{L}_i$, that $\bar{y}$ does not induce a field automorphism of $\bar{L}_i$ of order 2 when $\bar{L}_i \cong SL(2,q)$ and that $\bar{y}$ does not induce an outer automorphism of order 2 of $\bar{L}_i$ when $\bar{L}_i \cong \hat{A}_7$, $i = 1$ or 2.

Hence either (ii) holds or $\bar{y}$ induces a nontrivial inner automorphism of $\bar{L}_i$, $i = 1$ or 2. Suppose then that the latter case holds for, say, $i = 1$. If $\bar{L}_1\langle\bar{y}\rangle \cong SL(2,q) \times Z_2$, $\bar{y}$ would necessarily centralize $\bar{L}_1$, which is not the case, so $\bar{L}_1\langle\bar{y}\rangle \cong SL(2,q) * Z_4$. Thus $\bar{L}_1\langle\bar{y}\rangle = \bar{L}_1 * \langle\bar{v}\rangle$ for some element v of S of order 4 with $v^2 = x$. Furthermore, $y = t_1 v$ for some element t_1 of order 4 in Q_1. Replacing y by a conjugate in M, if necessary, we can assume without loss that $\langle t_1 \rangle \lhd N_S(Q_1)$. In particular, $C_{Q_1}(y)$ contains a cyclic subgroup V_1 of order 2^{n-1}, where $|Q_1| = 2^n$. But S/Q is isomorphic to a subgroup of D_8 and $\langle x \rangle = \Omega_1(V_1)$. Since $V_1^g \subseteq C_S(y)^g \subseteq S$, it follows therefore that either $n = 3$ or $x^g \in Q$. However, in the latter case, our fusion pattern again forces $x^g = x$, a contradiction.

Hence we are reduced to the case that $n = 3$, $Q_1 \cong Q_8$, and $V_1^g \cap Q = 1$. Since $r(G) \leq 4$ and $\bar{L}_2\langle \bar{v} \rangle$ centralizes $\bar{L}_1$, we have $r(\bar{L}_2\langle \bar{v} \rangle) = 2$. But $\bar{v} \notin \bar{L}_2$ as $y \notin Q$. Since $\bar{v}^2 = \bar{x}$, we conclude now from Proposition 6.11 that $\bar{L}_2\langle \bar{v} \rangle \cong SL^{(2)}(2, q_2)$, whence $Q_2\langle v \rangle$ is generalized quaternion. But then v lies in a quaternion subgroup W_2 of $Q_2\langle v \rangle$ and we have $Q_1 W_2 = Q_8 * Q_8$ with $y \in Q_1 W_2$. This in turn implies that $C_{Q_1 W_2}(y) = V \times \langle y \rangle$, where $V \cong D_8$ and $\langle x \rangle = Z(V)$. But now $V^g \subseteq S$ and if $V^g \cap Q \neq 1$, then $x^g \in Q$ as $V^g \cap Q \lhd V^g$. However, this leads to a contradiction exactly as before. Hence we must also have $V^g \cap G = 1$, whence $S/Q \cong D_8$. On the other hand, since we are in case (B), some element of $\bar{S}$ interchanges $\bar{L}_1$ and $\bar{L}_2$ under conjugation. Since $Q_1 \cong Q_8$, this forces also $Q_2 \cong Q_8$. Thus $S = QV^g$ is isomorphic to a split extension of $Q_8 * Q_8$ by D_8.

Observe next that $\tilde{Q} \cong E_{16}$ and V^g acts faithfully on $\tilde{Q}$ as $r(G) \leq 4$. Since $N_{\tilde{L}}(\tilde{Q}) = \tilde{Q}\tilde{D}$, where $\tilde{D} \cong Z_3 \times Z_3$ and $C_{\tilde{D}}(\tilde{Q}) = 1$, it follows now that $\tilde{D}\tilde{S}/\tilde{Q}$ is isomorphic to a Sylow 3-normalizer in $A_8 \cong L_4(2)$. Lemma 2.2(vii) of Part II now yields that $\tilde{S}$ is of type A_{10}, whence $\tilde{S} \cong D_8 \int Z_2$. It follows therefore that S contains subgroups $P_i \supset Q_i$ such that $\tilde{P}_i \cong D_8$, $i = 1, 2$ with $\tilde{P}_1\tilde{P}_2 = \tilde{P}_1 \times \tilde{P}_2$ and $\tilde{S} = \tilde{P}_1\tilde{P}_2\langle\tilde{t}\rangle$ for some element t of S with $\tilde{t}$ interchanging $\tilde{P}_1$ and $\tilde{P}_2$. Furthermore, it is immediate from the structure of a Sylow 3-normalizer in A_8 that we can choose t to be an involution in V^g. Thus $S = P_1P_2\langle t\rangle$ with t an involution which interchanges P_1, P_2 and with $P_1 \cong P_2 \cong Q_{16}$ or QD_{16}.

We argue next that $P_i \cong Q_{16}$, $i = 1, 2$. Indeed, if we set $\tilde{D}_1 = \tilde{D} \cap \tilde{L}_1$, we have that $\tilde{D}_1 \cong Z_3$ with $\tilde{D}_1$ acting nontrivially on $\tilde{Q}_1$ and trivially on $\tilde{Q}_2$. By the structure of $\tilde{D}\tilde{S}$, we see that $\tilde{P}_2$ is precisely equal to $C_{\tilde{S}}(\tilde{D}_1)$. But $\tilde{v}$ centralizes $\tilde{D}_1$ as $\tilde{v}$ centralizes $\tilde{L}_1$. Hence $\tilde{P}_2 = \tilde{Q}_2\langle\tilde{v}\rangle$ and consequently $P_2 = Q_2\langle v\rangle$. However, we have already shown that the latter group is generalized quaternion. Thus $P_1 \cong P_2 \cong Q_{16}$, as asserted.

Finally we have that $y = t_1v$ with $\langle t_1\rangle \triangleleft P_1 \subseteq N_S(Q_1)$ and $v \in P_2$. Hence $\langle\tilde{t}_1\rangle = Z(\tilde{P}_1)$ and as $\tilde{P}_2$ centralizes $\tilde{P}_1$, it follows that $\tilde{y}$ centralizes $\tilde{P}_1$. Thus y centralizes $P_1/\langle x\rangle$ and this implies that $V_1^* = C_{P_1}(y)$ is a maximal subgroup of P_1. Since $P_1 \cong Q_{16}$, we conclude that $V_1^* \cong Q_8$ or Z_8. But now $(V_1^*)^g \subseteq S$ and as $S/Q \cong D_8$, this forces $x^g \in Q$, giving the same contradiction as in the preceding cases. This completes the proof of (ii).

<u>Lemma 9.13</u>. If $|S/Q| = 4$, then $S/Q \cong Z_2 \times Z_2$.

<u>Proof</u>: Suppose instead that $S/Q \cong Z_4$, in which case yQ is a square in S/Q. But then $\tilde{y}$ must leave $\tilde{L}_i$ invariant and so $\tilde{L}_i\langle\tilde{y}\rangle \cong PGL(2,q)$, q odd, $i = 1, 2$, by the preceding lemma. As in Lemma 9.8, we obtain now that $C_Q(y) \cong Z_2 \times Z_2$ and that all involutions

of $Q\langle y\rangle - Q$ are conjugate to y. Moreover, we also see that $\Omega_1(C_Q(y)) \cong E_8$. Hence we can repeat the argument of Lemma 9.9 with $\Omega_1(C_S(y))$ in place of $C_S(y)$ to reach the same contradiction.

Next we prove

Lemma 9.14. If $|S/Q| = 4$ and $\tilde{y}$ leaves $\tilde{L}_i$ invariant, $i = 1, 2$, then S is of type $L_4(q)$ for some $q \equiv 7 \pmod 8$.

Proof: By Lemmas 9.12(ii) and 9.13, $S/Q \cong Z_2 \times Z_2$ and $\tilde{L}_1\langle\tilde{y}\rangle \cong PGL(2,q)$, $i = 1, 2$. Again the argument of Lemma 9.8 yields that every involution of $Q\langle y\rangle - Q$ is conjugate to y and as Q and $Q\langle y\rangle$ are normal in S, this implies that $S = QC_S(y)$. Thus $S = Q\langle y,v\rangle$ for some v which centralizes y and we must have that $\bar{v}$ interchanges $\bar{L}_1$ and $\bar{L}_2$, whence v interchanges Q_1 and Q_2. Since $SCN_3(S)$ is empty, we have that $n \geq 3$. Hence if v is an involution, S will be of type $L_4(q)$ for some $q \equiv 7 \pmod 8$, as asserted. Therefore, we assume that v cannot be taken to be an involution.

Again as in Lemma 9.8, $C_Q(y) = U = \langle u,x\rangle$ and $x \notin \Phi(C_S(y))$. Thus $C_S(y) = U\langle v\rangle \times \langle y\rangle$ with $v^2 = u$ or ux. Clearly then $U\langle v\rangle \cong Z_4 \times Z_2$ and $C_S(y) \cong Z_4 \times Z_2 \times Z_2$. If $\langle u_1\rangle$ denotes the unique cyclic normal subgroup of Q_1 of order 4 invariant under $N_S(Q)$ and if we set $u_2 = u_1^v$, then $\langle u_2\rangle$ is the unique cyclic normal subgroup of Q_2 of order 4 invariant under $N_S(Q)$. But then $\Omega_1(\langle u_1,u_2\rangle) = \langle x,u_1u_2\rangle$ is a normal four subgroup of S, so $U = \langle x,u_1u_2\rangle$. Hence without loss we can assume that $u = u_1u_2$. If $v^2 = u = u_1u_2$, we set $w = vu_1$; while if $v^2 = ux = u_1^{-1}u_2$, we set $w = yvu_1$. In both cases, w is an involution and $\langle y,w\rangle$ is a dihedral group of order 8 with center $\langle x\rangle$. Since w

also interchanges Q_1 and Q_2, we conclude that the structure of S is uniquely determined.

Since $Q_1^{wy} = Q_2$ and $(wy)^2 = x$, we check directly that $Q\langle wy\rangle - Q$ contains no involutions. Hence every involution of $S - Q$ is contained in $Q\langle w\rangle$ or $Q\langle y\rangle$. Moreover, as $Q_1^w = Q_2$, all involutions of $Q\langle w\rangle - Q$ are conjugate in $Q\langle w\rangle$ to w or wx. But $w^y = wx$ and so all involutions of $Q\langle w\rangle - Q$ are conjugate in S to w. On the other hand, we know already that every involution of $Q\langle y\rangle - Q$ is conjugate in S to y. Thus every involution of $S - Q$ is conjugate in S to w or y.

We claim next that wy is not conjugate in G to an element of $Q\langle y\rangle$. Indeed, if $(wy)^g = t \in Q\langle y\rangle$ for some g in G, then as $x = (wy)^2$, it would follow that $x^g = t^2 \in Q$, whence $t^2 = x$. But then $g \in M$ and so $wy \sim t$ in M, which is clearly not the case. This proves our assertion. But now [27, Lemma 16] yields $x = (wy)^2$ must be fused to an involution of $S - Q\langle y\rangle$. It follows therefore from the preceding paragraph that $x \sim w$ in G. But $x \sim y$ in G by our choice of y and we conclude now that every involution of $S - Q$ is conjugate in G to x. As a consequence, u is not conjugate in G to any involution of $S - Q$.

Finally we choose g in G so that $w^g = x$ and $C_S(w)^g \subseteq S$. We have that $Q = Q_1 * Q_2$ with $Q_1 \cong Q_{2^n}$ for some $n \geq 3$ and so $C_Q(w) = \langle x\rangle \times D$, where $D \cong D_{2^{n-1}}$. But then $C_Q(w)$ is generated by the set of involutions of $C_Q(w) - \langle x\rangle$. Since each such involution is conjugate to u, it follows from the preceding paragraph that $a^g \in Q$ for every involution a of $C_Q(w) - \langle x\rangle$ and we conclude that

$C_Q(w)^g \subseteq Q$. Thus $x^g \in Q$ and so $x^g = x$, which is a contradiction. This completes the proof of the lemma.

Lemma 9.15. If $|S/Q| = 4$ and $\tilde{y}$ interchanges $\tilde{L}_1$ and $\tilde{L}_2$, then S is of type $L_4(q)$, $q \equiv 7 \pmod 8$.

Proof: Let u_1 be as in the preceding lemma, but this time set $u_2 = u_1^y$. Again we have that $U = \langle x, u_1u_2\rangle$ and we suppose without loss that $u = u_1u_2$. If $Q_1 \cong Q_{2^n}$, we also have that $\tilde{Q}\langle\tilde{y}\rangle \cong D_{2^{n-1}} \int Z_2$ and consequently every involution of $\tilde{Q}\langle\tilde{y}\rangle - \tilde{Q}$ is conjugate in $\tilde{S}$ to $\tilde{y}$. Since $\tilde{Q}$ and $\tilde{Q}\langle\tilde{y}\rangle$ are normal in $\tilde{S}$, it follows that $\tilde{S} = \tilde{Q}C_{\tilde{S}}(\tilde{y})$. In particular, $C_{\tilde{S}}(\tilde{y}) = \langle\tilde{y}\rangle \times \tilde{V}$, where $\tilde{V}$ leaves $\tilde{L}_1$ and $\tilde{L}_2$ invariant and $|\tilde{V} : C_{\tilde{Q}}(\tilde{y})| = 2$. Since $\tilde{y}$ interchanges $\tilde{L}_1$ and $\tilde{L}_2$, $\tilde{L}_0 = C_{\tilde{L}_1\tilde{L}_2}(\tilde{y}) \cong \tilde{L}_1$. Moreover, $C_{\tilde{Q}}(\tilde{y})$ is a Sylow 2-subgroup of $\tilde{L}_0$, $\tilde{L}_0$ is $\tilde{V}$-invariant, and every element of $\tilde{V} - C_{\tilde{Q}}(\tilde{y})$ induces the same type of automorphism of both $\tilde{L}_1$ and $\tilde{L}_2$. Lemma 6.4(iii) and (iv) imply now that $\tilde{L}_0 \cong \tilde{L}_1 \cong L_2(q)$ and that no element of $\tilde{V} - C_{\tilde{Q}}(\tilde{y})$ induces a field or an inner automorphism of $\tilde{L}_0$. Hence $\tilde{L}_0\tilde{V} \cong PGL(2,q)$ or $PGL^*(2,q)$ and correspondingly $\tilde{V} \cong D_{2^n}$ or QD_{2^n}.

Consider the latter case first. In particular, $n \geq 4$. Furthermore, $Z(C_{\tilde{Q}}(\tilde{y})) = \langle\tilde{u}\rangle = \langle\tilde{u}_1\tilde{u}_2\rangle$ and $\tilde{V} - C_{\tilde{Q}}(\tilde{y})$ contains an element $\tilde{v}$ such that $\tilde{v}^2 = \tilde{u}$. We let v be a representative of $\tilde{v}$ in S. Then v normalizes $\langle x,y\rangle$ and so v^2 centralizes y. But also $v^4 \in \langle x\rangle$ as $\tilde{v}^4 = 1$. If $v^4 = x$, then $x \in \Phi(C_S(y))$ and this leads to a contradiction, as usual. Hence $v^4 = 1$ and so $v^2 = u$ or ux.

If every involution of $S - Q$ is conjugate to x in G, then Q contains every conjugate of u in G which lies in S. This leads to a contradiction exactly as in the final paragraph of the preceding lemma. Hence $S - Q$ must contain a conjugate t of u in G. Suppose every involution of $S - Q$ lies in $Q\langle y\rangle - Q$. Since the involution y interchanges Q_1 and Q_2, every involution of $Q\langle y\rangle - Q$ is conjugate to y or yx. In particular, $t \sim y$ or yx in G. But $y \sim x$ and $t \sim u$ in G, so we must have $t \sim yx$ and $y \not\sim yx$ in G. Since v normalizes $\langle y,x\rangle$ and centralizes x, this forces v to centralize y. On the other hand, $v^2 = u$ or ux and $u = u_1u_2 \in \Phi(C_S(y))$, as $n \geq 4$. This forces $v^2 = u$, since otherwise $x \in \Phi(C_S(y))$, which we have seen is not the case. Since $\tilde{Q}_i\langle\tilde{v}\rangle \cong QD_{2^n}$, $\tilde{v}$ inverts the unique cyclic subgroup of index 2 in $\tilde{Q}_i$, $i = 1,2$, and hence v inverts the corresponding subgroup U_i of Q_i, $i = 1,2$. Since $n \geq 4$ and $\langle u_i\rangle \lhd Q_i$, $u_i \in U_i$ and so v inverts u_i, $i = 1,2$. We now compute that $(yvu_1)^2 = (yvyv)u_2^{-1}u_1 = (y^2v^2)(xu) = uxu = x$.

By [27, Lemma 16], $yvu_1 \sim w$ in G for some w in $Q\langle y\rangle$. But then $x = (yvu_1)^2 \sim w^2$ in G and $w^2 \in Q$, so, as usual, $w^2 = x$ and $yvu_1 \sim w$ in M, which is clearly impossible. We thus conclude that $S - Q\langle y\rangle$ must contain an involution.

We continue the study of the distribution and conjugacy of involutions of $S - Q$. Set $\bar{S} = S/U$. Then $\bar{Q} = \bar{Q}_1 * \bar{Q}_2 \cong Q_{2^{n-1}} * Q_{2^{n-1}}$, $\bar{Q}_1^{\bar{y}} = \bar{Q}_2$, and $\bar{Q}_i\langle\bar{v}\rangle \cong D_{2^n}$, $i = 1,2$, with $\langle\bar{y},\bar{v}\rangle \cong Z_2 \times Z_2$. Hence $\bar{S}$ is of type $L_4(q)$ for some odd $q \equiv 7 \pmod 8$. It follows therefore that every involution of $\bar{S} - \bar{Q}$ is conjugate in $\bar{S}$ to $\bar{y}$, $\overline{yu_1}$, $\bar{v}$, $\overline{yv}$, or $\overline{yvu_1}$. (Note that $\langle\bar{u}_1\rangle = Z(\bar{S})$.) Hence if we set $V = \langle y,v,u_1,u_2\rangle$ and use the fact that $V \supset \langle u_1,u_2\rangle \supset U$, we conclude that every involution of $S - Q$ is conjugate in S to an involution of V. As in the preceding lemma, v

inverts the unique cyclic subgroup U_i of index 2 in Q_i and $u_i \in U_i$, so v inverts u_i, $i = 1,2$. Since $v^2 = u$ or ux, we see that $\langle v,u_1,u_2\rangle/\langle x\rangle \cong Z_4 \times Z_2$ and consequently $\langle v,u_1,u_2\rangle - \langle u_1,u_2\rangle$ contains no involutions. But as $S - Q\langle y\rangle$ contains an involution, $V - \langle y,u_1,u_2\rangle$ must contain an involution and this yields that $\langle yv,u_1,u_2\rangle - \langle u_1,u_2\rangle$ contains an involution.

We know that v normalizes $\langle x,y\rangle$ and so $v^{-1}yv = y$ or yx, whence $y^{-1}vy = v$ or vx. Since $v^2 = u$ or ux, $v^{-1} \neq v$ or vx and consequently y does not invert v. Thus yv is not an involution. Since $U \subseteq Z(V)$, it follows that yvx, yvu are also not involutions and that yvu_1, yvu_2 are both involutions. Thus

$$1 = (yvu_1)(yvu_1) = yvyvu_2^{-1}u_1 = (yv)^2xu ,$$

whence $(yv)^2 = ux$ and so

$$v^{-1}yv = v^2yux .$$

If v centralizes y, it follows that $v^2 = ux$. But then $ux \in \Phi(C_S(y))$. Since also $u \in \Phi(C_S(y))$ (as $\bar{u} \in C_{\bar{L}}(\bar{y})'$), we see that $x \in \Phi(C_S(y))$, giving the usual contradiction. Hence we must have $y^{-1}vy = vx$, which in turn forces $v^2 = u$.

Observe that as y interchanges Q_1 and Q_2, no element of $Q\langle y\rangle$ conjugates y into yx. Since $y^v = yx$ and $S = Q\langle y,v\rangle$ and $S = Q\langle y,v\rangle$, we see that no element of $S - Q\langle y\rangle$ centralizes y. Thus we also conclude that $C_S(y) \subseteq Q\langle y\rangle$.

We consider this remaining subcase. Since yvu_1 is an involution interchanging Q_1 and Q_2, it follows, as usual, that every involution of

$Q\langle yvu_1\rangle - Q$ is conjugate to yvu_1 or yvu_1x in $Q\langle yvu_1\rangle$. Note also that $S = Q\langle yvu_1, v\rangle$ and $(yvu_1)^v = yxvu_1^{-1} = yvu_1$, so yvu_1 and yvu_1x are not conjugate in S. On the other hand, we know already that $y \sim yx$ in S as $y^v = yx$.

Our argument yields finally that $S - Q$ has three conjugacy classes of involutions, represented by y, yvu_1, and yvu_1x. Since $y \sim x \not\sim u \sim t$ in G and $t \in S - Q$, we conclude that $u \sim yvu_1$ or yvu_1x. The argument being the same in either case, we can suppose without loss that $u \sim yvu_1$. If $x \sim yvu_1x$, then put $y' = yvu_1x$ and reason on y' exactly as we have done above on y. We shall then reach the conclusion that $C_S(y') \subseteq Q\langle y'\rangle$. However, as noted in the preceding paragraph, v centralizes yvu_1 and so centralizes y'. Thus $v \in C_S(y')$ and so $C_S(y') \not\subseteq Q\langle y'\rangle$. But by Thompson's tranfer lemma applied to $y' = yvu_1x$ and $Q\langle y\rangle$, we conclude that $yvu_1x \sim u$. In particular, the only conjugates of x which lie in S are contained in $Q\langle y\rangle$.

Finally we have $y^g = x$ and $C_S(y)^g \subseteq S$ for some g in G. Moreover, $C_S(y) = \langle y\rangle \times \langle x\rangle \times D$, where $D \subseteq Q$ and $D \cong D_{2^{n-1}}$. By the preceding paragraph $x^g \in Q\langle y\rangle$ and, as $x^g \neq x$, we have $x^g \in Q\langle y\rangle - Q$. But all involutions of $Q\langle y\rangle - Q$ are conjugate in S, so all are conjugate in G to x. This implies that no involution of $Q - \langle x\rangle$ is conjugate to any involution of $Q\langle y\rangle - Q$. But $D\langle x\rangle - \langle x\rangle$ is generated by involutions of $Q - \langle x\rangle$ and $(D\langle x\rangle)^g \subseteq S$. Since $x^g \notin Q$, it follows that for some involution d of $D\langle x\rangle - \langle x\rangle$, $d^g \notin Q$, whence $d^g \notin Q\langle y\rangle$. Thus $d^g \in S - Q\langle y\rangle$. Since $x^g \in Q\langle y\rangle - Q$, we conclude that $S = Q\langle x^g, d^g\rangle$. Since $\langle x^g, d^g\rangle = \langle x, d\rangle^g \cong Z_2 \times Z_2$, we see that S

splits over Q. Hence S must be of type $L_4(q)$ for some $q \equiv 7 \pmod 8$. Thus the lemma holds when $\tilde{L}_0\tilde{V} \cong PGL^*(2,q)$.

It therefore remains to treat the case in which $\tilde{L}_0\tilde{V} \cong PGL(2,q)$. Then $\tilde{V} - C_{\tilde{Q}}(\tilde{y})$ contains an involution $\tilde{v}$ and as $\tilde{v}$ induces the same type of automorphism on $\tilde{L}_i$ as it does on $\tilde{L}_0$, it follows that $\tilde{L}_i\langle\tilde{v}\rangle \cong PGL(2,q)$, $i = 1,2$. Since $SCN_3(S)$ is empty, we must also have $n \geq 3$.

If $C_S(y)$ contains an involution v which maps on $\tilde{v}$, then S splits over Q and again S is of type $L_4(q)$ for some $q \equiv 7 \pmod 8$ and the lemma holds. We can therefore assume that this is not the case. But then if V denotes the inverse image of $\langle\tilde{y},\tilde{v}\rangle$ in S, we have that $V \cong Z_4 \times Z_2$ or D_8 with $\langle x\rangle \in \Phi(V)$ in either case. However, if $V \cong Z_4 \times Z_2$, then $V \subseteq C_S(y)$ and so $x \in \Phi(C_S(y))$, giving the usual contradiction. Thus, in fact, $V \cong D_8$.

Let v be a representative of $\tilde{v}$ in V, so that either $v^2 = 1$ or x. In the first case, Thompson's transfer lemma yields that v is conjugate to an involution of $Q\langle y\rangle$. But as $y^v = yx$, every involution of $Q\langle y\rangle$ is conjugate in G to x or u. If $v \sim x$ in G, then v satisfies the same conditions as y did in Lemma 9.14 and so present lemma holds by that lemma. Hence we can assume that $v \sim u$ in G.

Applying [27, Lemma 16] we also have that either yv is conjugate to an element of $Q\langle y\rangle$ or $x = (yv)^2$ is conjugate to an involution of $S - Q\langle y\rangle$. But we compute from the structure of S that every involution of $S - Q\langle y\rangle$ is conjugate v. Since $v \sim u$, we see that the latter case cannot occur and so $(yv)^g = w$ for some g in G and

w in $Q\langle y\rangle$. Thus $x^g = w^2$, forcing $w^2 = x$ and $g \in M$. But clearly yv is not conjugate to w in M. We therefore conclude that $v^2 = x$.

Applying [27, Lemma 16] now to v and $Q\langle y\rangle$ and reasoning as in the preceding paragraph, it follows that $x \sim yv$ in G (using the fact that every involution of $S - Q\langle y\rangle$ is conjugate to yv in S in the present case). But every involution of $S - Q$ in conjugate to y or vy as $Q\langle v\rangle - Q$ contains no involutions and $y^v = yx$. We thus conclude that every involution of $S - Q$ is conjugate to x in G.

Again $y^g = x$ with $C_S(y)^g \subseteq S$ for some g in G. But $C_Q(y) \cong Z_2 \times D_{2^{n-1}}$ and $C_Q(y)$ is generated by involutions conjugate to u in G. Since $u \not\sim x$ in G, the preceding paragraph yields $C_Q(y)^g \subseteq Q$, whence $x^g \in Q$. Hence $x^g = x$. This contradiction proves the lemma.

By the preceding analysis, Proposition 9.6 holds if $|S/Q| \leq 4$. Hence it remains to treat the case in which $S/Q \cong D_8$. Let V_1 be the maximal subgroup of S containing Q such that $\tilde{V}_1$ leaves $\tilde{L}_i$ invariant, $i = 1,2$. By Lemma 6.4, $V_1/Q \cong Z_2 \times Z_2$. Also let V_2 be the maximal subgroup of S containing Q such that $V_2 \neq V_1$ and $V_2/Q \cong Z_2 \times Z_2$. Finally set $V_0 = V_1 \cap V_2$, so that $V_0/Q = Z(S/Q)$. To eliminate this case, we need the following property of V_1.

Lemma 9.16. $V_1 - V_0$ does not contain an element of order 4 whose square is x.

Proof: Observe first that $\tilde{L}_1\tilde{V}_0/\tilde{Q}_2 \cong \tilde{L}_2\tilde{V}_0/\tilde{Q}_1 \cong PGL(2,q)$ or $PGL^*(2,q)$, by Lemma 6.4(iv). This implies that if v_1 is an involution of $V_1 - V_0$, $\tilde{L}_i\langle\tilde{v}_i\rangle$ cannot be isomorphic to $PGL(2,q)$

for both $i = 1$ and 2. Indeed, in that case an element of $\tilde{V}_0\tilde{v}_1 - \tilde{Q}\tilde{v}_1$ would induce a field automorphism on both $\tilde{L}_1$ and $\tilde{L}_2$. But then it would easily follow that $r(G) \geq 5$, a contradiction. In view of Lemma 9.12(ii), our argument thus yields that no involution of $V_1 - V_0$ can be conjugate to x in G and so all conjugates of x in S lie in V_2.

Suppose now that the lemma is false and let $v \in V_1 - V_0$ with $v^2 = x$. By the preceding paragraph and [27, Lemma 16], $v \sim v_2$ in G for some v_2 in V_2, whence $x = v^2 \sim v_2^2$. But $v_2^2 \in Q$ as $V_2/Q \cong Z_2 \times Z_2$. It follows therefore from Lemma 9.7 that $v_2^2 = x$. Thus $v \sim v_2$ in M, which is clearly impossible.

We can now prove

Lemma 9.17. The case $S/Q \cong D_8$ does not occur.

Proof: Suppose false. Clearly $\tilde{Q}_2$ centralizes $\tilde{L}_1$ and, by Lemma 6.4(ii), $|C_{\tilde{V}_1}(\tilde{L}_1) : \tilde{Q}_2| \leq 2$. Suppose equality holds. Since Q_2 is generalized quaternion, it follows then that $V_1 - Q_2$ contains an element v with $v^2 \in \langle x \rangle$ such that $\tilde{v}$ centralizes $\tilde{L}_1$. Moreover, $v \notin V_0$ and so $v \in V_1 - V_0$. Hence by the preceding lemma, we have $v^2 = 1$. Thus $Q_1\langle v \rangle = Q_1 \times \langle v \rangle$. But Q_1 contains an element u_1 of order 4 with $u_1^2 = x$. Since $Q_1 \subseteq V_0$, we see that $u_1 v \in V_1 - V_0$ and that $(u_1 v)^2 = x$, again contradicting the preceding lemma. We therefore conclude that $C_{\tilde{V}_1}(\tilde{L}_1) = \tilde{Q}_2$.

Since $\tilde{V}_1/\tilde{Q} \cong Z_2 \times Z_2$, it follows now that $V_1 - Q$ contains an element v_1 such that $\tilde{v}_1$ induces a field automorphism of $\tilde{L}_1$ of order 2. As above, we can assume without loss that $v_1^2 \in \langle x \rangle$. By Lemma 6.4(iv),

$v_1 \notin V_0$, so $v_1^2 = 1$ by the preceding lemma. But $C_{Q_1}(v_1)$ is generalized quaternion and so contains an element u_1 of order 4 with $u_1^2 = x$. Again $u_1v_1 \in V_1 - V_0$ and $(u_1v_1)^2 = x$, so the preceding lemma yields a contradiction in this case as well.

This establishes the lemma and completes the proof of Proposition 9.6. Together Propositions 9.1 and 9.6 show that S is of type M_{12}, A_8, A_{10}, $\hat{A}_8$, or $L_4(q)$, $q \equiv 7 \pmod 8$, or $PSp(4,q)$, $G_2(q)$, $q \equiv 1,7 \pmod 8$. But now applying [17], [18], [19], [21], [36], [37], and keeping in mind the structure of M, we conclude that G is isomorphic to one of the groups listed in Theorem B. Combining this with Propositions 3.1, 4.1, 5.4, 5.7, and 8.1, it follows that Theorem B holds for each of the six possibilities for L_x listed in Section 3. Thus Theorem B is proved.

PART IV

A CHARACTERIZATION OF THE GROUP $D_4^2(3)$

1. Introduction. We continue the analysis of finite simple groups G of sectional 2-rank at most 4 begun in Parts I, II, and III. Again for any involution x of G, we set $L_x = L(C_G(x)/O(C_G(x))$, the latter group denoting by definition the unique largest normal semisimple subgroup of $C_G(x)/O(C_G(x))$ - equivalently the ultimate term of the derived series of $C_G(x)/O(C_G(x))$, as $r(G) \leq 4$. We also designate by $\mathcal{L}(G)$ the set of quasisimple components of L_x as x ranges over the involutions of G.

Consider a minimal counterexample to our Main Theorem, stated in the Introduction of the paper. Then, first of all, the nonsolvable composition factors of every proper subgroup of G are isomorphic to groups listed in the Main Theorem. Furthermore, the principal results of Parts I, II, and III imply that $\mathcal{L}(G)$ is nonempty and that for any involution x of G, L_x is not of sectional 2-rank 4 with each of its components nonsimple. Moreover, by Theorem D of Part II the centralizer of no involution of G is nonsolvable and 2-constrained. (Theorem C of Part II will not be needed until Part VI). Finally we note that by the principal classification theorems [5], [43], and [3], G must have 2-rank at least 3 and nonabelian Sylow 2-groups.

It will be convenient to incorporate these conditions into a single definition. Thus we shall call a group G an $\mathfrak{R}_1$-group provided it satisfies the following conditions:

(a) G has 2-rank at least 3, sectional 2-rank at most 4, and nonabelian Sylow 2-subgroups;

(b) The nonsolvable composition factors of the proper subgroups of G satisfy the conclusion of the Main Theorem;

(c) $\mathcal{L}(G)$ is nonempty;

(d) The centralizer of every involution of G is either solvable or non 2-constrained;

(e) For no involution x of G is L_x of sectional 2-rank 4 with each of its components nonsimple.

By the preceding discussion, we see that our Main Theorem will be completely proved provided we show that it holds for simple $\mathfrak{R}_1$-groups. This we shall do in Parts IV, V, and VI.

To state the principal result of this part, we require a further definition. We shall denote by $\mathcal{L}_c(G)$ the set of quasisimple components of L_x as x ranges over the central involutions of G. Clearly $\mathcal{L}_c(G) \subseteq \mathcal{L}(G)$.

Our goal in the next two parts will be to reduce to the case in which $\mathcal{L}_c(G)$ is empty. By condition (d) above this will correspond to the case in which the centralizer of every central involution of G is solvable. We shall here prove

Theorem A. If G is a simple $\mathfrak{R}_1$-group in which $\mathcal{L}_c(G)$ is nonempty and each element of $\mathcal{L}_c(G)$ is nonsimple, then $G \cong D_4^2(3)$; the triality twisted $D_4(3^3)$.

We shall establish Theorem A by showing that G must have Sylow 2-subgroups of type M_{12}. Once this is accomplished the main results of [6], [19] will imply that $G \cong M_{12}$, $G_2(q)$, or $D_4^2(q)$, $q \equiv 3,5 \pmod 8$. However, M_{12} and $G_2(3)$ are excluded as $\mathcal{L}_c(G)$ would then be empty. On the other hand, if $G \cong G_2(q)$ or $D_4^2(q)$ with $q > 3$, then for any involution x of G, $L_x \cong SL(2,q) * SL(2,q)$ or $SL(2,q) * SL(2,q^3)$ and so L_x has sectional 2-rank 4, contrary to the fact that G is an $\mathfrak{R}_1$-group. Thus the only possibility will be that $G \cong D_4^2(3)$. In this case, G is, in fact, an $\mathfrak{R}_1$-group in which every element of $\mathcal{L}_c(G)(= \mathcal{L}(G))$ is isomorphic to $SL(2,27)$.

2. Preliminary lemmas. We shall need a number of preliminary results for the proof of Theorem A. Many of these will also be needed in Parts V and VI.

In Lemmas 6.1 and 6.2 of Part III, we established a few properties of the groups $SL(2,q)$, $L_2(q)$, q odd, and $\hat{A}_7$. We first enlarge on this list.

First consider a group H with normal subgroup $L \cong SL(2,q)$, q odd, or $\hat{A}_7$ such that $R_1 = C_H(L)$ is a 2-group. Let S be a Sylow 2-subgroup of H and set $R_2 = S \cap L$. Clearly $R_1 \subseteq S$.

Lemma 2.1. The following conditions hold:

(i) S/R_1R_2 is abelian;

(ii) If $R_1 = Z(L)$, then $m(S) \leq 2$;

(iii) R_2 contains a cyclic subgroup T of index 2 with $T \triangleleft S$;

(iv) For any such subgroup T of R_2, $|C_S(T): TR_1| \leq 2$ and $C_S(T)/R_1$ is cyclic;

(v) If $|C_S(T):TR_1| = 2$ in (iv), then $S/C_S(T)R_2$ is cyclic.

Proof. Set $\overline{H} = H/R_1$, so that $\overline{H}$ is isomorphic to a subgroup of $P\Gamma L(2,q)$ or S_7. It is immediate from the structure of these groups that

S/R_1R_2 is abelian, so (i) holds. Next assume $R_1 = Z(L)$ and suppose (ii) is false. Let $A \subseteq S$ with $A \cong E_8$. Since R_2 is generalized quaternion, $|A \cap L| \leq 2$ and consequently $\bar{A}$ contains a four group disjoint from $\bar{L}$, contrary to Lemma 6.2 (iv) of Part III. Thus (ii) also holds. Since $R_2 \triangleleft S$ and R_2 is generalized quaternion, (iii) is also immediate.

Let T be a cyclic subgroup of index 2 in R_2 with $T \triangleleft S$. Suppose first that $L \cong SL(2,q)$. Clearly if (iv) and (v) hold in the special case that $\bar{H} \cong P\Gamma L(2,q)$, they will hold for arbitrary H. Hence without loss we can assume $\bar{H} \cong P\Gamma L(2,q)$, in which case $\bar{H}$ contains a normal subgroup $\bar{K} \cong PGL(2,q)$ with $\bar{K} \supseteq \bar{L}$ such that $\bar{S} = (\bar{S} \cap \bar{K})\bar{W}$, where $\bar{W} \cap \bar{K} = 1$, $\bar{W}$ is cyclic, and the elements of $\bar{W}$ are induced from field automorphisms of GF(q). Let R,W be the inverse images of $\bar{S} \cap \bar{K}$ and $\bar{W}$, respectively, in S, so that $R_2 \subseteq R$, $R_1 \subseteq W$, $|R:R_1R_2| \leq 2$, $S = RW$, and T lies in a cyclic subgroup V of index at most index 2 in R. Furthermore, every element of R-V inverts T, so $C_R(T) = R_1V$. On the other hand, we know the action of W on T and we check that no element of W inverts T or centralizes T. We conclude at once that $C_S(T) = VR_1$. Moreover, if $|V:T| = 2$, then $VR_1R_2 = R$ and consequently $S/VR_1R_2 \cong W$ is cyclic. Thus (iv) and (v) hold in this case.

If $L \cong \hat{A}_7$, then $|H : LR_1| \leq 2$ and we conclude readily from Lemma 6.1 (ii) of Part III that $C_S(T) = TR_1$, so (iv) and (v) hold in this case as well.

We next consider a group H with normal subgroup $L \cong L_2(q)$, q odd, or A_7 and with $C_H(L) = 1$. Then H is isomorphic to a subgroup of Aut(L). In the first case, it follows from the structure of $P\Gamma L(2,q)$ that H possesses a cyclic 2-subgroup W disjoint from L such that $|H:LW| = d$ or 2d, where d is odd, with the action of each element of W induced from an element of the Galois group of GF(q). Let $W = \langle w \rangle$.

Since $C_L(w) \subseteq C_L(w^2) \subseteq C_L(w^4) \subseteq \dots$ with each centralizer W-invariant, L possesses a W-invariant Sylow 2-subgroup R such that $C_R(w^i)$ is a Sylow 2-subgroup of $C_L(w^i)$ for all i. We let S be a Sylow 2-subgroup of H containing RW. On the other hand, if $L \cong A_7$, we conclude from the structure of S_7 that L has a complement W in H with $|W| \leq 2$. Then again L possesses a W-invariant Sylow 2-subgroup R with $C_R(W)$ a Sylow 2-subgroup of $C_L(W)$; and in this case, we set $S = RW$.

With this notation, we have the following omnibus lemma.

Lemma 2.2. The following conditions hold:

(i) $W \cap L = 1$;

(ii) $|S:RW| \leq 2$ with equality holding only if $L \cong L_2(q)$;

(iii) $C_L(W)$ contains a subgroup isomorphic to $L_2(3)$;

(iv) $\Omega_1(W)$ centralizes R;

(v) S' is cyclic;

(vi) Let $R \cong D_{2^n}$. If T is a cyclic subgroup of S, then $|T| \leq 2^n$. Moreover, if equality holds, then we have

(a) H contains a subgroup isomorphic to PGL(2,q);

(b) $C_S(T) = T$;

(c) If $T \triangleleft S$, then $|T \cap L| \cong 2^{n-1}$ and S/RT is cyclic;

(vii) If x is an involution of S - RW, then $L\langle x\rangle \cong PGL(2,q)$ and either $W = 1$ or $[x, \Omega_1(W)] = Z(R\langle x\rangle) \cong Z_2$;

(viii) If x is an involution of S - RW, then $C_S(x)$ is a Sylow 2-subgroup of $C_H(x)$ and $\Omega_1(C_S(x)) \subseteq R\langle x\rangle$.

(ix) If U is a four subgroup of S, then $C_H(U)$ has a normal 2-complement. Moreover, if $U \not\subseteq L$, then $N_H(U)$ has a normal 2-complement.

(x) If U is a four subgroup of R, then $N_H(U)/C_H(U) \cong A_3$ or S_3 and $U\Omega_1(W) = \Omega_1(Y)$ for some Sylow 2-subgroup Y of $C_H(U)$.

Proof. First, (i) and (ii) hold by our choice of W. Clearly (iii) and (iv) hold if $W = 1$; so we can suppose $W \neq 1$. If $L \cong L_2(q)$ and $|W| = 2^m$ then it follows from the action of W on L that $C_L(W) \cong PGL(2,q_0)$, where $q_0^{2^m} = q$ and that $C_R(W)$ is a Sylow 2-subgroup of $C_L(W)$. In particular, (iii) holds. Similarly if $W_1 = \Omega_1(W)$, then $C_L(W_1) \cong PGL(2,q_1)$ with $q_1^2 = q$ and $C_R(W_1)$ is a Sylow 2-subgroup of $C_L(W_1)$. But now a comparison of orders shows that W_1 centralizes R and so (iv) also holds. On the other hand, if $L \cong A_7$, $W = W_1 \cong Z_2$ and $C_L(W) \cong S_4$ or S_5. In either case, we obtain (iii) and (iv).

Observe next that if $L \cong A_7$, then $S \cong D_8$ or $D_8 \times Z_2$ and $S = RW$. In this case, it is clear that all assertions of (v), (vi), (vii), and (viii) hold. Hence in proving these parts, we can assume that $L \cong L_2(q)$. Likewise we can assume that $LS \ncong PGL^*(2,q)$. In this case, $S = QW$ with Q dihedral, $Q \lhd S$, $Q \cap W = 1$, and R of index at most 2 in Q . Then Q contains a W-invariant cyclic subgroup Q_o of index 2 in Q and by (iii), $Q = Q_oC_Q(W)$. It is immediate now that $S' \subseteq Q_o$, so S' is cyclic and (v) holds. If x is an involution of $S - RW$, it follows from [Part III, Lemma 6.2 (i)] that $L\langle x\rangle \cong PGL(2,q)$. Hence we can take $Q = R\langle x\rangle$ in this case. Now we check directly, if $W \neq 1$, that $[x, \Omega_1(W)] = Z(Q) \cong Z_2$ and we see that (vii) holds. Furthermore, if $|W| = 2^m$, we check that $C_S(x) = \langle x, Z(Q), rw\rangle$, where $r \in R$, $\langle r\rangle \lhd Q$, $|r| = 2^{m+1}$, and $W = \langle w\rangle$. In particular, using (iv), we see that $\Omega_1(C_S(x)) = \langle x, Z(Q)\rangle$. Likewise it follows easily that $C_S(x)$ is a Sylow 2-subgroup of $C_H(x)$. Thus (viii) also holds.

Now we establish (vi); so assume that $R \cong D_{2^n}$. Again with $|W| = 2^m$, we have that $S/R \cong Z_{2^m}$ or $Z_{2^m} \times Z_2$. Furthermore, $q = q_0^{2^m}$ for suitable prime power q_0 and if a Sylow 2-subgroup of $L_2(q_0)$ has order 2^a, we compute that $n = m+a$. Suppose now that T is a cyclic subgroup of S of order 2^{n+1}. Since S/R has exponent at most 2^m and R has exponent 2^{n-1}, we have $|TR/R| = 2^{m-k}$ for some k with $0 \leq k \leq m-2$. Hence $T_0 = T \cap L$ has order 2^{a+1+k}. In particular, T_0 is cyclic of order at least 8 and so T_0 lies in the unique cyclic subgroup V of index 2 in Q. But now considering the action of W on L and observing the order of T_0, we compute that $T \subseteq C_S(T_0) = V\mho^{k+1}(W)$. However, this is impossible as $|TR/R| = 2^{m-k}$, while $V\mho^{k+1}(W)R/R$ is of exponent 2^{m-k-1}. Hence no such T exists and so every cyclic subgroup of S has order at most 2^n.

Suppose next that T is a cyclic subgroup of S of order 2^n. This time $T_0 = T \cap L$ has order at least 2^a and as $a \geq 2$, again $T_0 \subseteq V$ and so $T \subseteq C_S(T) \subseteq C_S(T_0) \subseteq VW$. Set $V_0 = V \cap R$. We argue that $T \not\subseteq V_0W$. Indeed, $V_0 \cong Z_{2^{n-1}}$ as $R \cong D_{2^n}$ and so V_0W is a split extension of $V_0 \cong Z_{2^{n-1}}$ by $W \cong Z_{2^m}$. But one checks directly that such an extension has exponent at most 2^{n-1} and so cannot contain T. In particular, $V \supset V_0$, so $Q \supset R$ and (a) holds.

Now assume (b) is false. Since T is a maximal cyclic subgroup of S, it follows that $C_S(T) - T$ contains an involution t. We have that $t \in VW$ and so $t \in V\Omega_1(W)$. But $C_V(\Omega_1(W)) = V_0$ and hence $\Omega_1(V\Omega_1(W)) = \Omega_1(V) \times \Omega_1(W)$. Since $\Omega_1(V) \subseteq T_0 \subseteq T$ and $t \notin T$, we can therefore assume without loss that $t \in \Omega_1(W)$. Since $T \subseteq VW$, a generator of T has the form vw for some v in V and w in W. Then t centralizes vw and as t centralizes w, it follows that t centralizes v, whence $v \in V_0$ and so $T \subseteq V_0W$, contrary to what we have shown above. Hence (b) also holds.

Finally assume, in addition, that $T \triangleleft S$. Since $C_R(W)$ contains a four group, there exists an involution u in $Q-V$ which centralizes W. But then if again $T = \langle vw \rangle$, u inverts v and centralizes w, so $v^{-1}w \in T$ as $T \triangleleft S$. Hence also $v^2 \in T$. However, as $T \not\subseteq V_0W$, $v \in V - V_0$ and so $V = \langle v \rangle$, where $\langle v^2 \rangle = V_0$. Thus $T_0 = V_0$ has order 2^{n-1}. Likewise as $T \not\subseteq V_0W$, we see that $S/TR \cong W$ and so is cyclic. Hence (c) also holds and all parts of (vi) are proved.

Now we prove (ix) and (x). It is well-known that $C_H(u)$ has a normal 2-complement for every involution u of Q (when $L \cong L_2(q)$) and for every involution u of R (when $L \cong A_7$). Since correspondingly S/Q or S/R is cyclic, any four subgroup U of S intersects Q or R respectively nontrivially and so $C_H(U)$ has a normal 2-complement. Moreover, by Lemma 6.2 (iv) of Part III, any four subgroup U of S intersects L nontrivially. Hence if $U \not\subseteq L$, the involutions of U cannot be conjugate in H and consequently $N_H(U) = C_H(U)$ has a normal 2-complement. Thus (ix) also holds. On the other hand, if $U \subseteq R$, then $N_L(U)/C_L(U) \cong A_3$ or S_3 and U is a Sylow 2-subgroup of $C_L(U)$ by the structure of L. Moreover, U centralizes $\Omega_1(W)$ and we check directly from the action of S on L that no involution of $H - L\Omega_1(W)$ centralizes U, which implies at once that $U\Omega_1(W) = \Omega_1(Y)$ for any Sylow 2-subgroup Y of $C_H(U)$ containing $U\Omega_1(W)$. Therefore (x) holds and the lemma is proved.

Now we consider a group H with normal subgroup $L \cong L_3(q)$ or $U_3(q)$, q odd, and $C_H(L) = 1$. We again let S be a Sylow 2-subgroup of H and set $R = S \cap L$. Thus R is either quasi-dihedral or wreathed. Moreover,

H can be identified with a subgroup of $K = \mathrm{Aut}(L)$, the structure of which is well-known. First, K contains a subgroup $K_1 \cong PGL(3,q)$ or $PGU(3,q)$ with $|K_1 : L| = 3$ or 1 according as $q + \delta$ is or is not divisible by 3, where $\delta = -1$ if $L \cong L_3(q)$ and $\delta = +1$ if $L \cong U_3(q)$. Moreover, if W denotes the cyclic subgroup of K induced from the Galois group of $GF(q)$ or $GF(q^2)$ (according as $L \cong L_3(q)$ or $U_3(q)$), then $K = K_1W$ if $L \cong U_3(q)$ and $K = K_1W\langle x \rangle$ if $L \cong L_3(q)$, where x is an involution induced from the transpose-inverse map of $SL(3,q)$. In particular, the structure of our group H is completely determined as is the action of the elements of S on L. Moreover, we easily see that H/L has a normal 2-complement.

We note that if $L \cong U_3(q)$, $|W|$ is always even and the involution x of W can also be described as being induced from the transpose-inverse map of $SU(3,q)$. It will be convenient to define the groups $L_3^*(q)$ and $U_3^*(q)$ to be the extension of $L_3(q)$ respectively by the corresponding group $\langle x \rangle$.

Lemma 2.3. The following conditions hold:

(i) If R is wreathed, then the unique abelian subgroup A of index 2 in R is characteristic in S;

(ii) If R is wreathed, then $N_L(A)/C_L(A) \cong S_3$ and for some Sylow 3-subgroup P of $N_L(A)$, we have $S = AN_S(P)$;

(iii) If R is wreathed and $S \supset R$, then $r(S) \geq 4$;

(iv) If R is quasi-dihedral, then $S = RW$ where $R \cap W = 1$. W is cyclic and all involutions of $S - R$ are conjugate under the action of L;

(v) If R is quasi-dihedral with $W \neq 1$ and x is the involution of W, then $C_L(x) \cong PGL(2,q)$. Moreover, for a suitable choice of

R, $C_R(x) = \Omega_1(R)$ is a Sylow 2-subgroup of $C_L(x)$, and $[R,x] = Z(R)$;

(vi) If R is quasi-dihedral and y is an involution of $S - R$, then $L < y > \cong L_3^*(q)$ or $U_3^*(q)$;

(vii) If R is quasi-dihedral and U is a four subgroup of S, then $C_H(U)$ has a normal 2-complement. Moreover, if $U \not\subseteq L$, then $N_H(U)$ has a normal 2-complement;

(viii) If R is quasi-dihedral and U is a four subgroup of R, then $N_H(U)/C_H(U) \cong S_3$ and $U\Omega_1(W) = \Omega_1(Y)$ for some Sylow 2-subgroup Y of $C_H(U)$.

(ix) $Z(S) \subseteq R$.

(x) If z is an involution of L, then $C_L(z)/Z(C_L(z)) \cong PGL(2,q)$.

Proof: First, (ix) is immediate from the action of W on L, while (x) follows from the known structure of the centralizer of an involution in these groups. Now assume that R is wreathed (whence $q \equiv 1 \pmod 4$ if $L \cong L_3(q)$ and $q \equiv -1 \pmod 4$ if $L \cong U_3(q)$) and let A be the unique abelian subgroup of index 2 in R. Setting $N = N_L(A)$ and $C = C_L(A)$, we check that $C = A \times O(C)$ and that $N/C \cong S_3$. Since $A \lhd S$, S leaves C and N invariant. Hence by the Frattini argument, we can choose a Sylow 3-subgroup P of N such that $S = AN_S(P)$. In particular, (ii) holds. Since $C_A(P) = 1$ and $N/C \cong S_3$, we can write $S = A(R_0 \times V)$ where $R_0 \cong Z_2$ inverts $P/P \cap C$, $R = AR_0$, and V centralizes $P/P \cap C$. Moreover, we have $V \cap R = 1$.

Set $A = < a,b >$ and $R_0 = < c >$, so that $A = < a > \times < b >$ and $a^c = b$, $b^c = a$. Since V centralizes $P/P \cap C$ and c, for any v in V, we have $a^v = a^k$, $b^v = b^k$ for some number k. Since no outer automorphism of L of order 2 centralizes a Sylow 2-subgroup of L, we conclude that $C_V(A) = 1$. But now we can check that A is the unique abelian subgroup of its order in S and so A is characteristic in S, proving (i).

Setting $|a| = 2^n$, $a_1 = a^{2^{n-2}}$, and $b_1 = b^{2^{n-2}}$, we see that $< a_1 b_1 > \lhd S$. Hence if $\bar{S} = S/ < a_1^2 b_1^2 >$, then $\bar{V}$ and $\bar{c}$ both centralize $\bar{a}_1 \bar{b}_1$. Likewise $< a_1^2 , b_1^2 > = \Omega_1(A) \lhd S$ and so $\bar{V}$ and $\bar{c}$ also both centralize $\bar{a}_1^2$. Thus $\bar{V}$ centralizes $< \bar{a}_1 \bar{b}_1 , \bar{a}_1^2 , \bar{c} >$ with the latter group isomorphic to E_8 and disjoint from $\bar{V}$. Hence if $\bar{V} \neq 1$, we conclude that $r(\bar{S}) \geq 4$, so (iii) also holds.

Suppose now that R is quasi-dihedral. In proving (iv) and (v), we can assume that $S \supset R$. Then $S = RW$, where $R \cap W = 1$, $W = < x > \cong Z_2$ with x induced from the transpose-inverse map of $SL(3,q)$ if $L \cong L_3(q)$; while W is cyclic and induced from the Galois group of $GF(q^2)$ if $L \cong U_3(q)$. Setting $< x > = \Omega_1(W)$ in the latter case, we note that x is similarly induced from the transpose-inverse map of $SU(3,q)$. We can now check in either case that $C_L(x) \cong PGL(2,q)$ and for a suitable choice of R, $C_R(x) = \Omega_1(R)$ is a Sylow 2-subgroup of $C_L(x)$ and $[x,R] = Z(R)$, so (v) holds.

To prove the last assertion of (iv), we first verify (viii). If U is a four subgroup of R, then $N_L(U)/C_L(U) \cong S_3$ and U is a Sylow 2-subgroup of $C_L(U)$, by the structure of L. Moreover, U centralizes x and any involution of $C_H(U)$ clearly lies in $L < x >$. It follows at once that $U < x > = \Omega_1(Y)$ for any Sylow 2-subgroup Y of $C_H(U)$ containing $U < x >$. Thus (viii) holds. But now if we choose U in $C_L(x) - C_L(x)'$ and consider the groups $N_H(U)$ and $C_L(x)$, we easily conclude that all involutions of $\Omega_1(R) < x > - \Omega_1(R)$ are conjugate to x **under** L. Let

xa be an involution of $\langle x \rangle R - R$ with $a \notin \Omega_1(R)$. Then, since $x^a = az$, $\langle z \rangle = Z(R)$, we have that $a^2 = z$. Hence a is contained in the unique maximal generalized quaternion subgroup Q of R. On the other hand, one checks that $C_{L\langle x \rangle}(z)$ contains a subgroup L_0 isomorphic to $Z_4 * SL(2,q)$ with Sylow 2-subgroup $\langle xb \rangle * Q$, where $|b| = 4$ and $\langle b \rangle \triangleleft R$. Clearly all involutions of L_0 are conjugate to x or z in L_0. Hence $x \sim xa$ by the action of L. Thus (iv) holds.

Again if R is quasi-dihedral and y is an involution of $S - R$, then $R\langle y \rangle = R\langle x \rangle$ as $S/R \cong W$ is cyclic. Hence $L\langle y \rangle = L\langle x \rangle$ and (vi) follows.

Finally let the assumptions be as in (vii). If $U \subseteq L$, we see from the structure of the centralizer of an involution in L that $C_L(U) = U \times O(C_L(U))$. Since H/L has a normal 2-complement, it follows that $C_H(U)$ does as well. Suppose then that $U \not\subseteq L$. Then $U - (U \cap L)$ contains an involution u conjugate to x as all involutions of $x\langle L \rangle - L$ are conjugate under L by (iv). Setting $K = C_H(u)$, we have then that $K \cap L = PGL(2,q)$ by (v) and $K/K \cap L$ is $2'$-closed. Furthermore, as S/R is cyclic, R contains an involution v of U. But then $C_{K \cap L}(v)$ is a dihedral group and we conclude at once that $C_K(v) = C_H(U)$ has a normal 2-complement.

Again if $U \not\subseteq L$, we know that $U \cap L \neq 1$. Since $L \triangleleft H$, the involutions of U are not conjugate in H and therefore $N_H(U)/C_H(U)$ is a 2-group. Thus $N_H(U)$ has a normal 2-complement and all parts of (vii) hold.

Lemma 2.4 If S is a 2-group with $r(S) \leq 4$ which admits a fixed-point-free automorphism of order 3, then either $S \cong Z_{2^n} \times Z_{2^n}$ for some n

or $S/\Phi(S) \cong E_{16}$.

Proof: Let the given automorphism be α and suppose $\bar{S} = S/\Phi(S) \ncong E_{16}$. Since α acts fixed-point-free on $\bar{S}$ and $r(S) \leq 4$, the only possibility is that $\bar{S} \cong Z_2 \times Z_2$. Thus S is generated by two elements. On the other hand, as $|\alpha| = 3$ and α acts fixed-point-free on S, S has class at most 2. It follows at once from these conditions that S' is cyclic. But then α centralizes S', forcing $S' = 1$. Thus S is abelian and as S is generated by two elements, we conclude that $S \cong Z_{2^n} \times Z_{2^n}$.

Lemma 2.5. If S is a 2-group of order 64 which admits a fixed-point-free automorphism of order 3, then S is either isomorphic to E_{64}, $Z_4 \times Z_4 \times Z_2 \times Z_2$, or $Z_8 \times Z_8$ or S is of type $L_3(4)$ or $U_3(4)$.

Proof: If S is abelian, the lemma is clear; so we can assume S is nonabelian, whence $S' \neq 1$. This implies that S must now have class 2. Let α be the given automorphism of S. Clearly any α-invariant subgroup of S has order 2^{2k}, $0 \leq k \leq 3$. Hence if T is a non-trivial proper α-invariant subgroup of S, $|T| = 4$ or 16 and it is immediate that $T \cong Z_2 \times Z_2$, $Z_4 \times Z_4$, or E_{16}. In particular, $Z(S)$ is a nontrivial proper α-invariant subgroup of S. If $|Z(S)| = 16$, then $S = Z(S) < a,b >$ for suitable a,b and it follows that S' is cyclic. But then α centralizes S' and we have a contradiction. Hence $|Z(S)| = 4$ and so $Z(S) \cong Z_2 \times Z_2$. Since $S' \subseteq Z(S)$ and α does not centralize S', the only possibility is that $S' = Z(S)$. If $S/Z(S) \cong Z_4 \times Z_4$, then again $S = Z(S) < a,b >$ for suitable a,b and so S' is cyclic, which is not the case. Hence we must have $S/Z(S) \cong E_{16}$, whence also $\Phi(S) = Z(S)$. In other words, S is a special 2-group.

Since S is not abelian, S contains an element x of order 4. Setting $\bar{S} = S/Z(S)$, we have that $\bar{A} = \langle \bar{x}, \bar{x}^{\alpha} \rangle$ is an α-invariant group. Hence the inverse image A of $\bar{A}$ is α-invariant of order 16 and contains x. By the preceding paragraph, we have $A \cong Z_4 \times Z_4$. We can also write $\bar{S} = \bar{A} \times \bar{B}$, where $\bar{B} \cong Z_2 \times Z_2$ and $\bar{B}$ is α-invariant. It follows that $S\langle \alpha \rangle$ is isomorphic to an extension of $Z_4 \times Z_4$ by A_4. Moreover, as $C_S(A)$ is α-invariant and $A \not\subseteq Z(S)$, we have that $C_S(A) = A$. Hence this extension is faithful. But MacWilliams [35,Lemma 3] has determined all such possible extensions in which a 3-element acts fixed-point-free on the Sylow 2-group of the extension and we conclude that S is of type $L_3(4)$ or $U_3(4)$.

3. <u>The centralizer of a central involution.</u> Henceforth G will denote a simple $\mathfrak{R}_1$-group which satisfies the hypotheses of Theorem A. Thus $\mathcal{L}_c(G)$ is nonempty and each of its elements is nonsimple.

We first prove

<u>Proposition 3.1.</u> If z is a central involution of G for which $C_G(z)$ is nonsolvable, then $L_z \cong SL(2,q)$, q odd, or $\hat{A}_7$.

<u>Proof:</u> Since $C_G(z)$ is nonsolvable, $C_G(z)$ is not 2-constrained by condition (d) defining an $\mathfrak{R}_1$-group and so $L_z \neq 1$. By condition (b) defining an $\mathfrak{R}_1$-group, Theorem A and Corollary A of Part III are applicable since every component of L_z is nonsimple by hypothesis and $r(L_z) \leq 3$ by condition (e) defining an $\mathfrak{R}_1$-group, the theorem and corollary yield this proposition at once (note that, apart from $SL(2,q)$, q odd, and $\hat{A}_7$, the groups listed in Theorem A(II) of Part III each have sectional 2-rank 4).

We can sharpen this result slightly

<u>Proposition 3.2.</u> There exists a central involution z in G such that if $M = C_G(z)$ and $\overline{M} = M/O(M)$, then $\langle \bar{z} \rangle = Z(L(\overline{M})) = Z(L_z)$.

<u>Proof:</u> First of all, let u be an arbitrary central involution of G such that $H = C_G(u)$ is nonsolvable. Setting $\widetilde{H} = H/O(H)$, the preceding proposition yields that $L_u = L(\widetilde{H}) \cong SL(2,q)$, q odd, or $\hat{A}_7$. In particular, $\widetilde{Z} = Z(L(\widetilde{H})) \cong Z_2$ and $\widetilde{Z} \subseteq Z(\widetilde{H})$.

Let S be a Sylow 2-subgroup of H and let z be an involution of S such that $\langle \tilde{z} \rangle = \widetilde{Z}$. Then S is a Sylow 2-subgroup of G and $z \in Z(S)$, so z is also a central involution. In addition, $C_H(z)$ covers $L(\widetilde{H})$. Let L be a minimal S invariant normal subgroup of $C_H(z)$ which covers $L(\widetilde{H})$. Then L is perfect and $z \in L$. Setting $M = C_G(z)$, we have that $L \subseteq M$ and so M is also nonsolvable. Setting $\overline{M} = M/O(M)$, Proposition 3.1 yields likewise that $L_z = L(\overline{M}) \cong SL(2,r)$, r odd, or $\hat{A}_7$. By Proposition 2.1 of Part III, $\overline{M}/L(\overline{M})$ is solvable. Since L is perfect, it follows that $\overline{L} \subseteq L(\overline{M})$. But $z \in L$ and so $\bar{z}$ is an involution of $L(\overline{M})$. However, $Z(L(\overline{M}))$ contains the unique involution of $L(\overline{M})$ and we conclude that $\langle \bar{z} \rangle = Z(L(\overline{M}))$.

We now choose a central involution z of G in accordance with Proposition 3.2. Thus if we set $M = C_G(z)$, $\overline{M} = M/O(M)$, and $\overline{L} = L(\overline{M})$, we have that $\langle \bar{z} \rangle = Z(\overline{L})$. Moreover, $\overline{L} \cong SL(2,q)$, q odd, or $\hat{A}_7$. We also set $\overline{C} = C_{\overline{M}}(\overline{L})$, we let S be a Sylow 2-subgroup of M and we let R_1, R_2 be subgroups of S whose images in $\overline{M}$ are Sylow 2-subgroups of $\overline{C}$, $\overline{L}$ respectively. Then S is a Sylow 2-subgroup of G, R_1 centralizes R_2, $R_1 \cap R_2 = \langle z \rangle$ and R_1 , R_2 are normal in S .

Setting $\tilde{M} = \bar{M}/\langle \bar{z} \rangle$, we have that $\tilde{L} \cong L_2(q)$ or A_7 and that $\tilde{C} = C_{\tilde{M}}(\tilde{L})$. Thus $\tilde{M}/\tilde{C}$ is isomorphic to a subgroup of $\mathrm{Aut}(\tilde{L})$ and so we can apply Lemma 2.2 to $\tilde{M}/\tilde{C}$. We conclude at once that $\tilde{S}$ contains a subgroup $\tilde{W}$ with the following properties: $\tilde{W} \supseteq \tilde{R}_1$, $\tilde{W} \cap \tilde{L} = 1$, $|\tilde{S}:\tilde{R}_2\tilde{W}| \leq 2$ with equality holding only if $\tilde{L} \cong L_2(q)$, for any subgroup $\tilde{W}_0$ of $\tilde{W}$, $C_{\tilde{R}_2}(\tilde{W}_0)$ is a Sylow 2-subgroup of $C_{\tilde{L}}(\tilde{W}_0)$, $C_{\tilde{L}}(\tilde{W})$ contains a subgroup isomorphic to $L_2(3)$, and if $\tilde{W}_1$ denotes the inverse image in $\tilde{W}$ of $\Omega_1(\tilde{W}/\tilde{R}_1)$ then $\tilde{W}_1$ centralizes $\tilde{R}_2$. In particular, as $r(G) \leq 4$, these conditions imply that $r(\tilde{W}) \leq 2$. We let W, W_1 be the inverse images of $\tilde{W}$, $\tilde{W}_1$ in S. Since R_2 is generalized quaternion and normal in S, R_2 possesses a cyclic subgroup $\langle r_2 \rangle$ of order 4 which is normal in S. In particular, $r_2^2 = z$. Finally we let L be a minimal S-invariant normal subgroup of $C_M(R_1)$ which covers $\bar{L}$. Then L is perfect, R_2 is a Sylow 2-subgroup of L, $O(L) \subseteq O(M)$, and $L/O(L) \cong \bar{L}$. Moreover, for any subgroup W_0 of W, $C_{R_2}(W)$ is a Sylow 2-subgroup of $C_L(W_0)$.

We fix all this notation for the balance of the proof. We conclude this section by establishing some basic elementary properties of these various groups that we shall need for the proof of Theorem A.

Lemma 3.3. The following conditions hold:

(i) $r(W/\langle z\rangle) \leq 2$;

(ii) W centralizes r_2;

(iii) $Q_2 = C_{R_2}(W_1)$ is generalized quaternion of index 1 or 2 in R_2 according as $|W_1:R_1| = 1$ or 2;

(iv) Any involution of WR_2-W has the form wt, where $w \in W_1$, $t \in R_2^{\#}$, with w centralizing or inverting t and correspondingly $w^2 = z$ or 1;

(v) If y is an involution of $S - WR_2$, then $R_2\langle y\rangle$ is quasi-dihedral;

(vi) Every elementary abelian normal subgroup of S lies in WR_2;

(vii) $\Omega_1(Z(S)) \subseteq R_1$;

(viii) For any 2-subgroup X of M, $N_M(X)/C_M(X)$ is a $\{2,3\}$-group;

(ix) $C_{\bar{S}}(C_{\bar{L}}(\bar{W}_1)') = \bar{W}_1$.

Proof: By Lemma 2.2 (iii), $C_{\tilde{L}}(\tilde{W})$ contains a subgroup $\tilde{L}_0 \cong L_2(3)$ with $\tilde{R}_2 \cap \tilde{L}_0$ as Sylow 2-subgroup. In particular, as $r(G) \leq 4$ and $\tilde{R}_2 \cap \tilde{W} = 1$, $r(\tilde{W}) \leq 2$ and consequently (i) holds. Furthermore, it follows that $C_{\bar{L}}(\bar{W})$ contains a subgroup $\bar{L}_0 \cong SL(2,3)$ with $\bar{R}_2 \cap \bar{L}_0$ as Sylow 2-subgroup. Then $\bar{R}_2 \cap \bar{L}_0 \cong Q_8$. However, as $\langle r_2\rangle \triangleleft R_2$, r_2 lies in every quaternion subgroup of R_2 and so $\bar{r}_2 \in \bar{R}_2 \cap \bar{L}_0$. Thus $\bar{W}$ centralizes $\bar{r}_2$ and (ii) follows.

If $W_1 = R_1$, then $Q_2 = R_2$ and (iii) holds. Assume then that $W_1 \supset R_1$. If $\bar{L} \cong SL(2,q)$, then by Lemma 6.1 (i) of Part III $C_{\bar{L}}(\bar{W}_1) \cong SL(2,q_1)$, where $q_1^2 = q$, and $\bar{Q}_2 = C_{\bar{R}_2}(\bar{W}_1)$ is a Sylow 2-subgroup of $C_{\bar{L}}(\bar{W}_1)$ and we obtain (iii) at once. On the other hand, if $\bar{L} \cong \hat{A}_7$, then $W = W_1$ and (iii) follows similarly from Lemma 6.1 (ii) and (iii) of Part III.

Next let y be an involution of $WR_2 - W$, so that $y = wt$, $w \in W$ and $t \in R_2^{\#}$. Since $y^2 = wtwt = 1$, it follows that $w^{-1}tw = w^{-2}t^{-1}$. Since $R_2 \triangleleft S$, $w^{-1}tw \in R_2$ and so $w^{-2} \in R_2$. But $W \cap R_2 = \langle z\rangle$ and consequently $w^2 = 1$ or z. Correspondingly $w^{-1}tw = t^{-1}$ or zt^{-1}. In the first case (iv) holds; and to establish (iv) in the latter case, it remains only to prove that w then centralizes t. If $|t| = 4$, then $t^2 = z$ as R_2 is generalized quaternion. But then $zt^{-1} = t$ and so w

centralizes t. Since clearly $t \neq z$ in this case, we can assume that $|t| = 2^n$, $n \geq 3$. Then w inverts t^2 and $|t^2| \geq 4$. But $t^2 \in Q_2$ as $|R_2:Q_2| \leq 2$ and $w \in W_1$ as $w^2 = z$. Since W_1 centralizes Q_2 by (iii), this is a contradiction and (iv) is proved.

Suppose next that y is an involution of $S - WR_2$ which centralizes r_2. Then necessarily $\bar{L} \cong SL(2,q)$ and $\tilde{L} < \tilde{y} > \cong PGL(2,q)$, whence $\tilde{R}_2 < \tilde{y} >$ is dihedral. But $\bar{R}_2 < \bar{y} >$ is a nonsplit extension of $\tilde{R}_2 < \tilde{y} >$ by $< \bar{z} >$ as $\bar{R}_2$ is generalized quaternion. The only possibility therefore is that $\bar{R}_2 < \bar{y} >$, and hence also $R_2 < y >$, is quasi-dihedral, proving (v).

Next let A be an elementary abelian normal subgroup of S and suppose $A \not\subseteq WR_2$. Choosing a in A with $a \notin WR_2$, we have that $R_2 < a >$ is quasi-dihedral by (v). But then $[R_2, a]$ is cyclic of order at least 4, contrary to the fact that $[R_2, a] \subseteq [R_2, A] \subseteq A$ and A is elementary. This establishes (vi).

By (vi), $Z = \Omega_1(Z(S)) \subseteq WR_2$ and as Z is elementary, we have, in fact, $Z \subseteq W_1 R_2$. If $W_1 = R_1$, then W_1 centralizes R_2 and as R_2 is generalized quaternion with $Z(R_2) \subseteq W_1$, it is immediate that $Z \subseteq W_1 = R_1$. Suppose then that $W_1 \supset R_1$, in which case $|R_2| \geq 16$. We have $\tilde{W}_1 \tilde{R}_2 = \tilde{W}_1 \times \tilde{R}_2$ and $\tilde{R}_2 = D_{2^n}$ for some $n \geq 3$. This implies that $Z(\tilde{R}_2) = < \tilde{r}_2 >$. Since $\tilde{Z} \subseteq Z(\tilde{W}_1 \tilde{R}_2)$, it follows that $\tilde{Z} \subseteq Z(\tilde{W}_1) < \tilde{r}_2 >$. But r_2 is a square in R_2 and so $r_2 \in Q_2$. Thus $Z \subseteq W_1 Q_2$. However, W_1 centralizes Q_2 and no element of $Q_2 - < z >$ is in the center of Q_2. Together these conditions imply that $Z \subseteq W_1$. But by (iii) no element of $W_1 - R_1$ centralizes R_2. This forces $Z \subseteq R_1$ and so (vii) holds.

Clearly to establish (viii), we need only show that $N_{\tilde{M}}(\tilde{X})/C_{\tilde{M}}(\tilde{X})$ is a $\{2,3\}$-group. Since $\tilde{M}/\widetilde{LC}$ is isomorphic to a subgroup of $\mathrm{Aut}(\bar{L})/\mathrm{Inn}(\bar{L})$, it is abelian, and so any element of odd order in $N_{\tilde{M}}(\tilde{X})$ which centralizes $\tilde{X} \cap \widetilde{LC}$ will centralize $\tilde{X}$. Hence without loss we can suppose that $\tilde{X} \subseteq \widetilde{LC} = \tilde{L} \times \tilde{C}$. Let $\tilde{u}$ be a p-element of $N_{\tilde{M}}(\tilde{X})$ for any prime $p \geq 5$. Then as $r(\tilde{L}) = 2$, $r(\tilde{C}) \leq 2$ and so $\tilde{y}$ must centralize both $\tilde{X}/\tilde{X} \cap \tilde{L}$ and $\tilde{X}/\tilde{X} \cap \tilde{C}$. This implies that $\tilde{y}$ centralizes $\tilde{X}$ and (viii) holds.

Finally (ix) is obvious if $W = R_1$. In the contrary case, $W_1 \supset R_1$ and $C_{\bar{L}}(\bar{W}_1) \cong SL(2,q_0)$ with $q_0^2 = q$ if $\bar{L} \cong SL(2,q)$ and $C_{\bar{L}}(\bar{W}_1) \cong SL(2,3)$ or $SL(2,5)$ if $\bar{L} \cong \hat{A}_7$. But now (ix) follows at once from the structure of $\mathrm{Aut}(\bar{L})$. Thus all parts of the lemma are proved.

We next prove

Lemma 3.4. Either R_1 is cyclic, R_1 is of maximal class, or R_1 contains a normal four subgroup of S.

Proof: By Lemma 6.9 of Part III, $\tilde{R}_1$ is either cyclic, of maximal class, or $\Omega_1(\tilde{R}_1) \cong Z_2 \times Z_2$. If $\tilde{R}_1$ is cyclic, R_1 is either cyclic or abelian of rank 2. In the second case $\Omega_1(R_1)$ is then a normal four subgroup of S as $R_1 \lhd S$ and so the lemma holds in either case.

Suppose $\tilde{R}_1$ is nonabelian of maximal class and let R_0 be a maximal subgroup of R_1 containing $\langle z \rangle$ and normal in S. Then $\tilde{R}_0$ is cyclic and so R_0 is either cyclic or abelian of rank 2 and is normal in S. Hence in the latter case, $\Omega_1(R_0)$ is a normal four subgroup of S. We can therefore assume that R_0 is cyclic. Since $|R_1:R_0| = 2$ and $\tilde{R}_1$ is nonabelian of maximal class, a result of Schur implies that $\tilde{R}_1$ is dihedral and R_1 is of maximal class and again the lemma holds.

It therefore remains to treat the case that $\Omega_1(\tilde{R}_1) \cong Z_2 \times Z_2$. Let R_0 now denote the inverse image of $\Omega_1(\tilde{R}_1)$ in R_1. Then $R_0 \lhd S$ and

$R_0 \cong D_8$, Q_8, $Z_4 \times Z_2$, or $Z_2 \times Z_2 \times Z_2$. In the latter two cases, R_0 clearly contains a normal four subgroup of S, so we can assume that $R_0 \cong D_8$ or Q_8. If $R_1 = R_0$, the lemma holds, so we can also suppose that $R_1 \supset R_0$.

If Q_0 denotes the stabilizer in W of the chain $R_0 \supset Z(R_0) \supset 1$, then $Q_0 = R_0 C_{Q_0}(R_0)$ by [13, Lemma 5.4.6]. Since $R_0 \cap C_{Q_0}(R_0) = Z(R_0) = \langle z \rangle$, it follows that either $C_{Q_0}(R_0) = \langle z \rangle$ or else $r(\tilde{Q}_0) \geq 3$. However, the latter possibility is excluded by Lemma 3.3 (i) and we conclude that $Q_0 = R_0$. Since $\tilde{R}_0 \cong Z_2 \times Z_2$, this forces $\tilde{R}_1$ to be nonabelian of order 8, whence $\tilde{R}_1 \cong D_8$, contrary to the fact that $\Omega_1(\tilde{R}_1) = \tilde{R}_0 \cong Z_2 \times Z_2$ in the present case.

In analyzing the third possibility of Lemma 3.4, the following lemma will be critical.

<u>Lemma 3.5.</u> If R_1 contains a normal four subgroup of S, then any four subgroup of WR_2 intersects W nontrivially.

<u>Proof:</u> Let U be a normal four subgroup of S with $U \subseteq R_1$. Clearly we can assume that $z \in U$. Suppose, by way of contradiction, that WR_2 contains a four subgroup B such that $B \cap W = 1$. Thus $B = \langle w_1t_1, w_2t_2 \rangle$ with $w_i \in W$, $t_i \in R_2^{\#}$, $i = 1,2$. By Lemma 3.3 (iv), we have that $w_i \in W_1$, $i = 1,2$.

Observe next that as $\tilde{W}_1\tilde{R}_2 = \tilde{W}_1 \times \tilde{R}_2$ and $\tilde{W}_1\tilde{Q}_2$ is of index 2 in $\tilde{W}_1\tilde{R}_2$, $B \cap W_1Q_2 \neq 1$. Hence some involution of B has its "R_2-component" in Q_2. Without loss we can assume to begin with that $t_1 \in Q_2$.

Then as $w_1 \in W_1$, w_1 centralizes t_1 and by Lemma 3.3(iv), $w_1^2 = z$. Since w_1t_1 is an involution, it follows that also $t_1^2 = z$.

Consider first the case that $\langle w_1 \rangle = \langle w_2 \rangle$, whence also $w_2^2 = z$. Lemma 3.3 (iv) implies now that w_2 centralizes t_2. Thus $w_1t_1w_2t_2 = w_1w_2t_1t_2$ with $w_1w_2 \in \langle z \rangle$ and consequently $t' = t_1t_2$ or $z\, t_1t_2$ is the third involution of B. However, $B \cap W = 1$ by assumption and so $t' \neq z$. But $t' \in R_2$ and z is the unique involution of R_2. We conclude that $\langle w_1 \rangle \neq \langle w_2 \rangle$.

Now set $\overline{S} = S/R_2$, so that $\overline{B} = \langle \overline{w_1}, \overline{w_2} \rangle$. Since $R_2W/R_2 \cong W/\langle z \rangle$, it follows that $\overline{B}$ is a four group. On the other hand, as $U \triangleleft S$ and $U \cap R_2 = \langle z \rangle$, we have $|\overline{U}| = 2$ and $\overline{U} \triangleleft \overline{S}$, whence $\overline{U} \subseteq Z(\overline{S})$. Since $r(\overline{W}) \leq 2$ by Lemma 3.3 (i) and $\overline{B} \cong Z_2 \times Z_2$, this forces $\overline{U} \subseteq \overline{B}$. In addition, $\overline{w}_1 \in \overline{U}$ as $|w_1| = 4$, while $U \cong Z_2 \times Z_2$. Hence $\overline{B} = \langle \overline{w}_1, \overline{U} \rangle$ and so either $w_2 \in U$ or $w_2 \in w_1U$. However, in the latter case, the involution $w_1t_1w_2t_2$ has its "W-component" in U, so without loss we can assume that $w_2 \in U$. But then $w_2 \in R_1$ and so w_2 centralizes t_2. Hence $w_2^2 = z$ by Lemma 3.3 (iv), contrary to the fact that $w_2 \in U$ is an involution.

4. <u>The intersection of W and its conjugates.</u> In this section we shall establish the following basic result:

<u>Proposition 4.1</u> If g is an element of G such that $W \cap W^g \neq 1$ and $z^g \in S$, then $g \in M$.

We carry out the proof in a sequence of lemmas.

<u>Lemma 4.2.</u> If x is an involution of R_1, then z is isolated in $C_G(x)$.

<u>Proof:</u> Suppose false for some x and set $H = C_G(x)$ and $\widetilde{H} = H/O(H)$. Since $L \subseteq H$, H is nonsolvable and as G is an $\mathfrak{R}_1$-group, it follows that

$\tilde{H}$ is not 2-constrained. Hence by [Part III, Proposition 2.1], $H/L(H)$ is solvable and so $\tilde{L} \subseteq L(\tilde{H})$. The possible components of $L(\tilde{H})$ are listed in Theorem A of Part III. Moreover, again as G is an $\mathcal{R}_1$-group, either $L(\tilde{H})$ consists of a single component or of two components, one simple and one nonsimple. We easily check now that either $L(\tilde{H})$ possesses a component $\tilde{J} \cong L_3(r)$ or $U_3(r)$, r odd, $r \geq 5$, or else $\tilde{z} \in Z(L(\tilde{H}))$. However, in the latter case, z would be isolated in H, contrary to assumption. Hence such a component $\tilde{J}$ exists.

We argue next that, in fact, $\tilde{J} = L(\tilde{H})$. If false, then $L(\tilde{H})$ possesses a second component $\tilde{K} \cong SL(2,t)$, t odd, or $\hat{A}_7$. Since $\tilde{R}_2 \subseteq \tilde{L} \subseteq L(\tilde{H})$ and $z \in R_2'$, it is immediate that $\tilde{z}$ centralizes $\tilde{K}$. But then by the structure of M, L_z must cover $\tilde{K}$ and consequently $\tilde{L} = \tilde{K}$. Again $\tilde{z} \in Z(L(\tilde{H}))$ and z is isolated in H. This proves our assertion.

Since $\tilde{z} \in \tilde{J} \triangleleft \tilde{H}$ and $\tilde{J}$ has only one conjugacy class of involutions, we see that $z \in Z(T)$ for some Sylow 2-subgroup T of H. Since $T \subseteq M$, we can suppose without loss that $T \subseteq S$. Since every element of $\mathcal{L}_c(G)$ is nonsimple by assumption and $\tilde{J}$ is simple, it follows that x is not a central involution and so $T \subset S$.

Among all possible choices of x for which the lemma is false, we assume now that x and T are chosen so that T has maximal order. Define T_1, T_2 to be the subgroups of T with the property that $\tilde{T}_1$ is a Sylow 2-subgroup of $C_{\tilde{H}}(\tilde{J})$ and $\tilde{T}_2$ is a Sylow 2-subgroup of $\tilde{J}$. Since $T_1 \subseteq S$ and $\tilde{T}_1$ centralizes $\tilde{L}$, we have that $T_1 \subseteq R_1$. Furthermore, T_2 is either quasi-dihedral or wreathed and $T_1T_2 = T_1 \times T_2$.

Next set $U = \Omega_1(Z(T))$. By Lemma 2.3 (ix), we have $U \subseteq T_1T_2$. Clearly $\langle x,z \rangle \subseteq U$ and $U \cap T_2 = \langle z \rangle$, so $U = \langle z \rangle \times U_1$, where $U_1 = U \cap T_1$.

Suppose $U_1 \supset \langle x \rangle$. Choosing y in $N_S(T) - T$ with $y^2 \in T$, y normalizes U. But then $C_U(y)$ is noncyclic and so y centralizes some u in $U_1^{\#}$. Since $T\langle y \rangle \subseteq C_G(u)$, our maximal choice of x and T implies that z is isolated in $C_G(u)$. However, this is obviously impossible as $C_G(u)$ covers $\tilde{J}$ and $\tilde{z}$ is not isolated in $\tilde{J}$. We therefore conclude that $U_1 = \langle x \rangle$ and $U = \langle x,z \rangle$. Furthermore, with y as above, it follows for the same reason that $x^y = xz$, as $z^y = z$.

We argue next that y normalizes $\Omega_1(T_2)$; so assume false. Since all involutions of T_2 are conjugate in H and since $T_2{}^y \subseteq T$, it follows that $T - T_2$ contains an involution t which is conjugate to z. Setting $M_1 = C_G(t)$ and $\overline{M}_1 = M_1/O(M_1)$, this implies that $L(\overline{M}_1) \cong SL(2,q)$ and $\overline{t}$ is the unique involution of $L(\overline{M}_1)$. But now if $t \in T_1T_2 - T_2$, we can write $t = t_1t_2$ with $t_1 \in T_1{}^{\#}$ and $t_2 \in T_2$. If $t_2 = 1$, then $M_1 = C_G(t)$ covers $\tilde{J}$, which is clearly impossible as $\overline{M}_1/L(\overline{M}_1)$ is solvable. Hence $t_2 \neq 1$ and so by the structure of $\tilde{J}$, we have $\tilde{t}_2 \in C_{\tilde{J}}(\tilde{t})' \cong SL(2,q)$. This implies that t_2 lies in a perfect subgroup of M_1 and hence that $\overline{t}_2 \in L(\overline{M}_1)$, contrary to the fact that $\overline{t}_2 \neq \overline{t}$ and $\overline{t}$ is the unique involution of $L(\overline{M}_1)$. We therefore conclude that $t \in T - T_1T_2$. Applying Lemma 2.3 (v), it follows now that $C_{\tilde{J}}(\tilde{t}) \cong PGL(2,r)$. Thus M_1 possesses a perfect subgroup K_1 with $K_1/O(K_1) \cong L_2(r)$. However, this is clearly impossible as $\overline{K}_1 \subseteq L(\overline{M}_1) \cong SL(2,q)$. This contradiction proves our assertion.

If T_2 is quasi-dihedral, then all four subgroups of T_2 are conjugate in T_2 and as y normalizes $\Omega_1(T_2)$, it follows that yy_2 normalizes a four subgroup V_2 to T_2 for some y_2 in T_2. Since yy_2 has the same properties as y, we can suppose without loss that y normalizes V_2.

Assume, on the other hand, that T_2 is wreathed. Since y centralizes z, $y \in M$ and so $\bar{y}$ leaves $\bar{L}$ invariant. Hence y normalizes both $\Omega_1(T_2)$ and $T_2 \cap L$. However, it is immediate that these two groups generate T_2. Thus y normalizes T_2 and so normalizes the unique normal four subgroup V_2 of T_2.

Finally set $A = V_2 \times \langle x \rangle$, $N = N_G(A)$, and $D = C_G(A)$. We have that $A \cong E_8$ and hence that N/D is isomorphic to a subgroup of $GL(3,2)$. Again by the structure of J, $H \cap N/H \cap D \cong S_3$. Furthermore, clearly $z \in V_2$ and as y normalizes both U and V_2, it follows that $y \in N$. Since y does not centralize x, $y \notin H \cap N$ and consequently $N/D \ncong S_3$. On the other hand, $N/D \ncong GL(3,2)$, otherwise $x \sim z$ in N, contrary to the fact that x is a noncentral involution of G. Hence the only possibility is that $N/D \cong S_4$.

Since x is not isolated in N, we also have that $V_2 \lhd N$. Since $z \in V_2$ and the involutions of V_2 are conjugate in N, it follows that $z \in Z(P)$ for some Sylow 2-subgroup P of N. We have then that $P \subseteq M$. But the involutions of $A - V_2$ are all conjugate under the action of P. Since x is one of these and $x \in R_1$, we conclude now that $\bar{A} \subseteq \bar{C} = C_{\bar{M}}(\bar{L})$. Thus, in fact, $A \subseteq R_1$ and, in particular, V_2 centralizes R_2. However, $\tilde{R}_2 \subseteq \tilde{L} \subseteq \tilde{J}$ and by our choice of T, we have that $R_2 \subseteq T_2$. But as T_2 is quasi-dihedral or wreathed, the four subgroup V_2 cannot centralize the generalized quaternion group R_2. This completes the proof of the lemma.

As a corollary of the lemma, we have

<u>Lemma 4.3.</u> If g is an element of G such that $R_1 \cap R_1^g \neq 1$ and $z^g \in S$, then $g \in M$.

Proof: Let x_1 be an involution of R_1 such that $x = x_1^g \in R_1$. Setting $H = C_G(x)$, we have H contains both L and $J = L^g$. If $\tilde{H} = H/O(H)$, the preceding lemma yields that $\tilde{z} \in Z(\tilde{H})$. Hence $M \cap H$ covers $\tilde{H}$ and consequently $\tilde{L} = L(\tilde{H})$ with $\tilde{H}/\tilde{L}$ solvable. Since $\tilde{J}$ is perfect and $J/O(J) \cong L/O(L)$, it follows that $\tilde{J} = \tilde{L}$. Thus $Z(\tilde{J}) = Z(\tilde{L}) = \langle \tilde{z} \rangle$. Since the image of $\langle z^g \rangle$ in $\tilde{H}$ is the center of $\tilde{J}$, we see that $z^g \in O(H)\langle z \rangle$. But by assumption, $z^g \in S$. Since $O(H)\langle z \rangle \cap S = \langle z \rangle$, we conclude therefore that $z^g = z$, whence $g \in M = C_G(z)$, as asserted.

We shall next strengthen Lemma 4.2.

Lemma 4.4. If x is an involution of W, then z is isolated in $C_G(x)$.

Proof: Assume false for some x. Then $x \notin R_1$ by Lemma 4.2. Since x is an involution, we have $W_1 = R_1\langle x \rangle$. By our choice of W, $\bar{L}_0 = C_{\bar{L}}(\bar{x}) \cong SL(2,q_0)$, where $q_0^2 = q$ if $\bar{L} \cong SL(2,q)$ and $q_0 = 3$ or 5 if $\bar{L} \cong \hat{A}_7$ and, in addition, $\bar{R}_2 \cap \bar{L}_0$ is a Sylow 2-subgroup of $\bar{L}_0$.

Again we set $H = C_G(x)$ and $\tilde{H} = H/O(H)$. Consider first the case that $q_0 > 3$ and let L_0 be a minimal subgroup of $C_L(x)$ which maps on $\bar{L}_0$ and has $R_2 \cap L_0$ as a Sylow 2-subgroup. Then L_0 is perfect. We can now repeat the argument of the first part of Lemma 4.2 with L_0 in place of L to conclude that $L(\tilde{H}) = \tilde{J} \cong L_3(r)$ or $U_3(r)$ for suitable $r \geq 5$ and that for some subgroup T_2 of S, $\tilde{T}_2$ is a Sylow 2-subgroup of $\tilde{J}$. We have that $\tilde{L}_0 \subseteq \tilde{J}$ and that $\tilde{L}_0$ centralizes $\tilde{z}$, which together imply that, in fact, $\tilde{L}_0 = L(C_{\tilde{J}}(\tilde{z}))$, whence $r = q_0$. But now applying Lemma 2.3(x), it follows that T_2 contains an element t such that $\tilde{t}^2$ centralizes $\tilde{L}_0$ and $\tilde{L}_0\langle \tilde{t} \rangle/\langle \tilde{t}_2 \rangle \cong PGL(2,q_0)$. Since $t \in S$ and t centralizes x, we conclude that $\bar{L}_0\langle \bar{t} \rangle \subseteq C_{\bar{M}}(\bar{x})$ with $\bar{t}^2$ centralizing $\bar{L}_0$ and $\bar{L}_0\langle \bar{t} \rangle/\langle \bar{t}^2 \rangle \cong PGL(2,q_0)$.

We shall now contradict this conclusion. Indeed, as $C_{\overline{SL}}(\bar{x}) \subseteq \overline{WL}_0$, it is immediate that $C_{\bar{S}}(\bar{x})$ cannot contain an element $\bar{t}$ such that $\bar{L}_0 < \bar{t} > / < \bar{t}^2 > \cong PGL(2,q_0)$. Thus the lemma holds if $q_0 > 3$.

Thus we can assume that $q_0 = 3$, whence $\bar{L} \cong SL(2,9)$ or $\hat{A}_7$. We set $Q = C_{R_2}(x)$, so that $Q \cong Q_8$. Then $C_L(x)$ contains a cyclic 3-subgroup P such that $[Q,P] = Q$. We have $QP \subseteq H$. We argue first that H is not 2-constrained, so assume false. Since G is an $\mathfrak{R}_1$-group, H is therefore solvable. Set $\tilde{V} = O_2(\tilde{H})$. Since $O(\tilde{H}) = 1$ and $[\tilde{Q},\tilde{P}]=\tilde{Q}$, it is immediate from the structure of a solvable group of sectional 2-rank at most 4 that $\tilde{Q} \subseteq \tilde{V}$. Let U be a 2-subgroup of $C_H(z)$ containing Q whose image in $\tilde{H}$ is $C_{\tilde{V}}(\tilde{z})$. Since $U \subseteq M$ and U centralizes x, we can suppose without loss that $U \subseteq S$.

By assumption, z is not isolated in H. We argue first that z is not isolated in $H_0 = N_H(U)$. Indeed, if $\tilde{U} = \tilde{V}$, then H_0 covers $\tilde{H}$ and as $\tilde{z}$ is not isolated in $\tilde{H}$, our assertion is clear. On the other hand, if $\tilde{U} \subset \tilde{V}$, then certainly $\tilde{z}$ is not isolated in $N_{\tilde{V}}(\tilde{U})$ as $\tilde{U} = C_{\tilde{V}}(\tilde{z})$. Since H_0 covers $N_{\tilde{V}}(\tilde{U})$, our assertion is again clear. We conclude that $z^h \neq z$ for some z in H_0. But as $Q \subseteq U \subseteq S$, $z \in U' \cap Z(U)$ and so also $z^h \in U' \cap Z(U)$. However, $S' \subseteq R_1R_2$ by Lemma 2.1 (i) and so $z^h \in R_1R_2$. But z^h centralizers Q as $z^h \in Z(U)$. Since Q centralizes R_1 and $C_{R_2}(Q) = < z >$, it follows therefore that $z^h \in R_1$ Thus, $z^h \in R_1 \cap R_1^h$ and we conclude from the preceding lemma that $h \in M$. Hence $z^h = z$, contrary to our choice of h. This proves that H is not 2-constrained, as claimed.

Suppose $L(\tilde{H})$ possessed a component $\tilde{K} \cong SL(2,t)$ or $\hat{A}_7$. Since G is an $\mathfrak{R}_1$-group, $\tilde{K}$ is the only such component and so $\widetilde{QP}$ normalizes $\tilde{K}$.

Since $[\tilde{Q},\tilde{P}] = \tilde{Q} \cong Q_8$ and $\tilde{z} \in \tilde{Q}'$, it follows at once that $\tilde{z}$ centralizes $\tilde{K}$. But then $C_M(x)$ covers $\tilde{K}$, contrary to the fact that $C_M(x) = C_H(z)$ is solvable in the present case. Hence $\tilde{J} = L(\tilde{H})$ is simple and the possibilities for $\tilde{J}$ are again listed in Theorem A of Part III.

Set $\tilde{F} = C_{\tilde{H}}(\tilde{J})$, so that $\tilde{J}\tilde{F} = \tilde{J} \times \tilde{F}$. Since $\mathrm{Aut}(\tilde{J})/\mathrm{Inn}(\tilde{J})$ is 2'-closed in all cases, we have that $\tilde{Q} \subseteq \tilde{J}\tilde{F}$. Since $C_H(z)$ is solvable, clearly $\tilde{z} \notin \tilde{F}$. Hence either $\tilde{z} \in \tilde{J}$ or else $\tilde{J},\tilde{F}$ each contain quaternion subgroups $\tilde{Q}_0,\tilde{Q}_1$ respectively with $\tilde{Q}$ a "diagonal" of $\tilde{Q}_0\tilde{Q}_1$. Consider the latter case. Clearly $\tilde{x} \in \tilde{F}$ and $\tilde{x}$ centralizes $\tilde{Q}_0\tilde{Q}_1$. This forces $\tilde{x} \in \tilde{Q}_1$, as otherwise $r(\tilde{Q}_0\tilde{Q}_1 < \tilde{x} >) = 5$, contrary to $r(G) \leq 4$. But as $\tilde{z}$ centralizes $\tilde{Q}_1$, it follows that M contains a quaternion subgroup Q_1 with $x \in Q_1$ which maps on $\tilde{Q}_1$. However, $\overline{M}/\overline{LC}$ is abelian and as $x \in Q_1'$, we conclude that $\bar{x} \in \overline{LC}$, whence $x \in R_1R_2$, contrary to the fact that $x \in W_1 - R_1$. Our argument yields that $\tilde{z} \in \tilde{J}$.

Similarly if $\tilde{Q} \not\subseteq \tilde{J}$, it would follow that $\tilde{Q} \subseteq \tilde{Q}_0 \times \tilde{Q}^*$, where $\tilde{Q}_0$ is a quaternion subgroup of $\tilde{J}$ containing $\tilde{z}$ and $\tilde{Q}^*$ is a four subgroup of $\tilde{F}$ normalized, but not centralized by $\tilde{P}$. However, as $\tilde{x}$ centralizes $\tilde{P}$, $\tilde{x} \notin \tilde{Q}^*$ and so $r(\tilde{Q}_0\tilde{Q}^* < \tilde{x} >) = 5$, a contradiction. We thus conclude that $\tilde{Q} \subseteq \tilde{J}$. But now considering the various possibilities for $\tilde{J}$ and using the fact that $C_{\tilde{J}}(\tilde{z})$ is solvable, we see that we must have $\tilde{J} \cong L_3(3)$, $U_3(3)$, or M_{11}.

It follows now that z is in the center of a Sylow 2-subgroup T of H with $Q \subseteq T$. Since $\bar{Q}$ is a Sylow 2-subgroup of $\bar{L}_0 = C_{\bar{L}}(\bar{x})$, no loss of generality will occur if we assume that $T \subseteq S$. Again let T_2 be a subgroup of T such that $\tilde{T}_2$ is a Sylow-subgroup of $\tilde{J}$. As in the case

$q_0 > 3$, using Lemma 2.3 (x) and the known structure of the centralizer of an involution of M_{11}, we conclude now that for some t in T_2, $\bar{t}^2$ centralizes $\bar{L}_0$ with $\bar{L}_0\langle\bar{t}\rangle/\langle\bar{t}^2\rangle \cong PGL(2,3)$, and we reach the same contradiction as before. This completes the proof of the lemma.

Now we can establish Proposition 4.1. by essentially the same argument that yielded Lemma 4.3. Suppose then that $W \cap W^g \neq 1$ with $z^g \in S$ for some g in G and let x_1 be an involution of W such that $x = x_1^g \in W$. If both x_1 and x are in R_1, then $g \in M$ by Lemma 4.3. Hence we can assume this is not the case, so that $W_1 = R_1 \langle x \rangle$ or $R_1 \langle x_1 \rangle$ (or both). Let L_0 be the same subgroup of $C_L(\langle x, x_1 \rangle)$ that we defined in Lemma 4.4 for $C_L(x)$ and set $J_0 = L_0^g$. Then x centralizes both J_0 and L_0 and so each is contained in H.

By Lemma 4.4, z is isolated in H, so $M \cap H$ covers $\tilde{H}$. If $x \in R_1$, then $L \subseteq H$ and $\tilde{L} \triangleleft \tilde{H}$ with $\tilde{L}_0 \subseteq \tilde{L}$. In the contrary case, we obtain that $\tilde{L}_0 \triangleleft \tilde{H}$. Furthermore, correspondingly we have that $\tilde{H}/\tilde{L}$ or $\tilde{H}/\tilde{L}_0$ is solvable. Hence if L_0 is nonsolvable, it follows that $\tilde{J}_0 \subseteq \tilde{L}$ or $\tilde{L}_0$, as the case may be. But then $Z(\tilde{J}_0) = Z(\tilde{L}_0) = \langle \tilde{z} \rangle$ and the argument of Lemma 4.3 yields the desired conclusion $g \in M$.

It remains therefore to show that $g \in M$ also when L_0 is solvable. Let J_1 be a minimal subgroup of $M \cap H$ which covers $\tilde{J}_0$, so that $J_1/O(J_1) \cong \tilde{J}_0 \cong SL(2,3)$ and J_1 contains a normal subgroup $Q_1 \cong Q_8$. In addition, $z^g \in O(H)Z(Q_1)$ and $\langle z^g \rangle$ and $Z(Q_1)$ are each Sylow 2-subgroups of $O(H)Z(Q_1) \cap M$. Hence replacing J_1 by a conjugate by some element of $O(H) \cap M$, we can assume without loss that $\langle z^g \rangle = Z(Q_1)$.

Now we examine $\bar{M}$. Since $\bar{L} \cong SL(2,9)$ or $\hat{A}_7$ in the present case, we have that $\bar{M}/\bar{L}\,\bar{C}$ is a 2-group. Since $J_1 \subseteq M$ and J_1 has no normal subgroups of index 2, it follows that $\bar{J}_1 \subseteq \bar{L}\,\bar{C}$. In particular, $\bar{Q}_1 \subseteq \bar{L}\,\bar{C}$. But $\bar{L}\,\bar{C}/\bar{C} \cong L_2(9)$ or A_7 does not contain a quaternion subgroup and this implies that $Z(\bar{Q}_1) \subseteq \bar{C}$. Since $< z^g > = Z(Q_1)$ and $z^g \in S$ by assumption, we conclude that $Z(\bar{Q}_1) \subseteq \bar{R}_1$ and hence that $z^g \in R_1 \cap R_1^g$ and now Lemma 4.3 is applicable to yield the desired conclusion $g \in M$.

5. <u>The normal four subgroup case.</u> In this section we shall prove

<u>Proposition 5.1.</u> R_1 does not contain a normal four subgroup of S.

We assume false and argue to a contradiction in a sequence of lemmas. Let then U be a four subgroup of R_1 which is normal in S. We can assume that $U = < u,z >$

We shall make use of Goldschmidt's refinement [12] of Alperin's fusion theorem [1] to study the fusion of z in S with respect to G. To this end, we let $\mathfrak{T}$ denote the set of 2-subgroups T of S with the following properties:

(a) $N_S(T)$ is a Sylow 2-subgroup of $N = N_G(T)$,

(b) $C_S(T) \subseteq T$,

(c) T is a Sylow 2-subgroup of $O_{2',2}(N)$,

(d) Either N/T has cyclic or generalized quaternion Sylow 2-subgroups or $N/O_{2',2}(N)$ possesses a normal subgroup of odd index isomorphic to $L_2(2^n)$, $U_3(2^n)$, or $Sz(2^n)$ for some n.

Goldschmidt's theorem asserts that the fusion pattern of subsets of S in G is determined by the fusion pattern of subsets of T in $N_G(T)$ for T in $\mathfrak{T}$ in the sense that $\mathfrak{T}$ forms a conjugation family as this term is defined by Alperin.

Since G is simple, z is not isolated in S by Glauberman's

Z^*- theorem [11]. Hence by the above result, we have

Lemma 5.2. For some T in $\mathfrak{T}$, z is not isolated in $N_G(T)$.

We fix T in $\mathfrak{T}$ such that z is not isolated in $N = N_G(T)$. We also set $Z = \Omega_1(Z(T))$ and $\bar{N} = N/C_N(Z)$ and proceed to analyze the structure of T and $\bar{N}$.

Lemma 5.3. One of the following holds:

(i) $\bar{N} \cong A_5$ or $Z_3 \times A_5$;

(ii) $\bar{N}$ has a normal 2-complement and cyclic Sylow 2-subgroup.

Proof: Since $S \cap N$ is a Sylow 2-subgroup of N and $C_S(T) \subseteq T$, $Z(T)$ is a Sylow 2-subgroup of $C_N(T)$ and so $C_N(T) = O(N) \times Z(T)$. In particular, N is 2-constrained. Set $\tilde{N} = N/TC_N(T)$. Since T is a Sylow 2-subgroup of $O_{2',2}(N)$, it follows that $O_2(\tilde{N}) = 1$. This implies that $\tilde{N}$ acts faithfully on $\tilde{T} = T/\Phi(T)$. Since $r(G) \leq 4$, $|\tilde{T}| \leq 16$ and so $\tilde{N}$ is isomorphic to a subgroup of A_8. But as $C_N(T)T = O(N)T$, $\tilde{N}$ is a homomorphic image of N/T by a group of odd order. Hence applying condition (d) above and examining the various subgroups of A_8, we easily conclude that $\tilde{N}$ has one of the forms listed in (i) or (ii). But $C_N(Z) \supseteq C_N(T)T$ and so $\bar{N}$ is a homomorphic image of $\tilde{N}$. Hence $\bar{N}$ also satisfies (i) or (ii).

We next prove

Lemma 5.4. We have $T \subset S$.

Proof: Suppose $T = S$, in which case $Z = \Omega_1(Z(S))$. Lemma 6.1(iii) of Part III implies that $Z \subseteq R_1$. But as $Z \lhd N$, $z^x \in Z$ for any x in N and so $z^x \in R_1 \cap R_1^x$. Lemma 4.3 now yields that $x \in M$, whence $z^x = z$ for all x in N. Hence z is isolated in N, contrary to our choice of T.

Lemma 5.5. The following conditions hold:

(i) $\Omega_1(Z(S)) \subseteq Z$;

(ii) $Z \cap W = \langle z \rangle$;

(iii) $Z \cong Z_2 \times Z_2$ or E_8.

Proof: Since $C_S(T) \subseteq T$, $Z(S) \subseteq T$ and this implies (i). Set $W_0 = Z \cap W$, so that $z \in W_0$. Suppose $W_0 \supset \langle z \rangle$. If $|Z| \leq 8$ or if $|W_0| \geq 8$, then $W_0^x \cap W_0 \neq 1$ for any x in N. Since $z^x \in Z \subseteq S$, Proposition 4.1 implies that $x \in M$. Again we reach the contradiction that z is isolated in N. Thus $|Z| = 16$ and $|W_0| = 4$.

We can write $Z = W_0 \times Z_0$, where $Z_0 \cong Z_2 \times Z_2$. Suppose $Z_0 \subseteq WR_2$. Since R_1 contains the normal four subgroup U of S and Z_0 is a four group, Lemma 3.5 implies that $Z_0 \cap W \neq 1$. Since $W_0 = Z \cap W$, this is clearly impossible. Therefore Z_0 contains an involution y of $S - WR_2$.

On the other hand, the element r_2 of R_2 satisfies $r_2^2 = z$ and $\langle r_2 \rangle \lhd S$. Since $z \in T$, it follows that $r_2 \in N_S(T)$ and hence that $r_2 \in N$. But $R_2 \langle y \rangle$ is quasi-dihedral by Lemma 3.3 (v) and so y does not centralize r_2. Thus $r_2 \notin T$. Since $\langle r_2 \rangle \lhd S$, it follows now that $Z_1 = C_Z(r_2) \cong E_8$.

We have that $\bar{r}_2$ is an involution. Suppose $\bar{r}_2$ inverts an element $\bar{v}$ of odd order in $\bar{N}$. Then $\langle \bar{r}_2, \bar{r}_2^{\bar{v}} \rangle$ centralizes $Z_1 \cap Z_1^v \cong Z_2 \times Z_2$ or E_8 and $\bar{v} \in \langle \bar{r}_2, \bar{r}_2^{\bar{v}} \rangle$. Since $\bar{v}$ acts nontrivially on Z, we see that the only possibilities are that $|\bar{v}| = 3$ or 7. In particular, $|\bar{v}| \neq 5$ and so $\bar{N} \not\cong A_5$ or $Z_3 \times A_5$. Hence $\bar{N}$ has a normal 2-complement and has cyclic Sylow 2-subgroups by Lemma 5.2 and $O(\bar{N})$ is a 5'-group. Since $\bar{N}$ is isomorphic to a subgroup of A_8, we conclude that $O(\bar{N}) \cong Z_3$ or $Z_3 \times Z_3$.

Since $S \cap N$ is a Sylow 2-subgroup of N and $S \cap N$ centralizes z, it follows now that there exists a 3-element x in N such that $z^x \neq z$ with $\bar{r}_2$ centralizing or inverting $\bar{x}$. Suppose first that $\bar{r}_2$ centralizes $\bar{x}$, in which case x leaves Z_1 invariant. But by Lemma 3.3 (ii), W_0 centralizes r_2 and so $W_0 \subseteq Z_1$. Since $W_0 \cong Z_2 \times Z_2$ and $Z_1 \cong E_8$, it follows that $W_0 \cap W_0^x \neq 1$. Another application of Proposition 4.1 now yields $x \in M$, whence $z^x = z$, a contradiction. Thus $\bar{r}_2$ inverts $\bar{x}$. Again as $Z_1 \cong E_8$, $\bar{x}$ cannot act regularly on Z and so $Z = X_1 \times X_2$, where $X_1 = C_Z(\bar{x}) \cong Z_2 \times Z_2$ and $X_2 = [Z,\bar{x}] \cong Z_2 \times Z_2$. Moreover, $\bar{r}_2$ leaves both X_1 and X_2 invariant and $\bar{r}_2$ does not centralize X_2. This forces $\bar{r}_2$ to centralize X_1 and so $X_1 \subseteq Z_1$. Since X_1 and W_0 are four subgroups of Z_1, it follows that $W_0 \cap X_1 \neq 1$. Since x centralizes $W_0 \cap X_1$, $W_0 \cap W_0^x \neq 1$ and again Proposition 4.1 implies that $x \in M$, giving the same contradiction $z^x = z$. This establishes (ii).

Suppose now that (iii) is false, in which case $Z \cong E_{16}$. Then $Z \cap WR_2 \cong E_8$ or E_{16}, so $Z \cap WR_2$ contains a four subgroup Z^* with $z \notin Z^*$. But Lemma 3.5 implies that $Z^* \cap W \neq 1$. Now (ii) forces $Z^* \cap W = \langle z \rangle$, which is clearly impossible. Thus (iii) also holds.

As a consequence, we can prove

Lemma 5.6. The following conditions hold:

(i) $\bar{N} \cong S_3$;

(ii) T is a Sylow 2-subgroup of $C_N(Z)$.

Proof: We have seen in Lemma 5.3 that N is 2-constrained, that $\bar{N}$ is a homomorphic image of $\tilde{N} = N/O(N)T$, that $O_2(\tilde{N}) = 1$, and that either $\tilde{N} \cong A_5$ or $Z_3 \times A_5$ or $\tilde{N}$ has a normal 2-complement and cyclic Sylow 2-subgroup. However, in the first two cases, Theorem B of Part II would imply that $T \cong E_{16}$ or $Q_8 * D_8$. But then correspondingly $Z \cong E_{16}$ or Z_2, contrary to the preceding lemma. Hence the third case must occur.

Setting $V = S \cap N$, we have that V is a Sylow 2-subgroup of N. Since $T \subset S$ by Lemma 5.4, $\tilde{V} \neq 1$. Setting $\tilde{P} = O(\tilde{N})$, we have by the preceding paragraph, $\tilde{N} = \tilde{V}\tilde{P}$ and $\tilde{V}$ acts faithfully on $\tilde{P}$. Since $\tilde{N}$ is isomorphic to a solvable subgroup of A_8, it also follows that $\tilde{P} \cong Z_3$, $Z_3 \times Z_3$, Z_5 or $Z_3 \times Z_5$. However, as $|Z| = 4$ or 8, $O_5(\tilde{P})$ centralizes Z. Since $z \in Z$, the assumption that $O_5(\tilde{P}) \neq 1$ would imply that $|N_M(T)/C_M(T)|$ is divisible by 5, contrary to Lemma 3.3 (viii). Hence $O_5(\tilde{P}) = 1$ and so $\tilde{P} \cong Z_3$ or $Z_3 \times Z_3$. Furthermore, $\tilde{P}$ does not centralize Z, for then $\bar{N} = \bar{V}$ and z would be isolated in N, which is not the case. Hence if $\tilde{P} \cong Z_3$, $\tilde{P}$ acts faithfully on Z and we conclude that $\bar{N} \cong S_3$. In addition, $|\tilde{V}| = 2$ in this case, so $|V/T| = 2$ and hence $|\bar{V}| = 2$. Clearly these conditions force T to be a Sylow 2-subgroup of $C_N(Z)$, so both parts of the lemma hold.

Consider finally the case that $\tilde{P} \cong Z_3 \times Z_3$. Since $Z \cong Z_2 \times Z_2$ or E_8 and $\tilde{P}$ does not centralize Z, we have $\tilde{P}_0 = C_{\tilde{P}}(Z) \cong Z_3$. Furthermore, $\tilde{P}_0 \lhd \tilde{N}$. If $\tilde{P}_0$ is not inverted by an involution of $\tilde{N}$, then necessarily $\tilde{N}/\tilde{P}_0 \cong S_3$ and again we see that both parts of the lemma hold. Thus we can suppose that $\tilde{P}_0$ is inverted by an involution of $\tilde{N}$. Setting $T_0 = [T, \tilde{P}_0]$, we have that $T_0 \lhd N$. We shall reach a contradiction by showing that $T_0 \subseteq R_1$ or T_0 is nonabelian and $T_0' \subseteq R_1$. Indeed, assume this to be the case and let $x \in N$ be such that $z^x \neq \bar{z}$. By assumption, T_0 or T_0' is a nontrivial subgroup of R_1. Since each is normal in N, it follows that $R_1 \cap R_1^x \neq 1$ and once again Lemma 4.3 yields a contradiction.

By the Frattini argument, we can find a cyclic subgroup P_0 of N which maps an $\tilde{P}_0$ and which is inverted by a 2-element of $V = S \cap N$. Then P_0 normalizes T and clearly $T_0 = [T,P_0]$. We have that $P_0 \subseteq M$ and P_0 is inverted by a 2-element of S. Since $\overline{M}/\overline{L}\,\overline{C}$ is abelian, it follows that $\overline{P}_0 \subseteq \overline{L}\,\overline{C}$. Since $T_0 = [T_0,P_0]$, also $\overline{T}_0 \subseteq \overline{L}\,\overline{C}$ and consequently $T_0 \subseteq R_1R_2$. We can suppose $T_0 \not\subseteq R_1$ or else our assertion is clear. Considering $\tilde{M} = \overline{M}/\overline{C}$ and letting $\tilde{T}_0, \tilde{P}_0$ now denote the images of T_0, P_0 respectively in $\tilde{M}$, we have $\tilde{T}_0$ is dihedral and $[\tilde{T}_0,\tilde{P}_0] = \tilde{T}_0$. This forces $\tilde{T}_0 \cong Z_2 \times Z_2$ and consequently $T_0 \subseteq R_1Q$ for some quaternion subgroup Q of R_2.

If T_0 is nonabelian, then as R_1 centralizes R_2 and $Q' = \langle z \rangle$, it follows that $T_0' \subseteq R_1$, as required. Suppose that T_0 is abelian and set $B = \Omega_1(T_0)$. Since P_0 acts fixed-point-free on B, $z \notin B$ and as $r(G) \leq 4$, we must have $B \cong Z_2 \times Z_2$. But $B \triangleleft T$ as $T_0 \triangleleft T$ and as P_0 normalizes T, it follows that $B \subseteq Z(T)$. Thus $B \subseteq Z$. However, $B \subseteq R_1R_2 \subseteq WR_2$ and so $B \cap W \neq 1$ by Lemma 3.5. Since $z \notin B$ and $B \subseteq Z$, this yields $Z \cap W \supset \langle z \rangle$, contrary to Lemma 5.5 (ii).

We next prove

Lemma 5.7. $Z \cap WR_2$ is a normal four subgroup of N.

Proof: Suppose $Z \subseteq WR_2$. If $Z \cong E_8$, Z contains a four subgroup Z_0 with $z \notin Z_0$. Since $Z_0 \subseteq WR_2$, $Z_0 \cap W \neq 1$ by Lemma 3.5 and as $z \notin Z_0$, Lemma 5.5 (ii) yields a contradiction. Hence $Z \cong Z_2 \times Z_2$ by Lemma 5.5 (iii) and the lemma holds.

Hence we can suppose that $Z \cap WR_2 \subset Z$. Since $|S:WR_2| \leq 2$, it follows that $Z = (Z \cap WR_2) \times \langle y \rangle$ for some involution y of $S - WR_2$. It will suffice to prove that $z \in \Phi(T)$. Indeed, assume that this is the case. Then the normal closure Z_0 of z in N lies in $Z \cap \Phi(T)$. But $\Phi(T) \subseteq \Phi(S)$ and $\Phi(S) \subseteq WR_2$, so $Z_0 \subseteq WR_2$. On the other hand, as z is not isolated in N, $Z_0 \supset \langle z \rangle$. Since $|Z| \leq 8$ and $Z \not\subseteq WR_2$, the only possibility is that $Z_0 = Z \cap WR_2 \cong Z_2 \times Z_2$. Since $Z_0 \triangleleft N$, the lemma will therefore be proved.

Since $\langle r_2 \rangle$ is normal in S and $r_2^2 = z$, we have that $r_2 \in N$. Likewise as $U \triangleleft S$ and $z \in U$, also $U \subseteq N$. Set $U = \langle z,u \rangle$ for some involution u. Since $\overline{N} \cong S_3$ and T is a Sylow 2-subgroup of $C_N(Z)$ by Lemma 5.6, one of the three element u, r_2, or ur_2 lies in T. But $r_2^2 = z$ and $(ur_2)^2 = z$ (as $u \in R_1$ centralizes r_2). Hence $z \in \Phi(T)$ if either of these elements is in T, so we can suppose that $u \in T$. However, $u \notin Z$ as then $U \subseteq Z$, contrary to the fact that $Z \cap W = \langle z \rangle$. Since $U \triangleleft T$, it follows now that $\langle z \rangle = [U,T]$, whence $z \in T' \subseteq \Phi(T)$ in this case as well; and the lemma is proved.

By Lemma 5.6, there exists a 3-element x in N and an element v in $S \cap N$ such that $\langle \overline{x} \rangle = O(\overline{N}) \cong Z_3$, $\overline{v}$ is an involution inverting $\overline{x}$, and $\overline{N} = \langle \overline{x}, \overline{v} \rangle \cong S_3$. Since v centralizes z and z is not isolated in N, x must act regularly on $Z \cap WR_2$. We set $Z_1 = Z \cap WR_2$ and fix all this notation.

We next prove

<u>Lemma 5.8.</u> The following conditions hold:

(i) $C_{T \cap WR_2}(x) = 1$;

(ii) $|C_T(x)| \leq 2$ and $C_{\Phi(T)}(x) = 1$;

(iii) $|T: [T,x]| \leq 2$ and x acts fixed-point-free on $[T,x]$.

<u>Proof:</u> Suppose (i) holds. Since $|S:WR_2| \leq 2$, we have $|T:T \cap WR_2| \leq 2$ and, in particular, $\Phi(T) \subseteq WR_2$. Clearly these conditions imply (ii).

By (ii), x acts fixed-point-free on $\Phi(T)$ and so if we set $\overline{T} = T/\Phi(T)$, we have that $|C_{\overline{T}}(x)| \leq 2$ by (i). Hence x acts fixed-point-free on $[\overline{T},x]$ and $|\overline{T} : [\overline{T},x]| \leq 2$. Since $[T,x]$ maps on $[\overline{T},x]$ and x acts fixed-point-free on $\Phi(T)$, (iii) also follows.

Thus it will suffice to prove (i). We have that $Z_1 = \langle z,wt \rangle$ with $t \in R_2 - \langle z \rangle$ and $w \in W - \langle z \rangle$ as $Z_1 \cong Z_2 \times Z_2$ and $Z \cap W = \langle z \rangle$. Furthermore, by Proposition 4.1, $W \cap W^x = 1$, as otherwise $x \in M$ and $z^x = z$, which is not the case. In particular, $C_{W\cap T}(x)=1$. Thus if $C_{T \cap WR_2}(x) \neq 1$, we have that $C_T(x)$ contains an involution w_1t_1 for some $w_1 \in W - \langle z \rangle$ and $t_1 \in R_2 - \langle z \rangle$. Since $Z_1 \subseteq Z(T)$, $\langle wt, w_1t_1 \rangle$ is a four group and so by Lemma 3.5, we have that $(wt)(w_1t_1) \in W$. But then $w_1t_1 = w_0t$ for some w_0 in $W - \langle z \rangle$ and consequently we can assume without loss that $t_1 = t$. Since x acts regularly on Z_1, we can also suppose, by replacing x by x^{-1}, if necessary, that $(wt)^x = zwt$.

But now as $(t^{-1}w_1^{-1})^x = t^{-1}w_1^{-1}$, we have

$$(ww_1^{-1})^x = (wt)^x(t^{-1}w_1^{-1}) = (zwt)(t^{-1}w_1^{-1}) = zww_1^{-1}.$$

Thus $W \cap W^x \neq 1$, contrary to what we have shown above; and so (i) holds.

We now set $C_T(x) = T_0$ and $T_1 = [T,x]$ and study the structure of T_1 and T. We first prove

<u>Lemma 5.9.</u> The following conditions hold:

(i) $T = T_1$;

(ii) $T/\Phi(T) \cong E_{16}$;

(iii) $T' \subseteq Z(T) \subseteq \Phi(T)$ and $\Phi(T) \cong Z_{2^n} \times Z_{2^n}$ for some n;

(iv) T contains an x-invariant normal subgroup isomorphic to E_{16}.

<u>Proof:</u> We first verify (ii) and (iii) for T_1 in place of T. We shall then prove that $T = T_1$ and shall establish (iv).

We argue first that $|T_1| > 4$. Suppose false, in which case $T_1 = Z_1 \cong Z_2 \times Z_2$ and so $T = T_0T_1 = T_0 \times T_1 \cong Z_2 \times Z_2$ or E_8 as $|T_0| \leq 2$ by Lemma 5.8 (ii) . In the first case $T = Z = T_1$ and so $N_S(T) \cong D_8$. But then S is of maximal class and so is dihedral or quasi-dihedral, contrary to the fact that $m(G) = 3$ or 4 as G is an $\mathfrak{R}_1$-group. Thus $T \cong E_8$ and $T_0 = \langle y \rangle \cong Z_2$ with $y \in S - WR_2$ by Lemma 5.8 (i). But as in Lemma 5.7, both r_2 and $U = \langle z,u \rangle$ are in N and $r_2^2 = (ur_2)^2 = z$. Since $\overline{N} \cong S_3$ with T a Sylow 2-subgroup of $C_N(Z)$ and elementary abelian, the only possibility is that $u \in T$. But then $U \subseteq Z = T$, contrary to the fact that $Z \cap W = \langle z \rangle$. Thus $|T_1| > 4$, as asserted. Since U is a normal four subgroup of $S \cap N$ and x acts fixed-point-free on T_1 with $|T_1| > 4$, $\overline{u}$ cannot invert $\overline{x}$ and so $u \in T$.

Next set $E = \Omega_1(Z(T_1))$. We argue that $E = Z_1$. Since $Z_1 \subseteq Z(T_1)$, $Z_1 \subseteq E$ and so if false, then $E \cong E_{16}$. If $E \subseteq WR_2$, then $|E \cap W| \geq 8$ by Lemma 3.5. But then as x leaves E invariant, it follows that $W \cap W^x \neq 1$ and Proposition 4.1 yields the usual contradiction. Thus $E \not\subseteq WR_2$ and so $E \not\subseteq \Phi(T_1)$. Since x acts fixed-point-free on T_1 and $E \subseteq Z(T_1)$, it follows that $T_1 = T_1^* \times E_1$, where x leaves T_1^* and E_1 invariant and $E_1 \cong Z_2 \times Z_2$. Since $r(T_1) \leq 4$, this forces $r(T_1^*) = 2$ and now Lemma 6.10 of Part III yields that $T_1^* = Z_{2^n} \times Z_{2^n}$ for some n . In particular, T_1 is abelian. Since $Z \not\cong E_{16}$ by Lemma 5.5 (iii), this implies that $T \supset T_1$, whence $T_0 \cong Z_2$.

Now consider the element r_2 . We have that $|T:C_T(r_2)| \leq 2$ as $\langle r_2 \rangle \triangleleft S$. Since x acts fixed-point-free on T_1 and $|T_1| > 4$, $\overline{r}_2$ cannot then be an involution inverting $\overline{x}$. Thus $\overline{r}_2 = 1$ and $r_2 \in T$. Since T_1 is abelian and $T = T_0T_1$ with $T_0 \cong Z_2$, either $r_2 \in T_1$ or

$C_{T_1}(r_2) = C_{T_1}(T_0)$. However, in the latter case $|T_1 : C_{T_1}(T_0)| \leq 2$ and as x leaves $C_{T_1}(T_0)$ invariant and acts fixed-point-free on $T_1/C_{T_1}(T_0)$, this forces $T_1 = C_{T_1}(T_0)$, whence $T = T_1 \times T_0$. Since $u \in T$ and T_1 is abelian, $u \in Z = \Omega_1(Z(T))$, contrary to Lemma 5.5 (ii). Thus, in fact, $r_2 \in T_1$. But by Lemma 3.3 (v), no involution of $S - WR_2$ centralizes r_2. Thus $E \subseteq WR_2$, contrary to what we have shown above. This establishes that $E = Z_1$.

Now we argue that $U = \langle z, u \rangle$ and r_2 lie in T_1. As in the preceding paragraph, $r_2 \in T$ and as $U \triangleleft S$, it follows in the same way that $u \in T$. If T_1 is abelian, again the argument of the preceding paragraph yields that r_2 and u are in T_1. Hence we can suppose that T_1 is nonabelian and also clearly that $T_1 \subset T$ and that r_2 or $u \notin T_1$. Setting $\overline{T} = T/\Phi(T_1)$, we have that $\overline{T}_1 \cong Z_2 \times Z_2$ or E_{16}. Repeating the argument of the preceding paragraph with $\overline{T}_1, \overline{T}_0$ in place of T_1, T_0, we conclude similarly that $\overline{T}_0$ centralizes $\overline{T}_1$. Since $r(G) \leq 4$, this forces $\overline{T}_1 \cong Z_2 \times Z_2$. But now Lemma 2.4 yields that $T_1 \cong Z_{2^n} \times Z_{2^n}$ for some n, contrary to the fact that T_1 is nonabelian. Thus $\langle U, r_2 \rangle \subseteq T_1$, as asserted.

If $U \subseteq Z(T_1)$, then $U = E$ as $E = \Omega_1(Z(T_1)) = Z_1 \cong Z_2 \times Z_2$. But then $U \subseteq Z$, contrary to the fact that $Z \cap W = \langle z \rangle$. Thus $U \not\subseteq Z(T_1)$ and, in particular, T_1 is nonabelian. Since x acts fixed-point-free on T_1, it follows therefore that T_1 has class 2. Again Lemma 2.4 implies that $T_1/\Phi(T_1) \cong E_{16}$, proving (ii) for T_1. Since T_1 is of class 2, $T_1' \subseteq Z(T_1)$. Since $T_1/\Phi(T_1) \cong E_{16}$ and $r(T_1) \leq 4$, it also follows that $Z(T_1) \subseteq \Phi(T_1)$. Hence to prove (iii) for T_1, we need only show that

$\Phi(T_1) \cong Z_{2^n} \times Z_{2^n}$ for some n. We set $V = \Phi(T_1)$. We have $\Omega_1(Z(V)) \subseteq \Phi(S) \subseteq WR_2$. Hence if $\Omega_1(Z(V)) \cong E_{16}$, Lemma 3.5 implies that $|\Omega_1(Z(V)) \cap W| \geq 8$. Since x leaves $\Omega_1(Z(V))$ invariant, Proposition 4.1 once again yields a contradiction. Thus $Z(V) \cong Z_{2^n} \times Z_{2^n}$ for some n and so we need only show that V is abelian; so assume false.

If $V - Z(V)$ contains an involution v, then as x acts fixed-point-free on V, the image of $< v, v^x >$ in $V/Z(V)$ is an x-invariant four group, as $V/Z(V)$ is abelian. Setting $V_1 = < Z(V), v, v^x >$, it follows that V_1' is cyclic. But as V_1 is x-invariant, this forces $V_1' = 1$ and so V_1 is abelian. Since $v \in V_1 - Z(V)$, we see that $\Omega_1(V_1) \cong E_{16}$. Since $\Omega_1(V_1) \subseteq \Phi(S) \subseteq WR_2$, Lemma 3.5 and Proposition 4.1 again yield a contradiction. We conclude that $V - Z(V)$ contains no involutions and hence that $\Omega_1(V) \cong Z_2 \times Z_2$. But then V is a Suzuki 2-group and so V is of type $U_3(4)$ by [30]. In particular, this forces $Z(V) \cong Z_2 \times Z_2$.

Since $Z(T_1) \subseteq V = \Phi(T_1)$, it follows now that $Z(T_1) = Z(V)$. But $T_1' \subseteq Z(T_1)$ and so $T_1' \cong Z_2 \times Z_2$. Setting $\overline{T}_1 = T_1/T_1'$, we have that $\overline{T}_1$ is abelian and that $\overline{V} = \Phi(\overline{T}_1)$. Clearly then $\overline{V} = \mho^1(\overline{T}_1)$. However, as T_1' is elementary abelian and $T_1' \subseteq Z(T_1)$, we conclude at once from this last equality that $V = Z(T_1)$, contrary to the fact that V is of type $U_3(4)$. Our argument thus yields that $V = \Phi(T_1) \cong Z_{2^n} \times Z_{2^n}$ for some n and so (iii) holds for T_1.

Observe next that, as with V above, the group $< Z(T_1), u, u^x >$ is abelian and x-invariant. But then $A = EUU^x$ is elementary abelian and x-invariant. Since $U \neq E$, we have $A \cong E_{16}$. Furthermore, $E \triangleleft N$ and $U \triangleleft S \cap N$. In addition, $U^x \triangleleft T_1$.

Finally we prove $T = T_1$; so assume by way of contradiction that $T_0 \neq 1$, whence $T_0 \cong Z_2$. Note that as T_0 centralizes x and both E and U are normal in T, we have that T_0 normalizes A. Hence if we set $\overline{T} = T/E$, we have that $\overline{A} \cong Z_2 \times Z_2$ and that $\overline{T}_0$ centralizes $\overline{A}$. But $A \cap Z(T_1) = E$ and so $A \cap V = E$. If $V \supset E$, then $\overline{T}_0$ also centralizes $\Omega_1(\overline{V}) \cong Z_2 \times Z_2$. Since $\Omega_1(\overline{V}) \subseteq Z(\overline{T}_1)$ and $\Omega_1(\overline{V}) \cap \overline{A} = 1$, it follows that $\Omega_1(\overline{V})\overline{A}\ \overline{T}_0 \cong E_{32}$, contrary to $r(G) \leq 4$. Hence we must have $V = E$ and so $\overline{T}_1 \cong E_{16}$.

We now consider the subgroup $B = \langle E, r_2, r_2^x \rangle$ of T_1 which, as in the case of A, is abelian and x-invariant with $B/E \cong Z_2 \times Z_2$. Since $|r_2| = 4$, $B \cong Z_4 \times Z_4$ and so $\overline{B} \cong Z_2 \times Z_2$ with $\overline{B} \cap \overline{A} = 1$. But $\overline{T}_0$ also centralizes $\overline{B}$ and so $\overline{T} = \overline{B}\ \overline{A}\ \overline{T}_0 \cong E_{32}$, which is again a contradiction. Thus (i) holds and hence so do (ii) and (iii).

Furthermore, as $S \cap N = T_1 N_{S \cap N}(\langle x \rangle)$ by the Frattini argument, it follows that $\langle U, U^x, U^{x^2} \rangle$ is $S \cap N$ - invariant. Since $\langle U, U^x, U^{x^2} \rangle \subseteq A$, we see that A is invariant under $(S \cap N)\langle x \rangle = N$, proving (iv).

Finally we prove

Lemma 5.10. We have $S \subseteq N$.

Proof: Set $P = S \cap N$, so that P is a Sylow 2-subgroup of N and $|P : T| = 2$. To establish the lemma, we need only prove that T is characteristic in P, inasmuch as $P = N_S(T)$.

We have that $V = \Phi(T) \cong Z_{2^n} \times Z_{2^n}$ for some n. Suppose first that $n \geq 2$ and suppose T^* is a subgroup of P with $T^* \cong T$ and $T^* \neq T$. Since $|P:T| = 2$, we have $V^* = \Phi(T^*) \subseteq T$. But as $V^* \cong V$, it follows now that $\Omega_1(V^*) \subseteq \Omega_1(V)$, whence $\Omega_1(V^*) = \Omega_1(V) = E$. Since $E \subseteq Z(T)$, likewise $E \subseteq Z(T^*)$. But $P = TT^*$ as $T^* \neq T$ and $|P : T| = 2$. Hence $E \subseteq Z(P)$. However, as x acts regularly on E and $\bar{x}$ is inverted in $\bar{N}$ by an element of $\bar{P}$, no element of $P - T$ centralizes E. This contradiction shows that T is the unique subgroup of its form contained in P and so T is characteristic in P in this case.

Suppose then that $n = 1$, in which case $|T| = 64$. Since T contains an elementary abelian subgroup of order 16 and x acts fixed-point-free on T, Lemma 2.5 now yields that T is of type $L_3(4)$. But now Lemma 2.4 of Part II implies that P is of type $\hat{A}_8$. We easily check that every elementary subgroup of P of order 16 lies in T. Since T is generated by its subgroups of this form, T is characteristic in P in this case as well.

Now we can quickly derive a final contradiction. By Lemma 5.9 (iv), N contains a normal subgroup $A \cong E_{16}$. Since $S \subseteq N$, $A \lhd S$. Since $< y, R_2 >$ is quasi-dihedral for any involution y of $S - WR_2$, we have that $A \subseteq WR_2$. It follows therefore from Lemma 3.5 that $|A \cap W| \geq 8$. Since $(A \cap W)^x \subseteq A$, this implies that $W \cap W^x \neq 1$, whence $x \in M$ by Proposition 4.1, giving the usual contradiction $z^x = z$. This completes the proof of Proposition 5.1.

6. <u>The cyclic case.</u> In this section we shall prove

<u>Proposition 6.1.</u> R_1 is not cyclic.

We suppose false and derive a contradiction in a sequence of lemmas. We set $R_1 = < r >$ and $|R_1| = 2^n$, $n \geq 1$. Clearly $r^{2^{n-1}} = z$. Thus $R_1R_2 \cong Z_{2^n} * Q_{2^m}$.

<u>Lemma 6.2.</u> We have $n \geq 2$ and $S \supset R_1R_2$.

Proof: If $n = 1$, then $R_1 = \langle z \rangle$ and it follows from Lemma 2.1 that $m(S) \leq 2$, contrary to the fact that G is an $\mathfrak{R}_1$-group. Likewise if $S = R_1R_2$, then clearly $m(S) \leq 2$, giving the same contradiction.

Lemma 6.3. The following conditions hold:

(i) S' is abelian of rank at most 2;

(ii) All involutions of $R_1R_2 - \langle z \rangle$ are conjugate in M.

Proof: By Lemma 2.2, $(\tilde{S}/\tilde{R}_1)'$ is a cyclic subgroup of $\tilde{R}_1\tilde{R}_2/\tilde{R}_1$, where $\tilde{M} = \overline{M}/\langle z \rangle$. Hence $S' \subseteq R_1R_2$ and $S'R_1/R_1$ is cyclic. Since R_1 centralizes R_2 and R_1 is cyclic with z the unique involution of R_2, (i) follows at once. Furthermore, (ii) is an immediate consequence of the structure of M.

Now set $R_2 \cong Q_{2^m}$. If $m \geq 4$, then clearly $\langle r_2 \rangle$ is the unique subgroup of R_2 of order 4 that is normal in S. We claim the same holds when $m = 3$. Indeed, in this case $R_2 \cong Q_8$ and so $\tilde{L} \cong L_2(q)$, $q \equiv 3,5 \pmod 8$. Since $S \supset R_1R_2$, it follows that $\tilde{L}\tilde{S}/\tilde{R}_1 \cong PGL(2,q)$, whence $\tilde{S}/\tilde{R}_1 \cong D_8$ and so $\langle r_2 \rangle$ maps onto $Z(\tilde{S}/\tilde{R}_1)$. Thus $\langle r_2 \rangle$ is uniquely determined in this case as well.

We set $r_1 = r^{2^{n-2}}$, so that $|r_1| = 4$, and we put $u = r_1r_2$. Then u is an involution and as $\langle r_1 \rangle \triangleleft S$, we have that $U = \langle z,u \rangle$ is a normal four subgroup of S. By Lemma 6.1 (iii) of Part III, $Z(S)$ is cyclic, so $U \in \mathcal{U}(S)$.

Next set $T = C_S(u) = C_S(U)$, so that $|S:T| = 2$. Clearly $R_1 \subseteq T$ and $R_2 \cap T$ is cyclic of index 2 in R_2. We set $T_2 = R_2 \cap T$, so that $T \cap R_1R_2 = R_1T_2$ with $R_1 \cap T_2 = \langle z \rangle$. Thus if $|T_2| > |R_1|$, we can write $R_1T_2 = V \times T_2$, where V is cyclic of order $|R_1|/2$; while if $|T_2| \leq |R_1|$, we can write $R_1T_2 = R_1 \times V$, where V is cyclic of order $|T_2|/2$. In either case we note that R_1T_2 is not homocyclic abelian. Furthermore, as R_2 does not centralize U, we have $S = R_2T$ and hence also $T/R_1T_2 \cong S/R_1R_2$.

Finally by the structure of $\widetilde{M}$, $C_{\widetilde{M}}(\widetilde{U}) = C_{\widetilde{M}}(\widetilde{u})$ has a normal 2-complement and therefore so does $C_M(U) = C_G(U)$. Hence if we set $N = N_G(U)$ and $\overline{N} = N/O(N)$, we obtain that $\overline{T} \subseteq O_2(\overline{N})$.

Our goal now is to prove that $z \not\sim u$ in G. We argue by contradiction. Thus Lemmas 6.4-6.10 are proved under the assumption $z \sim u$ in G.

Lemma 6.4. The following conditions hold:

(i) $\overline{N}/\overline{T} \cong S_3$;

(ii) $|R_1| = |T_2|$;

(iii) If $\overline{x}$ is an element of $\overline{N}$ of order 3, then $\overline{R}_1\overline{R}_1^{\overline{x}} = \overline{R}_1 \times \overline{R}_1^{\overline{x}} \cong Z_{2^n} \times Z_{2^n}$, $\overline{R}_1\overline{R}_1^{\overline{x}} \lhd \overline{N}$, and $\overline{T}_2 \subseteq \overline{R}_1\overline{R}_1^{\overline{x}}$;

(iv) $\overline{T} = \overline{R}_1\overline{R}_1^{\overline{x}}\, C_{\overline{T}}(\overline{x})$ and $C_{\overline{T}}(\overline{x})$ is cyclic;

(v) $\overline{N} = \overline{R}_1\overline{R}_1^{\overline{x}}\, N_{\overline{N}}(\langle \overline{x} \rangle)$, and $N_{\overline{N}}(\langle \overline{x} \rangle) = C_{\overline{T}}(\overline{x}) \times \overline{F}$, where $\overline{F} \cong S_3$.

Proof: Since $U \in \mathcal{U}(S)$ and $Z(S)$ is cyclic, a standard argument yields that $N_G(U)/C_G(U) \cong S_3$, which immediately yields (i).

Let $\overline{x}$ be an element of $\overline{N}$ of order 3, so that $\overline{x}$ acts regularly on U. In particular, $z^{\overline{x}} \neq z$. Now set $D = T_2$ or $D = R_1$ according as $|T_2| > |R_1|$ or $|T_2| \leq |R_1|$. In either case, $D \lhd S$, so $\overline{D}$ and $\overline{D}^{\overline{x}}$ are each normal in $\overline{T}$. But $\overline{z}$ is the unique involution of $\overline{D}$ and $\overline{z}^{\overline{x}}$ the unique involution of $\overline{D}^{\overline{x}}$. Since $\overline{z} \neq \overline{z}^{\overline{x}}$, it follows that $\overline{D} \cap \overline{D}^{\overline{x}} = 1$ and as $[\overline{D},\overline{D}^{\overline{x}}] \subseteq \overline{D} \cap \overline{D}^{\overline{x}}$, we conclude that $\overline{D}\,\overline{D}^{\overline{x}} = \overline{D} \times \overline{D}^{\overline{x}}$. Let E be the inverse image of $\overline{D}^{\overline{x}}$ in S, so that $DE = D \times E$ with $E \cong D$.

Suppose first that $D = T_2$, whence $D \cong Z_{2^{m-1}}$. By Lemma 2.1 (iv) R_1T_2 has index at most 2 in $C_S(T_2)$ and $C_S(T_2) - R_1T_2$ contains an element of order 2^m. Since $E \cong Z_{2^{m-1}}$, it follows that $|R_1T_2| \geq |D|^2 = |T_2|^2$. However, as $D = T_2$, $|T_2| > |R_1|$ and as $R_1 \cap T_2 = \langle z \rangle$, we see that $|R_1T_2| < |T_2|^2$. This contradiction shows that we must have $D = R_1$. Thus $|R_1| \geq |T_2|$.

Since $E \cap D = E \cap R_1 = 1$, we have $\tilde{E} \cong E$ and $\tilde{E} \cap \tilde{R}_1 = 1$. It follows therefore from Lemma 2.2 (vi) that $|\tilde{E}| \leq 2|\tilde{T}_2|$, whence $|E| \leq |T_2|$. Since $|E| = |R_1| \geq |T_2|$, equality must hold and we obtain the conclusion of (ii).

Observe next that the same reasoning shows that also $\bar{D}^{\bar{x}^{-1}} = \bar{D}^{\bar{x}^2}$ centralizes both $\bar{D}$ and $\bar{D}^{\bar{x}}$, so $\bar{D}\,\bar{D}^{\bar{x}}\,\bar{D}^{\bar{x}^2}$ is an abelian subgroup of $\bar{T}$ invariant under $\bar{x}$. By Lemma 2.2 (vi,b), we have that $\Omega_1(\bar{D}\,\bar{D}^{\bar{x}}\bar{D}^{\bar{x}^2}) = \bar{U} \cong Z_2 \times Z_2$. Clearly this forces $\bar{D}^{\bar{x}^2} \subseteq \bar{D}\,\bar{D}^{\bar{x}}$ and so $\bar{x}$ leaves $\bar{D}\bar{E}$ invariant and acts regularly on it. Since $\bar{D} \lhd \bar{T}$ and $\bar{T}^{\bar{x}} = \bar{T}$, $\bar{D}\,\bar{E} \lhd \bar{T}$. Let $\bar{y}$ be an element of $\bar{S}$ which inverts $\bar{x}$. Then $\bar{D}^{\bar{x}\bar{y}} = \bar{D}^{\bar{y}\bar{x}^{-1}} \subseteq \overline{DD}^{\bar{x}^{-1}} = \overline{DE}$. Hence $\bar{D}\,\bar{E} \lhd \bar{N}$, as $\bar{N} = \langle \bar{T}, \bar{x}, \bar{y} \rangle$. Lemma 2.2 (vi,c) now yields $\bar{T}_2 \subseteq \overline{DE}$. Since $\bar{R}_1 = \bar{D}$, we obtain (iii).

By Lemma 2.2 (vi,c), we also have that S/ER_1R_2 is cyclic. Since $R_1T_2 \subseteq DE$ and $T/R_1T_2 \cong S/R_1R_2$, it follows now that T/DE is cyclic. Hence $\bar{x}$ centralizes $\bar{T}/\overline{DE}$ and so $\bar{T} = \overline{DE}C_{\bar{T}}(\bar{x})$ with $C_{\bar{T}}(\bar{x})$ cyclic, proving (iv).

We obtain now by the Frattini argument that $\bar{N} = \overline{DE}N_{\bar{N}}(\langle \bar{x} \rangle)$. Thus it remains to establish the final assertion of (v). Without loss we can assume $\bar{x}$ is chosen so that $N_{\bar{S}}(\langle \bar{x} \rangle)$ is a Sylow 2-subgroup of $N_{\bar{N}}(\langle \bar{x} \rangle)$, whence $\bar{S} = \overline{DE}N_{\bar{S}}(\langle \bar{x} \rangle)$. Setting $\tilde{S} = \bar{S}/\bar{D}\,\bar{E}$, we have that $\tilde{S} \cong N_{\bar{S}}(\langle \bar{x} \rangle)$

as $\bar{x}$ acts regularly on $\overline{DE}$ But $\tilde{S} = \tilde{R}_2\tilde{T}$ and $\tilde{R}_2 \not\subset \tilde{T}$. Furthermore $\tilde{R}_2 \cong Z_2$ as $\bar{T}_2 \subseteq \overline{DE}$ and $|R_2:T_2| = 2$. Since $\tilde{R}_2 \triangleleft \tilde{S}$, this implies that $\tilde{S} = \tilde{R}_2 \times \tilde{T}$. Thus $N_{\bar{S}}(<\bar{x}>) = C_{\bar{T}}(\bar{x}) \times <\bar{y}>$, where $\bar{y}$ is an involution which inverts $\bar{x}$. Since $C_{\bar{T}}(\bar{x})$ is cyclic, the final assertion of (v) follows and all parts of the lemma are proved.

With the notation as in the preceding proof we let y be an involution of S whose image in $\bar{N}$ inverts $\bar{x}$ and we let t be an element of S such that $<\bar{t}> = C_{\bar{T}}(\bar{x})$. In addition, N contains a 3-element x which centralizes t and maps on $\bar{x}$. We also put $T_1 = DE$. Clearly then $S = T_1 <y,t>$ with $T_1<y> \cong Z_{2^n} \int Z_2$, $T_1<y> \triangleleft S$, and $T_1<y> \cap <t> = 1$. Since $m(S) \geq 3$, $t \neq 1$ and so $|t| = 2^h$ for some $h \geq 1$. This is turn implies that $W \supset R_1$ and hence that $n \geq 3$. In particular, as a corollary of our analysis, we have

Lemma 6.5. S is a split extension of $Z_{2^n} \int Z_2$ by Z_{2^h}.

We next prove

Lemma 6.6. $SCN_3(S)$ is nonempty.

Proof: Since all involutions of U are conjugate in G, it follows in the contrary case from [29] and [35] that S is of type $G_2(q)$, $q \equiv 1,7 \pmod 8$, J_2, or $\hat{A}_{10}$. But then by [19], [16], and [18] we know the possibilities for G and in none of these cases does G possess an involution whose centralizer is of the form M.

As a consequence, we obtain

Lemma 6.7. $U<t^{2^{h-1}}>$ is the unique elementary abelian normal subgroup of S of order 8.

Proof: This is immediate from the structure of S and the fact that $SCN_3(S)$ is nonempty and $n \geq 3$.

We set $a = t^{2^{h-1}}$ and prove

Lemma 6.8. The following conditions hold:

(i) $C_S(a) = \mho^1(T_1)\langle y,t\rangle$;

(ii) $U \subseteq \Phi(C_S(a))$;

(iii) If $h > 1$, then $U \subseteq C_S(a))'$.

Proof: Since U, $U\langle a\rangle$, and $R_1 = \langle r\rangle$ are normal in S, we have $[a,R_1] \subseteq U \cap R_1 = \langle z\rangle$. Thus $r^a = r$ or rz and so a centralizes r^2. Thus a centralizes $\mho^1(R_1)$. Since x centralizes a and $T_1 = R_1 \times R_1^x$, it follows that a centralizes $\mho^1(T_1)$. Since $\langle y,t\rangle = \langle y\rangle \times \langle t\rangle$ is abelian, $C_S(a) \supseteq \mho^1(T_1)\langle y,t\rangle$. Again as x centralizes a and acts fixed-point-free on T_1, either equality holds or a centralizes T_1, in which case $a \in Z(S)$. However, $a \notin Z(S)$ as $Z(S)$ is cyclic.

We thus conclude that (i) holds.

Since $n \geq 3$ and $U = \Omega_1(T_1)$, we see that $U \subseteq \Phi(\mho^1(T_1))$, so (ii) follows from (i). Finally suppose $h > 1$, in which case $a = t_1^2$, where $t_1 = t^{2^{h-2}}$. Since t_1 centralizes x, either t_1 centralizes $\mho^1(T_1)$ or $[\mho^1(T_1),t_1] \supseteq U$. Since $\mho^1(T_1)$ and t_1 lie in $C_S(a)$, it follows in the latter case that $U \subseteq C_S(a)'$. Suppose, on the other hand, that t_1 centralizes $\mho^1(T_1)$. Then it is immediate that $[T_1,t_1] \subseteq \Omega_1(T_1) = U$. However, this implies that $[T_1,a] = [T_1,t_1^2] = 1$ and so a centralizes T_1, which is not the case. Thus (iii) also holds.

Now we can prove

Lemma 6.9. The following conditions hold:

(i) $h = 1$;

(ii) a is not conjugate in G to an element of T_1.

Proof: Suppose first that $a \sim b$ in G for some b in T_1. Then $b \in U = \Omega_1(T_1)$ and as the involutions of U are conjugate in G, we see that $a \sim z$ in G. Hence there exists v in G such that $a^v = z$ and $C_S(a)^v \subseteq S$.

If $h > 1$, $U \subseteq C_S(a)'$ by the preceding lemma and so $U^v \subseteq S'$. But $S' \subseteq T_1$ as $S = T_1 < y,t >$ with $< y,t >$ abelian. Since $U = \Omega_1(T_1)$, we conclude that $U^v = U$. However, as $a \notin U$, while $z \in U$, this is clearly impossible. Hence we must have $h = 1$, whence $t = a$ and $S/T_1 \cong Z_2 \times Z_2$, so that also $\Phi(S) \subseteq T_1$. But again by the preceding lemma, $U \subseteq \Phi(C_S(a))$, so $U^v \subseteq T_1$. Again we conclude that $U^v = U$, giving the same contradiction. Thus (ii) holds.

Suppose now that $h > 1$. Since every element of $S - T_1 < y,t^2 >$ has order at least 2^h, it follows from [27, Lemma 16] that $t \sim b$ in G for some b in $T_1 < y,t^2 >$. But then $a = t^{2^{h-1}} \sim c = b^{2^{n-1}}$ in G. However, as $< y,t^2 > \cong Z_2 \times Z_{2^{h-1}}$, it is immediate that $c \in T_1$. Since $a \sim c$ in G, (i) is thus contradicted. Hence (ii) also holds.

Now we obtain our objective.

Lemma 6.10. z is not conjugate to u in G.

Proof: Suppose false. All involutions of $T_1 < y > - T_1$ and $T_1 < ya > - T_1$ are conjugate in S. Since we have shown that a is conjugate to no involution of T_1, Thompson's transfer lemma applied to a and the maximal subgroups $T_1<y>$ and $T_1 < ya >$ of S implies that $a \sim y \sim ya$ in G. This yields that z has only 3 conjugates in S under G--namely, the elements of $U^{\#}$. Since $U \subseteq R_1R_2$, it follows now from Lemma 6.3 (ii) and Glauberman's Z*-theorem that all involutions of R_1R_2 are conjugate in G to z. However, this is impossible as R_1R_2 has more than 3 involutions.

By Lemmas 6.3 (ii) and 6.10 and the Z*-theorem, $z \sim t$ in G with $t \in S - R_1R_2$. We fix g in G such that $t^g = z$ and $C_S(t)^g \subseteq S$.

Lemma 6.11. The following conditions hold:

(i) $z \notin \Phi(C_S(t))$;

(ii) $\tilde{L}<\tilde{t}> \cong PGL(2,q)$ for some odd q;

(iii) $R_1 <t> \cong D_{2^{n+1}}$ or $QD_{2^{n+1}}$;

(iv) All involutions of $R_1R_2<t> - R_1R_2$ are conjugate in $R_1R_2<t>$.

Proof: We argue first that $z \notin C_S(t)'$. Suppose false. Since S/R_1R_2 is abelian, it follows that $z^g \in R_1R_2$. Lemmas 6.3 (ii) and 6.10 force $z^g = z$, which is not the case. This proves our assertion. If $\bar{t}$ induced a field automorphism of $\bar{L}$, then for a suitable conjugate $t_1 \in S - R_1R_2$ of t in M, we would have $z \in C_S(t_1)'$, contradicting the fact shown above. This proves (ii), as t is an involution.

Now we prove (i). Assume false, in which case $t' = z^g \in \Phi(S) \subseteq WR_2$. But then $\tilde{L} < \tilde{t}' > \not\cong PGL(2,q)$. On the other hand, $t' \neq z$ and so by Lemma 6.10, $t' \notin R_1R_2$. However, as t' is conjugate to z in G, we can apply the argument of the preceding paragraph with t' in place of t to conclude that $\tilde{L} < \tilde{t}' > \cong PGL(2,q)$. This contradiction establishes (i).

But now as $z \notin \Phi(C_S(t))$, we must have $C_{R_1}(t) = <z>$. Thus $R_1<t>$ is of maximal class and so $R_1<t> \cong D_{2^{n+1}}$ or $QD_{2^{n+1}}$, proving (iii).

Finally let v be an involution of $R_1R_2 <t> - R_1R_2$. Then $v = y_1y_2t$ with $y_i \in R_i$, $i = 1,2$. Since y_2 centralizes y_1 and $y_1^t = y_1^{-1}$ or $y_1^{-1}z$ (by the structure of $R_1 <t>$), it follows correspondingly that $(y_2t)^2 = 1$ or z. But $R_2 <t>$ is quasi-dihedral with R_2 generalized quaternion, so no element of $R_2 <t> - R_2$ is of order 4. Hence y_2t is an involution and $(y_2t)^{x_2} = t$ for some x_2 in R_2. Since x_2 centralizes y_1, this implies that $v^{x_2} = y_1t$. But as $R_1 <t>$ is dihedral or quasi-dihedral $(y_1t)^{x_1} = t$ or tz for some x_1 in R_1, whence $v^{x_2x_1} = t$ or tz. But $t \sim tz$ in $R_2\langle t \rangle$ and so $v \sim t$ in $R_1R_2\langle t \rangle$, proving (iv).

Lemma 6.12. The following conditions hold:

(i) $m(S) = 3$;

(ii) $C_{R_1R_2\langle t \rangle}(t) = \langle t \rangle \cong E_8$;

(iii) If $B \sim U\langle t \rangle$ in G with $B \subseteq S$, then $B \sim U\langle t \rangle$ in S.

Proof: Since $m(S) \geq 3$, either (i) holds or S contains a subgroup $E \cong E_{16}$. But then $\tilde{E} \cong E_8$ and so $|\tilde{E} \cap \tilde{R}_1\tilde{R}_2| \geq 4$. Therefore, $|E \cap R_1R_2| \geq 8$, contrary to the fact that $m(R_1R_2) = 2$. Thus (i) holds. Furthermore, (ii) follows immediately from the action of t on R_1 and R_2.

Finally set $A = U\langle t \rangle$ and $S_1 = R_1R_2\langle t \rangle$. Suppose that $A \sim B$ in G with $B \subseteq S$. To prove (iii), it will suffice to show that $B \subseteq S_1$. Indeed, since $m(R_1R_2) = 2$, we have $S_1 = R_1R_2\langle b \rangle$ for some b in B. By the preceding lemma, $b \sim t$ in S_1. Hence $C_{S_1}(b) \sim C_{S_1}(t) = A$. Since $B \subseteq C_{S_1}(b)$, it follows that $C_{S_1}(b) = B$ and that $B \sim A$ in S_1. Our argument yields, in fact, that any two elementary subgroups of S_1 of order 8 are conjugate in S_1.

Finally Lemmas 6.10 and 6.11 (iv) imply that z, u have precisely five and two conjugates in A respectively. Hence z, u have the same number of conjugates in B. But by Lemma 6.11 (ii), every conjugate of z in B must lie in $R_1R_2\langle t \rangle = S_1$. Since any five elements of B generate B, we conclude that $B \subseteq S_1$ and the lemma is proved.

Now we can quickly complete the proof of Proposition 6.1. Again we set $A = C_{S_1}(t)$, where $S_1 = R_1R_2\langle t \rangle$. By the preceding lemma, $A = U\langle t \rangle \cong E_8$ and any conjugate of A in G which lies in S is a conjugate of A in S. A standard argument now yields that if $z \sim a$ in G with a in A, then $z \sim a$ in $N_G(A)$.

We know that $A \cap R_1R_2 = U$, that $z \sim t$, and that all four involutions of $A - U$ are conjugate to t in S. Hence $z \sim a$ in G for a in $A - U$ and so $z \sim a$ in $N_G(A)$ for a in $A - U$. Since obviously $z \sim z$ in $N_G(A)$, we see that z has at least 5 conjugates in $N_G(A)$. On the other hand, $z \not\sim u$ or uz in G by Lemma 6.10 and so z has exactly 5 conjugates in $N_G(A)$. Hence $|N_G(A)/C_G(A)|$ must be divisible by 5, contrary to the fact that $N_G(A)/C_G(A)$ is isomorphic to a subgroup of $GL(3,2)$, which has order 168.

7. The maximal class case. By Propositions 5.1 and 6.1, R_1 is necessarily nonabelian of maximal class. In this section we treat this case and prove

Proposition 7.1. If R_1 is nonabelian of maximal class, then S is of type M_{12}.

We proceed by contradiction and divide the proof into a short sequence of lemmas.

Lemma 7.2. W is of maximal class and $|W/R_1| \leq 2$.

Proof: By Lemma 2.2 (iii), $r(\widetilde{W}) \leq 2$. We have that $\widetilde{R}_1 \lhd \widetilde{W}$ and that $\widetilde{R}_1 \cong R_1/\langle z \rangle \cong D_{2^n}$ for some n as R_1 is nonabelian of maximal class. If $n \geq 3$, then $\Omega_1(\widetilde{W}) \not\cong Z_2 \times Z_2$ and so $\widetilde{W}$ is of maximal class by Lemma 6.9 of Part III. Since $\widetilde{R}_1 \lhd \widetilde{W}$ and $\widetilde{W}/\widetilde{R}_1$ is cyclic, this forces $|\widetilde{W}/\widetilde{R}_1| \leq 2$. The lemma follows at once in this case.

Suppose then that $n = 2$, whence $R_1 \cong D_8$ or Q_8. Since $C_W(R_1) \cap R_1 = \langle z \rangle$, we have $C_W(R_1) = \langle z \rangle$, since otherwise $R_1C_W(R_1)/\langle z \rangle$ has sectional 2-rank at least 3. Contrary to $r(\widetilde{W}) \leq 2$. On the other hand if V denotes the subgroup of W which stabilizes the chain $R_1 \supset \langle z \rangle \supset 1$, we have $V = R_1 C_V(R_1)$ by [13, Lemma 5.4.6], so $V = R_1$. This implies that $C_{\tilde{W}}(\tilde{R}_1) = \tilde{R}_1$. Hence either $\tilde{W} = \tilde{R}_1$ or $\tilde{W} \cong D_8$. In either case, the desired conclusions follow.

We let $< r_1 >$ be a cyclic subgroup of R_1 of order 4 with $< r_1 > \triangleleft S$. We next prove

Lemma 7.3. z is not conjugate in G to an element of $R_1R_2 - < z >$.

Proof: Suppose $z \sim t$ in G for some $t \in R_1R_2 - <z>$. Then $t = y_1y_2$, where $y_i \in R_i$, $i = 1,2$, and either $y_1^2 = y_2^2 = z$ or $t = y_1$. Correspondingly $C_S(t)$ contains y_2 or R_2. Hence in either case $z \in \Phi(C_S(t))$. Now let $g \in G$ be such that $t^g = z$ and $C_S(t)^g \subseteq S$. Then $z^g \in \Phi(S)$. But as $|W/R_1| \leq 2$, S/R_1R_2 is elementary of order at most 4 and consequently $z^g \in R_1R_2$. But then by the structure of R_1R_2, $t \sim tz$ and $z^g \sim z^g z$ in S, whence also $z \sim zt$ in $S_1 = S^{g^{-1}}$. Setting $V = < z,t >$ and considering $N_S(V)$ and $N_{S_1}(V)$, we conclude now that $N_G(V)/C_G(V) \cong S_3$.

In particular, it follows that z is not isolated in $N = N_G(V)$. On the other hand, if $t = y_1 \in R_1$, then $L \subseteq C_G(V) \subseteq M$. By the structure of M, $O(C_G(V))< z >$ is then characteristic in $C_G(V)$ and consequently z is isolated in N, which is not the case. Thus $t \neq y_1$ and so $|y_i| = 4$, $i = 1,2$. But now by the structure of M, we see that $C_G(V)$ has a normal 2-complement. Setting $\bar{N} = N/O(N)$, our conditions imply that $O_2(\bar{N}) = C_{\bar{N}}(\bar{V})$ and that $\bar{N}/O_2(\bar{N}) \cong S_3$.

Next consider the case in which $V \in U(S)$. If $SCN_3(S)$ is empty, we reach a contradiction as in Lemma 6.5, so we have $SCN_3(S)$ nonempty. Let A be an elementary abelian normal subgroup of S of order 8 with $V \subset A$.

Suppose $\tilde{L}< \tilde{a} > \cong PGL(2,q)$, q odd, for some a in A. Then $R_2< a >$

is quasi-dihedral and so a inverts an element x_2 of $< a,R_2>$ of order 8. But then $|[a,x_2]| = 4$, contrary to the fact that $[a,x_2] \in [A,S] \subseteq A \cong E_8$. We thus conclude that $\tilde{A} \subseteq \widetilde{WR}_2$, so $A \subseteq WR_2$. Now assume $A \not\subseteq R_1R_2$, whence there exists a in A such that $a = t_2w$ with $t_2 \in R_2$ and $w \in W - R_2$. Then $W \supset R_1$ and so $|W| \geq 16$. Since W is of maximal class, w inverts an element w_1 of order 8 in W. But as $|W/R_1| = 2$, $\tilde{W}$ centralizes $\tilde{R}_2$ and consequently $[t_2,w_1] \in < z >$. It follows at once that $|[a,w_1]| = 4$, giving the same contradiction. We therefore conclude that $A \subseteq R_1R_2$.

Our argument yields that $SCN_3(R_1R_2)$ is nonempty. Since $R_1R_2 = R_1 * R_2$ with R_2 generalized quaternion and R_1 nonabelian of maximal class, this forces $R_i \cong Q_8$, $i = 1,2$, and $R_1R_2 \cong Q_8 * Q_8$. This in turn implies that $\bar{L} \cong SL(2,q)$, $q \equiv 3,5 \pmod 8$ and hence that $W = R_1$. If $S = R_1R_2$, then z would be isolated, which is not the case. Hence $S \supset R_1R_2$ and so $|S| = 64$.

On the other hand, we claim that S does not contain an elementary normal subgroup of order 16. Indeed, let B be such a subgroup of S. Suppose some b in B lies in $S - WR_2$. Then $R_2< b >$ is quasi-dihedral and so $[R_2,b]$ is not elementary, contrary to the fact that $[R_2,b] \subseteq [S,B] \subseteq B$. Thus $B \subseteq WR_2$. But W is of maximal class and $R_1 \cong Q_8$ is of index at most 2 in W. We conclude by the same argument that $B \subseteq R_1R_2$. Since $m(R_1R_2) = 3$, this is a contradiction and our assertion is proved. Therefore $SCN_4(S)$ is empty.

Now set $T = C_S(V)$, so that $|S:T| = 2$ and $\bar{T} = O_2(\bar{N})$. Since $V = < z,t >$ with $t = y_1y_2$, we have that $< \bar{y}_1,\bar{y}_2 > \subseteq \bar{T}$. Since $\bar{N}/\bar{T} \cong S_3$, $\bar{N}$ contains an element $\bar{x}$ of order 3 which does not centralize $\bar{V}$. Set $\tilde{T} = \bar{T}/\bar{V}$, so that $|\tilde{T}| = 8$. If $\bar{x}$ centralized $\tilde{y}_1$, then as $\bar{y}_1^2 = \bar{z}$, it would follow from the action of $\bar{x}$ on $\bar{V}<\bar{y}_1>$ that $\bar{x}$ centralized $\bar{z}$, which is not the case. Hence $\bar{x}$ does not centralize $\tilde{y}_1$. Since $|\tilde{T}| = 8$ and $|\tilde{y}_1| = 2$, this forces $\tilde{T} \cong E_8$ and $\tilde{T}_1 = [\tilde{T},\tilde{x}] = Z_2 \times Z_2$. Hence if $\bar{T}_1$ denotes the inverse image of $\tilde{T}_1$ in $\bar{T}$, we have that $|\bar{T}_1| = 16$, that $[\bar{T},\bar{x}] = \bar{T}_1$, and that $\bar{x}$ acts fixed-point-free on $\bar{T}_1$, whence $\bar{T}_1 \cong E_{16}$ or $Z_4 \times Z_4$. However, as $\bar{T}<\bar{x}> \triangleleft \bar{N}$, clearly $\bar{T}_1 \triangleleft \bar{S}$. Since $SCN_4(S)$ is empty, we conclude that $\bar{T}_1 \cong Z_4 \times Z_4$. Furthermore, we have $\bar{T} = \bar{T}_1 C_{\bar{T}}(\bar{x})$ with $C_{\bar{T}}(\bar{x}) \cong Z_2$ and so $\bar{N} = \bar{T}_1 N_{\bar{N}}(<\bar{x}>)$. Hence for a suitable choice of $\bar{x}$, if we set $\bar{S}_1 = N_{\bar{S}}(<\bar{x}>)$, we obtain that $\bar{S} = \bar{T}_1\bar{S}_1$ with $\bar{T}_1 \cap \bar{S}_1 = 1$ and $\bar{S}_1 \cong Z_2 \times Z_2$ or Z_4.

Consider the latter case first. Then $\bar{S}_1 = <\bar{u}_1>$ with $|\bar{u}_1| = 4$. Since $\bar{u}_1^2$ centralizes $\bar{x}$, we have $\bar{T} = \bar{T}_1 <\bar{u}_1^2>$. But then if $\bar{y} \in \bar{S} - \bar{T}$, it follows that $\bar{y}^2 \in \bar{T} - \bar{T}_1$. On the other hand, $\mho^1(R_1R_2) = <z>$ and R_1R_2 is generated by its elements of order 4. Since $\bar{z} \in \bar{T}_1$, our argument shows that no element of $\bar{R}_1\bar{R}_2$ of order 4 lies in $\bar{S} - \bar{T}$ and consequently $\bar{R}_1\bar{R}_2 \subseteq \bar{T}$. But $r(\bar{T}) = 3$, while $r(\bar{R}_1\bar{R}_2) = 4$. We therefore conclude that $\bar{S}_1 \cong Z_2 \times Z_2$.

Finally if $\bar{S}_1 \cap \bar{T} = C_{\bar{T}}(\bar{x})$ centralized $\bar{T}_1$, then $\bar{S}_1 \cap \bar{T} \subseteq Z(\bar{S})$ and so $Z(S) \cong Z_2 \times Z_2$, which is not the case. Thus $\bar{S}_1<\bar{x}> \cong Z_2 \times S_3$ is isomorphic to a subgroup of $Aut(\bar{T}_1)$. Now Lemma 2.1 (vi) of Part II yields that $\bar{S}$, and hence S is of type M_{12}. Since we are proving Proposition 7.1 by contradiction, our argument shows that $V \notin U(S)$.

Since all elements of order 4 in $\bar{R}_2$ are conjugate in $\bar{L}$, we can assume without loss that $y_2 = r_2$, whence $<y_2> \triangleleft S$. Hence $<y_1>$ is not normal in S. Set $\bar{S} = S/R_2$, so that $\bar{W} \cong W/<z>$ and $<\bar{t}> = <\bar{y}_1>$ is not normal in $\bar{S}$. But $\bar{W}$ is dihedral as W is of maximal class.

Furthermore, $|\overline{S}:\overline{W}| \leq 2$. It follows at once from these conditions that $\overline{t} \notin \overline{S}'$ and hence that $t \notin S'$. In particular, $t \notin C_S(V)'$. On the other hand, as $< y_1 > \not\triangleleft S$ and $|y_1| = 4$, R_1 is not dihedral, so R_1 is either generalized quaternion or quasi-dihedral. In either case, we see that $z \in C_{R_1R_2}(V)'$, whence $z \in C_S(V)'$.

Since $C_G(V) \subseteq M$, we can choose S to contain a Sylow 2-subgroup of $C_G(V)$, so without loss we can assume that $T = C_S(V)$ is a Sylow 2-subgroup of $C_G(V)$. (Actually the structure of M implies that this is the case for any Sylow 2-subgroup S of M.) But now we see that $\overline{z} \in \overline{T}'$, while $\overline{t} \notin \overline{T}'$. Hence $\overline{N}$ cannot possess an element of order 3 which cyclically permits the involutions $\overline{z}, \overline{t}, \overline{zt}$ of $\overline{V}$. This contradiction completes the proof of the lemma.

Lemma 7.3 together with the following results will establish Proposition 7.1.

<u>Lemma 7.4.</u> z is conjugate to an element of $R_1R_2 - < z >$.

<u>Proof:</u> Suppose false. By Glauberman's Z^*-theorem, $S - R_1R_2$ contains an involution t such that $t^g = z$ for some g in G with $C_S(t)^g \subseteq S$. Since $|W/R_1| \leq 2$, S/R_1R_2 is elementary. Hence if $z \in \Phi(C_S(t))$, it follows that $z^g \in R_1R_2$, whence $z^g = z$ by our assumption, contrary to the fact that $t^g = z$. Thus $z \notin \Phi(C_S(t))$ and consequently $\tilde{t} \notin \widetilde{WR_2}$. Hence $\tilde{L}< t > \cong PGL(2,q)$ for some odd q and $R_2< t >$ is quasi-dihedral.

We claim that $R_1< t >$ is of maximal class; so assume false, in which case $C_{R_1}(t) \supset < z >$. Since R_1 is of maximal class and $z \notin \Phi(C_{R_1}(t))$, the only possibility is that $C_{R_1}(t) = < z,v > \cong Z_2 \times Z_2$. But t inverts r_2 as $R_2 < t >$ is quasi-dihedral. Furthermore, as R_1 is nonabelian of maximal class, t also normalizes a cyclic subgroup

$\langle r_1 \rangle$ of order 4 in R_1. Since $C_{R_1}(t)$ is elementary, t must invert r_1. Since $r_1^2 = r_2^2 = z$, t thus centralizes r_1r_2. But v also inverts r_1, so vr_1 is an involution which centralizes r_2, whence $(vr_1r_2)^2 = z$. Since $vr_1r_2 \in C_S(t)$, we conclude that $z \in \Phi(C_S(t))$, which is not the case. Thus $R_1\langle t \rangle$ is of maximal class, as asserted. Hence $R_1\langle t \rangle$ is dihedral or quasi-dihedral. In either case, we compute as in Lemma 6.11 that all involutions of $R_1R_2\langle t \rangle - R_1R_2$ are conjugate to t in R_1R_2 and hence are conjugate to z in G.

Suppose first that $S = R_1R_2\langle t \rangle$. We have that $r_1r_2 \in C_S(t)$ and hence that $\langle r_1r_2, z \rangle^g \subseteq S$. But r_1r_2 and r_1r_2z are involutions of $R_1R_2 - \langle z \rangle$ and so are not conjugate to z in G. It follows therefore from the preceding paragraph that $(r_1r_2)^g$ and $(r_1r_2z)^g$ lie in R_1R_2. But then also $z^g \in R_1R_2$, whence $z^g = z$, which is a contradiction. Hence $S \supset R_1R_2\langle t \rangle$ and so $W \supset R_1$.

Since W is of maximal class, $W = R_1\langle w \rangle$, where $w^2 = 1$ or z. Since $\tilde{L}\langle \widetilde{tw} \rangle \cong PGL^*(2,q)$, $\tilde{R}_2\langle \widetilde{tw} \rangle \cong R_1R_2\langle tw \rangle/R_1$ is quasi-dihedral and consequently $R_1R_2\langle tw \rangle - R_1R_2$ contains no involution. Hence every involution of $S - R_1R_2$ lies in R_1R_2t or R_1R_2w.

Now all involutions of R_1R_2t are conjugate to z in G. Furthermore, w centralizes r_2, so $z \in \Phi(C_S(w))$. But now the argument of the first paragraph of the proof yields that $w \not\sim z$ in G.

Suppose first that w is an involution. By Thompson's transfer lemma, $w \sim v$ in G for some v in $R_1R_2\langle tw \rangle$. We can therefore choose Sylow 2-subgroups S_1, S_2 in G and $g \in G$ such that $C_S(w) \subseteq S_1$, $C_S(v) \subseteq S_2$, $S_1{}^g = S_2$, and $w^g = v$. Since $z \in \Phi(C_S(w))$, we have $z^g \in \Phi(S_2)$. But

$\Phi(S) \subseteq R_1R_2$ and by Lemma 7.2, z is conjugate in G to precisely one involution of R_1R_2 (namely, z itself). Hence z^g is conjugate in G to precisely one involution of $\Phi(S_2)$. On the other hand, since $v \in R_1R_2 < tw >$ and v is an involution, $v \in R_1R_2$ and consequently $z \in \Phi(C_S(v))$. Thus $z \in \Phi(S_2)$. Since obviously $z^g \sim z$ in G, we conclude that $z^g = z$. Therefore $g \in M = C_G(z)$. However, it is immediate from the structure of M that w and v are not conjugate in M.

Hence $w^2 = z$ and consequently $W - R_1$ contains no involutions. But now if u is an involution of $WR_2 - R_1R_2$, it follows from Lemma 3.3 (iv) that u centralizes an element of R_2 of order 4. Thus $z \in \Phi(C_S(u))$. Since $R_1R_2 < tw > = R_1R_2 < tu >$, we can repeat the argument of the preceding paragraph with u in place of w to conclude that $z \not\sim u$ in G. Thus z is not conjugate in G to an element of $WR_2 - R_1R_2$ and it follows now that z is not conjugate in G to an element of $S - R_1R_2 < t >$. Hence by [27, Lemma 16], $w^g = v$ for some v in $R_1R_2 < t >$ and some g in G. But then $z^g = (w^2)^g = v^2 \in R_1R_2$, whence $z^g = z$, again by Lemma 7.2, and so $g \in M$, giving the same contradiction as in the preceding case. This establishes the lemma.

By Proposition 7.1, G has Sylow 2-subgroups of type M_{12}. But now the discussion in the Introduction shows that $G \cong {}^2D_4(3)$ and the proof of Theorem A is complete.

PART V

CENTRAL INVOLUTIONS WITH NON 2-CONSTRAINED CENTRALIZERS

1. Introduction. In the Introduction of Part IV, we have noted that our Main Theorem, stated in the Introduction of the paper will be completely proved once it is established for simple $\mathfrak{R}_1$-groups. Here a group G is called an $\mathfrak{R}_1$-group provided:

(a) G has 2-rank at least 3, sectional 2-rank at most 4, and nonabelian Sylow 2-subgroups;

(b) The nonsolvable composition factors of the proper subgroups of G satisfy the conclusion of the Main Theorem;

(c) $\mathfrak{L}(G)$ is nonempty;

(d) The centralizer of every involution of G is either solvable or non 2-constrained;

(e) For no involution x of G is L_x of sectional 2-rank 4 with each of its components nonsimple.

In part IV, we have shown that a simple $\mathfrak{R}_1$-group in which $\mathfrak{L}_c(G)$ is nonempty and every element of $\mathfrak{L}_c(G)$ is nonsimple is necessarily isomorphic to $D_4^2(3)$. (Here, as in Part IV, $\mathfrak{L}_c(G)$ denotes the set of quasisimple components of L_x as x ranges over the central involutions of G). In particular, G satisfies the conclusion of the Main Theorem.

It will therefore be convenient to say that an $\mathfrak{R}_1$-group G is an $\mathfrak{R}_2$-group if either $\mathcal{L}_c(G)$ is empty or some element of $\mathcal{L}_c(G)$ is simple.

The preceding discussion together with our earlier results shows that the Main Theorem will be completely established once it is proved for $\mathfrak{R}_2$-groups.

We shall here prove

Theorem A. If G is a simple $\mathfrak{R}_2$-group, then the centralizer of every central involution of G is solvable.

As noted in Part IV, the centralizer of every central involution of G is solvable if and only if $\mathcal{L}_c(G)$ is empty. Under the assumption that Theorem A is false, our aim will be, as usual, to pin down the structure of a Sylow 2-subgroup of G. However, in the present instance, the applicable classification theorems will yield that there is no simple group satisfying the required conditions.

2. Initial reductions. Henceforth G will denote a simple $\mathfrak{R}_2$-group in which Theorem A is false and hence in which the centralizer of some central involution is nonsolvable and so has a simple component.

First of all, for any involution x of G, L_x is not of 2-rank 4 with each component nonsimple as G is an $\mathfrak{R}_2$-group. Hence by Theorem A and Corollary A of Part III, we have

Proposition 2.1. For any involution x of G, one of the following holds:

I. $L_x \cong L_2(q)$, $L_3(q)$, $U_3(q)$, $Re(q)$*, $SL(2,q)$, q odd, $L_2(8)$, $Sz(8)$, A_7, $\hat{A}_7$, M_{11}, or J_1 ; or

II. L_x is the direct product of components L_1, L_2 with $L_1 \cong SL(2,q)$, q odd, or $\hat{A}_7$, and $L_2 \cong L_2(q)$, q odd, $L_3(q)$, $q \equiv -1 \pmod 4$, $U_3(q)$, $q \equiv 1 \pmod 4$, A_7, or M_{11}.

By assumption, G possesses a central involution z such that $M = C_G(z)$ is nonsolvable. The possible structures of L_z are given by the preceding proposition. We first prove

Proposition 2.2. If z is a central involution of G, then L_z is not isomorphic to $Re(q)^*$, q odd, $L_2(8)$, $Sz(8)$, or J_1.

Proof: Suppose false for some z. Set $M = C_G(z)$, $\bar{M} = M/O(M)$, and put $\bar{L} = L_z$, so that $\bar{L} = L(\bar{M})$ is isomorphic to one of the listed groups. Also set $\bar{C} = C_{\bar{M}}(\bar{L})$, let S be a Sylow 2-subgroup of M and let R_1, R_2 be subgroups of S whose images in $\bar{M}$ are Sylow 2-subgroups of $\bar{C}$ and $\bar{L}$ respectively. By the structure of $Aut(\bar{L})$, we have that $|\bar{M}/\bar{L}\bar{C}|$ is odd and hence that $\bar{S} \subseteq \bar{C}\,\bar{L}$. It follows at once that $S = R_1R_2$ with R_1 centralizing R_2 and $R_1 \cap R_2 = 1$.

By the structure of $\bar{L}$, $m(R_2) = 3$. Since $r(G) \leqq 4$, this forces $r(R_1) = 1$ and so R_1 is cyclic. But then $R_2 \not\cong E_8$, since otherwise $S = R_1 \times R_2$ would be abelian, contrary to the fact that G is an $\mathfrak{R}_2$-group. Hence $\bar{L} \cong Sz(8)$. Since $\bar{R}_1$ is cyclic, C has a normal 2-complement and as $O(\bar{C}) \subseteq O(\bar{M}) = 1$, we have that $\bar{C} = \bar{R}_1$. We conclude now that M is <u>exceptional</u> in the sense of [24]. Since z is a central involution, [24, Proposition 4] now yields that $G \cong J_1$, whence S is abelian, which is not the case.

As a corollary, we have

Proposition 2.3. If z is a central involution of G such that

$C_G(z)$ is nonsolvable, then either $L_z \cong SL(2,q)$, q odd, or $\hat{A}_7$ or else L_z possesses a unique simple component isomorphic to $L_2(q)$, $L_3(q)$, $U_3(q)$, q odd, A_7, or M_{11}.

Since G is an $\mathfrak{R}_2$-group in which $\mathfrak{L}_c(G)$ is nonempty, some element of $\mathfrak{L}_c(G)$ must be simple. Hence we also have

Proposition 2.4. There exists a central involution z in G such that $C_G(z)$ is nonsolvable and L_z is not isomorphic to $SL(2,q)$, q odd, or $\hat{A}_7$.

Finally we can exclude certain possibilities as Sylow 2-subgroups of G.

Proposition 2.5. A Sylow 2-subgroup of G is not of type $L_3(4)$ or $L_3(4)^{(1)}$ and is not isomorphic to $Z_2 \times D_{2^n}$, $Z_{2^m} \times Z_{2^m} \times D_{2^n}$, $Z_{2^m} \times Z_{2^m} \times QD_{2^n}$, $D_{2^m} \times D_{2^n}$, $D_{2^m} \times QD_{2^n}$, or $QD_{2^m} \times QD_{2^n}$ for any m and n.

Proof: In the first case, $G \cong L_3(4)$ by [16, Theorem C] and so the centralizer of every involution of G is solvable, contrary to the preceding proposition. The second case is excluded as then G is not simple by [18, Lemma 9.4]. Since G is an $\mathfrak{R}_2$-group, a Sylow 2-subgroup of G is nonabelian. Hence in the third case a result of Harada [26] yields that G is not simple. In the remaining cases, we can apply the main results of [20], [38], [40], or [41] to conclude again that G is not simple. Thus we reach a contradiction in each case.

3. Theorem A; the wreathed case. By Proposition 2.4, there exists a central involution z of G such that $M = C_G(z)$ is nonsolvable and L_z is not isomorphic to $SL(2,q)$, q odd, or $\hat{A}_7$. Setting $\bar{M} = L(\bar{M})$, it follows therefore from Proposition 2.3 that $L_z = L(\bar{M})$ possesses a

unique simple component $\bar{L} \cong L_2(q)$, $L_3(q)$, $U_3(q)$, q odd, A_7, or M_{11}.

We let S be a Sylow 2-subgroup of M, so that S is a Sylow 2-subgroup of G. We also set $\bar{C} = C_{\bar{M}}(\bar{L})$ and let R_1, R_2 be subgroups of S whose images in $\bar{M}$ are Sylow 2-subgroups of $\bar{C}$, $\bar{L}$ respectively. Then $z \in R_1$ and R_1, R_2 are each normal in S with $R_1R_2 = R_1 \times R_2$. In particular, $Z(S)$ is noncyclic. Moreover, by the structure of $\bar{L}$, R_2 is either dihedral, quasi-dihedral, or wreathed.

Since $C_M(R_1)$ covers $\bar{L}$ and is S-invariant, $C_M(R_1)$ contains a perfect normal S-invariant subgroup L which maps on $\bar{L}$. Finally we set $\tilde{M} = \bar{M}/\bar{C}$, so that $\tilde{M}$ is isomorphic to a subgroup of $\mathrm{Aut}(\bar{L})$. We fix all this notation for the balance of part V.

In this section we eliminate the wreathed case.

<u>Proposition 3.1</u>. $\bar{L}$ is not isomorphic to $L_3(q)$, $q \equiv 1 \pmod 4$ or $U_3(q)$, $q \equiv -1 \pmod 4$.

<u>Proof</u>: By the structure of $\bar{L}$, R_2 is wreathed if and only if $\bar{L}$ is of one of these forms. Suppose the proposition false. Then $r(R_2) = 3$ and as $r(G) \leq 4$, it follows that $r(R_1) = 1$. Thus R_1 is cyclic. In particular, this implies that $\bar{C}$ has a normal 2-complement. Since $O(\bar{M}) = 1$, we conclude that $\bar{C} = \bar{R}_1$.

We claim that $\Omega_1(S) \subseteq R_1R_2$. Indeed, suppose false, in which case $S - R_1R_2$ contains an involution x. But then $\bar{L}\langle\bar{x}\rangle$ is isomorphic to a subgroup of $\mathrm{Aut}(\bar{L})$ containing $\bar{L}$ as a subgroup of index 2 and so $r(\bar{L}\langle \bar{x} \rangle) = 4$ by Lemma 2.3(iii) of Part IV. Hence $r(\bar{L}\langle\bar{x},\bar{z}\rangle) = 5$, which is a contradiction and our assertion is proved.

Let T be the abelian subgroup of index 2 in R_2 and let

$A = TR_1 = T \times R_1$, so that A is abelian. By Lemma 2.3(i) of Part IV, applied to $\tilde{M}$, we have that A is the unique abelian subgroup of its structure in S. Hence A is weakly closed in S with respect to G and so two elements of A conjugate in G are already conjugate in $N = N_G(A)$.

Finally we analyze the structure of N. We have $S \subseteq N$ as $A \triangleleft S$. By Glauberman's Z^*-theorem, z is conjugate in G to some involution y of $S - \langle z \rangle$. Then $y \in \Omega_1(S) \subseteq R_1R_2$ and so $y = y_1y_2$ with $y_i \in R_i$ and $|y_i| = 1$ or 2, $i = 1, 2$. Moreover $y_2 \neq 1$, since otherwise $y \in R_1$, whence $y = z$, which is not the case. But $\bar{L}$ has only one conjugacy class of involutions and so $y_2 \sim t$ in M for some involution t of T. Since $y_1 = 1$ or z, it follows that $y = y_1y_2 \sim a = y_1t$ in M, whence $z \sim a$ in G. But $a \neq z$ and $a \in A$, so $z \sim a$ in N by the preceding paragraph. We conclude that $N \not\subseteq M$.

On the other hand, since $\bar{C} = \bar{R}_1$, it follows from Lemma 2.3(ii) of Part IV, applied to $\tilde{M}$, that $N_M(T)$ contains a 3-subgroup P which centralizes R_1, does not centralize T, and is such that $S = TN_S(P)$. Setting $B = \Omega_1(A)$ and $\bar{N} = N/C_N(B)$, we have that $B = \Omega_1(T) \times \langle z \rangle \cong E_8$ and that $\bar{N}$ is isomorphic to a subgroup of $L_3(2)$. Since $B \not\subseteq Z(R_2)$, we have $\bar{R}_2 \neq 1$ and so $\bar{S} \neq 1$. Furthermore, $|\bar{P}| = 3$ and as $T \subseteq C_N(B)$, we also have that $\bar{S}$ normalizes $\bar{P}$. Since $\bar{S}$ is a Sylow 2-subgroup of $\bar{N}$, we conclude now from the structure of $L_3(2)$ that $\bar{N} = \bar{S}\bar{P} \cong S_3$. But now it follows that $\bar{N}$ centralizes $z \in B$ and so $N \subseteq M$, contrary to the preceding paragraph.

4. Preliminary results. By Proposition 3.1, $\bar{L} \cong L_2(q)$, q odd, $L_3(q)$, $q \equiv -1 \pmod 4$, $U_3(q)$, $q \equiv 1 \pmod 4$, A_7, or M_{11}; and, in particular, R_2 is either dihedral or quasi-dihedral. These conditions will prevail throughout the balance of Part V.

In this section, we shall establish a large number of properties of M which we shall need for our analysis. We begin with a list of consequences of Lemmas 2.2 and 2.3 of Part IV. If $\bar{L} \cong M_{11}$, then $\bar{M} = \bar{L}\bar{C}$ inasmuchas $\mathrm{Aut}(M_{11}) \cong M_{11}$. In this case, we set $W = R_1$, so that $S = WR_2$. In the remaining cases, as $\tilde{M} = \bar{M}/\bar{C}$ is isomorphic to a subgroup of $\mathrm{Aut}(\bar{L})$, the above-mentioned lemmas are applicable with $\tilde{M}$ in the role of the group H. We let $\tilde{W}$ be the subgroup of $\tilde{S}$ which corresponds to the subgroup W of these lemmas. Thus $\tilde{W}$ is a group of field automorphisms of $\tilde{L}$ if $\tilde{L} \cong L_2(q)$ or $U_3(q)$, $\tilde{W}$ is either trivial or generated by the image of the transpose-inverse map if $\tilde{L} \cong L_3(q)$, and $\tilde{W}$ is either trivial or is generated by an odd permutation of order 2 of S_7 if $\tilde{L} \cong A_7$. Furthermore, $C_{\tilde{R}_2}(\tilde{W})$ is a Sylow 2-subgroup of $C_{\tilde{L}}(\tilde{W})$.

We let W be the subgroup of S containing R_1 whose image in $\tilde{M}$ is $\tilde{W}$. In addition, let W_1 be the inverse image in W of $\Omega_1(\tilde{W})$. We also set $Q_2 = \Omega_1(R_2)$, so that Q_2 is dihedral of index 1 or 2 in R_2 according as R_2 is dihedral or quasi-dihedral. We fix all this notation. With the aid of the appropriate parts of these two lemmas from Part IV, we easily verify the following facts:

(4.1) $W \supseteq W_1 \supseteq R_1$, $r(W) \leq 2$, W/R_1 is cyclic, $\Omega_1(W/R_1) = W_1/R_1$ has order at most 2, $W \cap R_2 = 1$, $W = W_1$ if $\bar{L} \cong L_3(q)$ or A_7, and

$W = W_1 = R_1$ if $\bar{L} \cong M_{11}$;

(4.2) If $R_2 \cong Z_2 \times Z_2$, then $W = R_1$;

(4.3) W_1 centralizes Q_2 and $\Omega_1(WQ_2) = \Omega_1(W_1) \times Q_2$;

(4.4) If $S \supset WR_2$, then $\bar{L} \cong L_2(q)$. Moreover, $|S : WQ_2| \leq 2$ and if equality holds, then $\bar{L} \cong L_2(q)$, $L_3(q)$, or $U_3(q)$;

(4.5) If $S - WQ_2$ contains an involution y, then $\bar{L}\langle\bar{y}\rangle \cong PGL(2,q)$, $L_3^*(q)$, or $U_3^*(q)$;

(4.6) If $y \in S - WQ_2$ and $w \in W_1 - R_1$, then $\langle[y,w]\rangle \equiv Z(R_2) (\mathrm{mod}\ R_1)$. In particular, $C_W(y) \subseteq R_1$. Moreover, if R_2 is dihedral, then $\Omega_1(C_S(y)) \subseteqq (\Omega_1(C_{R_1}(y)) \times C_{R_2}(y))\langle y\rangle$;

(4.7) $Z(S) \subseteq W_1Q_2$; and if $S \supset WQ_2$, then $Z(S) \subseteq R_1Q_2$;

(4.8) If y is an involution of $WR_2 - R_1R_2$, then $C_L(y)/O(C_L(y)) \cong PGL(2,r)$ for some odd r.

Since W/R_1 is cyclic, we also have the following consequence of (4.1):

(4.9) If W is of maximal class, then $W = W_1$.

Finally as $r(W) \leq 2$ and $\Omega_1(W) = \Omega_1(W_1)$, Lemma 6.9 of Part III yields

(4.10) $\Omega_1(W) \cong Z_2$ or D_{2^m} for some m.

We next derive some further consequences of these results.

Lemma 4.1. Let T be a subgroup of S. Then we have

(i) If $m(T) = 4$, then $C_M(T)$ has a normal 2-complement;

(ii) If $m(T) = 4$ and $T \not\subseteq WR_2$, then $N_M(T)$ has a normal 2-complement;

(iii) If $T \cong E_8$ and $T \not\subseteq WR_2$, then $N_M(T)/C_M(T)$ is a 2-group;

(iv) If $T \cong E_{16}$ and R_2 is quasi-dihedral, then $T \subseteq W_1Q_2$;

(v) If $T \cong E_{16}$ and $T \subseteq WQ_2$, then $T = X \times Y$, where $X \subseteq W$, $Y \subseteq Q_2$, and $X \cong Y \cong Z_2 \times Z_2$.

Proof: We have that $\Omega_1(WQ_2) = \Omega_1(W) \times Q_2$ with $\Omega_1(W) \cong Z_2$ or D_{2^m} and $Q_2 \cong D_{2^n}$ for suitable m,n; and this immediately yields (v).

In proving (i), we can clearly assume that $T \cong E_{16}$. Since $m(R_1) \leq 2$, we have $m(\tilde{T}) \geq 2$. If $\tilde{L} \cong M_{11}$, then $\tilde{T} \subseteq \tilde{R}_2 = \tilde{S}$ and $\tilde{L} = \tilde{M}$, in which case it follows at once from the structure of M_{11} that $C_{\tilde{M}}(\tilde{T})$ is a 2-group. In the contrary case, we can apply Lemma 2.2(ix) or 2.3(vii) of Part IV to obtain that $C_{\tilde{M}}(\tilde{T})$ has a normal 2-complement.

In either case, let K be the inverse image of $O(C_{\tilde{M}}(\tilde{T}))$ in M and set $H = KT$. Then R_1 is a Sylow 2-subgroup of K and by the preceding paragraph $C_M(T) \subseteq H$. Setting $\bar{H} = H/O(H)$, we can apply Proposition 6.11 of Part III to $\bar{K}$ as $r(\bar{K}) = r(\bar{R}_1) \leq 2$. Since $\bar{z} \in Z(\bar{K})$, it follows that either $\bar{K} = \bar{R}_1$ or else $\bar{K}$ possesses a normal subgroup $\bar{J}$ of odd index with $\bar{J} \cong SL(2,r)$, r odd, $\hat{A}_7$, $SL^{(1)}(2,r)$, or $SL^{(2)}(2,r)$, where $SL^{(1)}(2,r)$ and $SL^{(2)}(2,r)$ are the nonsplit extentions of Z_2 by $PGL(2,r)$ having quasi-dihedral or generalized quaternion Sylow 2-subgroups respectively. In the first case, obviously $C_M(T) = C_H(T)$ has a normal 2-complement; so we can suppose $\bar{J}$ exists.

We have that $\bar{T} \cong T \cong E_{16}$ and that $\bar{z} \in \bar{T} \cap \bar{J}$. If $\bar{T} \cap \bar{J} \supset \langle \bar{z} \rangle$, then necessarily $\bar{T} \cap \bar{J} \cong Z_2 \times Z_2$ and $\bar{J} \cong SL^{(1)}(2,r)$ and we conclude at once from the structure of $P\Gamma L(2,q)$ that $C_{\bar{K}}(\bar{T} \cap \bar{J})$ has a normal 2-complement. Hence $C_M(T)$ has a normal 2-complement in this case. We reach the same conclusion if for some $\bar{t}$ in $\bar{T}$, either $\bar{J}'\langle\bar{t}\rangle \cong SL^{(1)}(2,r)$ or $\bar{t}$ induces a nontrivial inner automorphism of $\bar{J}'$. Hence we can suppose that none of these conditions holds.

We have that $\bar{T}_1 = C_{\bar{H}}(\bar{J})$ is a 2-group and so $\bar{T}_1 \subseteq \bar{T}$. Setting $\tilde{H} = \bar{H}/\bar{T}_1$, it follows under our present assumptions that $\tilde{J} \cap \tilde{T} = 1$ and $|\tilde{T}| \leq 2$. Moreover, if equality holds, either $\tilde{J} \cong A_7$ or $\tilde{T}$ induces a field automorphism of $\tilde{J} \cong L_2(r)$. We see then that in each remaining case $C_{\tilde{J}}(\tilde{T})$ contains a subgroup isomorphic to $L_2(3)$ disjoint from $\tilde{T}$ and this implies that $C_{\bar{J}}(\bar{T})$ contains a quaternion subgroup $\bar{Q}$ with $\bar{Q} \cap \bar{T} = \langle\bar{z}\rangle$. But then $r(\bar{Q}\bar{T}) = 5$, contrary to $r(G) \leq 4$. This completes the proof of (i).

Next assume $m(T) = 4$ and $T \not\subseteq WR_2$, whence $\bar{L} \cong L_2(q)$. Considering $\tilde{M}$ again, we have that $\tilde{L}\langle\tilde{t}\rangle \cong PGL(2,q)$ or $PGL^*(2,q)$ for some $\tilde{t}$ in $\tilde{T}$. We conclude again from the structure of $P\Gamma L(2,q)$, as in the proof of Lemma 2.2(ix) of Part IV, that $N_{\tilde{M}}(\tilde{T})/C_{\tilde{M}}(\tilde{T})$ is a 2-group. Thus $N_{\tilde{M}}(\tilde{T})$ has a normal 2-complement and now repeating the proof of (i), we see that $N_M(T)$ has a normal 2-complement in this case as well, so (ii) also holds.

Suppose next that (iii) is false, in which case $N_{\bar{M}}(\bar{T})$ contains an element $\bar{x}$ of odd order which does not centralize $\bar{T}$. Clearly $\bar{T} \cap \bar{C} = \bar{T} \cap \bar{R}_1$ has order at most 4 with equality holding only if $\bar{z} \in \bar{T} \cap \bar{C}$. Since $\bar{x}$ centralizes $\bar{z}$, it follows that $\bar{x}$ centralizes $\bar{T} \cap \bar{C}$. Hence $\tilde{x}$ acts nontrivially on $\tilde{T}$. But $\tilde{T}$ is elementary and $\tilde{L}\langle\tilde{t}\rangle \cong PGL(2,q)$ for some $\tilde{t}$ in $\tilde{T}$ as $T \not\subseteq WR_2$. Thus $\tilde{R}_2\langle\tilde{t}\rangle$ is dihedral and as $\tilde{T}$ centralizes $\tilde{t}$, we conclude that $\tilde{T} \cap \tilde{L}$ has order at most 2. Since $\tilde{M}/\tilde{L}$ is abelian, it is immediate now that $\tilde{x}$ must centralize $\tilde{T}$, which is not the case. This contradiction establishes (iii).

Finally we prove (iv). Thus R_2 is quasi-dihedral and $T \cong E_{16}$. Suppose $T \cap R_2 = T \cap Q_2 \cong Z_2 \times Z_2$. Since W_1 centralizes Q_2 and $S = WR_2$, it follows that $\Omega_1(C_S(T \cap Q_2)) \subseteq W_1Q_2$. Since T centralizes $T \cap Q_2$, and T is elementary, we conclude that $T \subseteq W_1Q_2$ and so (iv) holds in this case. We can therefore assume that $|T \cap R_2| \leq 2$, whence $T \cap R_2 = Z(R_2)$. Setting $\bar{S} = S/R_2'$, we have then that $\bar{R}_2 \cong Z_2 \times Z_2$, $\bar{T} \cong E_8$, $\bar{R}_2 \cap \bar{T} = 1$, and $\bar{R}_2 \lhd \bar{S}$. However, by the structure of the automomorphism group of a quasi-dihedral group, $\bar{T}$ necessarily centralizes $\bar{R}_2$, so $\bar{R}_2\bar{T} \cong E_{32}$, contrary to $r(G) \leq 4$. Hence (iv) holds and the lemma is proved.

We also have

<u>Lemma 4.2</u>. The following conditions hold:

(i) If Y is a four subgroup of R_2, then $N_L(Y)$ contains a 3-element which normalizes, but does not centralize Y;

(ii) $C_{R_2}(W)$ is a Sylow 2-subgroup of $C_L(W)$ and $C_L(W)$ contains a subgroup isomorphic to A_4. Moreover, either $S = R_1 \times R_2$ with $R_2 \cong Z_2 \times Z_2$ or $C_L(W)$ contains a subgroup isomorphic to S_4;

(iii) If $T = X \times Y \cong E_{16}$ with $X \subseteq W$ and $Y \subseteq R_2$, then $N_L(T)$ contains a 3-element u such that $[u,Y] = Y$ and $C_T(u) \cong Z_2 \times Z_2$;

(iv) In (iii), either $C_T(u) = X$ and $Y \subseteq C_L(X)'$ or $X \not\subseteq R_1$,

$Z(R_2) \cong Z_2$, $C_T(u) \subseteq XZ(R_2)$, and $Y \not\subseteq C_L(X)'$.

Proof: First, (i) is immediate from the structure of $\bar{L}$. We have noted the first assertion of (ii) above. The second and third assertions of (ii) are clear if $W = R_1$. In the contrary case, $\bar{L} \not\cong M_{11}$ and the desired conclusions follow from Lemmas 2.2(iii) and 2.3(v) of Part IV.

As for (iii) and (iv), if $X \subseteq R_1$, these follow from (i) with $C_T(u) = X$ and $Y \subseteq C_L(X)'$. Suppose then that $X \not\subseteq R_1$, whence $|R_2| \geq 8$ and so $Z(R_2) \cong Z_2$. Since $r(W) \leq 2$, we have $\tilde{X} \cong Z_2$. Set $\tilde{K} = \tilde{L}\tilde{X}$. We know that $C_{\tilde{K}}(\tilde{Y})$ has a normal 2-complement with $\tilde{T} = \tilde{X}\tilde{Y} \cong E_8$ as a Sylow 2-subgroup and, moreover, that $N_{\tilde{K}}(\tilde{Y})/C_{\tilde{K}}(\tilde{Y}) \cong S_3$ with $\tilde{P} = N_{\tilde{R}_2}(\tilde{Y})\tilde{X}$ as a Sylow 2-subgroup and $\tilde{P} \cap \tilde{R}_2 \cong D_8$. By the Frattini argument, there exists a 3-element $\tilde{u}$ in $N_K(\tilde{T})$ such that $[\tilde{u}, \tilde{Y}] = \tilde{Y}$. Then $C_{\tilde{T}}(\tilde{u}) = \langle \tilde{t} \rangle \cong Z_2$ with $\langle \tilde{t} \rangle$ normal in $\tilde{P}$. Hence $\tilde{t} \in Z(\tilde{P}) = Z(\tilde{R}_2) \times \tilde{X}$. Furthermore, it is clear from the structure of $C_{\tilde{L}}(\tilde{X})$ that $\langle \tilde{t} \rangle = \tilde{X}$ if and only if $\tilde{Y} \subseteq C_{\tilde{L}}(\tilde{X})'$. But now (iii) and (iv) follow at once in this case as well.

Finally we have the following four additional properties of M.

Lemma 4.3. The following conditions hold:

(i) If $S \supset WR_2$, then M has a normal subgroup of index 2 with Sylow 2-subgroup WR_2;

(ii) $\bar{M} = \bar{S}\bar{L}C_{\bar{M}}(\bar{R}_2)$;

(iii) If y is an involution of $S - WR_2$ and $C_{\bar{M}}(\bar{L})$ is a 2-group, then $m(C_S(y)) = m(C_M(y))$;

(iv) If $\bar{L} \cong L_2(q)$, then every involution of WR_2 is conjugate in M to an involution of $WZ(R_2)$.

Proof: Under the assumption of (i), $\bar{L} \cong L_2(q)$ and $\tilde{M}$ contains a normal subgroup isomorphic to PGL(2,q) or PGL*(2,q). Hence by the structure of $P\Gamma L(2,q)$, $\tilde{M}$ has a normal subgroup of index 2 with $\tilde{W}\tilde{R}_2$ as Sylow 2-subgroup and (i) follows.

As for (ii), we have $\bar{M} = \bar{L}N_{\bar{M}}(\bar{R}_2)$ by the Frattini argument. Moreover, $\bar{S} \subseteq N_{\bar{M}}(\bar{R}_2)$. If $\bar{R}_2 \cong D_{2^n}$ or QD_{2^n} with $n \geq 3$, then $Aut(\bar{R}_2)$ is a 2-group, which implies that $N_{\bar{M}}(\bar{R}_2) = \bar{S}C_{\bar{M}}(\bar{R}_2)$ and consequently $M = \overline{LSC_{\bar{M}}}(\bar{R}_2)$. The only other possibility is that $\bar{R}_2 \cong Z_2 \times Z_2$, whence $N_{\bar{L}}(\bar{R}_2)$ contains an element of order 3 which does not centralize $\bar{R}_2$. Hence $N_{\bar{M}}(\bar{R}_2) = N_{\bar{L}}(\bar{R}_2)\bar{S}C_{\bar{M}}(\bar{R}_2)$, so again $\bar{M} = \overline{LSC_{\bar{M}}}(\bar{R}_2)$. Thus (ii) also holds.

If y is as in (iii), then by (4.6), $\Omega_1(C_S(y)) = \Omega_1(C_{R_1}(y)) \times C_{R_2}(y) \times \langle y \rangle$, where $C_{R_2}(y) \cong Z_2$. An analogous result clearly holds for any Sylow 2-subgroup of M containing y. Hence to establish (iii), it will suffice to prove that $m(C_{\bar{R}_1}(\bar{y})) = m(C_{\bar{T}_1}(\bar{y}))$ for any $\bar{y}$-invariant Sylow 2-subgroup $\bar{T}_1$ of $\bar{C}$. This is obvious as $\bar{C}$ is a 2-group.

Finally in proving (iv), we clearly need only show that every involution of $\tilde{W}\tilde{R}_2$ is conjugate in $\tilde{M}$ to an involution of $\tilde{W}Z(\tilde{R}_2)$. However, this is immediate from the structure of $P\Gamma L(2,q)$.

All above results depended only upon M and not upon its embedding in G. Now we shall use this embedding to derive some additional properties of M and S.

Lemma 4.4. We have $Z(S) \cap R_2 \cong Z_2$.

Proof: Suppose false. Since $R_2 \lhd S$ and R_2 is either dihedral or quasi-dihedral, the only possibility is that $R_2 \cong Z_2 \times Z_2$ and $R_2 \subseteq Z(S)$. Then $W = R_1$ and $\bar{L} \cong L_2(q)$, $q \equiv 3, 5 \pmod 8$. Furthermore, $\tilde{M}$ does not contain a normal subgroup isomorphic to $PGL(2,q)$, so $S = WR_2 = R_1 \times R_2$. By Proposition 2.5, R_1 is neither abelian, dihedral, nor quasi-dihedral. On the other hand, if R_1 is generalized quaternion, a standard fusion argument shows that z is isolated in S with respect to G and then Glauberman's Z^*-theorem contradicts the simplicity of G. We conclude therefore from Lemma 6.9 of part III that $X = \Omega_1(R_1) \cong Z_2 \times Z_2$. Hence $A = X \times R_2 = \Omega_1(S)$.

Set $N = N_G(A)$, $C = C_G(A)$, and $\bar{N} = N/O(N)$. By Lemma 4.1(i), C has a normal 2-complement. Furthermore, $S \subseteq N$ and so $C = O(N)(S \cap C) = O(N)C_S(A)$. Thus $\bar{C} \subseteq \bar{S}$. Since $X \triangleleft R_1$, we also have that $|S:C_S(A)| \le 2$. Hence if $\tilde{N} = \bar{N}/\bar{C}$, it follows that $|\tilde{S}| \le 2$, whence $\tilde{N}$ has a normal 2-complement. In particular, $\bar{N}$ is solvable and so $\bar{N} = \bar{S}\bar{P}$, where $\bar{P}$ has odd order.

Since $\tilde{P} \triangleleft \tilde{N}$, $C_{\bar{A}}(\bar{P}) \triangleleft \bar{N}$. Hence if $C_{\bar{A}}(\bar{P}) \neq 1$, it would follow that some involution a of A is isolated in N. However, as $A = \Omega_1(S)$ is certainly weakly closed in S, two elements of A conjugate in G are already conjugate in N. Thus this would imply that a is isolated in G, again giving a contradiction. Hence $C_{\bar{A}}(\bar{P}) = 1$ and as $\bar{A} = \Omega_1(\bar{C})$, this yields that $C_{\bar{C}}(\bar{P}) = 1$. But by the Frattini argument, $\bar{N} = \bar{C}N_{\bar{N}}(\bar{P})$. Since $C_{\bar{C}}(\bar{P}) = 1$, $N_{\bar{N}}(\bar{P})$ is therefore a complement ot $\bar{C}$ in $\bar{N}$. In particular, it follows that $\bar{S}$ splits over $\bar{C}$ and hence that S splits over $C_S(A)$. However, all involutions of S lie in A. The only possibility therefore is that

$S = C_S(A)$. Thus $\bar{S} = \bar{C}$.

We have that $S = R_1 \times R_2$ and that S is nonabelian. Thus $S' = R_1' \neq 1$ and so $1 \neq \Omega_1(S') \subseteq X$. But $\bar{P}$ leaves $\Omega_1(\bar{S}')$ invariant and $C_{\bar{A}}(\bar{P}) = 1$. Since $\bar{X}$ is a four group, this forces $\bar{X} = \Omega_1(\bar{S}')$. Again by the action of $\bar{P}$ on $\bar{A}$, it follows now that $A = X \times B$, where $B \cong Z_2 \times Z_2$ and $\bar{B}$ is $\bar{P}$-invariant. Then $S = R_1 \times B$ and so $R_1 \cong S/B \cong \bar{S}/\bar{B} = \tilde{S}$. But $\bar{P}$ acts nontrivially on $\Omega_1(\tilde{S}) = \tilde{X} \cong Z_2 \times Z_2$ and so $\tilde{S}$ is necessarily a Suzuki 2-group. Since $r(R_1) = 2$ and $R_1 \cong \tilde{S}$, the only possibility is that $\tilde{S} \cong Z_{2^n} \times Z_{2^n}$, contrary to the fact that R_1 is nonabelian. This establishes the lemma.

For the balance of Part V we set $\langle z_2 \rangle = Z(S) \cap R_2$ and $Z = \langle z, z_2 \rangle$, so that Z is a four subgroup of $Z(S)$. We next prove

Lemma 4.5. No two of the involutions of Z are conjugate in G.

Proof: Suppose false and set $E = \Omega_1(Z(S))$, so that $Z \subseteq E$. By (4.7), $E \subseteq W_1 Q_2 = W_1 \times Q_2$. By the preceding lemma, $E \cap Q_2 = \langle z_2 \rangle$. Since $r(W_1) \leq 2$, it follows therefore that $E \cong Z_2 \times Z_2$ or E_8. Set $N = N_G(S)$. By Burnside's lemma, two elements of E conjugate in G are already conjugate in N. Hence for some element x of N of odd order, we have $(z')^x = z''$, where z', z'' are distinct involutions of Z.

Consider first the case that x fixes z_2. Since $|x|$ is odd, we must have $E \cong E_8$ and $E = \langle z_2 \rangle \times B$ for some x-invariant four subgroup B of E. If $B = \langle a, b \rangle$, then $a \sim b \sim ab$ and $az_2 \sim bz_2 \sim abz_2$ under the action of $\langle x \rangle$. Since x fixes z_2, clearly $z' \neq z_2$ and $z'' \neq z_2$, so $z'' = z'z_2$. If $z' \in B$, then also $z'' \in B$, whence $z_2 \in B$, which is not the case, so $z' \notin B$. Hence $z' = z_2b'$ for some b' in $B^{\#}$. By symmetry, we can clearly assume without loss that $b' = a$. Thus $z' = z_2a$ and so $z'' = z_2b$, whence $z'z_2 = z_2b$, forcing $z' = b$, which is not the case. We therefore conclude that $z_2^x \neq z_2$.

Since $r(S) \leq 4$, x induces an automorphism of S of order 3, 5, 7, or 15. However, as $m(E) \leq 3$, any automorphism of S of order 5 acts trivially on E. Hence without loss we can suppose that x induces an automorphism of S of order 3 or 7.

Since R_2 and R_2^x are each normal in S, the assumption $R_2 \cap R_2^x \neq 1$ would imply that $R_2 \cap R_2^x \cap E \neq 1$. But $R_2 \cap E = \langle z_2 \rangle$ by the preceding lemma and so $z_2 \in R_2^x \cap E$. However, $\langle z_2 \rangle^x = (R_2 \cap E)^x = R_2^x \cap E$ and so $z_2^x = z_2$, which is not the case. We thus conclude that $R_2 \cap R_2^x = 1$ and so $R_2R_2^x = R_2 \times R_2^x$. But also x^i does not centralize z_2 for any i for which x^i induces a nontrivial automorphism of S (as this automorphism has prime order). Hence by the same reasoning, also $R_2R_2^{x^i} = R_2 \times R_2^{x^i}$ for any such i. Conjugating these relations by appropriate powers of x, it follows now that $R_2^{x^i}R_2^{x^j} = R_2^{x^i} \times R_2^{x^j}$ for all i, j with $i \neq j$ and $0 \leq i, j \leq 2$ or 6, according as x induces an automorphism of S of order 3 or 7. Setting $T = R_2R_2^x$, it follows that $R_2^{x^j}$ centralizes T, where correspondingly $j = 2$ or $2 \leq j \leq 6$.

If R_2 is nonabelian, then $Z(T) \cong Z_2 \times Z_2$ and $r(T/Z(T)) = 4$. Hence $T \cap R_2^{x^j} \subseteq Z(T)$ and $r(TR_2^{x^j}/Z(T)) \geq 5$, contrary to $r(G) \leq 4$. Therefore R_2 must be abelian and so $R_2 \cong Z_2 \times Z_2$. Thus $T \cong E_{16}$. We reach the same contradiction if any $R_2^{x^j} \not\leq T$ (where j is in the appropriate range). Therefore our argument also yields that T is x-invariant. Since $R_2 \cong Z_2 \times Z_2$, (4.2) also yields that $W = R_1$. Since $R_2 \cap Z(S) = \langle z_2 \rangle$, this in turn implies that $S/R_1 \cong D_8$.

Assume next that $E = Z$, in which case $z^x \neq z$. Since $R_1 \triangleleft S$, the same reasoning as above yields that $R_1 R_1^x = R_1 \times R_1^x$. Hence $R_1 \cong R_1^x$ is isomorphic to a subgroup of D_8. Since $m(T) = m(S) = 4$, the only possibilities are $R_1 \cong Z_2 \times Z_2$ or D_8. However, in the latter case, $S = R_1 \times R_1^x \cong D_8 \times D_8$, contrary to Proposition 2.5. Hence $R_1 \cong Z_2 \times Z_2$. Then $C_S(R_2) = R_1 R_2 \cong E_{16}$ and so $T = R_1 R_2$. In addition, $|S : T| = 2$. Furthermore, by the structure of M, L contains a 3-element which centralizes R_1 and normalizes, but does not centralize R_2. Since x normalizes $R_1 R_2 = T$, we conclude easily in this case that $S = (Z_2 \times Z_2) \int Z_2$, contrary to [27, Lemma 18]. Thus $E \cong E_8$.

By the structure of S, we must have $E \cap R_1 \cong Z_2 \times Z_2$. Hence by (4.10), $E \cap R_1 = X = \Omega_1(R_1)$. It follows therefore that $C_S(T) = R_1 R_2 = R_1 \times R_2$. If $R_1 \cong Z_2 \times Z_2$, then $T = R_1 R_2$ and $|S : T| = 2$. Again we reach a contradiction as in the preceding paragraph. Thus $R_1 \not\cong Z_2 \times Z_2$ and so $R_1 \supset X$. This implies that $\mho^1(R_1) \neq 1$. But $\mho^1(R_1) = \mho^1(R_1 R_2)$ is normal in S as $R_1 \triangleleft S$. Hence $E_1 = \mho^1(R_1) \cap E \neq 1$. But E_1 is x-invariant as $R_1 R_2$ is. In particular, x does not act irreducibly on E. Since x acts nontrivially on $E \cong E_8$, x cannot induce an automorphism of S of order 7 and so x induces an automorphism of S of order 3.

If $X \subseteq \Phi(R_1) = \Phi(R_1 R_2)$, then x leaves X invariant. But $E = \langle z_2 \rangle \times X$ with $z \in X$ and we conclude easily in this case that no two involutions of Z can be conjugate under x. Hence $X \not\subseteq \Phi(R_1)$. Since $r(R_1) = 2$, this forces $R_1 \cong Z_{2^n} \times Z_2$ for some $n \geq 2$ (as $R_1 \cong Z_2 \times Z_2$). Thus $R_1R_2 \cong E_8 \times Z_{2^n}$.

Finally set $N = N_G(R_1R_2)$ and $\bar{N} = N/C_G(R_1R_2)$. Then $\langle S,x\rangle \subseteq N$ and $|\bar{S}| = |S/R_1R_2| = 2$. Thus $\bar{N}$ has a normal complement $\bar{P}$ and it is immediate from the structure of R_1R_2 that $|\bar{P}| = 3$. On the other hand, by the structure of M, L contains a 3-element d which acts trivially on R_1 and nontrivially on R_2. But then $d \in N$ and so $\bar{P} = \langle \bar{d}\rangle$. Hence M covers $\bar{N}$ so $N \subseteq M$. In particular, $x \in M$. However, it is immediate that no two of the involutions of Z are conjugate in M. This completes the proof of the lemma.

We next prove

Lemma 4.6. We have $m(S) = 4$.

Proof: Suppose false. Since W centralizes a four subgroup of R_2 by Lemma 4.2(ii), we must have $m(W) = 1$ and so W is either cyclic or generalized quaternion. Since $Z(S) \subseteq W_1R_2$ by (4.7) and $R_2 \cap Z(S) = \langle z_2\rangle$, we also have that $Z = \Omega_1(Z(S))$. Furthermore, $\Omega_1(WQ_2) = \Omega_1(W_1) \times Q_2$ by (4.3). But $\Omega_1(W_1) = \langle z\rangle$ in the present case. Hence all involutions of WR_2 lie in $\langle z\rangle \times Q_2$. However, by the structure of M, every involution of $\langle z\rangle \times Q_2$ is conjugate in M to an involution of Z. It follows therefore from the preceding lemma that z is not conjugate in G to an involution of $WR_2 - \langle z\rangle$. Since G is simple, Glauberman's Z*-theorem now yields that $S \supset WR_2$ and that $z \sim y$ in G for some involution y of $S - WR_2$. By (4.4), this in turn implies that $\bar{L} \cong L_2(q)$ and $\bar{L}\langle\bar{y}\rangle \cong PGL(2,q)$, whence $R_2\langle y\rangle$ is dihedral.

Since $y \sim z$ in G, there exists an element g in G such that $y^g = z$ and $S_1^g = S$, where S_1 is a Sylow 2-subgroup of $C_G(y)$ containing $C_S(y)$. Set $Z_1 = \Omega_1(Z(S_1))$, so that $y \in Z_1$. By the preceding paragraph, $Z_1 \cong Z_2 \times Z_2$. Now $C_S(y) \supseteq Z\langle y \rangle \cong E_8$. Since $m(S) = m(S_1) = 3$, it follows that $Z_1 \subseteq Z\langle y \rangle$ and consequently $Z_1 = \langle y, z' \rangle$ for some z' in $Z^{\#}$. Thus $y^g = z$ and $(z')^g \in Z = \Omega_1(Z(S))$. Lemma 4.5 now yields that $(z')^g = z'$, whence $z' = z_2$ or zz_2. On the other hand, as $R_2\langle y \rangle$ is dihedral, $y \sim yz_2$ in S. Hence $Z_1 \neq \langle y, yz_2 \rangle$, otherwise Lemma 4.5 would be contradicted. Thus we must have $z' = zz_2$.

We argue next that y centralizes R_1. Suppose false and set $V = C_{R_1}(y)$, so that $V \subset R_1$. Then there exists a y-invariant subgroup V_1 of R_1 containing V as a subgroup of index 2. Since V_1 is cyclic or generalized quaternion, it follows that $[V_1, y] = \langle z \rangle$, which implies that $y \sim yz$ in S. However, again as $R_2\langle y \rangle$ is dihedral, we have that $yz \sim yzz_2$ in S and consequently $y \sim yzz_2$ in S. But $Z_1 = \langle y, yzz_2 \rangle$ and Lemma 4.5 implies that no two of the involutions of Z_1 are conjugate in G. This contradiction establishes our assertion.

We have that $z \in S_1$ and so $z^g \in S$. If $z^g \in WR_2$, the first paragraph of the proof shows that $z^g = z$, contrary to $y^g = z$. Thus $z^g \in S - WR_2$. In particular, $z^g \notin \Phi(S)$. But $R_1 \subseteq C_S(y) \subseteq S_1$ and so $R_1^g \subseteq S$. Thus $z^g \notin \Phi(R_1^g)$. Since R_1 is cyclic or generalized quaternion with z as its only involution, we must have $R_1 = \langle z \rangle \cong Z_2$.

Since W/R_1 is cyclic, this in turn yields that W is cyclic. Furthermore, $R_2\langle y\rangle \lhd S$ as $y \not\sim yz$ in S.

Finally set $W = \langle w\rangle$ and $2^m = |w|$. Suppose $m \geq 2$. Applying [27], it follows now that for some x in G, w^x lies in the maximal subgroup $T = \langle w^2, y, R_2\rangle$ of S. But then $z^x \in \mho^{m-1}(T)$. However, $T/R_2 \cong Z_2 \times Z_{2^{m-1}}$ and so is of exponent 2^{m-1}. Clearly then we must have $z^x \in R_2$, contrary to the fact that z is not conjugate in G to an element of $WR_2 - \langle z\rangle$. Thus $m = 1$ and so $W = R_1 \cong Z_2$. Hence $S = R_1 \times R_2\langle y\rangle \cong Z_2 \times D_{2^n}$ for some n, contrary to Proposition 2.5.

Finally we prove

<u>Lemma 4.7</u>. W_1 is not cyclic or generalized quaternion.

<u>Proof</u>: Suppose false. By Lemma 4.6, there exists A in S with $A \cong E_{16}$. If $A \subseteq WR_2$, then $A \subseteq W_1Q_2 = W_1 \times Q_2$. Since Q_2 is dihedral, this forces $m(W_1) = 2$, contrary to assumption. Hence $A \not\subseteq WR_2$ and so $S - WR_2$ contains an involution a of A. In particular, $\bar{L} \cong L_2(q)$ and $R_2\langle a\rangle$ is dihedral. By (4.6), $A \subseteq \Omega_1(C_S(y)) \subseteq R_1 \times \langle z_2, y\rangle$, so $m(R_1) = 2$, contrary to the fact that $R_1 \subseteq W_1$.

We next establish a general result about the structure of $C_G(R_2)$. To this end, we set $K = C_G(R_2)WR_2$ and prove

<u>Lemma 4.8</u>. Either z is isolated in K or we have:

(i) $W \cong D_{2^n}$, QD_{2^n}, or $Z_{2^n} \times Z_{2^n}$ for some n;

(ii) K contains a characteristic subgroup J with $O(K) \subseteq J$ such that $R_2 \cap J = 1$ and $|WR_2 : WR_2 \cap JR_2| \leq 2$;

(iii) $J/O(K) \cong L_2(r)$, r odd, $L_3(r)$, $r \equiv -1 \pmod 4$, $U_3(r)$,

$r \equiv 1 \pmod 4$), A_7, M_{11}, or $PSL_m(2,3)$, $m \geq 2$.

Proof: We may assume that z is not isolated in K. Set $\bar{K} = K/R_2$ and suppose first that $\bar{z}$ is isolated in $\bar{K}$, whence $O(\bar{K})\langle\bar{z}\rangle \triangleleft \bar{K}$. But as every element of K of odd order centralizes R_2, it is immediate that $\overline{O(K)} = O(\bar{K})$ and consequently $O(K)R_2\langle z\rangle \triangleleft K$. On the other hand, all involutions of R_2 are conjugate in M. Since z is not conjugate to z_2 or zz_2 in G by Lemma 4.5, it follows that z is isolated in $N_G(R_2\langle z\rangle)$. However, $K = O(K)N_K(R_2\langle z\rangle)$ by the Frattini argument and so $K = O(K)C_K(z)$. Thus z is isolated in K, contrary to assumption. We conclude that $\bar{z}$ is not isolated in $\bar{K}$.

We have that $\bar{W}$ is a Sylow 2-subgroup of $\bar{K}$. Moreover, $\bar{W} \cong W$ as $W \cap R_2 = 1$. Hence $r(\bar{W}) = r(W) \leq 2$. We can therefore apply Proposition 6.11 of Part III to $\bar{K}/O(\bar{K})$ to obtain that $\bar{K}$ possesses a characteristic subgroup $\bar{J}$ with $\bar{J} \supseteq O(\bar{K})$ and $\bar{J}, \bar{W}$ isomorphic to one of the groups listed in (iii) and (i), respectively; and, in addition, that $|\bar{W} : \bar{W} \cap \bar{J}| \leq 2$.

Finally set $J_1 = O^2(K)$, so that $\bar{J}_1 \supseteq \bar{J}$. By the structure of $\bar{K}$, we have that $\bar{J}_1/\bar{J}$ is abelian of odd order with $\bar{J}_1 = \bar{J}$ if $\bar{J}$ is solvable. Furthermore, as $O^2(J_1) = J_1$, $|J_1 : J_1'|$ is of odd order. It follows therefore that $J_1 \cap R_2$ lies in both $Z(J_1)$ and J_1'. However, as WR_2 splits over R_2, we reach a contradiction (using Gaschütz' theorem) unless $J \cap R_2 = 1$. Hence this intersection must be trivial. But now setting $J = J_1$ if $\bar{J}$ is solvable and $J = J_1'$ in the contrary case, we immediately obtain all parts of the lemma.

We conclude this section with a somewhat unrelated result.

Lemma 4.9. Let H be a group in which $O(H) = 1$ and every component of $L(H)$ is isomorphic to $SL(2,q)$, q odd, or $\hat{A}_7$. If T is a Sylow 2-subgroup of H, then $Z(T) \subseteq O_2(H)$.

Proof: If $L(H) = 1$, then H is 2-constrained and the lemma is clear. Hence we can suppose $L(H) \neq 1$. Since $T \cap L(H)$ is a Sylow 2-subgroup of $L(H)$, $Z(T)$ must leave each compenent L_i of $L(H)$ invariant, $1 \leq i \leq m$. Since $L_i \cong SL(2,q_i)$, q_i odd, or $\hat{A}_7$, Lemma 6.1 (iii) of Part III implies that $Z(T)$ centralizes L_i, $i \leq i \leq m$, and hence centralizes $L(H)$. Since $O_2(H) \subseteq T$, $Z(T)$ also centralizes $O_2(H)$. But $C_H(L(H)O_2(H)) \subseteq O_2(H)$ by [23, Theorem 1] and so $Z(T) \subseteq O_2(H)$, as asserted.

5. Maximal elementary abelian 2-subgroups. By Lemma 4.6, $m(S) = 4$ and so S contains elementary subgroups of order 16. Let A and B be two such subgroups which are conjugate in G. In this section we examine the question of whether A and B are necessarily conjugate in S. Invoking Alperin's fusion theorem [1], we have, in any event, that one of the following holds:

(I) $A \sim B$ in S; or

(II) There exists a subgroup T of S containing A such that $N_S(T)$ is a Sylow 2-subgroup of $N_G(T)$ and for some g in $N_G(T)$, A^g is not conjugate to A in S.

In the latter case, we set $N = N_G(T)$, we let $A = A_1, A_2, \ldots, A_m$ be the conjugates of A in N, and we put $D = \langle A_i \mid 1 \leq i \leq m\rangle$. Since $A \subseteq T \triangleleft N$, we have $D \subseteq T$. Moreover, $D \triangleleft N$ and $D = \Omega_1(D)$. We fix this notation throughout the section.

Furthermore, by Lemma 4.4, $R_2 \cap Z(S) = \langle z_2\rangle \cong Z_2$. As in Section 4, we set $Z = \langle z, z_2 \rangle$ so that Z is a four subgroup of $Z(S)$. Throughout the balance of Part V, z_2 and Z will have these meanings.

We first prove

Proposition 5.1. If R_2 is quasi-dihedral, then $A \sim B$ is S.

Proof: Suppose then that R_2 is quasi-dihedral and the proposition is false, in which case (II) above holds. Combining Lemma 4.1(iv) and (4.3), we have that each $A_i \subseteq \Omega_1(W_1) \times Q_2$. Since Q_2 is dihedral and $A_i \cong E_{16}$, it follows now from (4.1) that $r(\Omega_1(W_1)) = 2$. But then Lemma 6.9 of Part III yields that $\Omega_1(W_1)$ is also dihedral. We conclude immediately that $D \cong D_{2^h} \times D_{2^k}$ for some h, k. Hence by the structure of D, any element of N of odd order necessarily normalizes each A_i. But as $S \cap N$ is a Sylow 2-subgroup of N, x is a product of elements of N of odd order and elements of $S \cap N$ and so $A^g = A^y$ for some y in $S \cap N$, contrary to assumption.

It may happen that S contains another central involution z' such that $L_{z'}$ also possesses a simple component of the same general form as L_z. If for any such z', the corresponding component has quasi-dihedral Sylow 2-subgroups, then clearly Proposition 5.1 applies with z' in the role of z. Hence we can suppose for the balance of the section that the component of any such z' has dihedral Sylow 2-subgroups. In particular, R_2 is dihedral and $\bar{L} \cong L_2(q)$ or A_7.

Among all such central involutions z', we assume now that z is chosen to satisfy the following conditions:

(a) If possible, $S = WR_2$;

(b) Subject to (a), $|R_2|$ is maximal.

In this case, we shall prove

Proposition 5.2. One of the following holds:

(i) $A \sim B$ in S;

(ii) (a) $A \not\subseteq WR_2$ and $B \not\subseteq WR_2$;

(b) $W \cong Z_{2^n} \times Z_{2^n}$ for some n;

(c) z is conjugate in G to an element of $R_1 - \langle z \rangle$.

Remark. By further argument, one can rule out the second possibility. However, we do not need to do so.

We let L be a minimal S-invariant normal subgroup of $C_M(R_1)$ which covers $\bar{L}$. Then L is perfect, R_2 is a Sylow 2-subgroup of L, $O(L) \subseteq O(M)$, and $L/O(L) \cong \bar{L}$. We fix all this notation as well. In addition, we assume that $A \not\sim B$ in S. Thus the conditions of (II) above hold. We shall argue in a sequence of lemmas that the the conclusions of Proposition 5.2(ii) are satisfied.

Lemma 5.3. The following conditions hold:

(i) $D \not\subseteq WR_2$;

(ii) $\bar{L} \cong L_2(q)$;

(iii) $S - WR_2$ contains an involution.

Proof: Since R_2 is dihedral, we have $R_2 = Q_2$. If $D \subseteq WR_2$, we can therefore repeat the argument of Proposition 5.1 to obtain a contradiction. Hence we have (i). Since $D = \Omega_1(D)$, (i) implies that some involution of D is not in WR_2 and so (iii) holds. Furthermore, (ii) follows at once from (iii) and (4.4).

Similarly we have

Lemma 5.4. z is not isolated in $N = N_G(T)$.

Proof: Suppose false. Since clearly $z \in A \subseteq T$, $O(N)$ centralizes z and so $N \subseteq M = C_G(z)$. But $m(D) = 4$ and $D \not\subseteq WR_2$, so by Lemma 4.1(ii), $N_M(D)$ has a normal 2-complement. Since $N \subseteq M$ and $D \lhd N$, it follows that N has a normal 2-complement. Thus $A^g = A^y$ for some y in $S \cap N$, contrary to the fact that (II) holds.

We next prove

Lemma 5.5. We have $A_i \cap W \cong Z_2 \times Z_2$, $1 \leq i \leq m$.

Proof: If $A_i \subseteq W_1R_2 = W_1 \times R_2$, the lemma is clear as R_2 is dihedral. In the contrary case, $A_i \not\subseteq WR_2$ and so A_i contains an involution a of $S - WR_2$. Hence by (4.6), $A_i \subseteq \Omega_1(C_S(a)) \subseteq \langle z_2, a\rangle \times C_{W_1}(a)$ and consequently $A_i \cap C_{W_1}(a) \cong Z_2 \times Z_2$. Thus the lemma holds in this case as well.

Lemma 5.6. If $\Omega_1(W) \cong Z_2 \times Z_2$, then Proposition 5.2(ii) holds.

Proof: Set $X = \Omega_1(W)$ and suppose $X \cong Z_2 \times Z_2$. Then $X = \Omega_1(W_1)$ and so $X \subseteq A_i$ for each i, $1 \leq i \leq m$, by the preceding lemma. This implies that $X \subseteq E = \Omega_1(Z(D))$. Since $D \not\subseteq WR_2$, it follows now from (4.6) that $X \subseteq R_1$. Furthermore, if $|E| = 16$, then as $r(G) \leq 4$, we must have $A = E$, contrary to the fact that A is not normal in N. Hence $|E| \leq 8$. However, $z_2 \in Z(S)$ and so also $z_2 \in A_i$ for all i, whence $z_2 \in E$. We conclude that $E = X\langle z_2\rangle = XZ \cong E_8$. We put $X = \langle z, z'\rangle$ for some involution z'. We make one further preliminary observation: namely, $z^g \neq z$. Indeed, in the contrary case, it would follow from Lemma 4.1(ii), as in Lemma 5.4, that $A^g = A^y$ for some y in S, which is not the case. Note also that $E \lhd N$.

By Lemma 4.5, no two of the involutions of Z are conjugate in G. In particular, $N_G(E)$ does not act transitively on $E^{\#}$ and so g can be taken to be a 3-element. The orbits of g on $E^{\#}$ are then of size 3, 3, 1, respectively, and each must contain one element of $Z^{\#}$. Since $Z \subseteq Z(S)$, it follows that all elements of $E^{\#}$ are central involutions of G. Furthermore, as g does not centralize z and $X \cap Z = \langle z \rangle$, g centralizes no element of $X^{\#}$ and so one of the orbits of g of size 3 contains at least two of the elements of $X^{\#}$.

Next set $M_1 = C_G(z')$ and $\bar{M}_1 = M_1/O(M_1)$. Since $X \subseteq R_1$, z' centralizes L and so $\bar{L} \subseteq \bar{L}_1$ for some simple component $\bar{L}_1$ of $L(\bar{M}_1)$. Since z' is a central involution, our present assumption implies that $\bar{L}_1$ does not have quasi-dihedral Sylow 2-subgroups. Hence the only possibility is that $\bar{L}_1 \cong L_2(r)$, r odd, or A_7. But clearly $\bar{L} = L(C_{\bar{L}_1}(\bar{z}))$ and consequently either $\bar{L} = \bar{L}_1$ or else we have $r = q^2$ or $q = 5$ and $\bar{L}_1 \cong A_7$. However, in either of the latter cases a Sylow 2-subgroup of $\bar{L}_1$ has order exceeding that of R_2, contrary to our maximal choice of z. Thus $\bar{L}$ is a component of $\bar{M}_1$. Similarly if $M_2 = C_G(zz')$, then also $\bar{L}$ is a component of $L(\bar{M}_2)$, where $\bar{M}_2 = M_2/O(M_2)$.

We use this to prove that g normalizes X. We know that g transforms one of the involutions of X into X. By the symmetry described in the preceding paragraph, we can assume, for simplicity, that $z^g = z'$, so that $M^g = M_1$. We have that X^g centralizes L^g. But by the preceding argument, $\overline{L^g} = \bar{L}$ in $\bar{M}_1$ and consequently X^g is a subgroup of E whose image in $\bar{M}_1$ centralizes $\bar{L}$. This forces $X^g = X$ and our assertion is proved.

Now we can show that W has the form asserted in Proposition 5.2(ii). Set $K = N_G(X)$ and $\tilde{K} = K/O(K)$. Since $C_G(X) \subseteq M$ and $C_G(X) \lhd K$, it is immediate that $\tilde{L} \lhd \tilde{K}$. But also $\langle S,g\rangle \subseteq K$ and so $\tilde{R}_1$ is a Sylow 2-subgroup of $\tilde{Y} = C_{\tilde{K}}(\tilde{L})$. Since the involutions of $\tilde{X}$ are all conjugate in $\tilde{Y}\langle\tilde{g}\rangle$ and since $\Omega_1(\tilde{R}_1) = \tilde{X}$, we conclude now from Proposition 6.11 of Part III that $\tilde{Y}\langle\tilde{g}\rangle/O(\tilde{Y}\langle\tilde{g}\rangle) \cong PSL_n(2,3)$ for some n. In particular, $R_1 \cong \tilde{R}_1 \cong Z_{2^n} \times Z_{2^n}$.

If $W = R_1$, then W has the required form. Hence we can assume that $W \supset R_1$, whence also $|R_2| \geq 8$. If $\tilde{Y} \cong PSL_n(2,3)$, choose $\tilde{y}$ of order 3 in $\tilde{Y}$. Then $\tilde{W}\langle\tilde{y}\rangle$ is a group and $[\tilde{X},\tilde{y}] = \tilde{X}$. We argue to the same conclusions if $\tilde{Y}$ is not of this form for some 3-element $\tilde{y}$ of $\tilde{K}$. Indeed, in this case, $\tilde{Y} = \tilde{R}_1$. Moreover, $\tilde{K}/\tilde{L}\tilde{Y}$ is abelian. Since $\tilde{R}_2$ is nonabelian dihedral, there thus exists an element $\tilde{y}$ of order 3 in $\tilde{K}$ which centralizes $\tilde{S}/\tilde{Y}$ and is such that $[\tilde{X},\tilde{y}] = \tilde{X}$. Clearly then $\tilde{W}\langle\tilde{y}\rangle$ is a group and so $\tilde{y}$ has the required properties. But also $\Omega_1(\tilde{W}) = \tilde{X}$ and hence we can again apply Proposition 6.11 of Part III to obtain that $W \cong Z_{2^m} \times Z_{2^m}$ for some m. Since $W \supset R_1$, this forces $m > n$. However, this is impossible as W/R_1 is cyclic.

Since the involutions of X are conjugate in G, all parts of Proposition 5.2(ii) will be established once we show that neither A nor B lies in WR_2. Since $W = R_1 \cong Z_{2^n} \times Z_{2^n}$ and $R_2\langle a\rangle$ is dihedral for some involution of D with $a \in S - WR_2$, any two elementary subgroups of WR_2 of order 16 are necessarily conjugate in S. Since $A \not\sim B$ in S, we can assume without loss that $A \not\subseteq WR_2$ and $B \subseteq WR_2$. But now by Alperin's theorem, we can suppose not only that (II) holds, but, in addition, that $A^g \subseteq WR_2$. However, the preceding analysis gives the structure of $K = N_G(X)$ and we see that K has a normal subgroup of index 2 with WR_2 as Sylow 2-subgroup. Since $g \in K$, the conditions $A \not\subseteq WR_2$ and $A^g \subseteq WR_2$ are therefore incompatible; and the proof is complete.

In view of the preceding lemma, we can assume henceforth that $\Omega_1(W)$ is not a four group. As a consequence, we have

Lemma 5.7. The following conditions hold:

(i) $W \cong D_{2^n}$ or QD_{2^n} for some $n \geq 3$;

(ii) $WR_2 = W_1R_2 = W_1 \times R_2$.

Proof: Since $r(W) \leq 2$, it follows under the present assumption from (4.10) that W is either cyclic or of maximal class. However, W is not cyclic or generalized quaternion by Lemma 4.7 and so (i) holds. Now (ii) follows from (4.3) and (4.9) (as $Q_2 = R_2$ in the present case).

We again set $E = \Omega_1(Z(D))$ and next prove

Lemma 5.8. The following conditions hold:

(i) $W \cap E = \langle z \rangle$;

(ii) $E = Z\langle y \rangle \cong E_8$ for some involution y of $S - WR_2$.

Proof: Set $X = W \cap E$. We have $z \in X$; so if (i) is false, $X \cong Z_2 \times Z_2$. Since D contains an involution y of $S - WR_2$, we have that $S = (W \times R_2)\langle y \rangle$ with y centralizing X. But then by the structure of W, we see that $Y = C_S(X) = X \times R_2\langle y \rangle$. But $D \subseteq Y$ as $X \subseteq Z(D)$. Since $R_2\langle y \rangle$ is dihedral, it follows therefore that $D \cong Z_2 \times Z_2 \times D_{2^r}$ for some r and again we reach a contradiction as in Proposition 5.1. Thus $X = \langle z \rangle$, proving (i).

By the structure of S, we have $Z = \langle z, z_2 \rangle = \Omega_1(Z(S))$. Since $Z \subseteq D$, also $Z \subseteq E$. As noted in Lemma 5.6, $|E| \neq 16$. So either $E \cong E_8$ or $E = Z$. However, in the latter case, Lemma 4.5 together with the fact that $E \lhd N$ yields that z is isolated in N, contrary to Lemma 5.4. Thus $E \cong E_8$.

Finally we argue that E is not contained in WR_2. Observe first that as $A_i \cap W \cong Z_2 \times Z_2$ for all i and $D = \langle A_i \mid 1 \leq i \leq m \rangle$, it follows from (i) that $A_i \cap W \neq A_j \cap W$ for some i, j, $i \neq j$. Hence by the structure of W, $D \cap W$ is either non-abelian dihedral or quasi-dihedral. Since E centralizes $D \cap W$, we conclude now that $E \cap WR_2 \subseteq \langle z \rangle R_2$ (using the fact that R_2 centralizes W). But now if our assertion is false, then $E \subseteq WR_2$ and so $E = \langle z \rangle \times (E \cap R_2)$ with $E \cap R_2 \cong Z_2 \times Z_2$. However, y centralizes $E \cap R_2$, which is impossible as $R_2\langle y \rangle$ is dihedral with $y \notin R_2$. Hence $E \not\subseteq WR_2$ and so we can choose y to be in E, whence $E = Z\langle y \rangle$, proving (ii).

Now we can easily complete the proof of the proposition. Set $D_1 = D \cap W$. Our goal will be to show that $D = D_1 \times \langle y, z_2 \rangle$, whence $D \cong D_{2^n} \times Z_2 \times Z_2$ for some n, which will yield the usual contradiction.

First of all, as $y \in Z(D)$ and $C_{R_2}(y) = \langle z_2 \rangle$, we have that $R_2 \cap D = \langle z_2 \rangle$. Furthermore, as W is of maximal class, $W = W_1$ by (4.9) and so $|W : R_1| \leq 2$. Moreover, if equality holds, then $[w,y] \equiv z_2$ (mod R_1) for any w in $W - R_1$ by (4.6). In particular, we conclude in either case that $W \cap D \subseteq R_1$.

Suppose now that D does not have the asserted structure, in which case $D \supset D_1 \times \langle y, z_2 \rangle$ and so there exists an involution d in D with

$d \notin D_1\langle y,z_2\rangle$. Then d or dy lies in WR_2 and, as dy has the same properties as d, we can assume without loss that $d \in WR_2$. Thus $d = wt$ with $w \in W^{\#}$, $t \in R_2^{\#}$. Since $WR_2 = W \times R_2$, w and t are, in fact, involutions. We have that $[w,y] \equiv 1$ or $z_2 \pmod{R_1}$ and that $[t,y] \in R_2$. Since y centralizes tw and $R_1R_2 = R_1 \times R_2$, it follows that $[w,y] = 1$ or z_2 and correspondingly that $[t,y] = 1$ or z_2. In the first case, $t \in C_{R_2}(y) = \langle z_2\rangle$, whence $t = z_2$ and consequently $w = dt^{-1} \in D \cap W = D_1$, contrary to our choice of d. Hence $[w,y] = [t,y] = z_2$. The first condition implies that $W \supset R_1$, whence $|R_2| \geq 8$. On the other hand, the second condition implies that y normalizes the four subgroup $R_0 = \langle t,z_2\rangle$ of R_2. However, as R_2 and $R_2\langle y\rangle$ are dihedral with $|R_2| \geq 8$, we have that $N_{R_2\langle y\rangle}(R_0)$ is dihedral of order 8 and lies in R_2. This contradiction completes the proof of Proposition 5.2.

We shall also need information concerning the structure of $N_G(A)$ for any elementary abelian subgroup A of S of order 16.

<u>Proposition 5.9</u>. $N_G(A)/C_G(A)$ is solvable of order 2^a3^b for some a,b.

<u>Proof</u>: Set $N = N_G(A)$, $C = C_G(A)$, and $\bar{N} = N/C$. Replacing A by a conjugate, if necessary, we can assume without loss that $N_S(A)$ is a Sylow 2-subgroup of N. We have that $Z \subset A$ and by Lemma 4.5 the involutions of Z lie in distinct orbits of $A^{\#}$ under the action of $\bar{N}$. In particular, $\bar{N}$ has at least 3 orbits in its action on $A^{\#}$. This implies that $\bar{N}$ does not contain a subgroup isomorphic to A_5, otherwise $\bar{N}$ would have at most 2 orbits on $A^{\#}$. Likewise if $|\bar{N}|$

is divisible by 7, then $\bar{N}$ has precisely 3 orbits on $A^{\#}$ of sizes 7, 7, and 1 respectively. Thus one of the involutions z' of Z is fixed by $\bar{N}$ and so lies in $Z(N)$. But then if $\bar{N}/O_2(\bar{N}) \cong L_3(2)$, $M_1 = C_G(z')$ would be nonsolvable as $N \subseteq M_1$. Since G is an $\mathfrak{K}_2$-group and z' is a central involution, it follows then that $L(\bar{M}_1) \neq 1$, where $\bar{M}_1 = M_1/O(M_1)$. But $\bar{M}_1/L(\bar{M}_1)$ is solvable by Proposition 2.1 of Part III and we conclude therefore from the structure of N that $L(\bar{M}_1)$ contains a 7-element which normalizes, but does not centralize an elementary subgroup of $L(\bar{M}_1)$ of order 8. However, this is clearly impossible as each component of $L(\bar{M}_1)$ has 2-rank one or two. Thus $\bar{N}$ is also not of this form and we conclude that $\bar{N}$ is solvable.

Suppose next that $|\bar{N}|$ is divisible by 14, whence $O_2(\bar{N}) \cong E_8$ and $\bar{N}$ contains an element $\bar{x}$ of order 7 which acts transitively on $O_2(\bar{N})^{\#}$. We have $A = A_0 \times A_1$, where $A_0 = C_A(\bar{x}) \cong Z_2$ and $A_1 = [A,\bar{x}] \cong E_8$. Because of the size of the orbits of $\bar{N}$ on $A^{\#}$, A_0 and A_1 must each be normal in N. But then the action of $\bar{x}$ on $O_2(\bar{N})$ and A_1 forces $O_2(\bar{N})$ to centralize A_1. Clearly $O_2(\bar{N})$ centralizes A_0 and so it centralizes A, contrary to the fact that $\bar{N}$ acts faithfully on A. Thus $|\bar{N}|$ is not divisible by 14. Likewise $|\bar{N}|$ is not divisible by 15, otherwise $\bar{N}$ would act transitively on $A^{\#}$, which is not the case.

Now consider the case that $A \subseteq WQ_2$. Since $R_2 \cap Z(S) \cong Z_2$, it follows directly from Lemma 4.2(ii) and (iii) that $N_M(A)/C_M(A)$ has order divisible by 6 and hence so does $\bar{N}$. But now we conclude from the preceding paragraph that $|\bar{N}|$ is not divisible by 7 or 5 and consequently $|\bar{N}| = 2^a 3^b$, as asserted.

Assume then that $A \not\subseteq WQ_2$. By Lemma 4.1(iv), $\bar{L} \cong L_2(q)$ and so $R_2 = Q_2$ is dihedral. Choose a in A with $a \notin WR_2$. Since also $R_2\langle a\rangle$ is dihedral, $A \cap R_2 = \langle z_2\rangle$ and it is immediate that some involution x of R_2 normalizes, but does not centralize A. Hence $\bar{N}$ has even order and so by the preceding argument $|\bar{N}|$ is not divisible by 7. Suppose $|\bar{N}|$ were divisible by 5, in which case $\bar{P} = O_5(\bar{N})$ has order 5 and $\bar{x}$ inverts $\bar{P}$. Considering the action of $\bar{P}\langle\bar{x}\rangle$ on A, it follows then that $C_A(x) \cong Z_2 \times Z_2$. However, $[A,x] \subseteq A \cap R_2 = \langle z_2\rangle$ and consequently $C_A(x) \cong E_8$, a contradiction. Thus $|\bar{N}|$ is not divisible by 5 and the proposition is proved.

In our final lemma, we assume, in addition, that whenever $A \sim B$ in G, then $A \sim B$ in S. Under this assumption, we shall show that, in fact, $N_S(A)$ is a Sylow 2-subgroup of $N_G(A)$. Indeed, we can prove

Lemma 5.10. The following conditions hold:

(i) $N_S(A)$ is a Sylow 2-subgroup of $N_G(A)$;

(ii) If $a,b \in A^{\#}$ with $a \sim b$ in G and if $a \in Z(S)$, then $a \sim b$ in $N_G(A)$.

Proof: Let T be a Sylow 2-subgroup of G such that $N_T(A)$ is Sylow 2-subgroup of $N_G(A)$. We have $T^g = S$ for some g in G, whence $A^g \subseteq S$ and so by our hypothesis on A, we have $A^{gx} = A$ for some x in S. But $N_S(A^g) = N_T(A)^g$ is a Sylow 2-subgroup of $N_G(A^g)$ and consequently $N_S(A) = N_S(A^{gx}) = N_T(A)^{gx}$ is a Sylow 2-subgroup of $N_G(A) = N_G(A^{gx})$, proving (i).

As for (ii), we have $b^g = a$ with $C_S(b)^g \subseteq S$ for some g in G. In particular, $A^g \subseteq S$ and so $A^{gx} = A$ for some x in S. Thus

$gx \in N_G(A)$ and $b^{gx} = a$. This completes the proof.

6. Fusion of involutions. We also require fusion information that occurs as a consequence of Glauberman's Z^*-theorem.

In the case that R_2 is quasi-dihedral, the desired result is easily obtained and is given by the following proposition:

Proposition 6.1. If R_2 is quasi-dihedral, then z is conjugate in G to an involution of $WQ_2 - \langle z \rangle$.

Proof: We have $S = WR_2$ in this case. Suppose the proposition false. By Glauberman's Z^*-theorem, $z \sim y$ in G for some $y \neq z$ with $y \in S$. Then $y \in S - WQ_2$. Since $\Omega_1(R_1R_2) = \Omega_1(R_1) \times Q_2 \subseteq W_1Q_2$, certainly $y \notin R_1R_2$. In particular, $W \supset R_1$ and so $\bar{L} \cong L_3(q)$ or $U_3(q)$. Correspondingly $\bar{L}\bar{W}_1 = \bar{L}\langle\bar{y}\rangle \cong L_3^*(q)$ or $U_3^*(q)$. But now Lemma 2.3(vi) of Part IV implies that all involutions of $\bar{L}\langle\bar{y}\rangle - \bar{L}$ are conjugate under the action of $\bar{L}$ and consequently $y \sim w$ in M for some involution w of W_1. Thus $z \sim w$ in G and $w \in WQ_2 - \langle z \rangle$, contrary to assumption. This establishes the proposition.

If R_2 is dihedral and $S = WR_2$, then $S = WQ_2$ and the Z^*-theorem immediately yields

Proposition 6.2. If R_2 is dihedral and $S = WR_2$, then z is conjugate in G to an involution of $WQ_2 - \langle z \rangle$.

Hence in analyzing the dihedral case, we can suppose that $S \supset WR_2$. Moreover, if z' is any central involution of G for which $L_{z'}$ has a simple component and if the conditions of Proposition 6.1 or 6.2 are satisfied in $C_G(z')$, then we can take z' as z and the required fusion properties will hold.

Thus we can suppose for the balance of the section that if z' is a central involution of G such that $M_1 = C_G(z')$ has a simple component, then $\overline{M}_1 = M_1/O(M_1)$ possesses a component $\overline{L}_1 \cong L_2(r)$, r odd, and $\overline{M}_1/C_{\overline{M}_1}(\overline{L}_1)$ contains a normal subgroup isomorphic to $PGL(2,r)$. In particular, we have that $R_2 = Q_2$ is dihedral. Furthermore, among all such possible choices of z', we assume z is chosen so that $|R_2|$ is maximal. We also have $Z(S) \subseteq R_1 \times R_2$ by (4.7).

In this case we shall prove

Proposition 6.3. There exist involutions x in R_1 and t in WR_2 with the following properties:

(i) $x \in Z(S)$;

(ii) $x \sim t$ in G, but $x \not\sim t$ in M.

We carry out the proof in a sequence of lemmas.

Lemma 6.4. If w is an involution of $Z(S) \cap W$, then z is not conjugate in $C_G(w)$ to an involution of $S - WR_2$.

Proof: Suppose false for some w and let y be an involution of $S - WR_2$ which is conjugate to z in $M_1 = C_G(w)$. Since $w \in Z(S)$, w centralizes y and so $w \in R_1$ by (4.6). Hence $L \subseteq M_1$ and so if we set $\overline{M}_1 = M_1/O(M_1)$, our conditions imply that $\overline{L}$ lies in a simple component $\overline{L}_1$ of $L(\overline{M}_1)$ with $\overline{L}_1 \cong L_2(r)$ for some odd r. Since $\overline{L} \subseteq C_{\overline{L}_1}(\overline{z})$, $\overline{z}$ induces a field automorphism of $\overline{L}_1$ and it follows as in Lemma 5.6 from our maximal choice of R_2 that $\overline{L} = \overline{L}_1$. Hence $\overline{z} \in \overline{C}_1 = C_{\overline{M}_1}(\overline{L}_1) \lhd \overline{M}_1$. But $\overline{L}_1\langle\overline{y}\rangle = \overline{L}\langle\overline{y}\rangle \cong PGL(2,q)$ and so certainly $\overline{y} \notin \overline{C}_1$. Clearly then y and z cannot be conjugate in M_1.

We next prove

Lemma 6.5. For any involution u of $Z(S)$, one of the following holds:

(i) z is not conjugate in $C_G(u)$ to an involution of $S - WR_2$; or

(ii) Some involution w of $Z(S) \cap W$ is conjugate in G to an involution of $WR_2 - \langle w \rangle$.

Proof: Suppose (i) is false. Set $M_1 = C_G(u)$ and let y be an involution of $S - WR_2$ with $z \sim y$ in M_1. By the preceding lemma, $u \notin W$. In any event, $u \in R_1R_2$ by (4.7) as $u \in Z(S)$.

Let K be the largest normal subgroup of M_1 of the form $K = O(M_1)(K \cap M)$ and set $T = S \cap K$. Then T is a Sylow 2-subgroup of K and $K \supseteq O_{2',2}(M_1)$. Consider first the case that $z \in K$. Setting $E = \Omega_1(Z(T))$, we have that $\langle z,u \rangle \subseteq E$. Since $z \sim y$ in M_1 and $K \triangleleft M_1$, clearly $y \in T$. But $M_1 = KN_{M_1}(T)$ by the Frattini argument and so $M_1 = (K \cap M)O(M_1)N_{M_1}(T)$. Since $K \cap M$ centralizes z, $z \sim y$ in $O(M_1)N_{M_1}(T)$ and consequently $z \sim y$ in $N_{M_1}(T)$. This forces $y \in E$, as E is characteristic in T. But then $[R_2,y] \subseteq E \cap R_2$ and so $[R_2,y]$ is an elementary abelian normal subgroup of R_2. However, as $R_2\langle y \rangle$ is dihedral, this implies that $R_2 \cong Z_2 \times Z_2$, whence $W = R_1$ and so $WR_2 = R_1 \times R_2$. In addition, $\langle z_2 \rangle = [R_2,y] \subseteq E$. Since y normalizes both R_1 and R_2, it is immediate that $C_{R_1R_2}(y) = \langle z_2 \rangle \times C_{R_1}(y)$. Since $y \in Z(T)$ and $S = R_1R_2\langle y \rangle$, we conclude also that $T = \langle z_2,y \rangle \times (R_1 \cap T)$.

Suppose first that T and hence $R_1 \cap T$ is not elementary abelian. Since $T \triangleleft S$, $\mho^1(R_1 \cap T)$ must then contain an involution w of $Z(S)$. But $y \notin \mho^1(T)$ and $z = y^x$ for some x in $N_{M_1}(T)$, so also $z \notin \mho^1(T)$.

Hence $z \neq w$. Since $w \in Z(S) \cap T$, we have $w \in E$ and so by the structure of T, it follows that $E = \langle z_2, y, z, w \rangle \cong E_{16}$ with $\langle w, z, z_2 \rangle = R_1R_2 \cap E \cong E_8$. We see then that $\langle w, z \rangle^x \cap (R_1R_2 \cap E) \neq 1$, which implies that $w_1^x \in R_1R_2$ for some $w_1 \neq 1$ in $\langle w, z \rangle$. If $w_1^x = w_1$, then $x \in C_G(w_1)$ and so $z \sim y$ in $C_G(w_1)$. However, as $w_1 \in Z(S) \cap W$, this contradicts the preceding lemma. Hence, in fact, $w_1^x \neq w_1$ and so $w_1^x \in R_1R_2 - \langle w_1 \rangle$. Therefore (ii) holds in this case.

We can therefore assume that T is elementary abelian, whence $T = E$. Since T is a Sylow 2-subgroup of K, either $N_K(T) = C_K(T)$ or $N_K(T) - C_K(T)$ contains an element a of odd order. However, as $K = O(M_1)(K \cap M)$, we can assume in the latter case that $a \in M$. But as $y \notin WR_2$, Lemma 4.1(ii) and (iii) imply that $N_M(T)/C_M(T)$ is a 2-group. Hence, in fact, $N_K(T) = C_K(T)$ and so K has a normal 2-complement by Burnside's transfer theorem. Since $O(K) \subseteq O(M_1)$, it follows that $K = O(M_1)T$. On the other hand, $C_{M_1}(T) \subseteq M$ and so by the maximality of K, we have $C_{M_1}(T) \subseteq K$. Hence if we set $\bar{M}_1 = M_1/O(M_1)$, we conclude that $C_{\bar{M}_1}(\bar{T}) = \bar{T}$. Furthermore, as $S \subseteq M_1$, our maximal choice of K also implies that $\bar{T} = O_2(\bar{M}_1)$.

Now $\langle z, z_2, y \rangle \subseteq T$ and $T \cong \bar{T} \cong E_8$ or E_{16}. Furthermore, as z is not isolated in M_1, $\bar{S} \subset \bar{M}_1$. On the other hand, as y does not centralize R_2, we have that $\bar{R}_2 \not\subseteq \bar{T}$. If $\bar{T} \cong E_8$, $|\bar{M}_1|$ is not divisible by 7 as then $z \sim z_2$ in G, contrary to Lemma 4.5. Since $O_2(\bar{M}_1) = \bar{T}$, it follows in this case that $\bar{M}_1/\bar{T} \cong S_3$, whence $S = TR_2 = \langle z \rangle \times R_2\langle y \rangle$, contrary to the fact that $m(S) = 4$. Thus $\bar{T} \cong E_{16}$ and now Proposition 5.9 implies that $\bar{M}_1$ is a $\{2,3\}$-group. Since

$O_2(\bar{M}_1) = \bar{T}$ and $\bar{u} \in \bar{T}$ lies in $Z(\bar{M}_1)$, we again conclude that $\bar{M}_1/\bar{T} \cong S_3$. Thus again $S = TR_2$, forcing $T \supseteq R_1 \cong Z_2 \times Z_2$ and $S = R_1 \times R_2\langle y \rangle \cong Z_2 \times Z_2 \times D_8$, contrary to Proposition 2.5.

This completes the proof of the lemma in the case that $z \in T$; so we can assume henceforth that $z \notin T$, whence $z \notin K$. Since $O_{2',2}(M_1) \subseteq K$, this implies that $z \notin O_{2',2}(M_1)$ and consequently M_1 is not 2-constrained. Moreover, if again $\bar{M}_1 = M_1/O(M_1)$, then $L(\bar{M}_1)$ is not isomorphic to $SL(2,r)$, r odd, or $\hat{A}_7$ by Lemma 4.9 and so $L(\bar{M}_1)$ possesses a simple component $\bar{L}_1$. Under the assumptions of the proposition, we know that $\bar{L}_1 \cong L_2(r)$ for some odd r and that $\bar{M}_1/\bar{C}_1$ contains a normal subgroup isomorphic to $PGL(2,r)$, where $\bar{C}_1 = C_{\bar{M}_1}(\bar{L}_1)$. Since z centralizes S, which is a Sylow 2-subgroup of M_1, (4.7) applied to M_1 now yields that $\bar{z} \in \bar{L}_1\bar{C}_1$. If $\bar{z} \in \bar{C}_1$, Proposition 6.11 of Part III would imply that $\bar{z} \in Z(\bar{C}_1)$, whence $z \in O_{2',2}(M_1)$, which is not the case. Hence we also have that $\bar{z} \notin \bar{C}_1$.

We first consider the case that $\bar{W}\bar{R}_2 \cap \bar{L}_1$ is noncyclic. Since $\bar{S} \cap \bar{L} \lhd \bar{S}$ and is dihedral, there thus exist distinct involutions u_1, u_2 in WR_2 with $u_1 \in Z(S)$ and $\bar{u}_1, \bar{u}_2$ in $\bar{L}_1$. Since $\bar{L}_1$ has only one conjugacy class of involutions, $u_1 \sim u_2$ in M_1. We argue that (ii) holds. This will clearly be the case if $u_1 \in W$ as then $u_1 \in Z(S) \cap W$ and $u_2 \in WR_2 - \langle u_1 \rangle$. Hence we can assume that $u_1 \notin W$. Since $u_1 \in Z(S) \cap WR_2$, it follows then that $u_1 = w_1z_2$ for some w_1 in W. But likewise our original involution u is in $Z(S) \cap WR_2$ with $u \notin W$, so $u = wz_2$ for some w in W. Since $\bar{u}_1 \in \bar{L}_1$, $u_1 \neq u$ and so uu_1 is a nontrivial element of $Z(S) \cap W$. But $u_1u \sim u_2u$ in M_1 as

$u_1 \sim u_2$ in M_1 and $u \in Z(M_1)$. Since $u_1 u \neq u_2 u$ and $u_2 u \in WR_2$, we conclude that (ii) holds. We can therefore assume for the balance of the proof that $\bar{W}\bar{R}_2 \cap \bar{L}_1$ is cyclic.

This forces $\bar{z} \notin \bar{L}_1$. Indeed, assume the contrary. Since $z \sim y$ in M_1 and $\bar{L}_1 \triangleleft \bar{M}_1$, also $\bar{y} \in \bar{L}_1$, whence $\bar{z}_2 \in [R_2, \bar{y}] \subseteq \bar{L}_1$. Thus $\bar{Z} = \langle \bar{z}, \bar{z}_2 \rangle \subseteq \bar{L}_1$, contrary to our assumption $\bar{W}\bar{R}_2 \cap \bar{L}_1$ cyclic. Thus $\bar{z} \notin \bar{L}_1$, as asserted.

Since $\bar{L}_1\bar{C}_1 = \bar{L}_1 \times \bar{C}_1$ with $\bar{z} \notin \bar{L}_1$ or $\bar{C}_1$, we can write $z = t_1 t_2$ with t_1, t_2 involutions of $Z(S)$, $\bar{t}_1 \in \bar{C}_1$ and $\bar{t}_2 \in \bar{L}_1$. Since $z^x = y$ with $x \in M_1$, it follows that $y = y_1 y_2$, where $t_i^x = y_i \in S$, $i = 1, 2$, $\bar{y}_1 \in \bar{C}_1$, and $\bar{y}_2 \in \bar{L}_1$. If $y_2 = t_2$, then x centralizes t_2 and so centralizes the four subgroup $\langle t_2, u \rangle$ of $Z(S) (\subseteq WR_2)$. It follows therefore as in the case $\bar{W}\bar{R}_2 \cap \bar{L}_1$ noncyclic that some involution w of $\langle t_2, u \rangle$ lies in W. Since $x \in C_G(w)$, $z \sim y$ in $C_G(w)$ and again Lemma 6.4 is contradicted. Thus $y_2 \neq t_2$. Since $\bar{W}\bar{R}_2 \cap \bar{L}_1$ is cyclic, this in turn implies that $y_2 \notin WR_2$. In particular, $R_2\langle y_2 \rangle$ is dihedral.

This last condition implies that $[\bar{R}_2\langle \bar{y}_2 \rangle, \bar{y}_2]$ is a maximal cyclic subgroup of $\bar{R}_2$. Moreover, it lies in $\bar{L}_1$. Hence if we set $Y_2 = [R_2\langle y_2 \rangle, y_2]\langle y_2 \rangle$, it follows that Y_2 is dihedral of the same order as R_2 and that $\bar{Y}_2 \subseteq \bar{L}_1$. But now our maximal choice of z forces $\bar{Y}_2$ to be a Sylow 2-subgroup of $\bar{L}_1$. Furthermore, $R_2 Y_2 = R_2\langle y_2 \rangle$. Hence if u_2 is any involution of R_2 other than z_2, we have that also $R_2 Y_2 = Y_2\langle u_2 \rangle$ with $Y_2\langle u_2 \rangle$ dihedral and $\bar{u}_2 \notin \bar{L}_1$. Note also that z_2 is necessarily the unique involution of $Z(S) \cap R_2\langle y_2 \rangle =$

$Z(S) \cap Y_2\langle u_2\rangle$ and therefore $t_2 = z_2$.

Since $W \cap R_2 = 1$ and $y_2 \in S - WR_2$, we have that $W \cap R_2\langle y_2\rangle = W \cap Y_2\langle u_2\rangle = 1$. But $W_1 \subseteq W$ and W_1 centralizes R_2. Hence if we set $\tilde{M}_1 = \bar{M}_1/\bar{C}_1$, we have that $\tilde{W}_1$ centralizes both $\tilde{R}_2 \cap \tilde{L}_1$ and $\tilde{u}_2$. Since $\tilde{M}_1$ is isomorphic to a subgroup of $P\Gamma L(2,r)$, it is immediate now from Lemma 2.2 of Part IV that $\tilde{W}_1 \subseteq \tilde{V}$, where $\tilde{V}$ is cyclic of order at most 4 and $\Omega_1(\tilde{V}) = \langle\tilde{z}_2\rangle$. (If $\tilde{V} = \langle\tilde{v}\rangle$ has order 4, then $\tilde{v}$ will be the product of a field automorphism of $\tilde{L}_1$ of order 2 and an element of order 4 in $\tilde{R}_2 \cap \tilde{L}$.) Hence if we let V be the inverse image of $\tilde{V}$ in S and denote by Y_1 the subgroup of S whose image in $\bar{M}_1$ is a Sylow 2-subgroup of $\bar{C}_1$, we have that V/Y_1 is cyclic of order at most 4 and that $W_1 \subseteq Y_1V = V$.

We use this conclusion to prove that R_1 is cyclic of order at most 4. Indeed, suppose false. Since $R_1 \subseteq W_1 \subseteq Y_1V$, it follows from the structure of V that $R_1 \cap Y_1 \neq 1$. But R_1 and Y_1 are each normal in S and so $R_1 \cap Y_1$ contains an involution z^* of $Z(S)$. Then $M^* = C_G(z^*)$ covers both $\bar{L}$ in $\bar{M}$ and $\bar{L}_1$ in $\bar{M}_1$. Hence if we set $\tilde{M}^* = M^*/O(M^*)$, then both $\tilde{L}$ and $\tilde{L}_1$ lie in the unique simple component $\tilde{L}^*$ of $\tilde{M}^*$. In particular $\tilde{R}_2\tilde{Y}_2 \subseteq \tilde{L}^*$. Since $R_2Y_2 \supset R_2$, our maximal choice of z is therefore contradicted. This proves our assertion.

Observe next that the situation is entirely symmetric with respect to M and M_1. Hence interchanging their roles, we conclude by an analogous argument that also Y_1 is cyclic of order at most 4.

Now we can complete the proof at once. Indeed, as u is clearly the unique involution of Y_1 and $u \notin W$, we have that $W_1 \cap Y_1 = 1$. Since $W_1 \subseteq Y_1V$, it follows therefore that W_1 is cyclic of order at most 4. However, this contradicts Lemma 4.7.

We make one further observation.

Lemma 6.6. Every involution of $Z(S) \cap W$ is isolated in M.

Proof: Let w be such an involution. Since $S \supset WR_2$, $w \in R_1$ by (4.6). Since $r(\bar{C}) \leq 2$ and $\bar{z} \in Z(\bar{C})$, Proposition 6.11 of Part III now yields that $\bar{w} \in Z(\bar{C})$. If $w = z$, our assertion is clear. In the contrary case, $\langle \bar{w}, \bar{z} \rangle$ is then a four subgroup of $Z(\bar{C})$. But $r(\bar{C}) \leq 2$, so $\langle \bar{w}, \bar{z} \rangle = \Omega_1(Z(\bar{C}))$, whence $\langle \bar{w}, \bar{z} \rangle \triangleleft \bar{M}$. Since $\bar{z} \in Z(\bar{M})$ and $w \in Z(S)$, it is immediate now that $\bar{w} \in Z(\bar{M})$ and the lemma holds.

Now we can establish Proposition 6.3. Suppose first that Lemma 6.5(i) is false. Then Lemma 6.5(ii) must hold and so for some involution w of $Z(S) \cap W$, we have that $w \sim t$ in G for some t in $WR_2 - \langle w \rangle$. By the preceding lemma, $w \not\sim t$ in M. Furthermore, $w \in R_1$ by (4.7) as $w \in Z(S)$ and $S \supset WR_2$. We see then that the conditions of the proposition are satisfied with w in the role of x. Hence it suffices to treat the case in which Lemma 6.5(i) holds and so we can assume that z is not conjugate in $C_G(u)$ to an involution of $S - WR_2$ for any involution u of $Z(S)$.

By Glauberman's Z^*-theorem, $z \sim y$ in G for some $y \neq z$ in S. Clearly y is not conjugate to z in M and y centralizes z. Hence if $y \in WR_2$, the proposition holds with $x = z$ and $t = y$. Thus we can also assume that $y \notin WR_2$.

Set $B = Z(y)$, so that $B \cong E_8$. We also have that $y^g = z$ for some g in G with $C_S(y)^g \subseteq S$. In particular, $B^g \subseteq S$. Consider first the case that $B^g \supseteq Z$, in which case $Z \cap Z^g \neq 1$. Hence there exists an involution u in Z such that $u^g \in Z$. Since no two involutions of $Z = \langle z, z_2 \rangle$ are conjugate in G by Lemma 4.5, we conclude that $u^g = u$. Hence $g \in C_G(u)$ and so $y \sim z$ in $C_G(u)$, contrary to our present assumption. Thus $Z \not\subseteq B^g$.

Since $m(S) = 4$, it follows that $B^g Z \cong E_{16}$ and consequently $B \subseteq A$ for some subgroup A of G with $A \cong E_{16}$. Since $z \in B$, $A \subseteq M$ and so we can assume $A \subseteq S$ by replacing y by a suitable conjugate in M. Since A centralizes $y \in B$, $A \subseteq R_1 R_2 \langle y \rangle$ by (4.6). Since $R_2 \langle y \rangle$ is dihedral, and $r(R_1) \leq 2$, it follows that $X = A \cap R_1 \cong Z_2 \times Z_2$. Thus $A = X \times \langle y, z_2 \rangle$. Clearly $z \in A$.

We now set $N = N_G(A)$, $T = S \cap N$, and $\bar{N} = N/C_G(A)$. If Proposition 5.2(ii) holds, then the present proposition also holds; so we can assume that this is not the case. Thus any G-conjugate of A which lies in S is already conjugate to A in S by Proposition 5.2(i). Lemma 5.10 now yields that T is a Sylow 2-subgroup of N. Moreover, by Proposition 5.9, $\bar{N}$ is a $\{2,3\}$-group.

Consider first the case that $|\bar{N}|$ is divisible by 9, in which case a Sylow 3-subgroup $\bar{P}$ of $\bar{N}$ is isomorphic to $Z_3 \times Z_3$. If no element of $\bar{P}^{\#}$ centralizes z, then z has 9 conjugates under the action of $\bar{P}$. Suppose none of these except z itself lies in $A \cap WR_2 = X\langle z_2 \rangle$. Then the 8 involutions of $A - X\langle z_2 \rangle$ each have 9 conjugates under $\bar{P}$ and so every involution of A which has 3 conjugates under $\bar{P}$ lies in $X\langle z_2 \rangle$. But $A = A_1 \times A_2$, where A_i

is a $\bar{P}$-invariant four subgroup of A and each involution of A_i has 3 conjugates under $\bar{P}$, $i = 1, 2$. Hence $A = A_1 \times A_2 \subseteq X\langle z_2 \rangle$, which is absurd. We thus conclude that $z \sim z'$ under $\bar{P}$ for some $z' \neq z$ in $A \cap WR_2$. Again the proposition holds with $x = z$ and $t = z'$.

We can therefore assume that $C_{\bar{P}}(z) \neq 1$. It follows now that $M = C_G(z)$ contains a 3-element which normalizes, but does not centralize A. However, as $y \in A$ and $y \in S - WR_2$, $N_M(A)/C_M(A)$ is a 2-group by Lemma 4.1(ii). Hence this case cannot arise and so we conclude that $|\bar{N}|$ is not divisible by 9. Thus $|\bar{N}| = 2^a$ or $2^a 3$. However, $z \sim y$ in N by Lemma 5.10 and as $T = S \cap N$ is a Sylow 2-subgroup of N, we must have $\bar{N} \supset \bar{T}$. Thus $\bar{N} = \bar{T}\bar{P}$, where $\bar{P}$ has order 3 and $z \sim y$ under the action of $\bar{P}$.

Suppose first that $\bar{T}_1 = O_2(\bar{N}) \neq 1$ and set $A_1 = C_A(\bar{T}_1)$. Since $\bar{T}_1 \subseteq \bar{T}$, we have $Z = \langle z, z_2 \rangle \subseteq A_1$. But $A_1 \lhd N$ as $\bar{T}_1 \lhd \bar{N}$. Since $z \in A_1$, we must also have $y \in A_1$. Since $\bar{T}_1$ does not centralize A, it follows that $A_1 = Z\langle y \rangle$. But $R_2 \cap N$ does not centralize y and so $\overline{R_2 \cap N} = \langle \bar{v} \rangle$ is not contained in $\bar{T}_1$. Hence $\tilde{N} = \bar{N}/\bar{T}_1 = \tilde{P}\langle \tilde{v} \rangle \cong S_3$ and consequently $C_{A_1}(\bar{P})$ is invariant under $\bar{v}$. But $C_{A_1}(\bar{P}) \cong Z_2$ as $A_1 \cong E_8$ and $\bar{P}$ does not centralize A_1. Thus $\bar{v}$ centralizes $C_{A_1}(\bar{P})$ and as $\bar{v}$ centralizes no element of $A_1 - Z$, it follows that $C_{A_1}(\bar{P}) = \langle u \rangle$ for some u in $Z^{\#}$. Since $z \sim y$ under $\bar{P}$, we conclude that $z \sim y$ in $C_G(u)$, contrary to assumption. Thus $\bar{T}_1 = O_2(\bar{N}) = 1$ and consequently $\bar{N} = \bar{P}\langle \bar{v} \rangle \cong S_3$.

Finally $A_2 = C_A(\bar{v}) = X\langle z_2 \rangle \cong E_8$ (as $X \subseteq R_1$ centralizes $\bar{v}$, the image of an element of R_2). Thus $\bar{P}$ does not act regularly on A. Hence $C_A(\bar{P})$ is a four subgroup of A. Since $\bar{V}$ does not centralize $A/C_A(\bar{P})$, it follows that $C_A(\bar{P}) \subseteq A_2$. But $Z \subset A_2$ and consequently $Z \cap C_A(\bar{P}) \neq 1$. Thus again $\bar{P}$ centralizes an involution u of Z, giving the same contradiction as in the preceding paragraph. This establishes Proposition 6.3.

7. Theorem A; the dihedral and quasi-dihedral cases. We now have sufficient information about the fusion of involutions and the conjugacy of elementary subgroups of order 16 in S to complete the proof of Theorem A in the cases we are considering.

Under the assumptions of Proposition 6.1 or 6.2, z is conjugate in G to an involution z' of $WQ_2 - \langle z \rangle$. Clearly then the conditions of Proposition 6.3 hold with $x = z$ and $t = z'$. Hence in all cases, it follows for a suitable choice of the involution z of $Z(S)$ that there exist involutions x, t in S which satisfy the conditions of Proposition 6.3. In particular, $\langle x,t \rangle$ is a four subgroup of WQ_2 and $\langle x,t \rangle \subseteq \Omega_1(WQ_2) = \Omega_1(W_1) \times Q_2$. By Lemma 4.7, $m(W_1) = 2$ and so by (4.10), $\Omega_1(W_1)$ is a dihedral group. Since also Q_2 is dihedral, it follows that $\langle x,t \rangle \subseteq A$ for some subgroup A of WQ_2 with $A \cong E_{16}$. We set $A \cap W = X$ and $A \cap Q_2 = Y$, so that X and Y are four groups. We also set $N = N_G(A)$, $T = S \cap N$, and $\bar{N} = N/C_G(A)$; and fix all this notation.

Our analysis will be based on the following result which is a consequence of what we have proved in the preceding three sections.

Lemma 7.1. The following conditions hold:

(i) $\bar{N}$ is a $\{2,3\}$-group;

(ii) T is a Sylow 2-subgroup of N;

(iii) $x \sim t$ in N;

(iv) $N \not\subseteq M$;

(v) $N \cap M$ contains a 3-subgroup D such that $[Y,D] = Y$ and $C_A(D) = \langle z,a \rangle$, where either $a \in X$ or $X \not\subseteq R_1$ and $a \in Xz_2$;

(vi) $\overline{N \cap M} \cong S_3$ or Z_3.

Proof: First, (i) is a restatement of Proposition 5.9. Since $A \subseteq WQ_2$, the results of Section 5 show that any G-conjugate of A which lies in S is conjugate to A in S. But now (ii) follows from Lemma 5.10. Since $\langle x,t \rangle \subseteq A$ and $x \in Z(S)$, Lemma 5.10 also yields (iii). Since $x \not\sim t$ in M, clearly $N \not\subseteq M$, so (iv) holds. Furthermore, (v) is a consequence of Lemma 4.2(iii) and (iv), while (vi) is immediate from the structure of M.

As an immediate corollary, we have

Lemma 7.2. $|\overline{N}|$ is divisible by 9.

Proof: Indeed, in the contrary case, the preceding lemma would imply that $\overline{N} = \overline{T}\overline{D}$, whence $\overline{N}$ centralizes z. But also $C_G(A)$ centralizes z since obviously $z \in A$. Hence $N \subseteq M$, which we have shown is not the case.

We now let P be a Sylow 3-subgroup of N containing D, so that $\overline{P} \cong Z_3 \times Z_3$ and $\overline{P} \supset \overline{D}$. We next prove

Lemma 7.3. $\overline{P}$ normalizes both X and Y.

Proof: We have that $A = Y \times \langle z,a \rangle$ with $[Y,D] = Y$ and $\langle z,a \rangle = C_A(D) \cong Z_2 \times Z_2$. Thus $\overline{P}$ leaves both Y and $\langle z,a \rangle$ invariant and so $z \sim a \sim za$ in N. If $a \in X$, then $\langle z,a\rangle = X$ and the lemma holds. Hence we can assume the contrary, whence $a \in Xz_2$ and $X \not\subseteq R_1$. Since X centralizes Q_2, it follows now from (4.8) that $L_0 = C_L(X)$ has Sylow 2-subgroup Q_2 and that $L_0/O(L_0) \cong PGL(2,r)$ for some odd r. We can suppose that $Y \not\subseteq L_0$, since otherwise we could choose D in $N_{L_0}(Y)$ and D would centralize X.

Let Y_0 be a four subgroup of $Q_2 \cap L_0'$ and set $A_0 = X \times Y_0$, $N_0 = N_G(A_0)$, $T_0 = S \cap N_0$, and $\bar{N}_0 = N_0/C_G(A_0)$. Note that $z_2 \in Y_0$ and as $a \in Xz_2$ in the present case, we have that $a \in A_0$. But $z \sim a \sim az$ in G and so all the preceding analysis applies with z as x, a or az as t, and A_0 as A. We then conclude that $z \sim a \sim az$ in N_0 and that $\bar{N}_0 = \bar{T}_0\bar{P}_0$, where $\bar{P}_0 \cong Z_3 \times Z_3$. On the other hand, L_0 contains a 3-subgroup D_0 such that $[D_0,Y_0] = Y_0$. Then D_0 centralizes X and so $D_0 \subseteq N_0$. Without loss, we can suppose that $\bar{D}_0 \subset \bar{P}_0$. But as $X = C_{A_0}(\bar{D}_0)$, $\bar{P}_0$ must leave X invariant. Since $z \in X$ and $\bar{T}_0$ centralizes z, it follows that all conjugates of z in N_0 lie in X. However, this is impossible as $z \sim a$ in N_0 and $a \notin X$.

A variation of the preceding argument will yield

Lemma 7.4. We have $X \subseteq R_1$.

Proof: Suppose false and let L_0 be as above. This time choose Y_0 to be a four subgroup of Q_2 with $Y_0 \not\subseteq L_0'$ and again set $A_0 = XY_0$. Again we have that $z \sim a$ in $N_0 = N_G(A_0)$, where now $\langle z,a\rangle = X$. By Lemma 4.2(iii) and (iv), $M \cap N_0$ contains a 3-subgroup D_0 such that $[D_0,Y_0] = D_0$ and $C_A(D_0) = X_0$ is a four group containing z, but distinct from X. This time we conclude that all conjugates of z in N_0 lie in X_0. But then $a \in X_0$, so $X = \langle z,a \rangle = X_0$, a contradiction.

As a corollary we have

Lemma 7.5. z is not conjugate in G to an element of $WQ_2 - R_1$.

Proof: Suppose $z \sim u$ in G for some u in $WQ_2 - R_1$. Then the preceding analysis applies with z,u as x, t respectively and so $z \sim u$ in N. But we have shown that all conjugates of z in N lie in X, so $u \in X$. Since $X \subset R_1$ and $u \notin R_1$, this is a contradiction.

Next we prove

Lemma 7.6. z is not isolated in $C_G(R_2)$.

Proof: Since P leaves both X and Y invariant and $\bar{P} \cong Z_3 \times Z_3$, it follows that $P_0 = C_P(Y)$ does not centralize X. We set $H = N_G(X)$, $C = C_G(X)$, and $\tilde{H} = H/O(H)$. Since $X \subseteq R_1$, $L \subseteq C$. Since $C \subseteq M$ and $O(\tilde{C}) = O(\tilde{H}) = 1$, it is immediate that $\tilde{L}$ is the unique simple component of $\tilde{C}$. Since R_2 is a Sylow 2-subgroup of L, it follows at once now by the Frattini argument that $N_H(R_2)$ covers H/C. Hence $N_H(R_2)$ contains a 3-element u which does not centralize X. In particular, $z^u \in X$ and $z^u \neq z$. If $|R_2| > 4$, then u centralizes R_2 as R_2 is dihedral or quasi-dihedral. But then $u \in C_G(R_2)$ and so z is not isolated in $C_G(R_2)$.

On the other hand, if $|R_2| = 4$, L contains a 3-element d such that $[R_2,d] = R_2$. Then $v = ud^i$ centralizes R_2 for some $i = 0,1$, or 2. But v normalizes, but does not centralize X as d centralizes X and we conclude in this case as well that z is not isolated in $C_G(R_2)$.

Setting $K = C_G(R_2)WR_2$, we now apply Lemma 4.8 to K. Since z

is not isolated in K, we conclude that

$$W \cong D_{2^n},\ QD_{2^n},\ \text{or}\ Z_{2^n} \times Z_{2^n}\ \text{for some}\ n;$$

and that K possesses a normal subgroup J containing $O(K)$ such that

$$J/O(K) \cong L_2(r),\ r\ \text{odd},\ L_3(r),\ r \equiv -1 \pmod 4,\ U_3(r),$$
$$r \equiv 1 \pmod 4,\ A_7,\ M_{11},\ \text{or}\ PSL_m(2,3),\ m \geq 2.$$

Furthermore, we have

$$R_2 \cap J = 1 \quad \text{and} \quad |WR_2 : WR_2 \cap JR_2| \leq 2.$$

We analyze the structure of K more closely.

Lemma 7.7. The following conditions hold:

(i) $R_1 \cap J$ is a Sylow 2-subgroup of J;

(ii) If $W \cong Z_{2^n} \times Z_{2^n}$, then $W = R_1 \subseteq J$.

Proof: We have shown in the preceding lemma that $C_G(R_2)$ contains a 3-element u which normalizes, but does not centralize X. On the other hand, if $\bar{K} = K/O(K)R_2$, $\bar{J} \cong J/O(K)$ and it is easily verified that $\bar{K}$ is isomorphic to a subgroup of $\mathrm{Aut}(\bar{J})$. In particular, $O^{2'}(\bar{K}') \subseteq \bar{J}$ and consequently $\bar{X} \subseteq \bar{J}$. Thus $A \subseteq JR_2$. It follows now from the structure of J that $J \cap N$ contains a 3-element v which centralizes Y and acts regularly on A/Y. By the structure of N, we must have $[X,v] = X$ and consequently $X \subseteq J$.

Setting $U = WR_2 \cap J$, we have that U is a Sylow 2-subgroup of J as WR_2 is a Sylow 2-subgroup of K. By the preceding paragraph $X \subseteq U$. But J has only one conjugacy class of involutions and so all involutions of U are conjugate to z. Now Lemma 7.5 forces all involutions of U to be in R_1. Hence if U is dihedral, then $U \subseteq R_1$ and (i) holds.

Suppose next that U is quasi-dihedral and $U \not\subseteq R_1$. Then $U = (U \cap R_1)\langle u \rangle$ with $U \cap R_1$ is dihedral of index 2 in U and $|u| = 4$. Moreover, $u \sim r_1$ for some element r_1 of order 4 in $U \cap R_1$ as J has only one conjugacy class of elements of order 4. Thus $r_1^t = u$ for some t in J. But as $z \in U$, we have $\langle z \rangle = Z(U)$ and consequently $r_1^2 = u^2 = z$.

It follows that $t \in M = C_G(z)$. But R_1 is a Sylow 2-subgroup of the inverse image C of $\bar{C} = C_{\bar{M}}(\bar{L})$ in M. Since $C \triangleleft M$, $u = r_1^t \in C \cap S = R_1$, which is a contradiction.

The only other possibility is that $W \cong Z_{2^n} \times Z_{2^n}$ for some $n \geq 2$. We conclude at once in this case that $\tilde{W} \subseteq \tilde{J}$ and that $\tilde{J} \cong PSL_n(2,3)$. In particular, $UR_2 = WR_2$. On the other hand, if we consider the action of W on L and apply Lemma 2.2(iii) and 2.3(v) of Part IV, we obtain that $C_L(W)$ contains a 3-element d which normalizes, but does not centralize a four subgroup Y_0 of R_2. Thus $[WY_0, d] = Y_0$. On the other hand, J contains a 3-element v which centralizes R_2 and acts fixed point-free on U. However, as $UR_2 = WR_2$ with Y_0 centralizing W and R_2 dihedral or quasi-dihedral, we see that $C_{UR_2}(Y_0) = WY_0$. Since v normalizes UR_2 and centralizes Y_0, it follows that v normalizes WY_0 and acts fixed-point-free on WY_0/Y_0. Note also that v normalizes $\mho^1(w)$ as this group is characteristic in $WY_0 = W \times Y_0$.

Now set $A_0 = XY_0$, $N_0 = N_G(A_0)$, and $\bar{N}_0 = N_0/C_G(A_0)$. Then $\langle d,v \rangle \subseteq N_0$ and, as usual, $\bar{N}_0$ is a $\{2,3\}$-group. But d acts trivially on X and nontrivially on Y_0; while v acts trivially on Y_0 and nontrivially on $X \subseteq \mho^1(W)$. Together these conditions imply that

$\langle \bar{d}, \bar{v} \rangle \cong Z_3 \times Z_3$. (Note that $\bar{N}_0$ has a normal 2-complement, as the Sylow 2-subgroup T_0/WY_0 of $\bar{N}_0$ is of order 2).

Observe next that $C_G(A_0)$ has a normal 2-complement by Lemma 4.1(i). Clearly then $O(C_G(A)) = O(N_0)$. Furthermore, as WY_0 is a Sylow 2-subgroup of $C_G(A_0)$, $C_G(A_0) = O(N_0)WY_0$. Hence if we set $\tilde{N}_0 = N_0/O(N_0)$, our argument yields that $\langle \tilde{d}, \tilde{v} \rangle$ induces a group of automorphisms of $\tilde{W}\tilde{Y}$ isomorphic to $Z_3 \times Z_3$. Since $\tilde{W} = C_{\tilde{W}\tilde{Y}_0}(\tilde{d})$, it follows that $\tilde{v}$ leaves $\tilde{W}$ invariant. However v normalizes WY_0, and so this in turn implies that v leaves W invariant. But v acts fixed-point-free on WY_0/Y_0 and so acts fixed-point-free on W. Since $v \in J$, this forces $W \subseteq J$ and we conclude that $W = U$.

Hence to complete the proof of both (i) and (ii), it thus remains only to show that $W = R_1$ in the case $W \cong Z_{2^n} \times Z_{2^n}$, $n \geq 2$. Clearly $X = \Omega_1(W)$ in this case. Setting $F = C_G(X)$, we have that $L \subseteq F$ as $X \subseteq R_1$. But $F \subseteq M$ as $z \in X$. Hence if we set $\tilde{F} = F/O(F)$, it follows that $\tilde{L}$ is the unique simple component of $\tilde{F}$ and that $\tilde{R}_1$ is a Sylow 2-subgroup of $C_{\tilde{F}}(\tilde{L})$. However, the element v above normalizes W and so acts on both F, $\tilde{F}$, and $C_{\tilde{F}}(\tilde{L})$. Since $\tilde{R}_1 = \tilde{W} \cap C_{\tilde{F}}(\tilde{L})$, we see that v normalizes R_1 as well as W. But W/R_1 is cyclic and as v acts fixed-point-free on W, the only possibility is that $W = R_1$ and the lemma is proved.

As an immediate corollary, we obtain

<u>Lemma 7.8</u>. We have $W = W_1$.

<u>Proof</u>: If $W \cong Z_{2^n} \times Z_{2^n}$, $n \geq 2$, then $W = R_1$ by the preceding lemma, so certainly $W = W_1$. On the other hand, if W is dihedral

or quasi-dihedral, the desired conclusion follows from (4.9).

We set $U = R_1 \cap J$ for the balance of the proof, so that U is a Sylow 2-subgroup of J by Lemma 7.7.

We next eliminate the quasi-dihedral case.

Proposition 7.9. We have $\bar{L} \cong L_2(q)$, q odd, or A_7.

Proof: Suppose false, in which case $R_2 \cong QD_{2^m}$ for some m. If $W = R_1$, then $S = R_1R_2 = R_1 \times R_2 \cong D_{2^n} \times QD_{2^m}$, $QD_{2^n} \times QD_{2^m}$, or $Z_{2^n} \times Z_{2^n} \times QD_{2^m}$, contrary to Proposition 2.5. Hence $W \supset R_1$. In particular, W is not homocyclic abelian by Lemma 7.7(ii) and so $W \cong D_{2^n}$ or QD_{2^n} for some $n \geq 3$ (as $R_1 \supset X \cong Z_2 \times Z_2$). Since $W = W_1$, this implies that $|W : R_1| = 2$. Since $|WR_2 : UR_2| \leq 2$ and $U = R_1 \cap J$ is a Sylow 2-subgroup of J, it follows now that $U = R_1$. Furthermore, by the structure of J and the fact that R_1 is a maximal subgroup of W, R_1 is dihedral.

We claim next that every involution of R_1R_2 is conjugate in G to z, z_2 or zz_2. By the structure of M and J, all involutions of R_1, R_2 are conjugate in J and M, respectively. Suppose then r_1r_2 is an involution with $r_i \in R_i^{\#}$, $i = 1, 2$. Then r_1 centralizes r_2 and so r_1, r_2 are each involutions. Since $J \subseteq C_G(R_2)$, $r_1r_2 \sim zr_2$ in J. But $zr_2 \sim zz_2$, in M, so $r_1r_2 \sim zz_2$ and our assertion is proved.

Consider first the case that W is dihedral. Then $W - R_1$ contains an involution w and $w \in S - R_1R_2$. Hence by Thompson's transfer lemma and the preceding paragraph $w \sim z'$ in G, where $z' = z, z_2$, or zz_2. Set $X_0 = \langle w, z \rangle$ and let Y_0 be a four subgroup of $C_L(X_0)'$, so that $A_0 = X_0 \times Y_0 \cong E_{16}$ and, as usual, $C_L(X_0)$ contains a 3-subgroup D_0 such that $[Y_0, D_0] = D_0$. Likewise if $N_0 = N_G(A_0)$, it follows as with A

and N that $w \sim z'$ in N_o and that all conjugates of w in N_o lie in X_o. Since z_2 and zz_2 are not in X_o, this forces $z' = z$, whence $w \sim z$ in G. However, this contradicts Lemma 7.5 as $w \in WQ_2 - R_1$.

Thus $W \cong QD_{2^n}$ and any element of $W - R_1$ is either of order 4 with z as its square or is of order 2^{n-1}. In particular, no element of $S - R_1R_2$ is an involution. Hence if we take w of order 4 in $W - R_1$, the extended form of Thompson's transfer lemma [27, Lemma 16] yields that $w^g = u \in R_1R_2$ for some g in G. But then $z^g = u^2$ and so by Lemma 7.5, $u^2 \in R_1$. Thus $u = r_1r_2$ with $r_1 \in R_1$ of order 4 and $r_2 \in R_2$ of order at most 2. Since R_1 is dihedral, it follows that $r_1^2 = z$. But then $u^2 = z$, whence $z^g = z$ and so $g \in M$. However, $\bar{R}_1\bar{R}_2$ is a Sylow 2-subgroup of $\overline{C}\overline{L}$, which is normal in $\bar{M}$ and consequently $w \in S - R_1R_2$ cannot be conjugate to $u \in R_1R_2$ in M. This contradiction establishes the proposition.

If for any central involution z' of S, $L_{z'}$ has a component with quasi-dihedral Sylow 2-subgroups, we can work with z' instead of z and Proposition 7.9 will yield a contradiction. Hence if z' is any involution of $Z(S)$ for which $L_{z'}$ has a simple component, then that component is isomorphic to $L_2(r)$, r odd, or A_7.

We next prove

<u>Proposition 7.10</u>. We have $S \supset WR_2$ and $\bar{L} \cong L_2(q)$.

<u>Proof</u>: Suppose false. If $W \cong Z_{2^n} \times Z_{2^n}$, then $W = R_1$ by Lemma 7.7(ii), whence $S = W \times R_2$. On the other hand, if $W \cong D_{2^n}$ or QD_{2^n}, the same conclusion follows from (4.3) as $W = W_1$. Thus

$S \cong Z_{2^n} \times Z_{2^n} \times D_{2^m}$, $D_{2^n} \times D_{2^m}$, or $QD_{2^n} \times D_{2^m}$, contrary to Proposition 2.5. Thus $S \supset WR_2$, forcing $\bar{L} \cong L_2(q)$.

By the proposition, $\tilde{M} = \bar{M}/\bar{C}$ contains a normal subgroup isomorphic to $PGL(2,q)$ or $PGL^*(2,q)$.

As with Proposition 7.9, we conclude now from Proposition 7.10 that if z' is an involution of $Z(S)$ such that $L_{z'}$ has a simple component, then $\bar{M}_1 = C_G(z')/O(C_G(z'))$ possesses a component $\bar{L}_1 \cong L_2(r)$, r odd, and that $\tilde{M}_1 = \bar{M}_1/C_{\bar{M}_1}(\bar{L}_1)$ contains a normal subgroup isomorphic to $PGL(2,r)$ or $PGL^*(2,r)$.

In the balance of the section we shall eliminate these cases as well. We first prove

<u>Lemma 7.11</u>. U is dihedral or homocyclic abelian.

<u>Proof</u>: Suppose false, in which case U is quasi-dihedral and $J/O(K) \cong L_3(r)$, $r \equiv -1 (\mathrm{mod}\ 4)$, $U_3(r)$, $r \equiv 1 \ (\mathrm{mod}\ 4)$, or M_{11}. Setting $M_1 = C_G(z_2)$, we have that $J \subseteq M_1$ and so M_1 is nonsolvable. Since z_2 is a central involution and G is an $\mathcal{R}_2$-group, we conclude, as usual, that $\bar{J}$ lies in a simple component $\bar{L}_1$ of $\bar{M}_1 = M_1/O(M_1)$. But then $\bar{L}_1$ has quasi-dihedral Sylow 2-subgroups, contrary to what we have shown above.

We require for our analysis three further results concerning $Z = \langle z, z_2 \rangle$.

<u>Lemma 7.12</u>. If $z' \in Z^{\#}$ and $z' \sim y$ in G for some y in $S - WR_2$, then $C_S(y)$ has 2-rank 4.

<u>Proof</u>: We argue essentially as in [19, Lemma 8.5]. Suppose false for some z' and y. We have $S = WR_2\langle y \rangle$. If $W = R_1$, then

$WR_2 = W \times R_2$, while if $W \supset R_1$, we reach the same conclusion from (4.3). Since $z_2 \in C_{R_2}(y)$, it follows that $m(C_W(y)) = 1$. On the other hand, y normalizes R_2 and so normalizes $U = R_1 \cap J$, which is a Sylow 2-subgroup of J. Since $U \subseteq W$, we have that $m(C_U(y)) = 1$. But $U \cong Z_{2^n} \times Z_{2^n}$ or D_{2^n} by the preceding lemma and so $C_U(y)$ is cyclic.

If $U \cong Z_{2^n} \times Z_{2^n}$, then $X = \Omega_1(U)$, in which case y normalizes X and $C_X(y) = \langle z \rangle$. Hence choosing u in $X - \langle z \rangle$, we have that $[u,y] = z$. On the other hand, if $U \cong D_{2^n}$, it again follows that U contains an element u such that $[u,y] = z$. Since $R_2\langle y \rangle$ is dihedral, there also exists an element r_2 in R_2 such that $[r_2,y] = z$. We conclude that u, r_2 and ur_2 normalize, but do not centralize $\langle y,z \rangle$, $\langle y,z_2 \rangle$, and $\langle y,zz_2 \rangle$ respectively. This implies that for z^* in $Z^{\#}$, $\langle y,z^* \rangle$ cannot be contained in the center of a Sylow 2-subgroup of G.

On the other hand, let S^* be a Sylow 2-subgroup of $C_G(y)$ containing $C_S(y)$. Since $z' \sim y$ in G, S^* is a Sylow 2-subgroup of G. Setting $Z^* = \Omega_1(Z(S^*))$, we have $y \in Z^*$. If $Z^* \subseteq \langle y,Z \rangle$, then $Z^* = \langle y,z^* \rangle$ for some z^* in $Z^{\#}$, contrary to the preceding paragraph. Hence $Z^* \not\subseteq \langle y,Z \rangle$ and consequently $B = \langle y,Z,Z^* \rangle \cong E_{16}$. But $B \subseteq C_G(Z) \subseteq M$. Since $y \in B \cong E_{16}$, while $m(C_S(y)) = 3$, Lemma 4.3(iii) now yields that $\bar{C} = C_{\bar{M}}(\bar{L})$ is not a 2-group. Since $r(\bar{C}) \le 2$ and $\bar{z} \in Z(\bar{C})$, we conclude, using Proposition 6.11 of Part III, that $C_{\bar{M}}(\bar{R}_2)$ contains a subgroup isomorphic to $SL(2,3)$. Clearly then Lemma 4.8 and Lemma 7.6 yield that $\bar{J} \cong M_{11}$, $L_3(r)$, $r \equiv -1 \pmod 4$, or $U_3(r)$, $r \equiv 1 \pmod 4$. Hence L_{z_2} has a simple component with quasi-dihedral Sylow 2-subgroup, contrary to our present assumption.

Lemma 7.13. If $z' \in Z^{\#}$, then $C_G(z')$ has a normal subgroup of index 2 with Sylow 2-subgroup WR_2.

Proof: Set $M_1 = C_G(z')$ and $\bar{M}_1 = M_1/O(M_1)$. If z is isolated in M_1, then $\bar{M}_1 = \overline{M \cap M_1}$ and Lemma 4.3(i) implies the desired conclusion. Thus we can assume that z is not isolated in M_1. In particular, $z' \neq z$.

Consider first the case that $\bar{M}_1$ has a simple component $\bar{L}_1$. Set $\bar{C}_1 = C_{\bar{M}_1}(\bar{L}_1)$ and let Y_1, Y_2 be subgroups of S whose images in $\bar{M}_1$ are Sylow 2-subgroups of $\bar{C}_1, \bar{L}_1$ respectively. In particular, $Y_i \triangleleft S$, $i = 1, 2$. Under our present assumptions, $\bar{L}_1 \cong L_2(r)$ for some odd r and so Y_2 is dihedral. Moreover, by our maximal choice of R_2, we have that $|Y_2| \leq |R_2| = 2^m$.

We wish to argue that $Y_2 \subseteq WR_2$; so assume the contrary, in which case $S - WR_2$ contains an element y of Y_2. Considering $\tilde{M} = \bar{M}/\bar{C}$, we have that $\tilde{S} = (\tilde{R}_2 \times \tilde{W})\langle\tilde{y}\rangle$ with $|\tilde{W}| \leq 2$ inasmuch as $W = W_1$. Furthermore, $\tilde{W} \cap \langle\tilde{y}\rangle = 1$ and $\tilde{R}_2\langle\tilde{y}\rangle \cong D_{2^{m+1}}$ or $QD_{2^{m+1}}$. Hence either $|\tilde{y}| = 2^m$ or $|\tilde{y}| \leq 4$. However, in the first case, $|y| \geq 2^m$ and as Y_2 is dihedral with $y \in Y_2$, it follows that $|Y_2| > |R_2|$, which is not the case. Thus $|\tilde{y}| \leq 4$ and so $[\tilde{R}_2, \tilde{y}]$ is cyclic of order 2^{m-1}. Since $[R_2, y] \subseteq Y_2$ and $|Y_2| \leq |R_2|$, this forces $|[R_2, y]| = 2^{m-1}$ and hence in turn $Y_2 = [R_2, y]\langle y\rangle$. Since Y_2 is dihedral and $y \notin [R_2, y]$, we conclude from this that y must be an involution. In particular, $R_2\langle y\rangle \cong D_{2^{m+1}}$. Since $R_2\langle y\rangle = R_2 Y_2$, it also follows that

$\bar{L}_1\bar{R}_2 \cong PGL(2,r)$. Note also that $z_2 \in [R_2,y]$ and so $z_2 \in Y_2$. Thus, in fact, $z' = zz_2$.

Since y is an involution, we see that $W \cap Y_2 = 1$. But $W = W_1$ centralizes R_2. Hence if we set $\tilde{M}_1 = \bar{M}_1/\bar{C}_1$, it follows now as in Lemma 6.5 that $\tilde{W} \subseteq \tilde{V}$, where $\tilde{V}$ is a cyclic subgroup of $\tilde{S}$ of order at most 4. Hence $W/W \cap Y_1$ is cyclic of order at most 4. If W is dihedral or quasi-dihedral, this implies the stronger conclusion that $|W/W \cap Y_1| \leq 2$ as such groups do not possess homomorphic images that are cyclic of order 4. On the other hand, we must have $R_1 \cap Y_1 = 1$, otherwise as in Lemma 6.5, the maximality of $|R_2|$ would be violated. Since $W = R_1$ if $W \cong Z_{2^n} \times Z_{2^n}$, we see that there is only one possibility for the structure of W: namely, $W \cong Z_2 \times Z_2$.

If y centralizes W, then $S = W \times R_2\langle y \rangle \cong Z_2 \times Z_2 \times D_{2^{m+1}}$, contrary to Proposition 2.5. Hence y does not centralize W and so $C_S(y) = \langle z,z_2,y \rangle \cong E_8$. On the other hand, $\bar{y} \sim \bar{z}_2$ in $\bar{L}_1$ and so $y \sim z_2$ in G. But as $y \in S - WR_2$, the preceding lemma implies that $C_S(y)$ has 2-rank 4, a contradiction. We have therefore shown that the assumption $Y_2 \not\subseteq WR_2$ leads to a contradiction in all cases and so we conclude that $Y_2 \subseteq WR_2$.

We shall argue next that $Z \cap Y_2 \neq 1$; so assume the contrary. Since $Y_2 \triangleleft S$, clearly this implies that $Z \subset \Omega_1(Z(S))$, whence $W \cong Z_{2^n} \times Z_{2^n}$ and $\Omega_1(Z(S)) = ZX$, where now $X = \Omega_1(W)$. Moreover, we have $W = R_1$ in this case. Under our present assumptions, we have that $\tilde{M}_1$ contains a normal subgroup isomorphic to $PGL(2,r)$ or $PGL^*(2,r)$. Hence by

(4.7), applied to M_1, we have that $X \subseteq Y_1 \times (Y_2 \cap Z(S))$, where $Y_2 \cap Z(S) \cong Z_2$. Thus $X \cap Y_1 \neq 1$ and consequently $R_1 \cap Y_1 \neq 1$. Again this leads to a contradiction as in Lemma 6.5. Thus $Z \cap Y_2 \neq 1$.

By Lemma 4.3(ii), applied to M_1, we have that $\bar{M}_1 = \bar{L}_1 C_{\bar{M}_1}(\bar{Y}_2)\bar{S}$. But by the preceding paragraph, $\bar{z} \in \bar{Y}_2\langle \bar{z}' \rangle$. Since clearly $\bar{z}' \in Z(\bar{M}_1)$, it follows that $C_{\bar{M}_1}(\bar{Y}_2\langle \bar{z}' \rangle) \subseteq C_{\bar{M}_1}(\bar{z}) = \overline{M \cap M_1}$ and so $\bar{M}_1 = \bar{L}_1(\overline{M \cap M_1})$. But $\bar{L}_1 \lhd \bar{M}_1$ and the Sylow 2-subgroup $\bar{Y}_2$ of $\bar{L}_1$ is contained in $\bar{W}\bar{R}_2$. Hence to prove the lemma in this case, we need only show that $\overline{M \cap M_1}$ has a normal subgroup of index 2 with Sylow 2-subgroup $\bar{W}\bar{R}_2$. However, this follows directly from Lemma 4.3(i). This completes the proof of the lemma in the case that $\bar{M}_1$ possesses a simple component.

We now consider the remaining case. Set $Q = S \cap O_{2',2}(M_1)$ and $B = \Omega_1(Z(Q))$. By Lemma 4.9, we have $\Omega_1(Z(S)) \subseteq B$. Again we first treat the case in which B contains an element y of $S - WR_2$. Since $[R_2, y] \subseteq B$ and B is elementary, we must have $[R_2, y] = \langle z_2 \rangle$. Since $[R_2, y]$ is cyclic of index 2 in R_2, it follows now that $R_2 \cong Z_2 \times Z_2$, whence also $W = R_1$. Furthermore, as $J \subseteq C_G(z_2)$, our maximal choice of z now yields, if J is nonsolvable, that also $U \cong Z_2 \times Z_2$. Since $|W : U| \leq 2$ and W is either dihedral, quasi-dihedral, or homocyclic abelian, we conclude, in addition, that $W \cong D_8$ or $Z_{2^n} \times Z_{2^n}$, $n \geq 1$.

Now y normalizes $W = R_1$. Hence if $C_W(y)$ is cyclic, it follows that $B = Z\langle y \rangle$ and that $m(C_S(y)) = 3$. But then Lemma 7.12 implies that z is not conjugate in G to any element of $B - Z$. Since z is not conjugate in G to any element of $Z - \langle z \rangle$ by Lemma 4.5, z is thus isolated in $N_{M_1}(B)$. Since $M_1 = O(M_1)N_{M_1}(B)$ by the Frattini argument, we conclude that z is isolated in M_1, which is not the case. Therefore $C_W(y)$ necessarily contains a four group. On the other hand, $S \neq W \times R_2\langle y\rangle \cong D_8 \times D_8$ or $Z_{2^n} \times Z_{2^n} \times D_8$ by Proposition 2.5, so y does not centralize W. In particular, W is not a four group.

Suppose $W \cong Z_{2^n} \times Z_{2^n}$, whence $n \geq 2$ and y centralizes $X = \Omega_1(W)$. Setting $W_0 = C_W(y)$, we have that $W_0 \subseteq Z(S)$ and so by Lemma 4.9, $W_0 \subset Q$. Since y acts on J with $J/O(K) = PSL_n(2,3)$ and since y centralizes X, we also have that W_0 is homocyclic abelian. Furthermore, as $[y,W] \subseteq B$, we must have $[y,W] = X$, which implies that $W/W_0 \cong Z_2 \times Z_2$. This in turn yields that W normalizes $\langle X,y\rangle$. Thus $B = X \times \langle y,z_2\rangle$, $Q = W_0 \times \langle y,z_2\rangle$ and $S/Q = WR_2\langle y\rangle/Q \cong E_8$. On the other hand, $F_1 = C_{M_1}(B)$ has a normal 2-complement by Lemma 4.1(i) as $F_1 \subseteq M$. Since Q is a Sylow 2-subgroup of $O_{2',2}(M_1)$ and of F_1, it follows that $O_2(\bar{M}_1/\bar{F}_1) = 1$ and $\bar{S}/\bar{Q} \cong E_8$. But $\bar{M}_1/\bar{F}_1$ is isomorphic to a subgroup of $N_G(B)/C_G(B)$, which by Proposition 5.9 is a $\{2,3\}$-subgroup of $L_4(2) \cong A_8$. Clearly A_8 contains no subgroups with these properties.

We thus conclude that $W \cong D_8$. We have that $K = WC_G(R_2)$ is S-invariant and that $J \triangleleft KS$. Setting $\tilde{K}\tilde{S} = KS/O(K)$, our conditions imply that $\tilde{J} \cong L_2(r)$ for some $r \equiv 3,5 \pmod 8$ and that $\tilde{J}\tilde{W} \cong PGL(2,r)$. But $\tilde{y}$ acts on $\tilde{J}\tilde{W}$ and so by the structure of $Aut(\tilde{J})$, we have $\tilde{J}\tilde{W}\langle\tilde{y}\rangle = \tilde{J}\tilde{W} \times \langle\tilde{y}'\rangle$ for some involution y' of S. Clearly $y' \in S - WR_2$ and y' centralizes W, so $S = W \times R_2\langle y'\rangle \cong D_8 \times D_8$, contrary to Proposition 2.5.

We have therefore shown that B contains no element of $S - WR_2$ and hence that $B \subseteq WR_2$. Now let B_1 be the normal closure of z in $N_{M_1}(B)$. Then $B_1 \subseteq B \subseteq WR_2$ and so $B_1 \subseteq R_1$ by Lemma 7.5. We use this to prove that $H = N_G(B_1)$ has a normal subgroup of index 2 with Sylow 2-subgroup WR_2. Since $N_{M_1}(B) \subseteq H$ and $M_1 = O(M_1)N_{M_1}(B)$, this will imply that M_1 has a normal subgroup of index 2 with Sylow 2-subgroup WR_2 and the lemma will be proved.

We have that $L \subseteq C_G(B_1) \subseteq M$ and that $C_G(B_1) \triangleleft H$. By the structure of M, $LO(C_G(B_1))$ is characteristic in $C_G(B_1)$. Hence if we set $\bar{H} = H/O(H)$, we have that $\bar{L} \triangleleft \bar{H}$ and that $\bar{L} \cong L_2(q)$. But then R_1 is a Sylow 2-subgroup of $C_{\bar{H}}(\bar{L})$. Since $\bar{R}_2$ is a Sylow 2-subgroup of $\bar{L}$, $\bar{R}_1\bar{R}_2$ is thus a Sylow 2-subgroup of $\bar{L}C_{\bar{H}}(\bar{L})$. On the other hand, by the structure of $P\Gamma L(2,q)$ and $W = W_1$, $\bar{H}/\bar{L}C_{\bar{H}}(\bar{L})$ is abelian and $\bar{S}/\bar{R}_1\bar{R}_2$ is elementary of order at most 4. Since WR_2 has index 2 in S, we conclude at once that H has a normal subgroup of index 2 with WR_2 as Sylow 2-subgroup. This completes the proof of the lemma.

<u>Lemma 7.14</u>. There exists an element y in $S - WR_2$ such that $y^2 \in Z$.

<u>Proof</u>: Suppose first that $S - WR_2$ contains an element y such that $y^2 \in R_2$. Then $\bar{L}\langle\bar{y}\rangle$ is a group and $|\bar{L}\langle\bar{y}\rangle : \bar{L}| = 2$. By definition of W, it follows that $\bar{L}\langle\bar{y}\rangle \cong PGL(2,q)$ or $PGL^*(2,q)$. Correspondingly there exists an element y_1 in $R_2\langle y\rangle$ such that $y_1^2 = 1$ or z_2. Taking y_1 as y, the lemma holds in this case. Hence we may assume that $S - WR_2$ contains no such element y.

Setting $\tilde{K}\tilde{S} = KS/O(K)R_2$, we have that $\tilde{S}$ normalizes both $\tilde{J}$ and

$\tilde{W}$ and that $\tilde{U} = \tilde{J} \cap \tilde{R}_1$ is a Sylow 2-subgroup of $\tilde{J}$. Our assumption implies that $\tilde{S} - \tilde{W}$ contains no involutions, which is possible only if $\tilde{W} \subseteq \tilde{J}$ and $\tilde{J}\tilde{S} \cong PGL^*(2,r)$. In this case, $S - WR_2$ contains an element y such that $\tilde{y}^2 = \tilde{z}$. This time we obtain that $\bar{L}\langle\bar{y}\rangle$ is a group and that $\bar{L}\langle\bar{y}\rangle/\langle\bar{z}\rangle \cong PGL(2,q)$ or $PGL^*(2,q)$. Correspondingly we conclude that $R_2\langle y\rangle - R_2$ contains an element y_1 such that $y_1^2 = z$ or $y_1^2 = zz_2$. Taking y_1 as y, lemma holds in this case as well.

We now begin to eliminate the various cases that remain. We first prove

Lemma 7.15. We can take y to be an involution.

Proof: If false, then $S - WR_2$ contains no involutions and our element y satisfies $y^2 = z' \in Z^{\#}$. The extended form of Thompson's fusion lemma [27, Lemma 16] yields that $y^g = u \in WR_2$ for some g in G. We know that $WR_2 = W \times R_2$ with $W \cong D_{2^n}$, QD_{2^n}, or $Z_{2^n} \times Z_{2^n}$ and $R_2 \cong D_{2^m}$. Furthermore, $|u| = 4$. If W is nonabelian, then clearly $u^2 \in \langle z\rangle R_2$. However, in the contrary case, we know that J contains a 3-element v which acts fixed-point-free on W and centralizes R_2. Hence $u^{v^i} \in WR_2$ and $(u^2)^{v^i} \in \langle z\rangle R_2$ for some i, so without loss we can assume that $u^2 \in \langle z\rangle R_2$. Similarly if R_2 is nonabelian, it follows that $u^2 \in Z$. In the contrary case, $R_2 \cong Z_2 \times Z_2$ and M contains a 3-element d which acts fixed-point-free on R_2 and centralizes W. Hence replacing u by u^{d^j} for some j, we can suppose without loss that $u^2 \in Z$.

If $y^2 = u^2$, then as $(y^2)^g = u^2$, g centralizes $y^2 = z'$ and so $g \in C_G(z')$. But then $y \sim u$ in $C_G(z')$. However this is impossible as $C_G(z')$ has a normal subgroups of index 2 with WR_2 as Sylow 2-subgroup, by Lemma 7.13. Thus $y^2 \neq u^2$. But $(y^2)^g = u^2$ and now Lemma 4.5 is contradicted.

Now let y be an involution of $S - WR_2$. We shall treat the cases $W = R_1$ and $W \supset R_1$ separately. In the first case, y normalizes R_1 and J and $\overline{R}_1\overline{J} \subseteq \overline{K} = K/O(K)$. Since $\overline{R}_1\overline{J} \cong PSL_n(2,3)$, $PSL(2,r)$, $PGL(2,r)$, $PGL^*(2,r)$, r odd, or A_7, we immediately obtain

Lemma 7.16. If $W = R_1$, then one of the following holds:

(i) $S - R_1R_2$ contains an involution y_1 which centralizes R_1;

(ii) $\overline{R}_1\overline{J} = \overline{J} \cong PSL(2,r)$ and $\overline{J}\langle\overline{y}\rangle \cong PGL(2,r)$;

(iii) $\overline{R}_1\overline{J} = \overline{J} \cong PSL_n(2,3)$; or

(iv) $\overline{R}_1\overline{J} \cong PGL(2,r)$ or $PGL^*(2,r)$ and $S - R_1R_2$ contains an involution y_1 which induces a field automorphism of order 2 on $\overline{J}$.

We next prove

Lemma 7.17. We have $W \supset R_1$.

Proof: Suppose false. Then one of the possibilities given in Lemma 7.16 must hold. (i) is immediately excluded by Proposition 2.5, as $\langle R_2, y_1\rangle$ is dihedral in this case.

Suppose that (ii) holds. By Thompson's transfer lemma, y is conjugate to an involution of R_1R_2 and so to one of $Z^{\#}$ as in the preceding lemma. But clearly $m(C_S(y)) = 3$, contrary to Lemma 7.12.

Suppose next that (iii) holds. Again y is conjugate in G to an involution of $R_1 \times R_2$ hence to one of $Z^{\#}$ as above. Thus by Lemma 7.12, $m(C_S(y)) = 4$ and so $C_{R_1}(y) \supseteq X = \Omega_1(R_1) \cong Z_2 \times Z_2$. Let $A_1 = X \times \langle y, z_2\rangle$.

We have $y^g = u \in Z^{\#}$ and $C_S(y)^g \subseteq S$ for some g in G. In particular, $X^g \subseteq A_1^g \subseteq S$ and so $u_1^g \in WR_2$ for some u_1 in X. But the elements of $X^{\#}$ are all conjugate in J and so by Lemma 7.5, we must have $u_1^g \in X$, whence $u_1^{gv} = u_1$ for some v in $N_J(R_1)$.

Considering $\tilde{K}\tilde{S}/O(K)R_2$ once again, we have that $\tilde{y}$ normalizes $\tilde{J}$ and centralizes $\tilde{X}$. Since $\tilde{R}_1 \cong Z_{2^n} \times Z_{2^n}$ is a Sylow 2-subgroup of $\tilde{J}$, $\tilde{y}$ thus centralizes $\tilde{J}/\tilde{R}_1 = \tilde{R}_1\langle\tilde{v}\rangle/\tilde{R}_1$ and so $\tilde{v}$ normalizes $\tilde{S} = \tilde{R}_1\langle\tilde{y}\rangle$. Consequently, we can choose v to normalize S as well as R_1R_2.

We have $u_1^{v^i} = z$ for some i. Setting $h = v^{-i}(gv)v^i$, it follows now that $z^h = z$. But $y^{v^i} \in S - WR_2$ as v normalizes S and R_1R_2 and $y \in S - WR_2$. Moreover, $(y^{vi})^h = y^{gv^{i+1}} \in WR_2$, as $u \in WR_2$ and v normalizes WR_2. However, $y^{v^i} \sim (y^{v^i})^h$ in $M = C_G(z)$, which is impossible as M has a normal subgroup of index 2 with WR_2 as a Sylow 2-subgroup.

Suppose finally that (iv) holds. Then $\tilde{S}\tilde{J}$ contains a normal subgroup of index 2 isomorphic to $PGL^*(2,r)$. Hence S contains a maximal subgroup T containing UR_2 such that $\Omega_1(T) \subseteq UR_2$. By Thompson's transfer lemma, every involution of $S - UR_2$ is conjugate in G to an involution of UR_2 and again, as above, to one of $Z^{\#}$. In particular, $y_1 \sim u$ in G, where $u \in Z^{\#}$. Choosing a suitable conjugate of y_1 in K, we may assume, by Lemma 2.2(iii) of Part IV, that $C_U(y_1)$ contains a four subgroup U_1 with $z \in U_1$ and that $C_J(y_1)$ contains a 3-element d such that $[d,U_1] = U_1$.

We now investigate the structure of $N_1 = N_G(A_1)$, where $A_1 = U_1 \times \langle y,z_2\rangle$. We know that $1 \neq \bar{d} \in \bar{N}_1 = N_1/O(N_1)$ with

$C_{\bar{A}_1}(\bar{d}) = \langle \bar{y}_1, \bar{z}_2 \rangle$. If $|\bar{N}_1|$ were divisible by 9, it would follow from the structure of $\bar{N}_1$ that N_1 also contains a 3-element d_1 which centralizes U_1 and normalizes, but does not centralize $\langle y_1, z_2 \rangle$. Since $z \in U_1$, $d_1 \in M$. On the other hand, since $A_1 \not\subseteq WR_2$, $N_M(A_1)/C_M(A_1)$ is a 2-group by Lemma 4.1(ii). This is clearly a contradiction. Therefore $|\bar{N}_1|$ is not divisible by 9. On the other hand, as $W = R_1 \not\cong Z_{2^n} \times Z_{2^n}$ in the present case, Proposition 5.2(i) must hold and so Lemma 5.10 is applicable. By Lemma 5.10(i) and Proposition 5.9, we conclude that $\bar{N}_1 = \bar{T}_1\langle \bar{d} \rangle$, where $T_1 \subseteq S$. But d leaves $U_1\langle z_2 \rangle = A_1 \cap WR_2$ invariant; consequently $A_1 \cap WR_2 \lhd N_1$, as $A_1 \cap WR_2 \lhd T_1$. However, this is impossible as $u \in A_1 \cap WR_2$, $y_1 \notin A_1 \cap WR_2$, and $y_1 \sim u$ in N_1 by Lemma 5.10(ii). This completes the proof of Lemma 7.17.

To complete the proof of Theorem A, it remains to eliminate the case $W \supset R_1$. Since $|W:U| \leq 2$ and $U \subseteq R_1$, we conclude that $R_1 = U$ and $|W:R_1| = 2$. This time we consider $\bar{M} = M/O(M)$. In the present case $S/R_1R_2 \cong Z_2 \times Z_2$ and so $\bar{L}\bar{S}/\bar{R}_1$ contains a normal subgroup of index 2 isomorphic to $PGL^*(2,q)$. We conclude now as in the previous lemma that every involution of $S - R_1R_2$ is conjugate to an involution of R_1R_2 and hence to one of $Z^{\#}$. If y induces an inner automorphism or a field automorphism of order 2 of $J/O(K)$, we can find an involution $y_1 \in S - WR_2$ with the property of y_1 in the previous case (iv), which leads to a similar contradiction. Therefore we have $J\langle y \rangle/O(K) \cong PGL(2,r)$ and so $R_1\langle y \rangle$ is dihedral. Since $m(C_S(y)) = 4$ by Lemma 7.12, $C_S(y)$ contains an involution wt, where $w \in W - R_1$ and $t \in R_2$. But then in $\tilde{M} = \bar{M}/\bar{C}$, $\langle \tilde{y}, \tilde{w}\tilde{t} \rangle$ is a four group disjoint from $\tilde{R}_2$, contrary to the structure of $P\Gamma L(2,q)$ given in Lemma 6.2(iv) of Part III. This contradiction completes the proof of Theorem A.

PART VI

A CHARACTERIZATION OF THE GROUP M_{12}

1. Introduction. In this part, we complete the proof of our Main Theorem and with it the classification of all finite groups whose 2-subgroups are generated by at most 4 elements.

In the Introduction of Part V, we have defined an $\mathfrak{R}_2$-group and have pointed out that the Main Theorem will be completely proved once it is shown to hold for simple $\mathfrak{R}_2$-groups. The principal result of Part V asserted that in a simple $\mathfrak{R}_2$-group G, the centralizer of every central involution of G is solvable. However, by Theorem C of Part II, the Main Theorem holds if the centralizer of every central involution of G is solvable and G possesses a nonsolvable 2-constrained 2-local subgroup.

In view of this, we define finally an $\mathfrak{R}_3$-group to be an $\mathfrak{R}_2$-group in which

(a) The centralizer of every central involution is solvable;

(b) Every 2-local subgroup is either solvable or non 2-constrained.

Thus the Main Theorem will be proved in its entirety once it is shown to hold for simple $\mathfrak{R}_3$-groups. This we shall carry out in the present part.

We shall prove

Theorem A. If G is a simple $\mathfrak{R}_3$-group, then $G \cong M_{12}$.

Again the object is to force a Sylow 2-subgroup of G to be of type M_{12}. Since $\mathcal{L}(G)$ is nonempty and the centralizer of every central involution of G is solvable, G clearly has at least two conjugacy classes of involutions and so a theorem of Brauer and Fong [6] will then yield that $G \cong M_{12}$.

The bulk of the long proof of Theorem A deals with the case in which every element of $\mathcal{L}(G)$ is simple of 2-rank 2. Included in this analysis is a study of arbitrary groups G which possesses an elementary abelian subgroup E of order 8 such that $N_G(E)/O(N_G(E))$ contains a self-centralizing normal subgroup isomorphic to $Z_4 \times Z_4$.

2. 2-Groups and their automorphism groups. We shall need a number of results concerning specific 2-groups and their automorphism groups, many of which are similar in character to the elementary results needed in Parts I and II.

We begin with some simple lemmas on 2-groups.

The next four lemmas can be established by straight forward calculation. We leave their proof to the reader. (Note that $T \cong D_{2^{n+1}} \wedge D_{2^{n+1}}$ in the first lemma. Use three-subgroup lemma for the proof of Lemma 2.4).

Lemma 2.1. If $T = (T_1 \times T_2)\langle t \rangle$, where $T_i \cong D_{2^n}$, $n \geq 3$, $T_i \lhd T$, $T/T_i \cong D_{2^{n+1}}$, $i = 1,2$, and t is an involution, then we have

(i) Every involution of $T - T_1T_2$ is conjugate in T to t;

(ii) The square of every element of T of order 4 lies in $Z(T)$.

Lemma 2.2. If $T = (T_1 \times T_2)\langle t\rangle$, where $T_1 \cong D_{2^n}$, $n \geq 3$, $T_i \lhd T$, $T/T_i \cong QD_{2^{n+1}}$, $i = 1, 2$, and t is an element of order 4 such that $t^2 \in Z(T) - Z(T_1) - Z(T_2)$, then every element of $T - T_1T_2$ of order 4 with square equal to t^2 is conjugate in T to t.

Lemma 2.3. If T is as in Lemma 2.1 or 2.2 and $S = T\langle y\rangle$, where $T \lhd S$, $y^2 = t$, and y interchanges T_1 and T_2 under conjugation, then every element of $S - T$ correspondingly has order at least 4 or 8.

Lemma 2.4. If $T = T_1T_2$ with $T_1 \cap T_2 = 1$, $T_i \cong D_{2^n}$ or QD_{2^n}, $n \geq 3$, $i = 1,2$, $Z(T) = Z(T_1)Z(T_2)$, and $T/Z(T) \cong D_{2^{n-1}} \times D_{2^{n-1}}$, then $[T,T] \cong Z_{2^{n-1}} \times Z_{2^{n-1}}$.

We now establish a number of results related to the automorphism groups of certain 2-groups. The first of these is the analogue of [Part II, Lemma 2.4] for 2-groups of type $U_3(4)$.

First of all, we denote by $U_3(4)^{(1)}$ and $U_3(4)^{(2)}$ the split extension of $U_3(4)$ by a field automorphism of order 2 or 4 respectively.

Lemma 2.5. Let A be a 2-group of type $U_3(4)$, set $B = \mathrm{Aut}(A)$, $C = O_2(B)$, and let D,E be respectively a Sylow 5-subgroup and a Sylow 3-subgroup of B. Then we have

(i) $C \cong E_{2^8}$ and C stabilizes the chain $A \supset Z(A) \supset 1$;

(ii) B/C is isomorphic to a Sylow 5-normalizer in A_8.

In particular, $|B| = 2^{10}\cdot 3\cdot 5$ and a Sylow 2-subgroup of B/C is isomorphic to Z_4;

(iii) $C_C(D) = 1$ and $N_B(D) \cong B/C$,

(iv) $C_C(E) \cong E_{16}$ and for a suitable choice of E, $N_B(E)$ is a split extension of $N_B(D)$ by $C_C(E)$;

(v) If f is an element of $N_B(D)$ of order 4, then $A<f>$ is of type $U_3(4)^{(2)}$ and $A<f^2>$ is of type $U_3(4)^{(1)}$;

(vi) A is characteristic in $A<f^2>$ and $A<f^2>$ is characteristic in $A<f>$;

(vii) f^2 inverts a subgroup of A of type $Z_4 \times Z_4$;

(viii) Every involution of $N_B(E) - C_C(E)$ is conjugate in $N_B(E)$ to f^2 and every element of $N_B(E) - C_C(E)$ of order 4 whose square is f^2 is conjugate in $N_B(E)$ to f or f^{-1};

(ix) Every involution of $A<f>$ or $A<f^2>$ not in A is conjugate in $A<f^2>$ to f^2.

Proof. A can be generated by elements $z_1, z_2, a_1, a_2, a_3, a_4$, subject to the relations

$$z_1^2 = z_2^2 = a_1^4 = a_2^4 = a_3^4 = a_4^4 = 1, a_1^2 = z_1, a_2^2 = a_3^2 = z_2, a_4^2 = z_1z_2,$$

$$[a_1,a_3] = z_1, [a_2,a_3] = [a_1,a_4] = z_1z_2, [a_2,a_4] = z_2,$$

with all other commutators of pairs of generators being trivial.

Clearly A is generated by a_1,a_2,a_3,a_4, $Z(A) = A' = \Omega_1(A) = <z_1,z_2> \cong Z_2 \times Z_2$ and $A/Z(A) \cong E_{16}$. Just as in the proof of Lemma 7.4 of part I, one argues that the stabilizer C_0 of the chain $A \supset Z(A) \supset 1$ is elementary abelian of order 2^8, that $C_0 \lhd B$, that $C_0 \subseteq C$, and that $\overline{B} = B/C_0$ is faithfully represented on $A/Z(A)$.

On the other hand, by the structure of the normalizer of a Sylow 2-subgroup of $U_3(4)$ in Aut $(U_3(4))$, one sees that $\overline{B}$ contains a subgroup isomorphic to a Sylow 5-normalizer in A_8 (which is of the form XY, where $X \cong Z_{15}$, $Y \cong Z_4$, and Y acts regularly on $O_5(X)$ and nontrivially on $O_3(X)$). Suppose $\overline{B}$ contained this subgroup properly. Since a Sylow 5-normalizer in A_8 is a maximal solvable subgroup of A_8, $\overline{B}$ would then be nonsolvable. But then B possesses a nontrivial perfect subgroup B_0. Since $Z(A) \cong Z_2 \times Z_2$, B_0 centralizes $Z(A)$. Since $\overline{B}_0$ is isomorphic to a subgroup of A_8, $|B_0|$ is divisible by 3. Thus B contains an element α of order 3 which centralizes $Z(A)$.

We shall show that B possesses no such element α. Indeed, one checks that A contains exactly five subgroups $A_i \cong Z_4 \times Z_4$, $1 \leq i \leq 5$, any two of which generate A. Since α permutes the A_i among themselves, α leaves at least two A_i invariant, say, A_1 and A_2. But $\Omega_1(A_i) \subseteq \Omega_1(A) = Z(A)$ and so α centralizes $\Omega_1(A_1)$ and $\Omega_1(A_2)$, whence α centralizes A_1 and A_2. Hence α centralizes $A = \langle A_1, A_2 \rangle$, contrary to the fact that α is an automorphism of A of order 3.

We conclude that $\overline{B} = B/C_0$ is isomorphic to a Sylow 5-normalizer in A_8. This in turn implies that $O_2(\overline{B}) = 1$, whence $C_0 = O_2(B) = C$. Thus (i) and (ii) hold.

To prove (iii), we need only show that $C_C(D) = 1$, for then the second assertion of (iii) will follow by the Frattini argument. Let $D = \langle \delta \rangle$. Since δ acts regularly and irreducibly on $\overline{A} = A/Z(A)$, we have $aa^{\delta}a^{\delta^2}a^{\delta^3}a^{\delta^4} = f(a) \in Z(A)$ for each a in A. Consider a coset $Z(A)a$ of $Z(A)$ in A. Then for z in $Z(A)$, we have that $f(az) = f(a)z^5 = f(a)z$ since δ clearly acts trivially on $Z(A) \cong Z_2 \times Z_2$. Hence we can choose z in $Z(A)$ such that $f(a)z^5 = 1$, whence $f(az) = 1$. We see then that each coset of $Z(A)$ in A contains an element a for which $f(a) = 1$.

Suppose now that δ centralizes a nontrivial element α of C. Since $\alpha \neq 1$, α does not centralize some coset of $Z(A)$ in A. Let a

be an element of this coset for which $f(a) = 1$. Then α does not centralize a and so $a^{\alpha} = a\epsilon$ for some $\epsilon \neq 1$ in $Z(A)$. Since $\alpha\delta = \delta\alpha$, this implies that

$$a^{\delta^i \alpha} = a^{\alpha\delta^i} = (a\epsilon)^{\delta^i} = a^{\delta^i}\epsilon$$

for all i. Hence

$$1 = f(a)^{\alpha} = a^{\alpha} a^{\delta\alpha} a^{\delta^2\alpha} a^{\delta^3\alpha} a^{\delta^4\alpha} = (a\epsilon)(a^{\delta}\epsilon)(a^{\delta^2}\epsilon)a^{\delta^3}\epsilon)(a^{\delta^4}\epsilon) = f(a)\epsilon^5 = \epsilon^5$$

and consequently $\epsilon = 1$. This contradiction proves that $C_C(D) = 1$ and so (iii) holds.

Next let $E = \langle \lambda \rangle$. Since λ corresponds to a 3-cycle in A_8, λ acts regularly on $A/Z(A)$. We have already shown that B contains no elements of order 3 which act trivially on $Z(A)$, so λ acts regularly on A. Let $A_1 = \langle x_1, x_1^{\lambda} \rangle$ and $A_2 = \langle x_2, x_2^{\lambda} \rangle$ be two λ-invariant subgroups of A with $A_1 \cong A_2 \cong Z_4 \times Z_4$, so that $A = \langle x_1, x_1^{\lambda}, x_2, x_2^{\lambda} \rangle$. Hence any element α of C which centralizes λ is determined by its effect on x_1 and x_2. On the other hand, if we define α in C by the conditions

$$x_1^{\alpha} = x_1\epsilon_1, \; x_2^{\alpha} = x_2\epsilon_2 \;, (x_1^{\lambda})^{\alpha} = x_1^{\lambda}\epsilon_1^{\lambda} \;, \text{ and } \; (x_2^{\lambda})^{\alpha} = x_2^{\lambda}\epsilon_2^{\lambda} \;,$$

with ϵ_1, ϵ_2 in $Z(A)$, we check that α centralizes λ. Conversely, any element of $C_C(\lambda)$ has such a form. Since ϵ_1 and ϵ_2 are arbitrary, we conclude that $C_C(\lambda) = C_C(E) \cong E_{16}$.

On the other hand, as $\bar{B} = B/C$ is isomorphic to a Sylow 5-normalizer in A_8, $\bar{E} \triangleleft \bar{B} = N_{\bar{B}}(\bar{D})$. Hence for a suitable choice of E, $N_B(E) \supseteq N_B(D)$ and $N_B(E)$ covers B/C. Since $C \cap N_B(D) = 1$, these results together with $C_C(E) \cong E_{16}$ yield (iv). Furthermore, we see that $N_B(E)/E$ is a split extention of E_{16} by a Frobenius group of order 20. In particular, the action of f on $C_C(E) \cong E_{16}$ is uniquely determined and it follows that $< C_C(E), f^2 > \cong Z_2 \times Z_2 \int Z_2$. From this we easily obtain (viii) .

Thus it remains to verify (v), (vi), (vii) and (ix) . First, if we identify A with a Sylow 2-subgroup of $U_3(4)$ and set $X = N_{U_3(4)^{(2)}}(A)$, we can regard $X/Z(A)$ as a subgroup of $\mathbf{B}$. Then for a suitable choice of D , we have that $D \subseteq X$ and $X = AD< f' >$, where f' normalizes D and induces a field automorphism of $U_3(4)$ of order 4. In particular, $A< f' >$ is a Sylow 2-subgroup of $U_3(4)^{(2)}$. On the other hand, using (viii) , it follows that all cyclic subgroups of $N_B(D)$ of order 2 or 4 are conjugate. Hence $A < f > \cong A < f' >$ and $A< f^2 > \cong A < (f')^2 >$, which together imply (v).

By the action of f on A, we have that $Z(A<f^2>) = Z(A)$ and $Z_2(A < f >) = Z(A)$. Hence $Z(A)$ is characteristic in both $A< f^2 >$ and $A < f >$. Setting $\bar{A}< \bar{f} > = A< f >/Z(A)$, we have that $\bar{A}< \bar{f}^2 > \cong Z_2 \times Z_2 \int Z_2$. But then $\bar{A}$ is the unique subgroup of its structure in $\bar{A}< \bar{f}^2 >$ and $\bar{A}< \bar{f}^2 >$ the unique one of its structure in $A< \bar{f} >$. From this, we immediately obtain (vi).

Finally we prove (vii) and (ix) simultaneously. With $\bar{A} < \bar{f} >$, as above, let F be the inverse image in A of $C_{\bar{A}}(\bar{f}^2)$. Then $|F| = 16$. Since $\bar{A}< \bar{f}^2 > \cong Z_2 \times Z_2 \int Z_2$, all involutions of $\bar{A} < \bar{f}^2 > - \bar{A}$ lie in $\bar{F} < \bar{f}^2 > - \bar{F}$ and consequently all involutions of $A< f^2 > - A$ lie in $F < f^2 > - F$. Thus $A < f^2 > - A$ contains at most 16 involutions. On the other hand, identifying f with a field automorphism of $U_3(4)$, we have that $C_{U_3(4)}(f^2) \cong L_2(4)$. In particular, $C_A(f^2) = Z(A)$ and so f^2 has 16 conjugates under the action of A . Hence (ix) holds. Furthermore, it also follows that F is normalized by an element x of order 3 of

$C_{U_3(4)}(f^2)$. Then x acts regularly on $Z(A)$ and as $\Omega_1(F) = \Omega_1(A) = Z(A)$, we must have $F \cong Z_4 \times Z_4$. Thus (vii) also holds and the lemma is proved.

In the next five lemmas, Y will denote a normal 2-subgroup of the group X such that $C_X(Y) \subseteq Y$ and $r(x) \leq 4$. We denote by S,P respectively a Sylow 2-subgroup and a Sylow 3-subgroup of X .

Lemma 2.6. If $Y \cong E_{16}$, $X/Y \cong A_4$, and $Z(S) \cong Z_2$, then S is of type A_8 .

Proof. We need only show that S splits over Y, for then by Gaschütz' theorem, X will split over Y in which case the desired conclusion follows from [Part II, Lemma 2.2 (iv)] .

In the present case, we have $X = SP$ with $S \lhd X$. Hence if we let $S = \langle Y, u_1, u_2 \rangle$ for suitable u_1, u_2, we can assume without loss that u_2 is conjugate to u_1 under the action of P. This implies that $C_Y(u_1) \cong C_Y(u_2)$. Setting $Y_i = C_Y(u_i)$, $i = 1,2$, and using the fact that u_i induces an automorphism of Y of order 2, it follows that $Y_i \cong Z_2 \times Z_2$ or E_8 , $i = 1, 2$. However, in the latter case, $Y_1 \cap Y_2 \cong Z_2 \times Z_2$ or E_8 , and $Y_1 \cap Y_2$ centralizes $\langle Y, u_1, u_2 \rangle = S$, contrary to the fact that $Z(S) \cong Z_2$. We conclude therefore that $Y_i \cong Z_2 \times Z_2$. This in turn implies that we can write $Y = A_i \times B_i$, where $A_i \cong B_i \cong Z_2 \times Z_2$ and u_i interchanges A_i and B_i under conjugation, $i = 1, 2$. But then by Lemma 6.8 of Part III, we can choose u_i to be an involution. Hence $Y\langle u_i \rangle \cong Z_2 \times Z_2 \int Z_2$, $i = 1, 2$ and now [Part II, Lemma 2.5] yields that S splits over Y, as required.

Lemma 2.7. If Y is of type A_8 and P has order 3, then

(i) $C_Y(P) \cong Z_2 \times Z_2$;

(ii) A Sylow 2-subgroup of $N_X(P)$ is abelian or dihedral of order 4 or 8;

(iii) If $X/Y \cong S_3$ and a Sylow 2-subgroup of $N_X(P)$ is elementary, then S is of type A_{10}.

Proof. We know that Y contains a characteristic subgroup $Q \cong Q_8 * Q_8$ and that any element of $Y - Q$ interchanges the two quaternion factors of Q. Since P centralizes $Y/Q \cong Z_2$, P does not centralize Q and our conditions imply that P acts regularly on $Q/Z(Q)$. Hence $N_Y(P) = C_Y(P)$ has order 4. If $C_Y(P) \cong Z_4$, then $Y - Q$ contains an element y with $\langle y^2 \rangle = Z(A) = Z(Y)$. However, in this case one can check directly that $Y - Q$ contains no involutions. But as Y is of type A_8, this contradicts the structure of Y. Thus (i) holds.

Since $r(X) \leq 4$, no 2-element of $X - Q$ induces an inner automorphism of Q. But $\mathrm{Aut}(Q)/\mathrm{Inn}(Q) \cong N_{A_8}(\langle (123),(456) \rangle)$ as noted before, so $X/Q \cong Z_2 \times Z_3$ or $Z_2 \times S_3$. Hence by (i) a Sylow 2-subgroup of $N_X(P)$ has order 4 or 8 and has the structure asserted in (ii).

Finally let the assumptions be as in (iii), so that a Sylow 2-subgroup of $N_X(P)$ is isomorphic to E_8 and X is a split extension of $Q \cong Q_8 * Q_8$ by $Z_2 \times S_3$. On the other hand, we know that $\mathrm{Aut}(Q)$ is isomorphic to a split extension of E_{16} by a Sylow 3-normalizer in A_8. Furthermore, the normalizer of P in $\mathrm{Aut}(Q)$ has Sylow 2-subgroup $T \cong Z_2 \times Z_2$ and $T(\mathrm{Inn}(Q))$ is of type A_8. Hence all complements to $\mathrm{Inn}(Q)$ in $T(\mathrm{Inn}(Q))$ are conjugate. This means that there is precisely one possibility for the structure of TQ and hence for that of a Sylow 2-subgroup of X. But

A_{10} contains a split extension of $Q_8 * Q_8$ by $Z_2 \times S_3$. Since this extension contains a Sylow 2-subgroup of A_{10}, we conclude that X has Sylow 2-subgroups of type A_{10}, proving (iii).

Lemma 2.8. If $Y \cong E_8$ and $X/Y \cong S_4$ and if we set $Q = [O_2(X), P]$, then one of the following holds:

(i) $Z(X) \cong Z_2$, $Q \cong Q_8 * Q_8$, and either S is of type A_8 or M_{12} or $\Omega_1(S) \subseteq Q$;

(ii) $Z(X) = 1$, $Q \cong E_{16}$ and either S is of type A_8 or $S/Q \cong Z_4$;

(iii) $Z(X) = 1$, $Q \cong Z_4 \times Z_4$, and either S is of type M_{12} or $S/Q \cong Z_4$.

(Hence there are 7 possible non-isomorphic structures for X and 5 for S).

Proof. Let X_1 be the subgroup of index 2 in X containing Y such that $X_1/Y \cong A_4$. The possible structures of X_1 have been determined in Lemma 3.2 of Part II. We set $S_1 = O_2(X)$, so that $S_1 = S \cap X_1$ is a Sylow 2-subgroup of X_1. If $Z(X) \neq 1$, then also $Z(X_1) \neq 1$ and so by that lemma $Z(X_1) \cong Z_2$ and $S_1 \cong Q_8 * Q_8$. Since P does not centralize Y or S_1/Y, clearly $S_1 = [S_1, P]$, so $S_1 = Q$. Let T be a Sylow 2-subgroup of $N_X(P)$. Then $T \cong Z_2 \times Z_2$ or Z_4 and $Z(Q) \subset T$. Moreover, any element of $T - Z(Q)$ either normalizes or interchanges the two quaternion factors of Q. Hence if $T \cong Z_2 \times Z_2$, it is immediate that correspondingly S is of type M_{12} or A_8. On the other hand, if $T \cong Z_4$, one checks easily that $S - Q$ contains no involutions. Thus (i) holds in this case.

Suppose next that $Z(X) = 1$. Since $X/X_1 \cong Z_2$ and $Z(X_1) \subseteq Y$, clearly also $Z(X_1) = 1$. Hence by the same lemma, there are three possible structures for S_1 and we conclude at once that $Q \cong E_{16}$ or $Z_4 \times Z_4$. Again if T is a Sylow 2-subgroup of $N_X(P)$, we have that $T \cong Z_2 \times Z_2$ or Z_4. Also $S = QT$ with $Q \cap T = 1$. Moreover, if $T \cong Z_2 \times Z_2$, then S is of type A_8 or M_{12} according as $Q \cong E_{16}$ or $Z_4 \times Z_4$ as split extensions of this form are uniquely determined (up to conjugacy) in $GL(4,2) \cdot E_{16}$ and $Aut(Z_4 \times Z_4) \cdot (Z_4 \times Z_4)$. Thus (ii) or (iii) holds in this case.

<u>Lemma 2.9</u> If $Y \cong E_{16}$, $X/Y \cong Z_3 \times Z_2 \times Z_2$, and $Y\langle u \rangle \cong Z_2 \times Z_2 \int Z_2$ for some u in $X - Y$, then S is of type $L_3(4)$.

<u>Proof.</u> Since P centralizes $Y\langle u \rangle/Y$, u leaves $Y_1 = [Y,P]$ invariant. If $Y_1 \cong Z_2 \times Z_2$, then u would be forced to centralize Y_1. But u also leaves $Y_2 = C_Y(P)$ invariant, and $Y_2 \cong Z_2 \times Z_2$ in this case. Hence $C_Y(u) = Y_1 \times C_{Y_2}(u) \cong E_8$, contrary to $Y\langle u \rangle \cong Z_2 \times Z_2 \int Z_2$. Hence we must have $Y_1 = Y$ and so P acts regularly on Y. Hence if T is a Sylow 2-subgroup of $N_X(P)$, $T \cong Z_2 \times Z_2$, $T \cap Y = 1$, and X splits over Y. Furthermore, a generator of P corresponds to a 3-cycle in $A_8 \cong GL(4,2)$. Since $C_{A_8}((123)) \cong Z_3 \times A_5$, a Sylow 2-subgroup of $C_{A_8}((123))$ is uniquely determined up to conjugacy. Thus the structure of $TY \cong S$ is uniquely determined. On the other hand, $PGL(3,4)$ contains such a split extension of E_{16} by $Z_3 \times Z_2 \times Z_2$ and so we conclude that S is of type $L_3(4)$.

<u>Lemma 2.10.</u> If $Y \cong E_{16}$, $X/Y \cong Z_2 \times S_3$, and $m(N_X(P)) \geq 3$, then $S \cong D_8 \times D_8$.

<u>Proof.</u> Set $A = C_Y(P)$ and let V be a Sylow 2-subgroup of $N_X(P)$. Thus $A = V \cap Y$. By our hypothesis, $|V| \geq 8$ and as $|X/Y|$ is not divisible by 8, it follows that $A \neq 1$. Hence $A \cong Z_2 \times Z_2$. Moreover,

by the Frattini argument, also $V/A \cong Z_2 \times Z_2$.

We argue first that it suffices to prove that V splits over A . Indeed, if this is the case, then V splits over Y and as VY is a Sylow 2-subgroup of X , X splits over Y by Gaschütz' theorem. Hence X can be identified with a subgroup of the holomorph of E_{16} and A_8 . Since P does not act regularly on Y , P is identified with a subgroup of A_8 which is generated by a product of two 3-cycles. But any two such subgroups of A_8 of order 3 are conjugate in A_8 and the normalizer in A_8 of any one of them is isomorphic to $Z_2 \times S_3$. Thus the structure of X is uniquely determined. Since $E_{16} \cdot N_{A_8}(P)$ has Sylow 2-subgroups isomorphic to $D_8 \times D_8$, as is easily checked,we conclude that $S \cong D_8 \times D_8$, as asserted.

To prove the required assertion, let v be an element of V whose image in X/Y generates $Z(X/Y)$. Then $v \in C_X(P)$ and v leaves $[Y,P] \cong Z_2 \times Z_2$ invariant, so in fact, v must centralize $[Y,P]$. Since $Y = A \times [Y,P]$, it follows that v does not centralize A. But $v^2 \in A$ as $V/A \cong Z_2 \times Z_2$ and consequently $\langle A,v \rangle \cong D_8$. Hence without loss we can choose v to be an involution.

By assumption, $m(V) \geq 3$ and so V contains a subgroup $W \cong E_8$. Since $|V| = 16$, $W \lhd V$. If $v \notin W$, v does not centralize W as V is nonabelian. Hence $C_W(v) \cong Z_2 \times Z_2$ and $V_1 = \langle v \rangle C_W(v) \cong E_8$. If $v \in W$, set $V_1 = W$, so in either case $v \in V_1 \cong E_8$. Since v does not centralize A and $|V| = 16$, we have that $A \cap V_1 \cong Z_2$. But then if B is a complement of $A \cap V_1$ in V , we have that $B \cong Z_2 \times Z_2$ and $A \cap B = 1$, so V splits over A and the lemma is proved.

We next prove two lemmas about 2-groups which admit an automorphism of order 7.

Lemma 2.11. If S is a 2-group with $r(S) \leq 4$ which admits a fixed-point-free automorphism of order 7, then either $S \cong Z_{2^n} \times Z_{2^n} \times Z_{2^n}$ for some n or S is of type $Sz(8)$.

Proof. Let the given automorphism be α. We need only prove that α acts transitively on the involutions of S, for then S is a Suzuki 2-group and the result follows from [30].

Let Z be a minimal α-invariant subgroup of $Z(S)$, so that Z is elementary abelian. Since α acts regularly on Z, we have $Z \cong E_8$. Since α induces a fixed-point-free automorphism of $\overline{S} = S/Z$ of order 7 and $r(\overline{S}) \leq 4$, the lemma holds by induction for $\overline{S}$. In particular, $\overline{W} = \Omega_1(\overline{S}) \cong E_8$. If W denotes the inverse image of $\overline{W}$ in S, we have that $Z \subseteq \Omega_1(S) \subseteq W$ and that W is α-invariant. Since α leaves $\Omega_1(S)$ invariant and α acts irreducibly an $\overline{W}$, either $\Omega_1(S) = Z$ or W. However, in the latter case each coset of Z in W would contain an involution, which would imply that every element of $W^{\#}$ is an involution. Hence $W \cong E_{64}$, contrary to $r(S) \leq 4$. Thus $\Omega_1(S) = Z \cong E_8$ and so α acts transitively on the involutions of S, as required.

Lemma 2.12. Let S be a 2-group with $r(S) \leq 4$ which admits an automorphism α of order 7. If α fixes an involution x of S such that $C_S(x) = \langle x \rangle \times T$, where α acts fixed-point-free on T, then one of the following holds:

(i) $S = [S,\alpha]\langle x \rangle$ and α acts fixed-point-free on $[S,\alpha]$; or

(ii) $\Omega_1(S) \cong E_{16}$.

Proof. If $x \in Z(S)$ then $S = \langle x \rangle \times T$ and the lemma clearly holds; so we can assume the contrary. Set $V = C_S(\alpha)$. If $V \supset \langle x \rangle$, then also

$C_V(x) \supset \langle x \rangle$. But $C_V(x) \subseteq C_S(x) \cap C_S(\alpha)$ and the latter intersection is equal to $\langle x \rangle$ by the assumed action of α on $C_S(x)$. This contradiction shows that $V = \langle x \rangle$.

Since $x \notin Z(S)$, it follows from the preceding paragraph that α acts fixed-point-free on $Z(S)$. We let W be a maximal α-invariant normal subgroup of S containing $Z(S)$ such that α acts fixed-point-free on W and we set $\bar{S} = S/W$. Since $C_S(\alpha)$ maps onto $C_{\bar{S}}(\alpha)$, we have that $C_{\bar{S}}(\alpha) = \langle \bar{x} \rangle$. By the maximality of W, α clearly centralizes $Z(\bar{S})$ and consequently $Z(\bar{S}) = \langle \bar{x} \rangle$. If $\bar{S} = \langle \bar{x} \rangle$, then $S = W\langle x \rangle$ and the first alternative of the lemma holds. Hence we can suppose that $\bar{S} \supset \langle \bar{x} \rangle$ and hence that $\tilde{S} = \bar{S}/\langle \bar{x} \rangle \neq 1$. It follows similarly that α acts fixed-point-free on $\tilde{S}$. Since $r(\tilde{S}) \leq 4$, the preceding lemma now yields that $\tilde{S} \cong Z_{2^m} \times Z_{2^m} \times Z_{2^m}$ for some m or that $\tilde{S}$ is of type $Sz(8)$.

For either structure of $\tilde{S}$, we have that $\tilde{R} = \Omega_1(\tilde{S}) \cong E_8$. If $\bar{R}$ denotes the inverse image of $\tilde{R}$ in $\bar{S}$, then $\bar{R}$ admits α as an automorphism of order 7. Since $\bar{R}/\langle \bar{x} \rangle \cong E_8$, the only possibility is that $\bar{R} \cong E_{16}$. In particular, $\bar{R} = \Omega_1(\bar{S})$. Furthermore, $\bar{R} = \langle \bar{x} \rangle \times \bar{R}_1$, where $\bar{R}_1$ is α-invariant and $\bar{R}_1 \cong E_8$. Since α acts fixed-point-free on $\bar{R}_1$, the maximality of W implies that $\bar{R}_1$ is not normal in $\bar{S}$. Let R, R_1 be the inverse images of $\bar{R}, \bar{R}_1$ in S. Then $R \triangleleft S$, $\Omega_1(S) \subseteq R$, $R = R_1\langle x \rangle$, and R_1 is not normal in S. The last condition implies that R_1 is not characteristic in R.

Since α acts fixed-point-free on R_1, the preceding lemma implies that $R_1 \cong Z_{2^n} \times Z_{2^n} \times Z_{2^n}$ for some n or that R_1 is of type $Sz(8)$. We have that $\Omega_1(S) \subseteq R$. Hence if $\Omega_1(R) = \langle x \rangle \times \Omega_1(R_1)$, then the second alternative of the lemma holds; so we can assume that this is not the case. Thus $R - R_1$ contains an involution y not in $\langle x \rangle \Omega_1(R_1)$. Then $y = xr$ for some r in R_1 with x inverting r and $|r| = 2^k$ for some $k \geq 2$. Since α centralizes x, it follows that x inverts r^{α^i} for all i and this implies that x inverts $\Omega_k(R_1)$. But now as $k \geq 2$, it is immediate that any element of $R - R_1$ has order less than that of a generator of R_1. Hence R_1 is the unique subgroup of R of its structure and consequently R_1 is characteristic in R. This contradiction establishes the lemma.

Lemma 2.18. If S is a 2-group of order 64 which admits a fixed-point-free automorphism of order 3, then S is either isomorphic to E_{64}, $Z_4 \times Z_4 \times Z_2 \times Z_2$, or $Z_8 \times Z_8$ or else S is of type $L_3(4)$ or $U_3(4)$.

Proof. If S is abelian, the lemma is clear; so we can assume S is nonabelian, whence $S' \neq 1$. It is well-known that S must now have class 2. Let α be the given automorphism of S. Clearly any α-invariant subgroup of S has order 2^{2k}, $0 \leq k \leq 3$. Hence if $T \neq 1$ is a proper α-invariant subgroup of S, $|T| = 4$ or 16 and it is immediate that $T \cong Z_2 \times Z_2$, $Z_4 \times Z_4$, or E_{16}. In particular, $Z(S)$ is a nontrivial α-invariant subgroup of S. If $|Z(S)| = 16$, then $S = Z(S)\langle a,b \rangle$ for suitable a,b and it follows that S' is cyclic. But then α centralizes S' and we have a contradiction. Hence $|Z(S)| = 4$ and so $Z(S) \cong Z_2 \times Z_2$. Since $S' \subseteq Z(S)$ and α does not centralize S', the only possibility is that $S' = Z(S)$. If $S/Z(S) \cong Z_4 \times Z_4$, then again $S = Z(S)\langle a,b \rangle$ for suitable a,b and so S' is cyclic, which is not the case. Hence we must have $S/Z(S) \cong E_{16}$, whence also $\Phi(S) = Z(S)$. In other words,

S is a special 2-group.

Since S is not abelian, S contains an element x of order 4. Setting $\bar{S} = S/Z(S)$, we have that $\bar{A} = \langle \bar{x}, \bar{x}^{\alpha} \rangle$ is an α-invariant four group. Hence the inverse image A of $\bar{A}$ is α-invariant of order 16 and contains x. By the preceding paragraph, we have $A \cong Z_4 \times Z_4$. We can also write $\bar{S} = \bar{A} \times \bar{B}$, where $\bar{B} \cong Z_2 \times Z_2$ and $\bar{B}$ is α-invariant. It follows that $S\langle \alpha \rangle$ is isomorphic to an extension of $Z_4 \times Z_4$ by A_4. Moreover, as $C_S(A)$ is α-invariant and $A \not\subseteq Z(S)$, we have that $C_S(A) = A$. Hence this extension is faithful. But MacWilliams [35, Lemma 3] has determined all such possible extensions in which a 3-element acts fixed-point-free on the Sylow 2-group of the extension and we conclude that S is of type $L_3(4)$ or $U_3(4)$.

In section 7, some explicit properties of the structure of $\mathrm{Aut}\,(L_3(3))$ and $\mathrm{Aut}(U_3(3))$ are needed.

<u>Lemma 2.19.</u> Let $X = \mathrm{Aut}(L_3(3))$ and $Y = \mathrm{Aut}(U_3(3))$. Then

(i) $|X:X'| = |Y:Y'| = 2$;

(ii) A Sylow 2-subgroup of Y is of type M_{12} and a Sylow 2-subgroup S of X is of order 32 and is generated by a,b,c,d,e with:

$$[c,d] = [b,e] = a, \quad [d,e] = b,$$
$$a^2 = c^2 = e^2 = 1, \quad b^2 = a, \; d^2 = b^{-1};$$

(iii) $Q = \langle ab^{-1}, cde \rangle$ is the unique normal subgroup of S isomorphic to Q_8;

(iv) S contains exactly two subgroups $R_1 = \langle ce, d \rangle$ and $R_2 = \langle e, cd \rangle$ isomorphic to QD_{16}. Moreover, $R_1 \cap R_2 = Q$.

(v) $SCN_3(S)$ is empty and S contains exactly two elementary abelian subgroups of order 8;

(vi) All involutions of $X - X'$ or $Y - Y'$ are conjugate in X or Y respectively;

(vii) If s,t are involutions of $X - X'$ or $Y - Y'$ respectively, then $C_{X'}(s) \cong C_{Y'}(t) \cong S_4$;

(viii) If x is an involution of X', then $C_{X'}(y) \cong GL(2,3)$ and $C_X(y)$ contains a normal subgroup of index 2 isomorphic to $Z_4 * SL(2,3)$;

(ix) If y is an involution of Y', then $C_{Y'}(y)/Z(C_{Y'}(y)) \cong S_4$ with $Z(C_{Y'}(y)) \cong Z_4$ and $C_Y(y)$ contains a normal subgroup of index 2 isomorphic $Q_8 * SL(2,3)$.

Proof. Some of these can be deduced from [Part IV, Lemma 2.3] and the rest can be directly checked. We omit the details.

Lemma 2.20. If the dihedral 2-group R is of index 2 in the group S, then $S - R$ contains an element s with $s^2 \in Z(R)$.

Proof. Suppose false, in which case clearly $|R| \geq 8$ and $\Omega_1(S) \subseteq R$. We set $Z(R) = \langle z \rangle$ and let x be an arbitrary element of $S - R$. If x^2 is an involution of $R - \langle z \rangle$, then $C_S(x^2) = \langle x,z \rangle \cong Z_4 \times Z_2$. Since $N_S(\langle x,z \rangle) \supset \langle x,z \rangle$, R contains an element y which normalizes $\langle x,z \rangle$. This forces $[y,x^2] = 1$, contrary to $C_R(x^2) = \langle x^2,z \rangle$. Hence x^2 can not be an involution of R and so $x^2 \in \langle a \rangle$, where $\langle a \rangle$ is the maximal cyclic subgroup of R. It follows that $|\langle x,a \rangle : \langle a \rangle| = 2$. Since $\Omega_1(S) \subseteq R$, $\langle x,a \rangle$ is not dihedral or noncyclic abelian. If $\langle x,a \rangle$ is quasi-dihedral or generalized quaternion, then necessarily $x^2 = z$ and the lemma holds. On the other hand, if $\langle x,a \rangle$ is cyclic, then $\langle x,a \rangle = \langle x \rangle$ and we see at once that S is dihedral, contrary to $\Omega_1(S) \subseteq R$. This establishes the lemma.

3. Some 2-groups associated with $\mathrm{Aut}(Z_4 \times Z_4)$. In [Part II, Lemma 2.1 (vi)], we have given a precise description of $B = \mathrm{Aut}(A)$, where $A \cong Z_4 \times Z_4$. In particular, if E is a Sylow 3-subgroup of B (and hence of order 3) and if F denotes a Sylow 2-subgroup of $N_B(E)$, we have shown that $F \cong D_8$ and have given an exact description of the action of F on A.

In this section we shall be concerned with a group X which possesses an elementary abelian subgroup W of order 8 such that

$$(*) \qquad N_X(W)/O(N_X(W)) \cong AEF^* ,$$

where F^* is a suitable subgroup of F. We shall consider three distinct possibilities for F^* and in each case our primary aim will be to show that the structure of a Sylow 2-subgroup of X is completely determined under these conditions.

The three candidates for F^* in which we shall be interested are:

(1) $F^* = Z(F)$;

(2) F^* is the unique cyclic subgroup of F order 4;

(3) F^* is the four subgroup of F which does not centralize E.

To distinguish these 3 cases, we shall denote these subgroups of F by F_1, F_2, F_3 respectively. In addition, we shall say, for brevity, that X is of type F_i, $i = 1, 2,$ or 3, if $(*)$ holds with $F^* = F_i$.

In each case, we let T be a Sylow 2-subgroup of $N_X(W)$ and let S be a Sylow 2-subgroup of X containing T. If X is of type F_1 or F_3, then correspondingly T is isomorphic to a split extention of $Z_4 \times Z_4$ by a group of order 2 inverting it (which, for brevity, we denote by D_{32}^*) or T is of type M_{12} by [Part II, Lemma 2.1]. Likewise the structure of T is uniquely determined if X is of type F_2. In particular, if $S = T$, the structure of a Sylow 2-subgroup of X is uniquely determined in each case. Hence it will suffice to consider the case in which $S \supset T$.

For simplicity, we shall identify T with AF_i, $i = 1,2$, or 3, as the case may be. Moreover, we shall use the same notation as in [Part II, Lemma 2.1] for AF. Thus $A = \langle \alpha,\beta \rangle = \langle \alpha \rangle \times \langle \beta \rangle$ and $F = \langle \delta,\epsilon \rangle$ with $\gamma = (\delta\epsilon)^2$ inverting A, ϵ interchanging α and β, $\alpha^\delta = \alpha^{-1}\beta^2$, $\beta^\delta = \beta\alpha^2$, and with $\langle \gamma,\delta \rangle$ centralizing E. In particular, we have $F_1 = \langle \gamma \rangle$, $F_2 = \langle \delta\epsilon \rangle$, and $F_3 = \langle \gamma,\epsilon \rangle$.

For each $i = 1,2,3$, $\langle \gamma,\alpha^2,\beta^2 \rangle$ is the unique elementary abelian normal subgroup of AEF_i of order 8. Hence in each case, we have $W = \langle \gamma,\alpha^2,\beta^2 \rangle$. We note also that $\langle \gamma,A \rangle - A$ has 16 involutions and four are conjugate to γ and 12 are conjugate to $\gamma\alpha$ in AEF_i for all $i = 1,2,3$. Hence if we set $N = N_X(W)$, it follows in all cases that $\langle \gamma,A \rangle - A$ has 16 involutions, four being conjugate to γ and 12 to $\gamma\alpha$ in N. We note also that the four conjugates of γ generate W. We set $V = \langle \gamma,A \rangle$ and check in all cases that A is characteristic in V and V is characteristic in T.

Furthermore, it follows by the Frattini argument that $N_N(A) \cap C_N(\gamma)$ contains a cyclic F_i-invariant 3-subgroup, which we again denote by E which acts fixed-point-free on A.

Next set $K = N_X(V)$. Since V is characteristic in T and $T \subset S$, $T \subset T_1$ for some Sylow 2-subgroup T_1 of K. Since A is characteristic in V, so is $\Omega_1(A)$. We now set $\bar{K} = K/O(K)\Omega_1(A)$ and $\bar{R} = [O_2(\bar{K}),\bar{E}]$. Since

$O_2(\bar{K}) \subseteq \bar{T}_1$, we can also consider the inverse image R of $\bar{R}$ in T_1.

We fix all this notation and first prove

Lemma 3.1. The following conditions hold:

(i) $|T_1 : T| = 4$;

(ii) $\bar{V} \cong E_8$, $C_{\bar{K}}(\bar{V}) = \bar{V}$, and $\bar{K}/\bar{V}$ is isomorphic to A_4 if X is of type F_1 and to S_4 if X is of type F_2 or F_3;

(iii) R is either isomorphic to $Z_8 \times Z_8$ or is of type $L_3(4)$ or $U_3(4)$;

(iv) $T_1 = RF_i$ with $R \cap F_i = 1$.

Proof. Since E normalizes A and centralizes γ, E normalizes V. Thus $E \subseteq K$ and as ET covers $N/O(N)$, so therefore does K. Hence by the preceding discussion either all involutions of V are conjugate in K or else the fusion of involutions of V in K is the same as that in N. However, in the latter case, all conjugates of γ under the action of T_1 would lie in W and as these conjugates generate W, it would follow that $T_1 \subseteq N_G(W) = N$. But this is impossible as $T_1 \supset T$ and T is a Sylow 2-subgroup of N. We conclude that all involutions of V are conjugate in K. This fact will enable us to prove (i) and (ii).

First of all, clearly $\bar{V} \cong E_8$. We let H be the inverse image of $C_{\bar{K}}(\bar{V})$ in K and also set $\tilde{K} = \bar{K}/\bar{V}$. Since $\bar{H}$ centralizes $\bar{\gamma}$ and $\Omega_1(A) \lhd K$, it follows that H normalizes $W = \Omega_1(A) \times \langle \gamma \rangle$. Thus $H \subseteq N$. Obviously $O(N) \cap K$ centralize V. But as $C_K(V) \subseteq N$, it is immediate from the structure of N that $C_K(V)$ has a normal 2-complement, which is clearly contained in $O(K)$. Hence $O(N) \cap H \subseteq O(C_K(V)) \subseteq O(K)$. We conclude that $\bar{H} \subseteq \overline{TE}$ and hence that $\tilde{H} \subseteq \tilde{F}_i\tilde{E}$. However, $\tilde{F}_i\tilde{E}$ acts faithfully on $\overline{V}$ by the structure of AF and so $\widetilde{H} = 1$. Thus $\overline{H} \subseteq \overline{V}$ and so $\overline{H} = C_{\overline{K}}(\overline{V}) = \overline{V}$. This in turn implies that $\widetilde{K}$ is isomorphic to a subgroup of $L_3(2)$.

Since every element of $\Omega_1(A)\alpha$ is of order 4, while every element of $\Omega_1(A)\gamma$ is of order $2, \overline{\alpha}$ and $\overline{\gamma}$ are not conjugate in $\overline{K}$ and so $|\widetilde{K}|$ is not divisible by 7. On the other hand, the involutions $\overline{\gamma}, \overline{\gamma\alpha}, \overline{\gamma\beta}, \overline{\gamma\alpha\beta}$, being the images of conjugate involutions of V, are all conjugate in $\overline{K}$. Thus $|\widetilde{K} : C_{\widetilde{K}}(\widetilde{\gamma})| \geq 4$. Since $C_{\widetilde{K}}(\widetilde{\gamma}) \supseteq \widetilde{F}_i\widetilde{E}$ and $\widetilde{F}_i\widetilde{E}$ is isomorphic to Z_3 if $i = 1$ and to S_3 if $i = 2$ or 3 and since $|\widetilde{K}|$ is not divisible by 7, we conclude now that $\widetilde{K} \cong A_4$ or S_4 with the latter holding if $i = 2$ or 3. Moreover, if $i = 1$, we claim that $\widetilde{K} \cong A_4$. Indeed, if H_1 denotes the inverse image of $C_{\widetilde{K}}(\widetilde{\gamma})$ in K, then as with H, we have that $H_1 \subseteq N$ and $O(N) \cap H_1 \subseteq O(K)$, whence $\overline{H}_1 = C_{\overline{K}}(\overline{\gamma}) = \overline{T}\overline{E}$. Therefore if $i = 1$, it follows that $C_{\overline{K}}(\overline{\gamma}) = \overline{V}\overline{E}$. We conclude easily from this that $C_{\widetilde{K}}(\widetilde{\gamma}) = \widetilde{E}$, which implies at once that $\widetilde{K} \cong A_4$ when $i = 1$. Thus (ii) holds.

Our argument shows that $|\widetilde{T}_1 : \widetilde{T}| = 4$ and consequently $|T_1 : T| = 4$, so (i) also holds.

Now let $\overline{R}_1 = O_2(\overline{K})$, so that $\overline{R} = [\overline{R}_1, \overline{E}]$. We have that $\overline{E}$ acts regularly on $\overline{A}$, $\overline{V} = \overline{A} \times \langle \overline{\gamma} \rangle$, and $\overline{R}_1/\overline{A} \cong E_8$ or Q_8 with $\overline{E}$ acting nontrivially on $\overline{R}_1/\overline{A}$. In the latter case, as the four involutions of $\overline{V} - \overline{A}$ are conjugate in $\overline{K}$, $\overline{R}_1$ would contain an element $\overline{r}$ such that $\overline{r}^{\,2} = \overline{\gamma}$. But then $\overline{r} \in C_{\overline{K}}(\overline{\gamma})$. However, we have shown above that $C_{\overline{K}}(\overline{\gamma}) = \overline{TE} = \overline{VF_i}\overline{E}$. Since $\overline{\gamma}$ is not a square in $\overline{TE}$, we reach a contradiction. Thus $\overline{R}_1/\overline{A} \cong E_8$ and so $[\overline{R}_1/\overline{A}, \overline{E}] \cong Z_2 \times Z_2$. This implies that $\overline{R} \cong E_{16}$ or $Z_4 \times Z_4$ with $\overline{E}$ acting regularly on $\overline{R}$.

Since E acts fixed-point-free on $\Omega_1(A)$ and $R/\Omega_1(A) \cong \overline{R}$, we conclude now by the Frattini argument that $N_K(R)$ contains an F_i-invariant cyclic 3-subgroup E^* which induces a fixed-point-free automorphism of R and which centralizes γ. Since $A \subset R$, R is not elementary abelian. It follows therefore from Lemma 2.14 that $R \cong Z_4 \times Z_4 \times Z_2 \times Z_2$ or $Z_8 \times Z_8$ or else R is of type $L_3(4)$ or $U_3(4)$. Hence to complete the proof of (iii), it remains only to rule out the first possibility.

In this case, $R = A \times R_0$, where $R_0 \cong Z_2 \times Z_2$. Then γ normalizes $\Omega_1(R) = \Omega_1(A) \times R_0$ and so every conjugate of γ under the action of R lies in $\Omega_1(R)\gamma$. In particular, γ is not conjugate to $\alpha\gamma$ under the action of R. But by the structure of $\overline{K}$, we clearly have $\overline{K} = \overline{R}C_{\overline{K}}(\overline{\gamma})$ and consequently γ is not conjugate to $\alpha\gamma$ in K, contrary to the fact that all involutions of V are conjugate in K. Thus (iii) also holds.

Finally as E^* acts fixed-point-free on R and E^* is F-invariant, we have that $R \cap F_i = 1$. Thus $|RF_i| = |T_1|$ and we conclude that $T_1 = RF_i$. Therefore (iv) also holds and the lemma is proved.

We shall treat these three possibilities separately. We recall first that $L_3(4)^{(1)}$ denotes the split extension of $L_3(4)$ by the image of the transpose-inverse map of $SL(3,4)$ and that $L_3(4)^{(1)}$ has Sylow 2-subgroups isomorphic to the group T_2 of [18, Lemma 7.1].

<u>Lemma 3.2.</u> If R is of type $L_3(4)$, then X is of type F_1 or F_3 and correspondingly S is of type $L_3(4)^{(1)}$ or $\hat{A}_{10}$.

<u>Proof</u>: First of all, if X is of type F_2 (in which case γ is a square in T_1), it follows from [Part II, Lemma 2.4 (ii)] that γ stabilizes the chain $R \supset Z(R) \supset 1$. But then we see that γ cannot be conjugate to

$\gamma\alpha$ in K, a contradiction. Thus X is necessarily of type F_1 or F_3. Since $C_R(\gamma) = Z(R) = \Omega_1(A)$, it follows now from [Part II, Lemma 2.4 (iv)] that T_1 is correspondingly of type $L_3(4)^{(1)}$ or $\hat{A}_{10}$. Hence to establish the lemma we need only show that T_1 is, in fact, a Sylow 2-subgroup of X.

Suppose false, in which case $N_X(T_1) - T_1$ possesses a 2-element y such that $y^2 \in T_1$. It is easy to see, using the structure of a 2-group of type $L_3(4)^{(1)}$ or $\hat{A}_{10}$, that R is the only subgroup of T_1 of type $L_3(4)$ and so R is characteristic in T_1. If $i = 1$, then $T_1 = R<\gamma>$ and so certainly y normalizes $R<\gamma>$. On the other hand, if $i = 3$, it follows from [Part II, Lemma 2.4 (iv)] that no two of the groups $R<\gamma>$, $R<\varepsilon>$, and $R<\varepsilon\gamma>$ are isomorphic. Hence $R<\gamma>$ is characteristic in T_1 and so y normalizes $R<\gamma>$ in this case as well. But we can also check that every involution of $R<\gamma> - R$ is conjugate in $R<\gamma>$ to γ. Hence replacing y by yr for suitable r in R, we can assume without loss that y centralizes γ. But then y normalizes $Z(R)<\gamma> = \Omega_1(A)<\gamma> = W$ and so $y \in N$. Since $T \subset T_1$, $<T,y>$ is thus a 2-subgroup of N and as T is a Sylow 2-subgroup of N, this forces $y \in T$. Thus $y \in T_1$, which is not the case.

We next prove

<u>Lemma 3.3.</u> If R is of type $U_3(4)$, then X is of type F_1 or F_2 and correspondingly S is of type $U_3(4)^{(1)}$ or $U_3(4)^{(2)}$.

<u>Proof:</u> Suppose $i = 3$, in which case $F_3 = <\varepsilon,\gamma>$ acts on R. By Lemma 2.5 (ii), either ε, $\varepsilon\gamma$, or γ must stabilize the chain $R \supset Z(R) \supset 1$. However, as $\alpha^{\varepsilon} = \beta$ and $\alpha^{\varepsilon\gamma} = \beta^{-1}$ neither ε nor $\varepsilon\gamma$ stabilizes this chain, so γ must stabilize this chain. But now it follows as before that γ cannot be conjugate to $\gamma\alpha$ in K, a contradiction. Therefore $i = 1$

or 2 and correspondingly $T_1 = RF_1$ or RF_2 with $F_1 \cong Z_2$ and $F_2 \cong Z_4$. If E^* is as in Lemma 3.1, E^* is F_i-invariant, E^* centralizes γ, and E^* induces a fixed-point-free group of automorphisms of R of order 3. We conclude therefore from Lemma 2.5 (v) and (viii) that T_1 is correspondingly of type $U_3(4)^{(1)}$ or $U_3(4)^{(2)}$.

Furthermore, by Lemma 2.5 (vi), we have that R and $R\langle\gamma\rangle$ are characteristic in T_1 and that every involution of $R\langle\gamma\rangle - R$ is conjugate in $R\langle\gamma\rangle$ to γ. We conclude therefore exactly as in the preceding lemma that T_1 is, in fact, a Sylow 2-subgroup of X, and the lemma follows.

To state our results in the case $R \cong Z_8 \times Z_8$, we first introduce some general terminology. Let $Z = \langle a,b \rangle \cong Z_{2^n} \times Z_{2^n}$, $n \geq 2$. A split extension of Z by a group $\langle c \rangle \cong Z_2$ will be said to be of <u>type</u> $R_{2^{2n+1}}$ provided:

(a) c inverts $\mho^1(Z) = \langle a^2,b^2 \rangle$; and

(b) Either c inverts Z or every involution of $Z\langle c \rangle$ lies in $\mho^1(Z)\langle c \rangle$.

Moreover, a split extension of Z by a group $\langle c,d \rangle$ of order 4 will be said to be of <u>type</u> $R_{2^{2n+2}}$ provided:

(a) $Z\langle c \rangle$ is of type $R_{2^{2n+1}}$; and

(b) Either $\langle c,d \rangle \cong Z_4$ or $\langle c,d \rangle \cong Z_2 \times Z_2$ and d interchanges a and b under conjugation.

We note that when $n \geq 3$, every 2-group of type $R_{2^{2n+1}}$ or $R_{2^{2n+2}}$ has no nontrivial normal abelian subgroups of rank 3. In addition, we see that our Sylow 2-subgroup T of $N = N_X(W)$ is of type R_{2^5} if X is of type F_1 and of type R_{2^6} if X is of type F_2 or F_3.

We now prove

<u>Lemma 3.4.</u> If $R \cong Z_8 \times Z_8$, then S is of type $R_{2^{2n+1}}$ if X is of type F_1 and of type $R_{2^{2n+2}}$ if X is of type F_2 or F_3 for some $n \geq 3$. Moreover, $SCN_3(S)$ is empty and the involutions of the unique element of $U(S)$ are conjugate in X.

<u>Proof</u>: Let $R = \langle a,b \rangle$. Then $A = \mho^1(R) = \langle a^2, b^2 \rangle$ and so γ inverts $\mho^1(R)$. Since $T_1 = RF_i$ with $R \cap F_i = 1$, it follows from the definition that T_1 is of type R_{2^7} if $i = 1$ and of type R_{2^8} if $i = 2$. Moreover, if $i = 3$, then ϵ inverts the 3-subgroup E^* which acts fixed-point-free on R. It is immediate from this that a and a^γ generate R. Hence we can take $b = a^\gamma$ and we see that T_1 is of type R_{2^8} in this case as well.

We now let A^* be an F_i-invariant 2-subgroup of X containing R and of maximal order such that

(a) $A^* = \langle \alpha^*, \beta^* \rangle = \langle \alpha^* \rangle \times \langle \beta^* \rangle \cong Z_{2^n} \times Z_{2^n}$;

(b) A^*F_i is of type $R_{2^{2n+1}}$ if $i = 1$ and of type $R_{2^{2n+2}}$ if $i = 2$ or 3;

(c) A^* is normalized by an F_i - invariant cyclic 3-subgroup (which we again denote by E^*) which acts fixed-point-free on A^* and centralizes γ.

We set $T^* = A^*F_i$ and argue that T^* is a Sylow 2-subgroup of X. We assume false and proceed as in Lemma 3.1. We set $V^* = A^* \langle \gamma \rangle$, $K^* = N_X(V^*)$, and let T_1^* be a Sylow 2-subgroup of K^* containing T^*. One checks directly that A^* is characteristic in V^* and V^* is characteristic in T^*. Since T^* is not a Sylow 2-subgroup of X, it follows therefore that $T^* \subset T_1^*$ for some Sylow 2-subgroup T_1^* of K^*.

Setting $V_1^* = \mho^1(A^*) < \gamma >$. Since γ inverts $\mho^1(A^*)$ by (b), we check that every involution of $V_1^* - \mho^1(A^*)$ is conjugate to γ in $V_1^* E^* < \gamma >$. We claim that there exists t in $N_{T_1^*}(T^*)$ such that $\gamma^t \in V^* - V_1^*$. Suppose false. Since V^* is characteristic in T^*, it follows that $\gamma^t \in V_1^*$ for each such element t. Since t also normalizes $\mho^1(A^*)$, this implies that $\gamma^{ta} = \gamma$ for some a in $V_1^* E^* < \gamma >$. But then ta centralizes γ. Since ta also normalizes $\Omega_1(A^*) = \Omega_1(A)$, ta thus normalizes $W = \Omega_1(A) \times < \gamma >$ and so $ta \in N$. Since $a \in V_1^* E^* < \gamma >$ and $E^* < \gamma > \subseteq N$, this in turn yields that $tv_1 \in N$ for some v_1 in V_1^*. Hence $tv_1 \in N \cap T_1^* = T$ and so $t \in Rv_1^{-1} \subseteq T^*$. We thus conclude that $N_{T_1^*}(T^*) = T^*$, contrary to the fact that $T^* \subset T_1^*$. This proves our assertion.

Taking t in T_1^* such that $\gamma^t \in V^* - V_1^*$, we have that $\gamma^t = a\gamma$ for some a in $A^* - V_1^*$. Since γ^t is an involution, it follows now that γ inverts a and also that $|a| = 2^n$. Since E^* centralizes γ and acts fixed-point-free on A^* and since $|a| = 2^n$, this in turn implies that $<a, a^\gamma> = A^*$. Hence γ inverts A^*.

Now we set $\overline{K}^* = K^*/O(K^*)\mho^1(A^*)$. As in Lemma 3.1, we conclude easily that $\overline{V}^* \cong E_8$, that $\overline{K}^*/\overline{V}^* \cong A_4$ if $i = 1$ and S_4 if $i = 2$ or 3, and that $\overline{R}^* = [O_2(\overline{K}^*), \overline{E}^*] \cong E_{16}$ or $Z_4 \times Z_4$. Moreover, if R^* denotes the inverse image of $\overline{R}^*$ in T_1^*, we find that $T_1^* = R^* F_i$ with $F^* \cap F_i = 1$ and that $N_{K^*}(R^*)$ possesses an F_i-invariant cyclic 3-group (which we again denote by E^*) which induces a fixed-point-free group of automorphisms of R^* of order 3 and which centralizes γ.

Setting $\tilde{R}^* = R^*/\mho^2(A^*)$, we have that $\tilde{R}^*$ is of order 64 and admits E^* as a fixed-point-free group of automorphisms of order 3. Hence by

Lemma 2.13, $\tilde{R}^* \cong Z_8 \times Z_8$ or $Z_4 \times Z_4 \times Z_2 \times Z_2$ or $\tilde{R}^*$ is of type $L_3(4)$ or $U_3(4)$. In the first or last case $\Omega_1(\tilde{R}^*) = \Omega_1(\tilde{A}^*) \cong Z_2 \times Z_2$ and it follows at once from the structure of A^* that also $\Omega_1(R^*) = \Omega_1(A^*) \cong Z_2 \times Z_2$. Hence E^* acts transitively on the involutions of R^* and so R^* is a Suzuki 2-group. Clearly the only possibility is that $R^* \cong Z_{2^{n+1}} \times Z_{2^{n+1}}$. Considering the action of F_i an R^*, we see now that R^* is of type $R_{2^{2n+3}}$ if $i = 1$ and of type $R_{2^{2n+4}}$ if $i = 2$ or 3. Since E^* acts fixed-point-free on R^* and $R^* \supset A^*$, our maximal choice of A^* is therefore contradicted.

If $\tilde{R}^* \cong Z_4 \times Z_4 \times Z_2 \times Z_2$, we conclude easily, as in Lemma 3.1, that γ cannot be conjugate in K^* to any element of $V^* - V_1^*$, which is a contradiction. Hence $\tilde{R}^*$ must be of type $L_3(4)$. To eliminate this case, we use the fact that by the action of E^* an R^*, R^* is necessarily of class at most 2. Since $[\tilde{R}^*, \tilde{R}^*] = \mho^1(\tilde{A}^*) = Z(\tilde{R}^*)$, it follows from the structure of A^* that $[R^*, R^*] = \mho^1(A^*)$. But then $\mho^1(A^*) \subseteq Z(R^*)$ as R^* has class at most 2. Since $R^*/\mho^1(A^*) \cong E_{16}$, this in turn implies that $[R^*, R^*]$ is elementary abelian, whence $[R^*, R^*] = \Omega_1(A^*)$. We conclude that $\Omega_1(A^*) = \mho^1(A^*)$, contrary to the fact that $A^* \cong Z_{2^n} \times Z_{2^n}$ with $n \geq 3$.

Thus T^* is a Sylow 2-subgroup of X, as asserted. By condition (b), T^* and hence S is of type $R_{2^{2n+1}}$ if $i = 1$ and of type $R_{2^{2n+2}}$ if $i = 2$ or 3. Moreover, $n \geq 3$ as $A^* \supseteq R \cong Z_8 \times Z_8$ and so the first assertion of the lemma holds. Since $n \geq 3$, $SCN_3(T^*)$ is empty, as we have noted above. Clearly $\Omega_1(A^*)$ is the unique element of $U(T^*)$ and the involutions of $\Omega_1(A^*)$ are conjugate in X by condition (c). Hence the second assertion of the lemma also holds.

If $S = T$, then S is of type R_{2^5} or R_{2^6}, as we have noted above. On the other hand, if $S \supset T$, the structure of S is given by Lemmas 3.2, 3.3 and 3.4. Hence together these results yield the following proposition, which is our principal result in this connection.

Proposition 3.5. If X is a group of type F_i, $i = 1,2$, or 3, then a Sylow 2-subgroup of X is of type $\hat{A}_{10}$, $L_3(4)^{(1)}$, $U_3(4)^{(1)}$, $U_3(4)^{(2)}$, $R_{2^{2n+1}}$, or $R_{2^{2n+2}}$ for some $n \geq 2$.

We can sharpen this result in the case that X is fusion-simple.

Proposition 3.6. If X is a fusion-simple group of Type F_i, $i = 1, 2$, or 3, then a Sylow 2-subgroup of X is of type $G_2(q)$, q odd, or $\hat{A}_{10}$.

Proof. Let the notation be as above. Suppose first that S is of type $R_{2^{2n+1}}$ or $R_{2^{2n+2}}$, $n \geq 3$. Then $SCN_3(S)$ is empty and the involutions of the unique element of $U(S)$ are conjugate in X by Lemma 3.4. Hence by the main results of [29] and [35], S is necessarily of type $G_2(q)$, $q \equiv 1, 7 (\text{mod } 8)$, J_2, or L. However, one checks that a group of type J_2, or L is not of type $R_{2^{2n+1}}$ or $R_{2^{2n+2}}$, $n \geq 3$, while a group of type $G_2(q)$, $q \equiv 1, 7 (\text{mod } 8)$ is of type $R_{2^{2n+2}}$ for suitable $n \geq 3$. Hence S is of type $G_2(q)$ for some $q \equiv 1, 7 (\text{mod } 8)$ in this case and the proposition holds.

If $S = T$, then S is of type D^*_{32}, R_{2^6}, or M_{12} according as X is of type F_1, F_2, or F_3. Since a group of type M_{12} is also of type $G_2(q)$ for $q \equiv 3, 5 (\text{mod } 8)$, the proposition holds in the third case. Moreover, in the second case $T/A = S/A \cong Z_4$ and we see that $SCN_3(S)$ is empty. Likewise $\Omega_1(A)$ is the unique element of $U(S)$ and its involutions are conjugate in X. But now the main result of [35] yields a contradiction and so this case cannot arise.

Observe next that we have already noted in Proposition 2.7 of Part I that the proof of [18, Lemma 9.6] yields that there is no fusion-simple group with Sylow 2-subgroups of type $L_3(4)^{(1)}$. Thus also S is not of this form.

Combining the preceding argument with Proposition 3.5, we conclude now that either this proposition holds or else S is of type D^*_{32}, $U_3(4)^{(1)}$, or $U_3(4)^{(2)}$. Hence we need only prove that X cannot be fusion-simple in any of these cases.

We shall obtain this as a consequence of a slightly more general result. Here D^*_{64} denotes a 2-group of type R_{64} in which the complement $\langle c,d \rangle \cong Z_4$ with the involution c inverting the normal subgroup $Z \cong Z_4 \times Z_4$.

Proposition 3.7. If X is an arbitrary group with Sylow 2-subgroups of type R_{32}, D^*_{64}, $U_3(4)^{(1)}$, or $U_3(4)^{(2)}$, then X is not fusion-simple.

Proof. Again let S be a Sylow 2-subgroup of X. Then $S = RF$, where $R \lhd S$, $R \cap F = 1$, $R \cong Z_4 \times Z_4$ in cases one and two and R is of type $U_3(4)$ in cases three and four, and where $F \cong Z_2$ in cases one and three and $F \cong Z_4$ in cases two and four. We set $F = \langle f \rangle$ and $\langle u \rangle = \Omega_1(F)$, so that $u = f$ or f^2.

We shall derive a contradiction from the assumption that X is fusion-simple. Suppose first that S is of type R_{32} with $\Omega_1(S) = \Omega_1(R)\langle u \rangle$. Then $W = \Omega_1(S)$ is weakly closed in S with respect to X. By Thompson's fusion lemma, $u \sim z$ in X for some z in R. But then $z \in R \cap W$ as z is an involution. Since W is weakly closed, it follows that $u \sim z$ in $N = N_G(W)$. But clearly $S \subseteq N$ and $W = C_S(W)$, so $\bar{N} = N/C_G(W)$ has

Sylow 2-subgroups isomorphic to $Z_2 \times Z_2$. Since $\overline{N}$ is isomorphic to a subgroup of $GL(3,2)$ and u is not isolated in N, the only possibility is that $\overline{N} \cong A_4$. In particular, $\overline{S} \triangleleft \overline{N}$ and so $R \cap W = C_W(\overline{S}) \triangleleft N$. Since $u \notin R \cap W$, we conclude that u cannot be conjugate to z in N, which is a contradiction.

Hence if S is of type R_{32}, then $\Omega_1(S) \supset \Omega_1(R)\langle u \rangle$ and it follows from the definition of a group of type R_{32} that u inverts R. Thus S is in fact of type D^*_{32}. If S is of type D^*_{64}, we also have that u inverts R. On the other hand, if S is of type $U_3(4)^{(1)}$ or $U_3(4)^{(2)}$, it follows from Lemma 2.5 (vii) that u inverts a subgroup V of R with $V \cong Z_4 \times Z_4$. For uniformity of notation, we put $V = R$ if S is of type D^*_{32} or D^*_{64}. Then u inverts V and $\Omega_1(V) = \Omega_1(R)$ in all cases.

If $F \cong Z_2$, then $u \sim z$ in X for some z in R by Thompson's fusion lemma. We claim that the same conclusion holds if $F \cong Z_4$. In this case $S - R\langle u \rangle$ contains no involutions and so the extended form of Thompson's fusion lemma [27] yields that f is conjugate in X to some element of $R\langle u \rangle$. Since $u = f^2$ and $\mho^1(R\langle u \rangle) \subseteq R$, it follows that $u \sim z$ in X for some z in R, as asserted.

First consider the case that $F \cong Z_2$, in which case $\Omega_1(R) \neq Z(S)$, whence $z \in Z(S)$. Hence $u^x = z$ with $C_S(u)^x \subseteq S$ for some x in X. Since $C_S(u) = Z(S)\langle u \rangle \cong E_8$ and $S/R \cong Z_2$ with $\Omega_1(R) \cong Z_2 \times Z_2$, it follows at once that $\langle u, z_0 \rangle^x = \Omega_1(R)$ for some z_0 in $\Omega_1(R)^\#$. On the other hand, as u inverts V and $\Omega_1(V) = \Omega_1(R)$, we have that $v^2 = z_0$ for some v in V. Since u inverts v, we see that v normalizes, but does not centralize $\langle u, z_0 \rangle$. Thus v^x normalizes, but does not centralize $< u,z_0>^x = \Omega_1(R)$. However, as $\Omega_1(R) = Z(S)$, any 2-element of X which normalizes $\Omega_1(R)$ necessarily centralizes it, so we reach a contradiction. Therefore $F \cong Z_4$.

In this case, $Z(S) = < z_0 > \cong Z_2$. We claim that we can take z to be z_0. Suppose false, in which case certainly u is not conjugate to z_0 in X. Since z_0 is not isolated in S, z_0 is conjugate to some involution $v \neq z_0$ with $v \in S$. Since $\Omega_1(R) = < z,z_0>$ with $z^f = zz_0$, our assumption implies that $v \in S - R$. Likewise, combined with Lemma 2.5 (ix), it implies that R is not of type $U_3(4)$ and so $S \cong D^*_{64}$. Let S_1 be a Sylow 2-subgroup of $C_G(v)$ containing $C_S(v)$. Then $C_S(v) \supseteq < v,z,z_0>$ and v lies in the unique subgroup $R_1 \cong Z_4 \times Z_4$ of S_1. Since S_1/R_1 is cyclic, it follows that $< z,z_0> \cap R_1 \neq 1$. Thus $\Omega_1(R_1) = < v,z'>$ where $z' = z,z_0$ or zz_0. Since v inverts $R \cong Z_4 \times Z_4$ and $\Omega_1(R) = < z,z_0>$, $v \sim vz$, vz_0, and vzz_0 in S. Hence v has at least two conjugates in $\Omega_1(R_1)$. But $\Omega_1(R) \sim \Omega_1(R_1)$ and $v \sim z_0$ in X, so $\Omega_1(R)$ contains at least two conjugates of z_0. However, this contradicts our present assumptions.

Hence without loss we can assume that $< z > = Z(S)$. Then $C_X(u)$ contains a Sylow 2-subgroup S_1 of X, which without loss we can take to contain $C_S(u)$. We have $S_1 = R_1F_1$, where R_1,F_1 are appropriate conjugates of R,F respectively. Observe next that $C_S(u) = \Omega_1(R)F$ and consequently $z \in C_S(u)'$, whence $z \in S_1'$. However, as S_1/R_1 is abelian, $z \in R_1$ and so $z \in \Omega_1(R_1)$. Thus $\Omega_1(R_1) = < z,u>$ with $< u > = Z(S_1)$. Since $u \sim z$ in X and $z \sim zu$ in S_1, all involutions of $\Omega_1(R_1)$ are conjugate in X and hence so are all involutions of $\Omega_1(R)$.

If S is of type $U_3(4)^{(2)}$, then $SCN_3(S)$ is empty and $\Omega_1(R)$ is the unique element of $U(S)$. Again the main result of [35] yields a contradiction. Therefore S must be of type D^*_{64}

Again we have that $u^x = z$ and $C_S(u)^x \subseteq S$ for some x in X. Since $z^x \in (C_S(u)^x)'$, it follows as in the preceding paragraph that $z^x \in R$, whence $\Omega_1(R) = \langle z, z^x \rangle$. But then $C_S(z^x) = R\langle u \rangle$. Since $f \in C_S(u)$ and f^x centralizes z^x, this implies that $f^x \in R\langle u \rangle$. But $R\langle u \rangle - R$ contains no elements of order 4 as u inverts R, so $f^x \in R$. On the other hand, f does not centralize $\Omega_1(R)$. Thus if we set $z_1 = z^x$, we have that $z_1^x \in C_S(u)^x \subseteq S$ and that z_1^x does not centralize f^x. Hence $z_1^x \in R\langle u \rangle - R$ and so z_1^x inverts f^x. This implies that $u^x \in (C_S(u)^x)'$. However, as $\langle z \rangle = C_S(u)'$, this forces $z^x = u^x$, whence $z = u$, which is not the case. This completes the proof of the proposition.

4. <u>Initial reductions.</u> Henceforth G will denote a simple $\mathfrak{K}_3$-group. In this short section we establish a few general properties of G.

<u>Proposition 4.1.</u> Every element of $\mathcal{L}(G)$ is isomorphic to $L_2(q)$, $L_3(q)$, $U_3(q)$, $Re(q)^*$, q odd, $L_2(8)$, $Sz(8)$, A_7, M_{11}, or J_1. In particular, every element of $\mathcal{L}(G)$ is simple.

<u>Proof.</u> Since G is an $\mathfrak{K}_2$-group, Proposition 2.1 of Part V implies that every element of $\mathcal{L}(G)$ is either of one of the above forms or else is isomorphic to $SL(2,q)$, q odd, or $\hat{A}_7$. Hence for some involution x of G, if we set $H = C_G(x)$ and $\bar{H} = H/O(H)$, we can assume that $L(\bar{H})$ possesses a component $\bar{L} \cong SL(2,q)$, q odd, or $\hat{A}_7$. Let T be a Sylow 2-subgroup of H and let S a Sylow 2-subgroup of G containing T. Then $Z = Z(S) \subseteq T$. Since $\bar{Z}$ centralizes $\bar{T} \cap L(\bar{H})$, which is a Sylow 2-subgroup of $L(\bar{H})$, $\bar{Z}$ leaves each component of $L(\bar{H})$ invariant and, in particular, leaves $\bar{L}$ invariant. But as $\bar{Z}$ centralizes $\bar{T} \cap \bar{L}$ which a

Sylow 2-subgroup of $\bar{L}$, Lemma 4.9 of Part V implies that $\bar{Z}$ centralizes $\bar{L}$. Since $C_H(Z)$ covers $C_{\bar{H}}(\bar{Z})$, it follows that $C_G(z)$ is nonsolvable for any involution z of Z. However, as any such z is a central involution, this contradicts the fact that G is an $\mathfrak{R}_3$-group.

We also have

Proposition 4.2. A Sylow 2-subgroup of G is not of type $L_4(q)$, $q \equiv 7 \pmod 8$, $PSp(4,q)$, q odd, $G_2(q)$, $q \equiv 1, 7 \pmod 8$, $L_3(4)$, A_8, A_{10}, $\hat{A}_{10}$, J_2, or L.

Proof. Note that a 2-group of type A_{10} is isomorphic to $D_8 \int Z_2$ and one of type $\hat{A}_{10}$ is also of type L. Hence if a Sylow 2-subgroup of G is of any of the above types, we can apply one of the results [16], [17], [18], [19], [21], [36], and [37] to obtain that G possesses a nonsolvable 2-constrained 2-local subgroup, contrary to the fact that G is an $\mathfrak{R}_3$-group.

As a consequence we also have

Proposition 4.3. If $SCN_3(2)$ is nonempty in G and U is an element of $U(2)$, then the involutions of U are not conjugate in G.

Proof. In the contrary case, the main results of [29] and [35] yield that a Sylow 2-subgroup of G is of type $G_2(q)$, $q \equiv 1, 7 \pmod 8$, J_2, or L and the preceding proposition is contradicted.

Finally we have

Proposition 4.4. Theorem A holds if a Sylow 2-subgroup of G is of type M_{12}.

Proof. As noted in the Introduction, G must have more than one conjugacy class of involutions, so $G \cong M_{12}$ by the main result of [6]. Hence Theorem A holds in this case.

5. Elimination of the rank 3 case. By hypothesis, G possesses an involution x such that $M = C_G(x)$ is nonsolvable (and hence non 2-constrained). Among all such involutions, we let x be any one for which a Sylow 2-subgroup of M has maximal order. The goal of the long analysis that follows is to pin down the exact structure of M.

We again let T be a Sylow 2-subgroup of M and S a Sylow 2-subgroup of G containing T. By our hypothesis, x is noncentral and so $T \subset S$. Next set $\overline{M} = M/O(M)$. By Proposition 4.1, each component of $\overline{M}$ is simple and so $L(\overline{M})$ is the direct product of its components, each of which has 2-rank at least 2. If $L(\overline{M})$ had more than one factor, then $m(L(\overline{M})\langle \bar{x} \rangle) \geq 5$, contrary to $r(G) \leq 4$. On the other hand, as $\overline{M}$ is not 2-constrained, $L(\overline{M}) \neq 1$. Thus $L(\overline{M})$ is, in fact, a nontrivial simple group. For brevity, we put $\overline{L} = L(\overline{M})$.

Again by Proposition 4.1, $\overline{L}$ is isomorphic to one of the following groups:

(*) $L_2(q)$, $L_3(q)$, $U_3(q)$, $Re(q)$,* q odd, $L_2(8)$, $Sz(8)$, A_7, M_{11}, or J_1.

We also set $\overline{C} = C_{\overline{M}}(\overline{L})$ and $\widetilde{M} = \overline{M}/\overline{C}$. Thus $\widetilde{M}$ is isomorphic to a subgroup of $Aut(\widetilde{L})$. We fix all this notation for the balance of the paper.

The following lemma will be useful throughout Part VI.

Lemma 5.1. If H is a 2-local subgroup of G in which a Sylow 2-subgroup has order greater than $|T|$, then H is solvable.

Proof. Let R be a Sylow 2-subgroup of H and set $Q = R \cap O_{2',2}(H)$. Then Q is a Sylow 2-subgroup of $O_{2',2}(H)$ and $Q \neq 1$ as H is a

2-local subgroup. By condition (b) in the definition of an $\mathfrak{R}_3$-group, either the lemma holds or H is not 2-constrained.

Consider the latter possibility. Then $C_H(Q)$ is certainly nonsolvable. But as $Q \triangleleft R$ and $Q \neq 1$, Q contains a central involution z of R. Since $C_H(Q) \subseteq C_G(z)$, $C_G(z)$ is nonsolvable. But $R \subseteq C_G(z)$ and $|R| > |T|$ by assumption. Since T is a Sylow 2-subgroup of $M = C_G(x)$, this contradicts our maximal choice of the involution x.

The main result of this section is the analysis of the case in which $\bar{L}$ has 2-rank 3. We prove

Proposition 5.2. $\bar{L}$ is not isomorphic to $L_2(8)$, $Sz(8)$, J_1, or $Re(q)^*$, q odd.

We suppose false and proceed in a series of lemmas to reach a contradiction. Let R_1, R_2 be subgroups of T such that $\bar{R}_1 = C_{\bar{T}}(\bar{L})$ and $\bar{R}_2 = \bar{L} \cap \bar{T}$. Then $\bar{R}_1$ is a Sylow 2-subgroup of $\bar{C} = C_{\bar{M}}(\bar{L})$ and $\bar{R}_2$ is a Sylow 2-subgroup of $\bar{L}$. In particular, $R_1R_2 = R_1 \times R_2$ and either $R_2 \cong E_8$ or R_2 is of type $Sz(8)$.

We first prove

Lemma 5.3. The following conditions hold.

(i) $T = R_1 \times R_2$;

(ii) $R_1 = \langle x \rangle$.

Proof. In each case, $|\tilde{M}:\tilde{L}|$ is odd by the structure of $\mathrm{Aut}(\tilde{L})$ and this implies that $T = R_1R_2 = R_1 \times R_2$. Furthermore, as $r(T) \leq 4$, we have $r(R_1) = 1$, so R_1 is cyclic. Hence $Z(T) = R_1 \times Z(R_2)$ with $Z(R_2) \cong E_8$ and $\langle x \rangle = \Omega_1(R_1)$. Since T is not a Sylow 2-subgroup of G, $\langle x \rangle$ cannot be characteristic in $Z(T)$. This forces $\langle x \rangle = R_1$

since otherwise $\langle x \rangle = \Omega_1(\mho^1(Z(T)))$ would be characteristic in $Z(T)$. Thus both parts of the lemma hold.

We set $A = Z(T)$ and $Z = Z(R_2)$. By the lemma, we have $A = Z \times \langle x \rangle \cong E_{16}$.

We next prove

Lemma 5.4. The following conditions hold:

(i) $\bar{L} \cong Sz(8)$;

(ii) $|S| = 2^{10}$;

(iii) $A = \Omega_1(S)$ and $Z = \Omega_1(Z(S))$;

(iv) S is normalized by a 7-element b such that $C_S(b) = \langle x \rangle$;

(v) All noncentral involutions of S are conjugate in S to x;

(vi) $A/Z = Z(S/Z)$

Proof. By the structure of $N_{\bar{L}}(\bar{R}_2)$, $N_M(T)$ contains a 7-element a such that $C_T(a) = \langle x \rangle$. Let V be a maximal 2-subgroup of G containing T such that $N = N_G(V)$ possesses a 7-element b such that $C_V(b) = \langle x \rangle$. We argue first that V is a Sylow 2-subgroup of G.

Suppose false, in which case a Sylow 2-subgroup W of N contains V properly. But then as $V \supseteq T$, $|W| > |T|$ and so N is solvable by Lemma 5.1. Since $r(N) \leq 4$ and $|N/C_G(V)|$ is divisible by 7, we conclude from [Part I, Lemma 4.4] that $W \subseteq O_{2',2}(N)$, whence $O_{2',2}(N) = O(N)W$. We have that x centralizes $O(N)$ and that $O(N)W \lhd O(N)W\langle b \rangle$. Hence by the Frattini argument, $N_N(W)$ contains a 7-element c which centralizes x, but not W. Since T is a Sylow 2-subgroup of $M = C_G(x)$ and $T \subseteq W$, we have $T = C_W(x)$. But now Thompson's $A \times B$-lemma [13, Theorem 5.3.4] yields that c does not centralize T. Since $T/\langle x \rangle \cong R_2 \cong E_8$ or is of type $Sz(8)$, c must act fixed-point-free on $T/\langle x \rangle$ and so $C_T(c) = \langle x \rangle$. Setting $W_0 = C_W(c)$, it follows that $C_{W_0}(x) = \langle x \rangle$, whence, $W_0 = \langle x \rangle$. Thus W satisfies the same conditions as V and as $W \supset V$, our maximal choice of V is contradicted. Therefore V is a Sylow 2-subgroup of G, as asserted.

Since S is, by definition, an arbitrary Sylow 2-subgroup of G containing T, we can suppose without loss that $S = V$. Then b is 7-element which normalizes S and satisfies $C_S(b) = \langle x \rangle$, so (iv) holds.

Observe next that as $r(S) \leq 4$, b^7 centralizes S, again by [Part I, Lemma 4.4]. Hence b induces an automorphism of S of order 7. Furthermore, b leaves $T = C_S(x)$ invariant and $C_T(b) = \langle x \rangle$, so $T = \langle x \rangle \times [T,b]$ with b acting fixed-point-free on $[T,b]$. Hence Lemma 2.12 is applicable and yields that either $S_1 = [S,b]$ is of index 2 in S with b acting fixed-point-free on S_1 or that $\Omega_1(S) \cong E_{16}$.

Consider the first possibility, in which case $S = S_1\langle x \rangle$ with $x \notin S_1$. By Lemma 2.11, we have that $S_1 \cong Z_{2^n} \times Z_{2^n} \times Z_{2^n}$ for some n or that S_1 is of type $Sz(8)$. Since x centralizes b, x must centralize $\Omega_1(S_1) \cong E_8$ and so every involution of S_1 is in the center of S. But then as x is a noncentral involution of G, x cannot be conjugate in G to any involution of S_1. Now Thompson's fusion lemma yields that G has a normal subgroup of index 2, contrary to the simplicity of G.

We therefore conclude that $\Omega_1(S) \cong E_{16}$. Since $A = Z \times \langle x \rangle \subseteq T \subseteq S$ and $A \cong E_{16}$, it follows that $A = \Omega_1(S)$. Furthermore, as $b \in M$ and b normalizes A, we must have that b normalizes Z. Since $A \cap Z(S)$ is nontrivial and b-invariant, we see that either $x \in Z(S)$ or

$Z \subseteq Z(S)$. However, $x \notin Z(S)$ as x is noncentral. We thus conclude that $Z = \Omega_1(Z(S))$ and so (iii) also holds.

Since b induces an automorphism of S of order 7 and $C_S(b) = \langle x \rangle \subseteq T$, we have that $|S:T| = 2^{3k}$ for some k. Moreover, $k \geq 1$ as x is noncentral. Since $T = C_S(x)$, x thus has 2^{3k} conjugates in S. However, as each of these conjugates must be in $A - Z$ it follows that $k = 1$. Hence $|S:T| = 8$ and all involutions of $A - Z$ are conjugate in S to x. In view of (iii), we conclude at once that (v) holds.

It remains to prove (i), (ii), and (vi). Since $|S:T| = 8$, (ii) will follow from (i). If (i) is false, then $R_2 = Z \cong E_8$ and $|S| = 2^7$. Since b acts nontrivially on S/Z, the only possibility is that $S/Z \cong E_{16}$. But then $S_1 = [S,b]$ has index 2 in S and b acts fixed-point-free on S_1. This leads to a contradiction as above. Hence (i) holds. Similarly if $\overline{S} = S/Z$ and $Z(\overline{S}) \supset \overline{A}$, then b would normalize, but not centralize a subgroup $\overline{Z}_1$ of $Z(\overline{S})$ with $\overline{Z}_1 \cong E_8$. But then $\overline{S}/\overline{Z}_1 \cong E_{16}$ and again it would follow that $S_1 = [S,b]$ is of index 2 in S with b acting fixed-point-free on S_1. Again we reach a contradiction. Hence $A/Z = Z(S/Z)$. Hence (vi) also holds and the lemma is proved.

We now derive a contradiction by showing that the 2-group S of order 2^{10} described in Lemma 5.4 does not exist. Suppose false and let a be an element of order 4 in $C_S(x) = \langle x \rangle \times R_2$. Since b acts fixed-point-free on S/A, $S/A \cong Z_4 \times Z_4 \times Z_4$ or S/A is of type $Sz(8)$ by Lemma 2.11. In any case, $\langle \overline{a}, \overline{x} \rangle \lhd \overline{S} = S/Z$. Hence S contains a maximal subgroup $S_1 \supset \langle x \rangle \times R_2$ with $[\overline{S}_1, \langle \overline{a}, \overline{x} \rangle] = 1$. Let d be an element of S_1. Then $a^d = az$, $z \in Z$. Hence $[a,d^2] = 1$ and so $[a,\Phi(S_1)] = 1$. On the other hand, $\Phi(S_1/A)$ is of order at least 4. Hence $\Phi(S_1)A$ is not contained in $C_{R_2}(a)A \cong Z_4 \times Z_2 \times Z_2 \times Z_2$. As $\Phi(S_1)A \subseteq R_2A$, this contradicts $[a, \Phi(S_1)] = 1$. Thus Proposition 5.2 is proved.

6. <u>The major reduction.</u> In this long section, we shall establish the following basic result:

<u>Proposition 6.1</u> One of the following three cases holds:

(I) $\overline{L} \cong L_2(q)$, $q \not\equiv 3,5 \pmod 8$, A_7, $L_3(3)$, or M_{11} and $\overline{C} = \langle \overline{x} \rangle \cong Z_2$;

(II) $\overline{L} \cong L_2(q)$, $q \equiv 3,5 \pmod 8$, and $\overline{C} \cong Z_2$ or $Z_2 \times Z_2$;

(III) $\overline{L} \cong U_3(3)$ and $\overline{C} \cong Z_2$ or Z_4.

We shall argue in a sequence of lemmas. First of all, we set $\overline{R}_1 = C_{\overline{T}}(\overline{L})$, $\overline{R}_2 = \overline{T} \cap \overline{L}$, and let R_i be subgroups of T which map an $\overline{R}_i$, $i = 1, 2$. Thus $\overline{R}_1$, $\overline{R}_2$ are Sylow 2-subgroups of $\overline{C}$, $\overline{L}$ respectively. Since $\overline{L}$ is simple, $\overline{LC} = \overline{L} \times \overline{C}$ and so $R_1R_2 = R_1 \times R_2$. Since $\overline{L}$ and $\overline{C}$ are normal in $\overline{M}$, likewise R_1 and R_2 are normal in T. Furthermore, as $S \supset T$, $N_S(T) - T$ contains an element y with $y^2 \in T$. Note also that $Z(S) \subseteq T$.

The following result is critical for the ensuing analysis.

<u>Lemma 6.2.</u> We have $R_1R_1^y = R_1 \times R_1^y$.

<u>Proof.</u> Since y normalizes T and $R_1 \triangleleft T$, also $R_1^y \triangleleft T$. Hence $W = R_1 \cap R_1^y \triangleleft T$. Moreover, as $y^2 \in T$, y normalizes W. Hence $T\langle y \rangle \subseteq H = N_G(W)$ and so a Sylow 2-subgroup of N has order exceeding $|T|$. Furthermore, $\overline{L} \subseteq C_{\overline{M}}(\overline{W})$ as $\overline{W} \subseteq \overline{R}_1 \subseteq \overline{C}$. Since $C_M(W)$ covers $C_{\overline{M}}(\overline{W})$, it follows that $C_M(W)$, and hence H, is nonsolvable. These conditions force $W = 1$, since otherwise H would be a nonsolvable 2-local

subgroup of G violating the hypothesis of Lemma 5.1. Since $W = R_1 \cap R_1^y$, this establishes the lemma.

If $\overline{L} \cong L_2(q)$, q odd, then $\widetilde{M}$ is isomorphic to a subgroup of $P\Gamma L(2,q)$, q odd. If some element of $\widetilde{M}$ induces a field automorphism of $\widetilde{L}$ of order 2, then this automorphism can be induced by an involution $\widetilde{u}$ of $\widetilde{M}$, which without loss we can assume lies in $\widetilde{T}$. We know that $[\widetilde{u},\widetilde{R}_2] = 1$. Hence if we choose a representative u of $\widetilde{u}$ in T such that $u^2 \in R_1$, we then have $R_1R_2\langle u\rangle = R_1\langle u\rangle \times R_2$.

Likewise if $\overline{L} \cong A_7$ and $\widetilde{M} \cong S_7$, there exists an element u in T which satisfies the same conditions and such that $\widetilde{M} = \widetilde{L}\langle \widetilde{u}\rangle$. In all remaining cases, we set $u = 1$ for uniformity. Then the above factorization of $R_1R_2\langle u\rangle$ again holds. Since $r(R_2) \geq 2$, we have in all cases $r(R_1\langle u\rangle) \leq 2$. We state these facts as a lemma.

<u>Lemma 6.3.</u> The following conditions hold:

(i) $R_1R_2\langle u\rangle = R_2 \times R_1\langle u\rangle$;

(ii) $r(R_1\langle u\rangle) \leq 2$.

We shall now systematically eliminate various possibilities for the structure of $\overline{M}$. We first prove

<u>Lemma 6.4.</u> If $\overline{L} \cong L_3(q)$ or $U_3(q)$, q odd, then $q = 3$.

<u>Proof.</u> Suppose $q \geq 5$. By [Part III, Lemma 6.3], $Z(\overline{T}) \subseteq \overline{R}_1 \times Z(\overline{R}_2)$. By the structure of $\overline{L}$, the centralizer of every involution of $\overline{L}$ contains a subgroup isomorphic to $SL(2,q)$ and so is nonsolvable. Since $\overline{R}_1$ centralizes $\overline{L}$, it follows that $C_{\overline{L}}(\overline{z})$ is nonsolvable for every involution z of $Z(T)$ and so also $C_G(z)$ is nonsolvable. But as $Z(S) \subseteq T$, we have $Z(S) \subseteq Z(T)$ and so we can choose z to lie in $Z(S)$, contrary to the fact that the centralizer of every central involution of G

is solvable by assumption.

Lemma 6.5. If $\bar{L} \cong L_2(q)$, $q \equiv 3,5 \pmod 8$, then $\bar{C} \cong Z_2$ or $Z_2 \times Z_2$.

Proof. In this case, $R_2 \cong Z_2 \times Z_2$ and $T/R_1 \cong Z_2 \times Z_2$ or D_8. Since $R_1^y \triangleleft T$ and $R_1 \cap R_1^y = 1$, it follows that R_1^y, and hence also R_1, is isomorphic to a subgroup of D_8. Thus $R_1 \cong Z_2$, $Z_2 \times Z_2$, Z_4, or D_8. Since $\bar{R}_1$ is a Sylow 2-subgroup of $\bar{C}$ and $\bar{x}$ is isolated in $\bar{C}$, we conclude in each case that $\bar{C}$ has a normal 2-complement. But $O(\bar{C}) \subseteq O(\bar{M}) = 1$ and so $\bar{C} = \bar{R}_1$. Hence the lemma holds if $R_1 \cong Z_2$ or $Z_2 \times Z_2$. We can therefore assume that neither is the case, whence $R_1 \cong Z_4$ or D_8 and necessarily $T/R_1 \cong D_8$.

Suppose first that $R_1 \cong Z_4$. Since $x^y \neq x$ and $R_1R_2 \cong Z_4 \times Z_2 \times Z_2$, clearly $R_1^y \not\subseteq R_1R_2$. Hence $T = R_1^y R_1 R_2$. Since R_1 centralizes R_1^y by Lemma 6.2, it follows that $R_1 \subseteq Z(T)$, whence $Z(T) \cong Z_4 \times Z_2$. But then $< x > = \mho^1(Z(T))$ and so $x^y = x$, which is not the case. We thus conclude that $R_1 \cong D_8$.

It follows now that $T = R_1 \times R_1^y \cong D_8 \times D_8$. Let U be an element of $U(S)$. Since $x \in T' \subseteq S'$, x centralizes U and so $U \subseteq T = C_S(x)$. But by the structure of $T< y >$, $Z(T)$ is the unique normal four subgroup of $T< y >$ and therefore $U = Z(T)$. In particular, $x \in U$ and so $T \subseteq C_S(U) \subseteq C_S(x) = T$. Thus $C_S(U) = T$ and we conclude that $S = T< y >$. Now Lemma 6.8 of Part III yields $S \cong D_8 \int Z_2$, contrary to Proposition 4.2.

By Lemma 6.5, Proposition 5.5 and Lemma 6.4, we can henceforth assume: $\bar{L} \cong L_2(q)$, $q \equiv 1,7 \pmod 8$, A_7, $L_3(3)$, $U_3(3)$ or M_{11}. In particular, $R_2 \cong D_{2^n}$, $n \geq 3$, QD_{16}, or R_2 is wreathed or order 32. Hence $Z(R_2) \cong Z_2$ or Z_4.

The following fact will be important and is a direct consequence of the structure of $\operatorname{Aut}(\bar{L})$.

Lemma 6.6 We have $C_T(R_2) = Z(R_2) \times R_1\langle u \rangle$.

We next prove

Lemma 6.7 If $R_1^y \cap R_2 = 1$, then either $\overline{C} \cong Z_2$ or $\overline{L} \cong U_3(3)$ and $\overline{C} \cong Z_4$.

Proof. Suppose $R_1^y \cap R_2 = 1$ and set $W = C_T(R_2)$. Since $R_1^y, R_2 \lhd T$, $[R_1^y, R_2] = 1$ and so by Lemmas 6.2 and 6.3, $R_1 \times R_1^y \subseteq W$. But $W = Z(R_2) \times R_1\langle u \rangle$ and $r(R_1\langle u \rangle) \leq 2$ by Lemmas 6.3 and 6.6. Since $Z(R_2)$ is cyclic, it follows that $r(W) \leq 3$. Since $R_1 \times R_1^y \subseteq W$, this forces $r(R_1) = 1$, whence R_1 is cyclic.

If $\overline{L} \not\cong U_3(3)$, then $Z(R_2) \cong Z_2$ and consequently $\mho^1(W) \subseteq R_1$. Hence $\mho^1(R_1^y) \subseteq R_1$. But now lemma 6.2 forces $\mho^1(R_1^y) = 1$, whence $R_1^y \cong R_1 \cong Z_2$. On the other hand, if $\overline{L} \cong U_3(3)$, then $Z(R_2) \cong Z_4$. Arguing now with $\mho^2(W)$, we conclude similarly that $R_1 \cong Z_2$ or Z_4. Since $O(\overline{C}) = 1$, it follows in each case that also $\overline{C} = \overline{R}_1$, and so one of the two alternatives of the lemma holds.

The lemma shows that if $R_1^y \cap R_2 = 1$, then one of the alternatives of Proposition 6.1 holds. Therefore we can also assume henceforth that

$$R_1^y \cap R_2 \neq 1 .$$

Using this inequality, we can prove

Lemma 6.8. We have $R_2R_2^y = R_2 \times R_2^y$.

Proof. By the inequality, some involution of R_2 is conjugate to one in R_1. But all involutions of R_2 are conjugate in G as $\overline{L}$ has only one conjugacy class of involutions. Since $C_G(R_1)$ is nonsolvable, it follows that $C_G(r)$ is nonsolvable for every involution r of R_2.

Set $W = R_2 \cap R_2^y$ and suppose that $W \neq 1$. Since y normalizes T with $y^2 \in T$ and since R_2, R_2^y are each normal in T, y normalizes W

and $W \triangleleft T$. Hence y centralizes an involution w of $W \cap Z(T)$. But then $C_G(w)$ is nonsolvable and contains $T\langle y \rangle$. Since $T\langle y \rangle \supset T$, this contradicts Lemma 5.1. Thus $W = 1$ and the lemma follows.

As an immediate corollary, we have

Lemma 6.9. $\bar{L}$ is not isomorphic to $U_3(3)$.

Proof. In the contrary case R_2 is wreathed and so $R_2 \times R_2^y$ has sectional 2-rank 6, contrary to $r(G) \leq 4$.

We set $R = R_2 \times R_2^y$. Our previous results imply that either $R \cong D_{2^n} \times D_{2^n}$, $n \geq 3$, or $R \cong QD_{16} \times QD_{16}$. We also set $z = xx^y$ and fix this notation. We have that $z \in Z(T)$ with y centralizing z, so that $T\langle y \rangle \subseteq C_G(z)$. Hence by Lemma 5.1, we obtain

Lemma 6.10. $C_G(z)$ is solvable.

We also need the following fact.

Lemma 6.11. We have $\langle x \rangle = Z(R_2^y)$.

Proof. By the structure of R, $m(R) = 4$. Since $r(T) \leq 4$, this forces $\Omega_1(Z(T)) \subseteq R$, whence $x \in Z(R) = \langle x_2 \rangle \times \langle x_2 \rangle^y$, where $\langle x_2 \rangle = Z(R_2)$. Since y does not centralize x, $x \neq x_2 x_2^y$ and consequently $x = x_2$ or x_2^y. However, in the first case, $x \in R_1 \cap R_2 = 1$, which is a contradiction. Thus $x = x_2^y$ and so $\langle x \rangle = Z(R_2^y)$, as asserted.

We next abstract a portion of the argument of Lemma 6.5.

Lemma 6.12. If every normal four subgroup of T or $T\langle y \rangle$ lies in R, then $S = T\langle y \rangle$.

Proof. Let $U \in \mathcal{U}(S)$. Since $x \in (R_2^y)'$ by the preceding lemma, we have $U \subseteq T$, whence U is normal in both T and $T\langle y \rangle$. Hence $U \subseteq R$ by our hypothesis and so $U \triangleleft R\langle y \rangle$. However, it is immediate from the structure of $R\langle y\rangle$ that $Z(R) = \langle x, x^y\rangle$ is the unique normal four subgroup of $R\langle y\rangle$. Thus $U = \langle x, x^y\rangle$, whence $x \in U$ and $U \subseteq Z(T)$. Together these conditions imply that $T \subseteq C_S(U) \subseteq C_S(x) = T$. Thus $T = C_S(U)$ and so $S = T\langle y\rangle$, as asserted.

Next let B be an arbitrary four subgroup of R_2 and set $A = B \times B^y$, so that $A \cong E_{16}$. The following lemma will be used several times.

Lemma 6.13. If y is an involution, then $N = N_G(A)$ has the following properties:

(i) $y \in N$;

(ii) $N/C_G(A)$ is isomorphic to a Sylow 3-normalizer in A_8;

(iii) x and z have 6 and 9 conjugates respectively in A under the action of N;

(iv) $C_A(y)$ is a four group containing z and $C_N(y)$ contains a 3-element which acts nontrivially on $C_A(y)$.

Proof. Since y is an involution, $A^y = A$ and so $y \in N$. By [Part V, Lemma 4.2 (ii)], $N_{\tilde{L}}(\tilde{B})$ contains a subgroup $\tilde{P}$ of order 3 which acts regularly on $\tilde{B}$, centralizes $\tilde{B}\langle\tilde{u}\rangle/\tilde{B}$ and is inverted by an involution of $\tilde{R}_2$. Since $\bar{L}$ centralizes $\bar{R}_1$, it follows that M possesses a cyclic 3-subgroup P which acts fixed-point-free on B, centralizes $BR_1\langle u\rangle/B$, and is inverted by a 2-element r of R_2. But $A = B \times B^y \subseteq B \times R_2^y \subseteq BR_1\langle u\rangle$ by Lemmas 6.6 and 6.8 together with the fact that $Z(R_2) \subseteq B$. Hence P centralizes A/B and so $P \subseteq N$.

Now set $\bar{N} = N/C_G(A)$. Then $\bar{P}$ is of order 3, acts regularly on B, and trivially on A/B. Furthermore, by the structure of $R = R_2 \times R_2^y$, we see that $N_R(A) = R \cap N = \langle A, r, r^y\rangle$, whence $\langle \bar{r}, \bar{r}^y\rangle \cong Z_2 \times Z_2$ and $\langle r, \bar{y}\rangle \cong D_8$. Since $r^y \in R_2^y \subseteq BR_1\langle u\rangle$, we also have that $\bar{P}$

centralizes $\bar{B}\langle \bar{r}^{\bar{y}}\rangle/\bar{B}$ and is inverted by $\bar{r}$. Likewise we have that $\bar{P}^{\bar{y}}$ acts regularly on B^y and trivially on A/B^y.

On the other hand, by **Lemma 6.11**, $x \in B^y$, **whence** $x^y \in B$ and so $z \in A$. Therefore $C_G(A)$ is solvable by Lemma 6.10 and so certainly $C_N(A)$ is solvable. Since A is contained in a Sylow 2-subgroup of $O_{2',2}(N)$, this in turn implies that N is 2-constrained. Since G is an $\mathfrak{R}_3$-group, we conclude that N and hence also $\bar{N}$ is solvable. Thus $\bar{N}$ is isomorphic to a solvable subgroup of $GL(4,2) \cong A_8$.

Suppose that $\bar{Q} = O_2(\bar{N}) \neq 1$. Let V be a Sylow 2-subgroup of N containing $\langle A,r,y \rangle$. Then $A_0 = C_A(\bar{Q})$ is a nontrivial normal subgroup of V and so $A_0 \cap Z(V) \neq 1$. But $\langle A,r,y \rangle \cong D_8 \int Z_2$ and has center $\langle z \rangle$. Since $\Omega_1(Z(V)) \subseteq A$, it follows that $\langle z \rangle = \Omega_1(Z(V))$ and we conclude that $z \in A_0$. Since $z = xx^y$ with $x^y \in B$ and A_0 **is $\bar{P}$-invariant, the action of $\bar{P}$ on A forces $B \subseteq A_0$.** Since $y \in V$ and $A_0 \triangleleft V$, this in turn yields that $A = B \times B^y \subseteq A_0$. Thus $\bar{Q}$ centralizes A, which is not the case. Hence $\bar{Q} = O_2(\bar{N}) = 1$.

But now examining the solvable subgroups of A_8, we see that there is only one possibility for the structure of $\bar{N}$: namely, that asserted in (ii). Then (iii) follows at once from the action of $\bar{N}$ on A. Moreover, $C_{\bar{N}}(\bar{y}) \cong Z_2 \times S_4$ and $\langle A,y \rangle \cong Z_2 \times Z_2 \int Z_2$, which together imply that a 3-element of $C_{\bar{N}}(\bar{y})$ acts fixed-point-free on A. Now we immediately obtain (iv).

We draw some consequences of this result. Again assume y is an involution and set $R_0 = C_R(y)$, so that R_0 is the "diagonal" of $R_2 \times R_2^y$. Hence $R_0 \cong R_2$. We prove

Lemma 6.14. If y is an involution and we set $R_0 = C_R(y)$, then we have

(i) All involutions of R_0 are conjugate in $C_G(y)$;

(ii) $C_G(y)$ is nonsolvable;

(iii) y is not conjugate in G to x or z.

Proof. Every four subgroup B_0 of R_0 is clearly equal to $C_A(y)$ for some elementary abelian subgroup A of R of the form $A = B \times B^y$ for a suitable four subgroup B of R_2. By Lemma 6.13 (iv), $z \in B_0$ and all involutions of B_0 are conjugate in $H = C_G(y)$ to z. Since every involution of R_0 is contained in a four subgroup of R_0, we conclude that all involutions of R_0 are conjugate in H, proving (i).

We have that $R_0 \cong D_{2^n}$, $n \geq 3$. But it is easily seen that no solvable group can possess a nonabelian dihedral 2-subgroup all of where involutions are conjugate. Therefore H must be nonsolvable and (ii) holds.

Since $C_G(z)$ is solvable by Lemma 6.10 and $H = C_G(y)$ is nonsolvable, y and z cannot be conjugate in G. Suppose that $y \sim x$ in G, in which case $M = H^g$ for some g in G. Setting $\bar{H} = H/O(H)$ and $\bar{L}_1 = L(\bar{H})$, we have that $\bar{L}_1 \cong \bar{L}$. Moreover, $\bar{L}_1$ is the ultimate term of the derived series of $\bar{H}$ by [Part III, Proposition 2.1]. Since $\bar{H}' \supseteq \bar{R}_0' \neq 1$ and all involutions of $\bar{R}_0$ are conjugate in $\bar{H}$, we have $\bar{R}_0 \subseteq \bar{H}'$. Repeating this argument, we conclude that $\bar{R}_0 \subseteq \bar{L}_1$. Thus the image of R_0^g in $\bar{M}$ lies in $\bar{L} = L(\bar{M})$. But $z = xx^y \in R_0$ and z is a central involution. This forces every involution of $\bar{L}$ to be the image of a central involution of G. However, this is impossible as $x^y \in R_2$ by Lemma 6.11, x^y is a noncentral involution, and $\bar{R}_2 \subseteq \bar{L}$.

This enables us to prove

Lemma 6.15. We have $T \supset R_1 \times R_2$.

Proof. Suppose false. By Lemmas 6.2 and 6.8, R_1^y is isomorphic to a subgroup of $T/R_1 \cong R_2$ and R_2^y to a subgroup of $T/R_2 \cong R_1$. Clearly this is possible only if

$$R_1 \cong R_2 \quad \text{and} \quad T = R_2 \times R_2^y = R .$$

By Lemma 6.12, we have $S = T\langle y \rangle$, whence by Lemma 6.8 of Part III $S \cong D_{2^n} \int Z_2$, $n \geq 3$, (and so S is of type $L_4(q)$ for suitable $q \equiv 7 \pmod 8$) or $S \cong QD_{16} \int Z_2$. In particular, we can choose y to be an involution.

The first case is excluded by Proposition 4.2. Hence $S \cong QD_{16} \int Z_2$ and $T = R \cong QD_{16} \times QD_{16}$. In this case we apply Lemmas 6.13 and 6.14. Let B and A be as in Lemma 6.13. By the structure of T , every elementary abelian subgroup of T of order 16 is conjugate in T to A . Since every involution of T lies in such an elementary abelian subgroup, it follows now from Lemma 6.13 (iii) that every involution of T is conjugate to x or z . Hence by Thompson's fusion lemma and the simplicity of G , $y \sim x$ or z in G . However, this contradicts Lemma 6.14(iii).

We can now prove

Lemma 6.16. $\bar{L}$ is not isomorphic to $L_3(3)$ or M_{11} .

Proof. Suppose false. By the preceding lemma, $T \supset R_1R_2$, whence $\tilde{M} \supset \tilde{L}$. Since $\mathrm{Aut}(M_{11}) \cong M_{11}$, it follows that $\bar{L} \not\cong M_{11}$ and hence that $\bar{L} \cong L_3(3)$. Since $|\mathrm{Aut}(\bar{L}):\mathrm{Inn}(\bar{L})| = 2$ in this case, we have $|T:R_1R_2| = 2$. Again using Lemmas 6.2 and 6.8 we conclude now that

$$2|R_1| \geq |R_2| \quad \text{and} \quad 2|R_2| \geq |R_1| .$$

If $|R_1| = 2|R_2|$, then $R_1^y \cong T/R_1$ and consequently R_1^y , and hence also R_1, is isomorphic to a Sylow 2-subgroup of $\mathrm{Aut}(L_3(3))$. But then $m(R_1)=3$ by Lemma 2.19(vii) and so $m(R_1R_2) = 5$, against $r(G) \leq 4$. This forces that

$2|R_2| > |R_1|$ and so $|R_2| \geq |R_1|$.

In the present case, $u = 1$ and so $W = C_T(R_2) = Z(R_2) \times R_1$. Since $R_2^y \subseteq W$ and is disjoint from $Z(R_2)$ by Lemma 6.8, we conclude now that also $W = Z(R_2) \times R_2^y$, whence $R_2 \cong R_2^y \cong R_1$ and $R = R_1 \times R_2$.

There exists an element t in T with $t^2 \in R_1$ such that $\tilde{t}$ induces the transpose-inverse map on $\tilde{L}$. But then $C_{R_2}(t) \cong D_8$ by Lemma 2.19(**vii**). Since $R_1 < t >$ centralizes $C_{R_2}(t)$, it follows that $r(R_1 < t >) \leq 2$. However, as $R_1 \cong QD_{16}$ by the preceding paragraph, this is impossible by Lemma 6.9 of Part III.

In view of the preceding lemma, we have $\bar{L} \cong L_2(q)$, $q \equiv 1,7 \pmod 8$ or A_7. Hence also $R_2 \cong D_{2^n}$ for some $n \geq 3$.

Our goal now will be to prove that $u \neq 1$. We shall derive a contradiction from the contrary assumption. Thus in Lemmas 6.17-6.22, we assume that $u = 1$.

<u>Lemma 6.17.</u> The following conditions hold:

(i) $\bar{L} \cong L_2(q)$;

(ii) No element of $\bar{M}$ induces a field automorphism of $\bar{L}$ or order 2;

(iii) $T/R_1 \cong D_{2^{n+1}}$ or $QD_{2^{n+1}}$.

<u>Proof.</u> If $\bar{L} \cong A_7$, then $\tilde{M} = \tilde{L}$ by definition of u and so $T = R_1R_2$, contrary to Lemma 6.15. Thus $\bar{L} = L_2(q)$. Since $T \supset R_1R_2$ by Lemma 6.15, (ii) and (iii) follow at once now from the structure of $P\Gamma L(2,q)$ and the fact that $u = 1$.

We next prove

<u>Lemma 6.18.</u> The following conditions hold:

(i) $R = R_1 \times R_2$ with $R_1 \cong R_2$;

(ii) $Z(R_2) \times R_1 = Z(R_2) \times R_2^y$;

(iii) $S = T\langle y \rangle$ and $Z(T) = \langle x, x^y \rangle$ is the unique element of $U(S)$.

Proof. As in the preceding lemma, we have that $2|R_2| \geq |R_1|$ and $2|R_1| \geq |R_2|$; and if $2|R_2| = |R_1|$, we conclude that $R_1^y \cong T/R_1$, whence $R_1 \cong D_{2^{n+1}}$ or $QD_{2^{n+1}}$. This inturn implies that $T = R_1 \times R_1^y$. But then $\langle x, x^y \rangle = Z(T)$ is the unique normal four subgroup of T and so $S = T\langle y \rangle$ by Lemma 6.12. Thus $S \cong D_{2^{n+1}} \int Z_2$ or $QD_{2^{n+1}} \int Z_2$ and y can be taken to be an involution by Lemma 6.8 of Part III. The first case being excluded by Proposition 4.2, we have, in fact, $R_1 \cong QD_{2^{n+1}}$.

Since $R = R_2 \times R_2^y \cong D_{2^n} \times D_{2^n}$, we see that T has only one conjugacy class of elementary abelian subgroups of order 16, that they all lie in R, and that every involution of T lies in one of them. We can therefore apply Lemmas 6.13 and 6.14 exactly as we did in Lemma 6.15 to reach the same contradiction.

Thus $|R_2| \geq |R_1|$ and we conclude now as in the corresponding case of Lemma 6.16 that $R_2 \cong R_1$, that $R = R_1 \times R_2$, and that $Z(R_2) \times R_1 = Z(R_2) \times R_2^y$. **It is immediate now from the structure of $T\langle y \rangle = R\langle t,y \rangle$** that $Z(T) = \langle x, x^y \rangle$ is the unique normal four subgroup of $T\langle y \rangle$. Hence $S = T\langle y \rangle$ by Lemma 6.12 and so all parts of the lemma hold.

According as T/R_1 is dihedral or quasi-dihedral, we can find an element t in $T - R_1R_2$ such that $t^2 \in R_1$ or $\langle t^2 \rangle R_1 = Z(R_2)R_1$. In either case, we have $S = \langle R, t, y \rangle$. The next step in the proof is to make a judicious choice of the elements t and y. We shall prove

Lemma 6.19. We can choose y and t to satisfy the following conditions:

(i) $[t,y] = 1$;

(ii) $t^2 = 1$ or z ;

(iii) $y^2 = 1$ or t .

Proof. We have that $T = R\langle t \rangle$ and $R = R_1R_2$. Moreover, $t^2 \in Z(R_2) \times R_1 = Z(R_2) \times R_2^y$. Hence $T/R_1 \cong T/R_2{}^y$. But then also $T/R_2 \cong T/R_1$ and so $T/R_2 \cong D_{2^{n+1}}$ or $QD_{2^{n+1}}$. In the first case, $t^2 \in R_1$ by our choice of t and it follows that $R_2 \cap R_1\langle t \rangle = 1$, whence $R_1\langle t \rangle \cong T/R_2 \cong D_{2^{n+1}}$. Hence in this case we can choose t to be an involution. In the second case, $t^2 \in Z(R_2)R_1 - R_1$ by our choice of t and it follows that $Z(R_2)R_1\langle t \rangle/Z(R_2) \cong T/R_2 \cong QD_{2^{n+1}}$. In this case we see that we can choose t so that $\langle t^2 \rangle$ and $Z(R_1)$ have the same images. Hence $t^2 \in Z(R_2)Z(R_1) = Z(T)$ and $t^2 \notin Z(R_i)$, $i = 1, 2$, whence $t^2 = xx^y = z$.

According as $|t| = 2$ or 4, Lemmas 2.1 (i) or 2.2 implies that $T - R$ possesses only one conjugacy class of elements of order 2 or 4 with square t^2. Since y centralizes t^2, it follows that $t^y = t^r$ for some $r \in R$. Hence replacing y by yr^{-1} , if necessary, we can assume without loss that y centralizes t . By the structure of T , we have that $C_T(t) = Z(T)\langle t \rangle$ and so $C_S(t) = Z(T)\langle t,y \rangle$. This implies that $y^2 \in Z(T)\langle t \rangle$. If $y^2 \notin Z(T)$, then $y^2 = tz_1$ for some z_1 in $Z(T)$. We see that tz_1 has the same properties as t and so we can suppose in this case that $y^2 = t$. On the other hand, if $y^2 \in Z(T)$, then as y centralizes y^2 and $Z(T) = \langle x,x^y \rangle$, we have that either $y^2 = 1$ or $y^2 = xx^y = z$. However, in the latter case, $(yx)^2 = 1$. Since yx centralizes t, we can assume that $y^2 = 1$.

Therefore we have shown that t and y can be chosen so that $t^2 = 1$ or z , y centralizes t, and either $y^2 = t$ or $y^2 = 1$. Thus y and t satisfy the condition of the lemma.

Note that if $t^2 = 1$, T is isomorphic to the crown product $D_{2^{n+1}} \wedge D_{2^{n+1}}$, as this term is defined in [21].

We next prove

Lemma 6.20. We have $y^2 = t$ and $S = R\langle y \rangle$.

Proof. Since $S = R\langle t, y \rangle$, the second assertion clearly follows from the first. Suppose, by way of contradiction, that $y^2 \neq t$, whence $y^2 = 1$. If $t^2 = 1$, then S is of type $Psp(4,q)$ for some odd q by [21], contrary to Proposition 4.2. Hence $t^2 = z$, and consequently all involutions of T lie in $R = R_1R_2$, as can be directly checked.

We examine the conjugacy classes in G of the involutions of R. By the structure of M, all involutions of $R_2 \times \langle x \rangle$ are conjugate in G to x or z. Hence the same is true of the involutions of $R_2^y \times \langle x^y \rangle$. Furthermore, by Lemma 6.13 (iii), the involutions of R of the form rr^y with $r \in R_2 - Z(R_2)$ are also conjugate in G to x or z. There thus remain the involutions of the form rr^{yt} with $r \in R_2 - Z(R_2)$. However, we note that $y' = ytx$ is also an involution of $S - T$ and so we can apply Lemma 6.13 with y' in place of y. If B is a four subgroup of R_2 containing r and we set $A' = B \times B^{y'}$, we have that $r^{y'} = r^{yt} \in A'$ and so $rr^{yt} \in A'$. But then by Lemma 6.13 (iii) every involution of this form is also conjugate to x or z in G.

We thus conclude that every involution of T is conjugate to x or z in G. But then by Thompson's fusion lemma, y must be conjugate to x or z in G, contrary to Lemma 6.14 (iii). Hence $y^2 \neq 1$ and so $y^2 = t$, as asserted.

Observe now that Lemma 2.3 implies that every element of $S - T$ has order at least 4 or 8 according as $t^2 = 1$ or z. Hence by [27, Lemma 16], y must be conjugate to an element v of T. But the elements of $T - R$ either have order 2^n or have order 2 or 4 according as $|y| = 4$ or 8 (that is, according as $t^2 = 1$ or z), **again by direct calculation. Since $n \geq 3$, we conclude that either $v \in R$ or $n = 3$ and $|y| = 8$.**

We next prove

Lemma 6.21. One of the following holds:

(i) $|y| = 8$, $v^4 = z$ and $y \sim v$ in $C_G(z)$; or

(ii) $|y| = 4$ and $t \sim x$ in $C_G(z)$.

Proof. Suppose first that $|y| = 8$, whence $|v| = 8$. Whether $v \in R$ or $v \in T - R$ one checks that $v^4 \in Z(T)$. Since $y^g = v$ for some g in G and $y^4 = t^2 = z$, we must have $v^4 = z$, since otherwise $z^g = (y^4)^g = v^4 = x$ or x^y, contrary to the fact that $z \not\sim x$ in G. Thus $g \in C_G(z)$ and so $y \sim v$ in $C_G(z)$. Therefore (i) holds in this case.

Assume then that $|y| = 4$, whence $t^2 = 1$ and $v \in R$. Now Lemma 2.1 (ii) yields that $v^2 \in Z(T)$. Suppose first that $v^2 = z$. By the structure of T, we find that $C_T(v) \cong Z_{2^n} \times Z_{2^{n-1}}$ with $\Omega_1(C_T(v)) = Z(T)$. Moreover, $S - T$ contains no involutions, so we have, in fact, that $\Omega_1(C_S(v)) = Z(T)$.

We claim that v is extremal in S. Indeed, we check that every element of $S - T$ of order 4 is conjugate in S to y or y^{-1}. Since $C_S(y) \subseteq C_S(t)$, we see that $C_S(y) = \langle z, y \rangle \cong Z_4 \times Z_2$. In particular, we have $|C_S(v)| > |C_S(y_1)|$ for any element y_1 of order 4 in $S - T$ (as $n \geq 3$). Hence if v is not extremal in S, there exists v_1 in R of order 4 with $v \sim v_1$ in G and $|C_S(v_1)| > |C_S(v)|$. Here we have used the fact that $T - R$ contains no element of order 4 **as** $T \cong D_{2^{n+1}} \wedge D_{2^{n+1}}$. **However, this is possible only if** $v_1^2 = x$ **or** x^y **. But then** $z = v^2 \sim x$ or x^y , which is not the case. This establishes our assertion.

Thus $y^g = v$ with $C_S(y)^g \subseteq C_S(v)$ for some g in G . Since $\Omega_1(C_S(y)) = \langle y^2, z \rangle = \langle t,z \rangle$ and $\Omega_1(C_S(v)) = Z(T)$, we must have $\langle t,z \rangle^g = \langle x,z \rangle$. But $t^g = z$ inasmuch as $y^g = v$, $y^2 = t$, and $v^2 = z$. Hence $z^g = x$ or x^y, which is not the case. **We conclude that** $v^2 \neq z$.

Since $v^2 \in Z(T)$ and $v^2 \neq z$ with $|v| = 4$, we have $v^2 = x$ or x^y. Without loss we can assume that $v^2 = x$. But x is extremal in S as $|S:C_S(x)| = |S:T| = 2$ and x is noncentral. Since $t = y^2 \sim v^2 = x$, we thus have $t^h = x$ with $C_S(t)^h \subseteq T$ for some h in G . But $\Omega_1(\Phi(C_S(t))) = \langle z,t \rangle$, while $\Omega_1(\Phi(T)) = Z(T)$, as is easily checked. Hence $\langle z,t \rangle^h = Z(T)$. Again our fusion pattern forces $z^h = z$, whence $h \in C_G(z)$ and $t \sim x$ in $C_G(z)$. Thus (ii) holds in this case.

We shall now eliminate the case $u = 1$, which we have been considering, by contradicting the conclusions of Lemma 6.21. This will yield our objective:

<u>Lemma 6.22.</u> We have $u \neq 1$.

<u>Proof.</u> Setting $K = C_G(z)$, we need only derive a contradiction from the assumption that either $|y| = 8$, $v^4 = z$, and $y \sim v$ in K or $|y| = 4$ and $t \sim x$ in K .

Since G is an $\mathcal{R}_3$-group and $z \in Z(S)$, we have that K is solvable and $S \subseteq K$. Set $\bar{K} = K/O(K)$ and $\tilde{K} = \bar{K}/\langle \bar{z} \rangle$. Because of the specified conjugacy in K , certainly K does not have a normal 2-complement and so $\bar{K} \supset \bar{S}$. Clearly $O(\tilde{K}) = 1$. Since $\langle z,x \rangle \lhd S$, also $\tilde{x} \in Z(\tilde{S})$. Since K is solvable, it follows that $\tilde{x} \in O_2(\tilde{K})$ and hence that $\bar{x} \in \bar{Q} = O_2(\bar{K})$.

Consider first the case that $t \sim x$ in K. Then also $\bar{t} \in \bar{Q}$ and so the normal closure of $\bar{t}$ in $\bar{S}$ lies in $\bar{Q}$. But if T_0 denotes the normal closure of t in $T = R\langle t \rangle \cong D_{2^{n+1}} \wedge D_{2^{n+1}}$, we check that $T_0 = \langle t, r_1, r_2 \rangle$, where $r_i \in R_i$, $|r_i| = 2^{n-1}$, and t inverts r_i, $i = 1, 2$. In particular, $|T_0 : C_{T_0}(t)| \geq 4$. Since $\bar{T}_0 \subseteq \bar{Q}$, it follows that also $|\bar{Q} : C_{\bar{Q}}(\bar{t})| \geq 4$. Since $\bar{Q} \triangleleft \bar{K}$ and $\bar{t} \sim \bar{x}$ in $\bar{K}$, this in turn implies that $|\bar{Q} : C_{\bar{Q}}(\bar{x})| \geq 4$. On the other hand, as $\langle x,z \rangle \triangleleft S$, $\langle \bar{x}, \bar{z} \rangle \triangleleft \bar{Q}$, whence $|\bar{Q} : C_{\bar{Q}}(\bar{x})| \leq 2$, a contradiction. Thus we must have $|y| = 8$, $v^4 = z$, and $y \sim v$ in K.

Since $y^2 = t$, we also have that $t \sim v^2$ in K. Suppose that $\bar{v}^2 \notin \bar{Q}$. We have that $v^2 \in S'$ as $v^2 \in \Phi(T) = T'$. Since $\bar{K}/\bar{Q}$ is isomorphic to a solvable subgroup of $GL(4,2) \cong A_8$, $\bar{S}/\bar{Q}$ is isomorphic to a subgroup **of** D_8 and so our conditions imply that $\bar{S}/\bar{Q} \cong D_8$ with the image of $\bar{v}^2$ central in $\bar{S}/\bar{Q}$. This in turn yields that $|\bar{K}|$ and $|\tilde{K}|$ are divisible by 9 and that $\tilde{K}$ contains a subgroup $\tilde{P}$ of order 3 such that $\tilde{v}^2$ inverts $\tilde{P}\tilde{Q}/\tilde{Q}$ with $\tilde{P}$ acting regularly on $\tilde{Q}/\Phi(\tilde{Q}) \cong E_{16}$. Clearly these conditions imply that $|\tilde{Q} : C_{\tilde{Q}}(\tilde{v}^2)| \geq 4$. On the other hand, as $v \in T$ with $v \in R$ if $n > 3$ and **as** $v^4 = z$, we check directly from the structure of T that $\langle \tilde{v}^2, \tilde{x} \rangle \triangleleft \tilde{S}$, whence $|\tilde{Q} : C_{\tilde{Q}}(\tilde{v}^2)| \leq 2$, a contradiction. We therefore conclude that $\bar{v}^2 \in \bar{Q}$.

Now it follows that also $\bar{t} \in \bar{Q}$. Again $\bar{T}_0 \subseteq \bar{Q}$ and $|\bar{Q} : C_{\bar{Q}}(\bar{t})| \geq 4$. On the other hand, as $\langle \tilde{v}^2, \tilde{x} \rangle \triangleleft \tilde{S}$, it is immediate that $|\bar{Q} : C_{\bar{Q}}(\bar{v}^2)| \leq 4$, whence $|\bar{Q} : C_{\bar{Q}}(\bar{t})| \leq 4$. Hence equality must hold and so $n = 2$ and $\bar{T}_0 \cong D_{32}^*$. If $\bar{Q} = \bar{T}_0$, then $\langle \bar{r}_1, \bar{r}_2 \rangle$ is characteristic in $\bar{Q}$. Since $\langle x,z \rangle = \Omega_1(\langle r_1, r_2 \rangle)$, it follows that $\langle \bar{x}, \bar{z} \rangle \triangleleft \bar{K}$, whence $C_{\bar{K}}(\langle \bar{x}, \bar{z} \rangle)$ is normal in $\bar{K}$ of index 2 and has $\bar{T}$ as a Sylow 2-subgroup. Clearly this is impossible as then $\bar{y}$ and $\bar{v}$ could not be conjugate in $\bar{K}$. This contradiction yields that $\overline{Q} \supset \overline{T}_0$.

Finally as $C_S(t) = \langle y,x,z \rangle$ and $|\overline{Q}:C_{\overline{Q}}(\overline{t})| = 4$, the only possibility is that $\overline{Q} = \overline{T}_0 \langle \overline{y} \rangle$. Again we conclude that $\langle \overline{r}_1,\overline{r}_2 \rangle$ is characteristic in $\overline{Q}$ and we reach the same contradiction as in the preceding paragraph. This completes the proof of the lemma.

We have therefore reduced ourselves to the case $u \neq 1$, which we now proceed to analyze.

Lemma 6.23. The following conditions hold:

(i) $R_1 \langle u \rangle \cong D_{2^m}$ or QD_{2^m}, where $m = n$ or $n + 1$;

(ii) $|T:R_1R_2 \langle u \rangle| = 2$;

(iii) $R_1 \cap \langle R_1, u \rangle^y = 1$.

Proof. As usual, $R_2^y \subseteq C_T(R_2) = Z(R_2) \times R_1 \langle u \rangle$ with $R_2^y \cap Z(R_2) = 1$, so R_2^y is isomorphic to a subgroup of $R_1 \langle u \rangle$, whence $2|R_1| \geq |R_2|$. Moreover, as $r(R_1 \langle u \rangle) \leq 2$, [Part III, Lemma 6.9] yields that $R_1 \langle u \rangle \cong D_{2^m}$ or QD_{2^m}. Since $2|R_1| \geq |R_2|$ and $|R_2| = 2^n$, we have, in fact, $m \geq n$.

Suppose next that T contains an element u_1 such that $\tilde{u}_1$ induces a field automorphism of $\tilde{L}$ or order 4. Without loss we can suppose that $\tilde{u}_1 \in \tilde{T}$. But then $\tilde{u}_1^2$ has the same properties as $\tilde{u}$ and so without loss we can assume that $\tilde{u}_1^2 = \tilde{u}$. Moreover, we can choose the representative u of $\tilde{u}$ so that $u = u_1^2$ for some representative u_1 of $\tilde{u}_1$ in T. Considering the action of $\tilde{u}_1$ on $\tilde{L}$ (which is necessarily isomorphic to $L_2(q)$ in this case), we see that $C_{R_2}(u_1)$ contains a four group and hence also $r(R_1 \langle u_1 \rangle) \leq 2$. Now another application of [Part III, Lemma 6.9] yields that $R_1 \langle u_1 \rangle \cong D_{2^{m+1}}$ or $QD_{2^{m+1}}$. However,

$R_1 \lhd R_1\langle u_1\rangle$ with the factor group isomorphic to Z_4, so we reach a contradiction in either case. Since $\tilde{M} \cong S_7$ or to a subgroup of $P\Gamma L(2,q)$ containing $PSL(2,q)$, we conclude now that $|T: R_1R_2\langle u\rangle| = 1$ or 2 with the latter occurring only if $\bar{L} \cong L_2(q)$ and $\tilde{M}$ contains a subgroup isomorphic to $PGL(2,q)$.

Suppose next that $T = R_1R_2\langle u\rangle$. Since $R_1R_2\langle u\rangle = R_2 \times R_1\langle u\rangle$ by Lemma 6.3, we have that $T \cong D_{2^n} \times D_{2^m}$ or $D_{2^n} \times QD_{2^m}$. In the latter case, considering the possible quasi-dihedral subgroups of T, we see that y must transform the center $\langle x\rangle$ of $R_1\langle u\rangle$ into itself, whence $y \in T$, which is not the case. We reach the same contradiction as in the former case if $m > n$. Hence the only possibility is that $T \cong D_{2^n} \times D_{2^n}$. Since $R = R_2 \times R_2^y$ has the same structure, we conclude that $T = R$. Now Lemma 6.12 yields that $S = T\langle y\rangle$, whence $S \cong D_{2^n} \int Z_2$ by Lemma 6.8 of Part III. Once again Proposition 4.2 yields a contradiction. Thus $|T: R_1R_2\langle u\rangle| = 2$ and (ii) is proved.

Suppose next that $R_1 \cap \langle R_1,u\rangle^y \neq 1$. Since $R_1 \lhd T$, it follows that $R_1 \cap Z(\langle R_1,u\rangle^y) \neq 1$. But as $\langle R_1,u\rangle \cong D_{2^m}$ or QD_{2^m} with $m \geq n \geq 3$ and $R_1^y \lhd \langle R_1,u\rangle^y$, we see that $Z(\langle R_1,u\rangle^y) \subseteq R_1^y$, whence $R_1 \cap R_1^y \neq 1$, contrary to Lemma 6.2. Thus $R_1 \cap \langle R_1,u\rangle^y = 1$ and (iii) holds.

By (iii), $\tilde{T} = T/R_1$ must contain a subgroup isomorphic to $R_1\langle u\rangle$. But $|\tilde{T}| = 4|R_2|$ by (ii) and $m(\tilde{T}) \geq 3$ as $\tilde{T} \supseteq \tilde{R}_2 \times \langle\tilde{u}\rangle$. Since $m(R_1\langle u\rangle) = 2$, we see that $\tilde{T}$ must contain a proper subgroup isomorphic to $R_1\langle u\rangle$. We thus conclude that $2|R_2| \geq |R_1\langle u\rangle|$, whence $|R_2| \geq |R_1|$. Since $2^m = |R_1\langle u\rangle| = 2|R_1|$ and $2^n = |R_2|$ and since $m \geq n$, the only possibilities are $m = n$ or $m = n + 1$. Therefore (i) also holds and the lemma is proved.

As an immediate corollary, we have

Lemma 6.24 $\bar{L}$ is not isomorphic to A_7.

Proof: In the contrary case, $\tilde{M} \cong S_7$, whence $|\tilde{T}| = 2|\tilde{R}_2| = 2|R_2|$. However, we have shown above that $|\tilde{T}| = 4|R_2|$.

Thus $\bar{L} \cong L_2(q)$. Moreover, our argument shows that $\tilde{M}$ contains a subgroup isomorphic to $PGL(2,q)$ and so $\tilde{T}$ contains a subgroup of index 2 of the form $\tilde{R}_2 < \tilde{t} > \cong D_{2^{n+1}}$ with $\tilde{t}$ an involution. Moreover, by the structure of $R_1 < u >$, R_1 is either dihedral, generalized quaternion, or cyclic. Since $\tilde{t}$ is an involution, $R_1 < t >$ is a group with $t^2 \in R_1$ for any representative t of $\tilde{t}$ in T. If R_1 is generalized quaternion or cyclic, then we can either choose t to be an involution or else $R_1 < t >$ is cyclic or generalized quaternion. Since x is then the unique involution of R_1, we see that in any case we can choose t so that either $t^2 \in < x >$ or $< t^2 > = R_1$. On the other hand, if R_1 is dihedral, we can choose t so that $t^2 \in Z(R_1) = \Omega_1(Z(R_1))$ by Lemma 2.20.

By the structure of $T< y >$, we have that $Z(T) = < x, x^y >$ is the unique normal four subgroup of $T< y >$, whence $S = T< y >$ by Lemma 6.12. This gives

Lemma 6.25. We have $S = R_1R_2 < t,u,y >$.

We shall treat the cases $m = n$ and $m = n + 1$ separately.

Lemma 6.26. If $m = n$, then the following conditions hold:

(i) $R_1 < u > \cong D_{2^n}$;

(ii) $R = R_1R_2 < u >$ and $T = R < t >$;

(iii) We can choose u to be an involution.

Proof. As noted at the beginning of the proof of Lemma 6.23, $R_2^y \subseteq Z(R_2) \times R_1 < u >$ and is isomorphic to a subgroup of $R_1 < u >$. Under our present assumptions, $|R_2^y| = |R_1 < u >|$, so we must have $R_1 < u > \cong R_2^y$, which gives (i). Furtheremore, $Z(R_2) \times R_1 < u > = Z(R_2) \times R_2^y$ and so also $R_2 R_1 < u > = R_2 R_2^y = R$, which implies (ii). Finally by (i), $R_1 < u > - R_1$ contains an involution, which we can take to be u, so (iii) also holds.

In particular, we see in this case that R_1, being of index 2 in $R_1 < u >$, is either isomorphic to $Z_{2^{n-1}}$ or $D_{2^{n-1}}$. We next prove

Lemma 6.27. If $m = n$, then $R_1 < t > \cong Z_{2^n}$.

Proof: By the discussion following Lemma 6.24 and our choice of t, either this lemma holds or $t^2 \in \Omega_1(Z(R_1))$; so we consider the latter case. We argue first that $t^2 \in Z(R_2^y)$. Since $Z(R_2^y) = < x >$ by Lemma 6.11, this is clearly the case if $t^2 \in < x >$ and, in particular, if $< x > = \Omega_1(Z(R_1))$. Suppose then that $t^2 \notin < x >$, which is possible only if $R_1 \cong D_4 \cong Z_2 \times Z_2$ and $R_1 = < t^2, x >$. Note that $R_1 < t > \cong Z_4 \times Z_2$ in this case.

Since $< x^y > = Z(R_2)$, it follows from Lemma 2.2 (vii) of Part IV that $[u,t] = x^y r$ for some r in R_1, whence $t^u = t x^y r$. Since $x^y \in Z(T)$ and $R_1 < t >$ is abelian with R_1 elementary, this implies that $(t^2)^u = t^2$, whence $< x, t^2, u > = R_1 < u >$ is abelian, contrary to the fact that $R_1 < u > \cong D_8$ in this case.

Our argument yields that the image of t in T/R_2^y is an involution. Since $T = R_2 R_2^y < t >$ by Lemma 6.26 (ii) and since we know the action of t on R_2, it follows that $T/R_2^y \cong D_{2^{n+1}}$, whence also $T/R_2 \cong D_{2^{n+1}}$.

But $T = (R_2 \times R_1\langle u\rangle)\langle t\rangle$ with $t^2 \in \langle x\rangle \subseteq R_1\langle u\rangle$ and $[t,u] \equiv x^y$ (mod R_1). Hence $R_2 \cap R_1\langle u,t\rangle = \langle x^y\rangle = Z(R_2)$ and so also $R_1\langle u,t\rangle/Z(R_2) \cong D_{2^{n+1}}$. However, as $n \geq 3$, no element of this factor group outside of $Z(R_2)R_1\langle u\rangle/Z(R_2)$ has order 4. Hence the image of t must be an involution and so $t^2 \in Z(R_2) = \langle x^y\rangle$. Hence, in fact, $t^2 = 1$. This in turn implies that $R_2^y\langle t\rangle \cong D_{2^{n+1}}$ and we conclude that

$$T = (R_2 \times R_2^y)\langle t\rangle \cong D_{2^{n+1}} \wedge D_{2^{n+1}} .$$

But now as in the proof of Lemma 6.19, we can choose y to centralize t and such that either $y^2 = t$ or $y^2 = 1$. In the latter case, S is of type $PSp(4,q)$ for some odd q and Proposition 4.2 is contradicted. On the other hand, in the former case we argue first as in Lemma 6.21 (ii) that $t \sim x$ in $C_G(z)$ and then repeat the first part of the argument of Lemma 6.22 to reach a contradiction.

We can now prove

Lemma 6.28. We have $m = n + 1$.

Proof: Assume false, in which case $m = n$. By the preceding lemma, we have that $R_1\langle t\rangle \cong Z_{2^n}$, whence $\langle t^2\rangle = R_1$ and $R_2 \cap \langle t\rangle = 1$. Since also $R_2 \cap R_1\langle u\rangle = 1$, it follows that $\langle t,u\rangle/Z(R_2) = R_1\langle u,t\rangle/Z(R_2)$ contains subgroups isomorphic to Z_{2^n} and D_{2^n}. (As noted earlier, $[t,u] \equiv x^y$ (mod R_1) and so $R_2 \cap R_1\langle u,t\rangle = Z(R_2)$). We conclude from this that $\langle t,u\rangle/Z(R_2) \cong D_{2^{n+1}}$ or $QD_{2^{n+1}}$. Correspondingly our conditions yield that $t^u = x^y t^{-1}$ or $t^u = x^y t^{-1} x$. Since $T = R_2\langle t,u\rangle$ with $R_2 \cap \langle t,u\rangle = Z(R_2)$, we also have that $T/R_2 \cong \langle t,u\rangle/Z(R_2)$.

Consider first the dihedral case, whence $T/R_2 \cong D_{2^{n+1}}$ and so also

$T/R_2^y \cong D_{2^{n+1}}$. But $t^u = x^y t^{-1}$ in this case and so $(tu)^2 = x^y$. Since $x^y \in R_2$ and $R_2 \cap R_2^y = 1$, it follows that the image of tu in T/R_2^y is of order 4. However, as $T/R_2^y \cong D_{2^{n+1}}$ with $n \geq 3$, all elements of order 4 in this group lie in $\Phi(T/R_2^y)$. Hence $tu \in \Phi(T)$, which is clearly false. We thus conclude that $\langle t,u \rangle/Z(R_2) \cong T/R_2 \cong QD_{2^{n+1}}$ and that $t^u = x^y t^{-1} x = t^{-1}z$.

To handle this case, we set $t_1 = tu$ and, as $(tu)^2 = z$, we have that $t_1^2 = z$. Moreover, as $T = R\langle t \rangle$ with $u \in R$ by Lemma 6.26 (ii), we have that $T = R\langle t_1 \rangle = R_2R_2^y\langle t_1 \rangle$. In addition, $T/R_2 \cong T/R_2^y \cong QD_{2^{n+1}}$. We see then that T has exactly the same structure as we encountered in Lemmas 6.19 - 6.22 in the case that $t^2 = z$, but now with t_1 in place of t. Reasoning as in Lemma 6.19, we again reduce to the case that y centralizes t_1 and either $y^2 = 1$ or $y^2 = t_1$.

In the first case, applying Lemma 6.13 (iii) to both y and the involution yt_1x, we again find that every involution of T is conjugate in G to x or z. But then $y \sim x$ or z in G, contrary to Lemma 6.14 (iii). In the second case, we again apply the extended form of Thompson's transfer lemma [27] to obtain that $y^g = v$ for some v in T and again find that $v^4 = z$ and hence that $g \in C_G(z)$. These are exactly the conditions we reached in Lemma 6.21 (i). But now we can repeat the argument of Lemma 6.22 for this case to again reach a contradiction.

Therefore to complete the proof of Proposition 6.1, it remains only to treat the case $m = n + 1$ of Lemma 6.23. By that lemma, we have that $R_1\langle u \rangle \cong D_{2^{n+1}}$ or $QD_{2^{n+1}}$ and that $R_1 \cap \langle R_1,u \rangle^y = 1$.

But by the structure of $P\Gamma L(2,q)$ and our choice of t, we have that $R_1R_2\langle t\rangle/R_1 \cong D_{2^{n+1}}$, $R_1R_2\langle tu\rangle/R_1 \cong QD_{2^{n+1}}$, and $R_1R_2\langle u\rangle/R_1 \cong D_{2^n} \times Z_2$. Hence $\langle R_1,u\rangle^y \equiv R_2\langle t\rangle$ or $R_2\langle tu\rangle \pmod{R_1}$. Since $T = R_1\langle u\rangle R_2\langle t\rangle$, it follows that also $T = R_1\langle u\rangle \langle R_1,u\rangle^y$. Since $|T| = 2^{2n+2}$, while $|R_1\langle u\rangle| = 2^{n+1}$, this in turn implies that $R_1\langle u\rangle \cap \langle R_1,u\rangle^y = 1$.

We also know that $[t,u] \equiv x^y \pmod{R_1}$. Since $R_1\langle u\rangle$ centralizes R_2, this implies that $\langle R_1,u,x^y\rangle = Z(T)R_1\langle u\rangle \lhd T$. But then also $Z(T)R_1\langle u\rangle \cap Z(T)\langle R_1,u\rangle^y = Z(T)$ and therefore $[R_1\langle u\rangle, \langle R,u\rangle^y] \subseteq Z(T)$. But $S = T\langle y\rangle$ and $T = R_1\langle u\rangle \langle R_1,u\rangle^y$. It follows therefore from Lemma 6.8 of Part III that $S/Z(T) \cong D_{2^n} \int Z_2$. In particular, we can therefore choose y so that $y^2 \in Z(T)$. Since y centralizes y^2, we have, as usual, that $y^2 = 1$ or z. However, in the latter case $(xy)^2 = 1$ and so we can assume without loss that y is an involution.

We argue next that $C_S(y) = \langle y\rangle \times C_T(y)$ with $C_T(y) \cong D_{2^n}$, $D_{2^{n+1}}$, or $QD_{2^{n+1}}$ and $\langle z\rangle = Z(C_T(y))$. Indeed, once again set $R_0 = C_R(y)$, so that $R_0 \cong R_2 \cong D_{2^n}$ and $\langle z\rangle = Z(R_0)$. Setting $\bar{S} = S/Z(T)$, we have that $\bar{T} \cong D_{2^n} \times D_{2^n}$ and that $\bar{y}$ interchanges two of the factors of $\bar{T}$, so that $C_{\bar{T}}(\bar{y}) \cong D_{2^n}$. Hence if T_0 denotes the inverse image of $C_{\bar{T}}(\bar{y})$ in T, we have that $C_T(y) \subseteq T_0$, that $|T_0| = 2^{n+2}$, and that $T_0/Z(T) \cong D_{2^n}$. But y does not centralize $Z(T) = \langle x,x^y\rangle$ and consequently $C_T(y) \subset T_0$. On the other hand, $R_0 \subseteq C_T(y)$, whence either $R_0 = C_T(y)$ or

$|C_T(y)| = 2^{n+1}$. Since $Z(T) \not\subseteq C_T(y)$, it follows in the latter case that $T_0 = Z(T)C_T(y)$, whence $C_T(y)$ covers $C_{\bar{T}}(\bar{y})$. Thus $C_T(y)$ is an extension of D_{2^n} by $\langle z\rangle = C_{Z(T)}(y)$. But $\langle z\rangle = Z(R_0)$ and $R_0 \cong D_{2^n}$ with $n \geq 3$. Since $R_0 \subseteq C_T(y)$, we see that this extension cannot split and so $C_T(y) \cong D_{2^{n+1}}$ or $QD_{2^{n+1}}$. Clearly $\langle z\rangle = Z(C_T(y))$ in all cases. Since $C_S(y) = \langle y\rangle \times C_T(y)$, our assertion is proved.

Now we are in a position to eliminate this case by arguments similar to those we used in Lemmas 6.21 and 6.22 to eliminate the case $u = 1$. By the structure of $\bar{S} = S/Z(T)$, every involution of $\bar{S} - \bar{T}$ is conjugate to the image of y and so every involution of $S - T$ is conjugate in S to an involution of $Z(T)y$. But as $Z(T)\langle y\rangle \cong D_8$, every such involution is conjugate in S to y. In particular, $yz \sim y$ in S.

Suppose first that $S - T$ contains an extremal involution of S. Then by the preceding paragraph, y is extremal in S. By the Thompson transfer lemma, $y \sim v$ in G for some v in T. Hence $v^g = y$ with $C_S(v)^g \subseteq C_S(y)$ for some g in G. Since $Z(T) \subseteq C_S(v)$, it follows that $Z(T)^g \subseteq C_S(y)$. But $C_S(y) = \langle y\rangle \times C_T(y)$ with $C_T(y)$ dihedral or quasi-dihedral and with $\langle z\rangle = C_T(y)$. Since $\langle y,z\rangle = Z(C_S(y))$ and $m(C_S(y)) = 3$, we conclude now that $Z(T)^g \cap \langle y,z\rangle \neq 1$.

If x^g or $(x^y)^g$ is in $\langle y,z\rangle$, then as x and z are not conjugate, it follows that x^g or $(x^y)^g = y$ or yz. But then as $y \sim yz$ in S, we see that $y \sim x$ in G, contrary to Lemma 6.14(iii). Hence $z^g \in \langle y,z\rangle$. But again by Lemma 6.14(iii), $z^g \neq y$ or yz, so $z^g = z$. Thus

$g \in K = C_G(z)$ and so $y \sim v$ in K.

Suppose, on the other hand, that all extremal involutions of S lie in T. Then again by Thompson's transfer lemma, $y \sim v$ in G for some v in T with v extremal in S. Thus $y^h = v$ and $C_S(y)^h \subseteq C_S(v)$ for some h in G. In particular, $R_0^h \subseteq C_S(v)$. But all involutions of R_0 are conjugate in G to z by Lemma 6.14(i), while all involutions of $S - T$ are conjugate to y. Since z and y are not conjugate by Lemma 6.14(iii), we conclude that $R_0^h \subseteq C_T(v)$. But then $z^h \in \Omega_1(T')$. On the other hand, since $T = R_1\langle u\rangle \langle R_1, u\rangle^y$ with $T/Z(T) \cong D_{2^n} \times D_{2^n}$, Lemma 2.4 implies that $T' \cong Z_{2^{n-1}} \times Z_{2^{n-1}}$ with $\Omega_1(T') = Z(T)$. Thus $z^h \in Z(T)$ and as $z \not\sim x$ in G, we must have $z^h = z$. Therefore $h \in K$ and again $y \sim v$ in K.

Hence to complete the proof, we need only show that the assumption $y \sim v$ in K leads to a contradiction. As in Lemma 6.22, set $\bar{K} = K/O(K)$ and $\tilde{K} = \bar{K}/\langle\bar{z}\rangle$. Again we have that $\bar{K} \supset \bar{S}$, $O(\bar{K}) = O(\tilde{K}) = 1$, and $\bar{K}$ is solvable. Again we set $\bar{Q} = O_2(\bar{K})$ and as $\langle x, z\rangle \triangleleft S$, it again follows that $\bar{x} \in \bar{Q}$. If $\langle\bar{x}, \bar{z}\rangle \triangleleft \bar{K}$, then $C_{\bar{K}}(\langle\bar{x}, \bar{z}\rangle)$ would be normal in $\bar{K}$ of index 2 with Sylow 2-subgroup $\bar{T}$, in which case $\bar{y}$ and $\bar{v}$ could not be conjugate in $\bar{K}$. Thus $\langle\bar{x}, \bar{z}\rangle$ is not normal in $\bar{K}$. Since $\bar{z} \in Z(\bar{K})$ and $\bar{x} \in \bar{Q}$, this implies that $\bar{x}$ is conjugate in $\bar{K}$ to some involution $\bar{a}$ of $\bar{Q} - \langle\bar{x}, \bar{z}\rangle$. But $x \not\sim y$ in G by Lemma 6.14(iii), while all involutions of $S - T$ are conjugate to y in S. These conditions clearly force $\bar{a}$ to lie in $\bar{T}$.

Finally as $\tilde{x} \in Z(\tilde{S})$, $\tilde{x} \in Z(\tilde{Q})$ and so $\tilde{a} \in Z(\tilde{Q})$. Hence if $\bar{T}_0$ denotes the normal closure of $\bar{a}$ in $\bar{S}$, we have that $\bar{T}_0 \subseteq \bar{Q}$ and $\tilde{T}_0 \subseteq Z(\tilde{Q})$, whence $\tilde{T}_0$ is abelian. On the other hand, we have that $\bar{S}/\langle\bar{x},\bar{z}\rangle \cong D_{2^n} \wr Z_2$ and $\bar{T}/\langle\bar{x},\bar{z}\rangle \cong D_{2^n} \times D_{2^n}$. Furthermore, the inverse image in $\bar{T}$ of $Z(\bar{T}/\langle\bar{x},\bar{z}\rangle)$ is isomorphic to $Z_4 \times Z_4$, so $\bar{a}$ maps on to a noncentral involution of $\bar{T}/\langle\bar{x},\bar{z}\rangle$. It follows therefore from the structure of $\bar{S}/\langle\bar{x},\bar{z}\rangle$ that $\langle\bar{x},\bar{z}\rangle \triangleleft \bar{T}_0$ and $\bar{T}_0/\langle\bar{x},\bar{z}\rangle \cong D_{2^{n-1}} \times D_{2^{n-1}}$. But $\bar{T}_0/\langle\bar{x},\bar{z}\rangle \cong \tilde{T}_0/\langle\tilde{x}\rangle$ and so is abelian. This forces $n = 3$ and $\bar{T}_0/\langle\bar{x},\bar{z}\rangle \cong \tilde{T}_0/\langle\tilde{x}\rangle \cong E_{16}$. Thus $\bar{T}_0$ is abelian of order 32. But $\bar{T}_0$ is the normal closure in $\bar{S}$ of $\langle\bar{a}\rangle$ and so is generated by involutions. Hence $\bar{T}_0$ is elementary and so $\bar{T}_0 \cong E_{32}$, contrary to $r(G) \leq 4$. Proposition 6.1 is therefore finally proved.

7. The non-dihedral case. We must now analyze the various cases to which Proposition 6.1 has reduced us. In this section we shall eliminate the possibilities that $\bar{L} \cong M_{11}, L_3(3)$, or $U_3(3)$. Thus we shall prove

Proposition 7.1. $\bar{L}$ is isomorphic to $L_2(q)$, q odd, or A_7.

We proceed by contradiction in a sequence of lemmas. We let R_1, R_2, T, y, and $\tilde{M}$ have the same meanings as in the preceding section. In particular, either $R_2 \cong QD_{16}$ or W_{32}. By Proposition 6.1, $R_1 \cong Z_2$ or Z_4 with the latter occurring only if $R_2 \cong W_{32}$ and $\bar{L} \cong U_3(3)$.

We first prove

Lemma 7.2. If R_2 is quasi-dihedral, then $T \supset R_1R_2$.

Proof: Suppose false, in which case $T = R_1 \times R_2$ with $R_1 \cong Z_2$, $R_2 \cong QD_{16}$, and $\bar{M} = \bar{L} \times \bar{R}_1$ with $\bar{L} \cong L_3(3)$ or M_{11}, by Proposition 6.1 and the structure of Aut$(\bar{L})$. Since R_2 contains only one conjugacy class of four groups, T contains only one conjugacy class of elementary abelian 2-subgroups of order 8. Hence if A is one of these, we can assume, as usual, that y normalizes A. Since $Z(S) \subseteq T$, we see that $Z(S) \subset A$. Clearly also $x \in A$.

Now set $N = N_G(A)$ and $\bar{N} = N/C_G(A)$, so that $\bar{N}$ is isomorphic to a subgroup of Aut$(A) \cong L_3(2)$. If $\bar{N}$ contained an element of order 7, all involutions of A would be conjugate in N. But then as x and $Z(S)$ are in A, x would be a central involution, which is not the case. Therefore $\bar{N}$ is solvable. By Lemma 2.19(vii), $N_M(A)/C_M(A) \cong S_3$ and so the image of $N_M(A)$ in $\bar{N}$ is isomorphic to S_3. But as $y \in N$, x is not isolated in N. We conclude at once from these conditions that $\bar{N} \cong S_4$, that $\bar{N}$ leaves invariant a four subgroup A_1 of A, and that x has 4 conjugates in A under the action of N which span the set $A - A_1$.

Since $C_G(A) \subseteq M$, it is immediate from the structure of M that $C_G(A)$ has a normal 2-complement which is clearly equal to $O(N)$. But now if we let W be a Sylow 2-subgroup of N, it follows from Lemma 2.8, applied to $N/O(W)$, that W is either of type A_8 or M_{12}. (By Lemma 2.19 (vii), $N_M(A)$ and hence W contains an involution outside of N'. From this, it follows that the other alternatives of Lemma 2.8 can not arise in the present situation.) In either case W does not contain a subgroup of the form $T \cong Z_2 \times QD_{16}$ and consequently W is not a Sylow 2-subgroup of G. Hence

$|S| \geq 2^7$ and our argument yields that T is not a maximal subgroup of S as $|T| = 2^5$.

Now let $U \in U(S)$. Since $|S:C_S(U)| \leq 2$, it follows for each u in $U^{\#}$ that a Sylow 2-subgroup of $C_G(u)$ has order at least 2^6. Since T is a Sylow 2-subgroup of $M = C_G(x)$, x cannot be conjugate in G to any u in $U^{\#}$. In particular, $x \notin U$. If $x \in C_S(U)$, then $U \subseteq T$ and so $U\langle x\rangle$ is an elementary abelian normal subgroup of T of order 8. But as $T \cong Z_2 \times QD_{16}$, $SCN_3(T)$ is empty. Thus $x \notin C_S(U)$ and we conclude that $S = C_S(U)\langle x\rangle$ with $C_S(U)$ maximal in S.

Hence by Thompson's transfer lemma, $x \sim v$ in G for some v in $C_S(U)$. By the preceding argument, $v \notin U$ and so $B = U\langle v\rangle \cong E_8$. Since M contains only one conjugacy class of E_8's, it follows that $A \sim B$ in G. But then also $N_G(B)/C_G(B) \cong S_4$ with $N_G(B)$ having a normal four subgroup B_1 of B and with v having 4 conjugates in B under the action of $N_G(B)$, which span the set $B - B_1$. On the other hand, v is not conjugate in G to an element of U since x is not. Thus, in fact, we have $B_1 = U$. We conclude therefore from the action of $N_G(B)/C_G(B)$ on B that the involutions of U are conjugate in $N_G(B)$.

Finally we argue that $SCN_3(S)$ is empty. Indeed, if false and

$E \in SCN_3(S)$, then $E_1 = C_E(x)\langle x\rangle$ would be abelian of 2-rank at least 3 and E_1 would be normal in T, contrary to the fact that $SCN_3(T)$ is empty. This proves our assertion. But now Proposition 4.3 yields a contradiction and the lemma is proved.

As an immediate corollary, we have

Lemma 7.3. $\bar{L}$ is not isomorphic to M_{11}.

Proof: Indeed, since $\mathrm{Aut}(M_{11}) = \mathrm{Inn}(M_{11})$, we would necessarily have $T = R_1R_2$ in this case, contrary to the preceding lemma.

We next prove the analogue of Lemma 7.2 in the wreathed case.

Lemma 7.4. If R_2 is wreathed, then $T \supset R_1R_2$.

Proof: Suppose false. In this case $R_1 \cong Z_2$ or Z_4. Let R_0 be the unique subgroup of R_2 isomorphic to $Z_4 \times Z_4$ and set $A = \Omega_1(R_0 \times R_1)$. Then $m(A) = 3$ and we see that $\Omega_1(Z(S)) \subseteq A$. Setting $N = N_G(A)$ and $\bar{N} = N/C_G(A)$, we conclude as in Lemma 7.2, again using the structure of M, that $\bar{N} \cong S_4$. Since $R_0R_1 \subseteq C_G(A)$, thus in turn implies that $|N|$ is divisible by $2^7|R_1|$. Hence the same is true of S and as $T = R_1R_2$ has order $2^7|R_1|$, it again follows that T is not a maximal subgroup of S.

We shall now contradict this fact by showing that $S = T\langle y\rangle$. We remark that portions of this argument will also be used again in the analysis of the case $T \supset R_1R_2$. By the structure of W_{32}, R_2 possesses a unique normal quaternion subgroup Q. Then $Q \triangleleft T$ and hence also

$Q^y \lhd T$. We set $P = QQ^y$, so that y normalizes P. We shall determine the structure of P. Set $Q_1 = Q \cap Q^y$. If $|Q_1| = 8$, then $P = Q$. If $|Q_1| = 4$, then $|P| = 16$ and we check directly from the list of groups of order 16 that Q_{16} and $Q_8 \times Z_2$ are the only ones generated by two normal quaternion subgroups. However, in the first case, $P \cap R_1 = 1$ as the unique subgroup of order 2 in P (namely $Z(Q) = Z(P)$) is disjoint from R_1. Thus $\tilde{T} = T/R_1 \supseteq \tilde{P} \cong Q_{16}$. Since $\tilde{T} \cong W_{32}$, this is a contradiction. We conclude therefore that $P \cong Q_8 \times Z_2$ and hence that $P = QC_P(Q)$ with $C_P(Q)$ a four group. Since $C_{R_2}(Q) = Z(R_2) \cong Z_4$, $C_T(Q) = Z(T) = Z(R_2)R_1$ and consequently $C_P(Q) = Q'\langle x\rangle = \Omega_1(Z(T))$. Thus if $|Q_1| = 4$, $P = Q \times \langle x\rangle$.

Suppose next that $|Q_1| = 2$, whence $|P| = 2^5$ and $Q_1 = Q'$. Since $[Q,Q^y] \subseteq Q_1 = Q'$ (as Q and Q^y are normal in P), Q^y stabilizes the chain $Q \supset Q' \supset 1$. But then we have that $P = QC_P(Q)$. As we observed above, $C_T(Q) = Z(T)$ and so $C_P(Q)$ is abelian. But now it is immediate that every quaternion subgroup of P lies in $Q\Omega_1(C_P(Q))$. Since $\Omega_1(C_P(Q)) \subseteq \Omega_1(Z(T)) = Q'\langle x\rangle$, it follows that $|P| = 16$, which is not the case. Finally if $Q_1 = 1$, then $P \cong Q_8 \times Q_8$. However, T contains no subgroups of this form. We thus conclude that $P = Q$ or $Q\langle x\rangle$. Since y normalizes P and $Z(T)$, it follows in either case that y normalizes $V = QZ(T) = QC_T(Q)$.

We next set $H = N_G(V)$ and $K = C_G(V)V$ We have that $T\langle y\rangle \subseteq H$, that $V \subseteq K$, and that $K \lhd H$. Observe that $\Omega_1(Z(V)) = Q'\langle x\rangle \lhd H$. Since y does not centralize x and since $C_G(\Omega_1(Z(V))) \subseteq C_G(x) \subseteq M$, it follows at once that $|H:H \cap M| = 2$. In particular, this implies that both $H \cap M$ and $K \cap M$ are normal in H. By Lemma 2.19 (viii) we have that $H \cap M/K \cap M \cong S_3$. Since $|H/K \cap M| = 2|H \cap M/K \cap M|$, we conclude now that $\bar{H} = H/K \cap M = Z(\bar{H}) \times (\overline{H \cap M}) \cong Z_2 \times S_3$. But V is clearly a Sylow 2-subgroup of $K \cap M$ and so our argument yields that $R = T\langle y\rangle$ is a Sylow 2-subgroup of H.

Furthermore, if W denotes the subgroup of index 2 in R containing V whose image in $\bar{H}$ is $Z(\bar{H})$, then $\bar{W} \not\leq \bar{T}$ and consequently $R = WT$. We also have that $Q = [V, O_3(\bar{H})]$ and so $\bar{W}$ leaves Q invariant. Then $\bar{W} \times O_3(\bar{H})$ acts an Q with $O_3(\bar{H})$ acting nontrivially, so by Thompson's $A \times B$-lemma, $\bar{W}$ centralizes Q. Since $V = QC_V(Q)$ and $W/V \cong \bar{W}$, it follows therefore that $W = QC_W(Q)$. We set $W_1 = C_W(Q)$, so that $W = QW_1$. Since $W \not\leq T$, also $W_1 \not\leq T$, and so $R = TW_1$. Since $R = T\langle y\rangle$, we can therefore assume without loss that $y \in W_1$.

We next analyze the structure of W_1. Since $W_1 \cap V = C_V(Q) = Z(T)$, we have that $W_1 = Z(T)\langle y\rangle$. Suppose first that $R_1 \cong Z_4$, whence $Z(T) = R_1 \times Z(R_2) \cong Z_4 \times Z_4$. Since y does not centralize x, it does not centralize $\Omega_1(Z(T))$. We conclude from this that y must interchange some pair of generators of $Z(T)$. But then $W_1 \cong W_{32}$ by Lemma 6.8 of Part III and so $W \cong Q_8 * W_{32}$. Since W_{32} contains a subgroup isomorphic to $Z_4 * Q_8$ and since $Q \cap W_1 = Q'$, it follows at once that $r(W) \geq 5$, contrary to $r(G) \leq 4$. We thus conclude that $R_1 = \langle x\rangle \cong Z_2$ and that $Z(T) \cong Z_2 \times Z_4$.

Observe that $Q \lhd R$ as $Q \lhd T$ and y centralizes Q. Hence also $Q' \lhd R$. We set $\bar{R} = R/Q'$ and argue that $\bar{W}_1$ is abelian of order 8.

Indeed, set $\langle z\rangle = Z(R_2)$, so that $|z| = 4$, $Z(T) = \langle z,x\rangle \cong Z_4 \times Z_2$, and $\langle z^2\rangle = Q' = \Omega_1(Z(T))$. Since $Z(T) \subseteq W_1$, W_1 is therefore not of maximal class, which implies that $|W_1:W_1'| > 4$. But $W_1 = Z(T)\langle y\rangle$ is of order 16 and as y centralizes z^2 and does not centralize x, we must have $x^y = xz^2$. We thus conclude that $\langle z^2\rangle = W_1'$, whence $\bar{W}_1$ is abelian of order 8, as asserted. Since $\bar{W} = \bar{Q} \times \bar{W}_1$ and $m(\bar{W}) \leq 4$, this in turn forces $\bar{W}_1 \not\cong E_8$. Hence $\bar{W}_1 = Z_2 \times Z_4$ and, in particular, $|\bar{y}| = 4$.

Now we argue that $W_1 = \langle y,x\rangle$ with $|y| = 8$ and x centralizing y^2. We have that $\bar{W}_1 = W_1/Q' \cong Z_2 \times Z_4$ and so $\mho^1(W_1) \subseteq Z(W_1)$. If $\bar{y}^2 = \bar{x}$, then $x \in \langle Q',y^2\rangle$. But as $y^2 \in \mho^1(W_1)$, $y^2 \in Z(W_1)$ and so $x \in Z(W_1)$. Thus x centralizes y, which is not the case. Hence $\bar{W}_1 = \langle\bar{y},\bar{x}\rangle$ and as $Z(T) = Z(R_2) \times \langle x\rangle$ with $Z(R_2) = Z_4$, we conclude at once that $W_1 = \langle y,x\rangle$ with $|y| = 8$. Since $y^2 \in Z(W_1)$, also x centralizes y^2 and our assertion is proved.

In particular, the structure of $W = QW_1$ is uniquely determined and we compute now that $V = \Omega_2(W)$. Thus V is characteristic in W.

Finally we argue that W is normal in $N_S(R)$. Suppose false and choose a in $N_S(R)$ with $W^a \neq W$. Since $Z(R) \subseteq Z(T)$ and y does not centralize x, clearly $\Omega_1(Z(R)) = Q'$ and so Q' is characteristic in R. Therefore a centralizes Q' and so a normalizes $\bar{R} = R/Q'$ and $\bar{W}^a \neq \bar{W}$. Note also that R_2, being isomorphic to W_{32}, contains an element v such that $\langle v^2\rangle = Z(R_2)$ with $v \notin QZ(R_2)$ and $\langle v,Q\rangle/\langle v^2\rangle \cong D_8$. Then $v \notin QW_1 = W$ and so $R = W\langle v\rangle$, whence $\bar{R} = \bar{W}\langle\bar{v}\rangle$. In particular, this implies that $Z(\bar{R}) \subseteq \bar{W}$ and $\bar{Q} \not\subseteq Z(\bar{R})$. We use this conclusion to derive a contradiction.

Now $\bar{W} = \bar{Q} \times \bar{W}_1 \cong Z_2 \times Z_2 \times Z_2 \times Z_4$ and $\bar{W}$ is of index 2 in $\bar{R}$. Hence $\bar{R} = \bar{W}\bar{W}^a$ and $\bar{W} \cap \bar{W}^a$ is abelian of order 16 and lies in $Z(\bar{R})$.

Since $Z(\overline{R}) \subseteq \overline{W}$ and $\overline{Q} \not\subseteq Z(\overline{R})$, this implies that $\overline{W} \cap \overline{W}^a = Z(\overline{R}) \cong Z_2 \times Z_2 \times Z_4$ and $\overline{W} = \overline{Q}Z(\overline{R})$. Since a normalizes $Z(\overline{R})$, the last equality yields that $\overline{Q}^a \not\subseteq \overline{W}$, otherwise $\overline{W}^a = \overline{W}$. Likewise a does not normalize $\Omega_1(\overline{W})$, otherwise a normalizes $\overline{W} = C_{\overline{R}}(\Omega_1(\overline{W}))$. Together these conditions imply that for some involution $\overline{b}$ of $\overline{Q}$, $\overline{c} = \overline{b}^a$ lies in $\overline{R} - \overline{W}$. In particular, $\overline{D} = \overline{Q}\langle\overline{c}\rangle \cong D_8$.

Finally set $H = C_G(Q')$ and $\overline{H} = H/Q'$. Then $\overline{D} \subseteq \overline{H}$ and as $a \in H$, $\overline{b} \sim \overline{c}$ in $\overline{H}$. On the other hand, by Lemma 2.19(viii), the involutions of $\overline{Q}$ are conjugate in $\overline{H}$. Hence all involutions of $\overline{D}$ are conjugate in $\overline{H}$ and so $\overline{H}$ must be nonsolvable. But as $Q' \cong Z_2$ is normal in R and $R \supset T$, it follows from Lemma 5.1 that H is solvable. This contradiction establishes that $W \triangleleft N_S(R)$, as asserted.

Since V is characteristic in W, V is normal in $N_S(R)$. Hence $R = N_S(R) = S = T\langle y\rangle$, which is the contradiction we have been seeking.

Lemma 7.5. $\overline{L}$ is not isomorphic to $L_3(3)$.

Proof: Suppose false. By Lemma 7.2, we have $T \supset R_1R_2$, whence $\widetilde{T} \supset \widetilde{R}_2$. Lemma 2.19(i) now yields that $\widetilde{M} \cong Aut(L_3(3))$ and that $|T:R_1R_2| = 2$. By the same lemma, $\widetilde{T} - \widetilde{R}_2$ contains an involution and so $T - R_1R_2$ contains an element u such that $u^2 \in R_1$. In the present instance $R_1 = \langle x\rangle$ by Proposition 6.1 and so either $u^2 = 1$ or x.

We first eliminate the second possibility. Indeed, in that case $T/R_2 \cong Z_4$ and so $\Omega_1(T) \subseteq R_1R_2$, whence all elementary abelian subgroups of T order 8 lie in R_1R_2. As observed in Lemma 7.2, all such subgroups of R_1R_2 are conjugate in R_1R_2. By Lemma 2.19(vi) and (vii), $C_{R_2}(u) \cong D_8$ and consequently $C_T(u) \cong D_8 \times Z_4$. In particular, $C_T(u)$ contains an elementary abelian subgroup A of order 8. Our argument yields that $A^y = A^r$ for some

r in R_1R_2 and so without loss we can suppose that y normalizes A. However, as $C_S(x) = T$ and $x \in A$, it is immediate that $C_S(A) = A\langle u\rangle \cong Z_2 \times Z_2 \times Z_4$ with $\langle x\rangle = \mho^1(C_S(A))$. Since $y \in N_S(A)$, y normalizes $C_S(A)$ and so y centralizes x, which is not the case.

We thus conclude that $u^2 = 1$. Hence if we set $R_2^* = R_2\langle u\rangle$, we have that $T = R_1 \times R_2^*$ with R_2^* isomrophic to a Sylow 2-subgroup of $Aut(L_3(3))$. By Lemma 2.19(iii), R_2^* contains a normal quaternion subgroup Q, $Z = C_{R_2^*}(Q) \cong Z_4$, and $Z\langle v\rangle \cong D_8$ for some involution v of $R_2 - Q$. We set $V = QZR_1$ and argue next that y normalizes V. Indeed, by Lemma 2.19(iv), R_2^* contains exactly two quasi-dihedral subgroups, namely, R_2 and a second one R_3 with $R_2 \cap R_3 = Q$. It follows that $(R_1R_2)^y = R_1R_2$ or R_1R_3. In the first case y normalizes R_1Q as this is the unique subgroup of R_1R_2 isomorphic to $Z_2 \times Q_8$. In the second case, as R_1R_2 is normal in T, y normalizes $R_1R_2 \cap R_1R_3 = R_1Q$. Since $C_T(R_1Q)R_1Q = R_1QZ = V$, we conclude in either case that y normalizes V, as asserted.

We are now in a position to emulate portions of the argument of the preceding lemma. We observe first that our group V has the identical structure as the subgroup V of that lemma in the case that $R_1 \cong Z_2$. Furthermore, if we again set $H = N_G(V)$ and $K = C_G(V)V$, it follows from Lemma 2.19(vii) that $H \cap M/K \cap M \cong S_3$. Setting $\bar{H} = H/K \cap M$, we again conclude that $\bar{H} \cong Z_2 \times S_3$ and that $R = T\langle y\rangle$ is a Sylow 2-subgroup of H. Moreover, if W again denotes the subgroup of R containing V such that $\bar{W} = Z(\bar{H})$, we again obtain that $W = QW_1$, where $W_1 = C_W(Q)$, that y can be chosen to lie in W_1, that $W_1 = \langle y,x\rangle$ with $|y| = 8$ and x centralizing y^2, and that $V = \Omega_2(W)$ is characteristic in W.

We argue next that also W is normal in $N_S(R)$. (However, we cannot copy the corresponding argument of Lemma 7.4 as the element v of that lemma was of order 8, while the present element v is an involution.) Setting $\bar{R} = R/Q'$, it again suffices to prove that $\bar{W}$ is characteristic in $\bar{R}$. Arguing by contradiction, it follows as in Lemma 7.4 that $\bar{W} \supseteq Z(\bar{R}) \cong Z_2 \times Z_2 \times Z_4$ and hence that $\bar{W}_1 \subseteq Z(\bar{R})$. In particular, $\bar{v}$ centralizes $\bar{W}_1$. Since $y \in W_1$, it follows that $y^v \equiv y (\mathrm{mod}\ Q')$, which implies that v centralizes y^2. But as $W_1 = \langle y,x\rangle$, we see that $\langle y^2,x\rangle = W_1 \cap T = R_1Z$. Thus v centralizes Z, contrary to the fact that $Z\langle v\rangle \cong D_8$. Therefore W is normal in $N_S(R)$. Since V is characteristic in W, we conclude at once that R is a Sylow 2-subgroup of G.

Our argument yields that $R = S = T\langle y\rangle$. In particular, $|S| = 2^7$. Note also that $SCN_4(S)$ is empty. Indeed, if false and we take B in $SCN_4(S)$, then $B \not\subseteq T$ as $SCN_4(T)$ is empty by Lemma 2.19(v). Therefore $S = TB$ and $m(T \cap B) = 3$. Since $C_S(x) = T$, $x \notin B$ and so $(B \cap T)\langle x\rangle$ is an abelian normal subgroup of T of rank 4, contrary to the fact that $SCN_4(T)$ is empty.

Finally as u is an involution, $C_T(u) \cong D_8 \times Z_2 \times Z_2$. Thus $C_T(u)$ contains two elementary abelian subgroups A, A_1 of order 16. Since $C_{\bar{L}}(\bar{u}) \cong PGL(2,3)$, we can suppose that $\bar{A} \subseteq C_{\bar{L}}(\bar{u})'$ and $\bar{A}_1 \not\subseteq C_{\bar{L}}(\bar{u})'$. By Lemma 2.19(v) A, A_1 are the only E_{16}'s in T and are conjugate in T. Clearly each is also normal in $C_T(u)$. Hence if we consider $N_S(T)$, it follows at once that $S_1 = N_S(A) \supset C_T(u)$. Since $|S| = 2^7$, we have that $|S_1| = 2^6$ or 2^7. However, as $SCN_4(S)$ is empty, $N = N_G(A)$ does not contain a Sylow 2-subgroup of G. Hence $|S_1| = 2^6$ and S_1 is a Sylow 2-subgroup of N.

We argue next that $S_1 \cong D_8 \times D_8$ by applying Lemma 2.10. Since $C_G(A) \subseteq M$, it follows from the structure of M that $C_G(A)$ has a normal 2-complement. Thus $O(N)A = C_G(A)$, whence $\bar{A} \cong E_{16}$ and $C_{\bar{N}}(\bar{A}) \subseteq \bar{A}$. Moreover, by Lemma 2.19(vii) and our choice of A, $C_M(\langle u,x \rangle)$ contains a cyclic 3-subgroup P which normalizes, but does not centralize $A \cap R_2$ and which is inverted by an involution r of $C_T(u)$. Then $P\langle r \rangle \subseteq N$ and $\bar{P}\langle\bar{r}\rangle \cong S_3$. Moreover, $\langle \bar{r},\bar{u},\bar{x} \rangle \cong E_8$ and $\langle \bar{r},\bar{u},\bar{x} \rangle$ normalizes $\bar{P}$, whence $m(N_{\bar{S}_1}(\bar{P})) \geq 3$. Since $C_G(A)$ has a normal 2-complement and G is a $\mathfrak{R}_3$-group, we also have that N is solvable. Thus $\tilde{N} = \bar{N}/\bar{A}$ is isomorphic to a solvable subgroup of A_8 with Sylow 2-subgroup $\tilde{S}_1$ of order 4.

Suppose that $O_2(\tilde{N}) \neq 1$. Then $O_2(\tilde{N}) \cong Z_2$ and hence $O_2(\tilde{N})\tilde{P}\langle\tilde{r}\rangle \cong Z_2 \times S_3$. But now Lemma 2.10 yields that $S_1 \cong D_8 \times D_8$, as asserted. Assume then that $O_2(\tilde{N}) = 1$. Since x is a noncentral involution and $\tilde{N}$ is solvable, 15 does not divide $|\tilde{N}|$. Hence the only possibility is that $O(\tilde{N}) \cong Z_3 \times Z_3$. As usual, this implies that S_1 is isomorphic to a subgroup of a 2-group of type A_{10}. Since $|\tilde{S}_1| = 4$ and A is not characteristic in S_1, we conclude at once in this case as well that $S_1 \cong D_8 \times D_8$.

On the other hand, if we set $S_0 = S_1 \cap W$, we have that $|S_0| \geq 2^5$ as $|S:W| = 2$. But S_0 has exponent 4 as S_1 does and $V = \Omega_2(W)$, so, in fact, $S_0 \subseteq V$. Since $|V| = 2^5$, this forces $S_0 = V$. However, $V = QZR_1 \cong (Q_8 * Z_4) \times Z_2$ and it is immediate that $D_8 \times D_8$ contains no subgroup of this form. This contradiction establishes the lemma.

Finally we prove

Lemma 7.6. $\bar{L}$ is not isomorphic to $U_3(3)$.

Proof: Suppose false. By Lemma 7.3, we have $T \supset R_1R_2$ and consequently $\tilde{M} \cong \mathrm{Aut}(U_3(3))$ by Lemma 2.19(i). Thus $|T:R_1R_2| = 2$, $\tilde{T} = T/R_1$ is of type M_{12}, and $\tilde{T} - \tilde{R}_2$ contains an involution. Hence $T - R_1R_2$ contains an element u such that $u^2 \in R_1$. Also $R_1 \cong Z_2$ or Z_4.

We argue first that $R_1 = \langle x \rangle \cong Z_2$ and that $u^2 = x$. We need only prove that $R_1\langle u \rangle$ is cyclic. Indeed, if this is the case, $R_1 \subseteq Z(T)$. But by the structure of $\tilde{T}$, we see that $Z(T) = (Z(T) \cap R_2) \times R_1$ with $Z(T) \cap R_2 \cong Z_2$. But then we must have $R_1 = \langle x \rangle \cong Z_2$ as otherwise $\langle x \rangle$ would be characteristic in T, and so y centralizes x, a contradiction. In particular, this yields that either $u^2 = 1$ or x. However, in the first case, $T = R_2\langle u \rangle \times R_1$ and $R_2\langle u \rangle \cong \tilde{T}$ is of type M_{12}. But then $r(R_2\langle u \rangle) = 4$ and so $r(T) \geq 5$, contrary to $r(G) \leq 4$. Therefore also $u^2 = x$.

Suppose then that $R_1\langle u \rangle$ is noncyclic, whence $R_1 \cong Z_4$ and $\langle x \rangle = \Phi(R_1\langle u \rangle)$. But now setting $\bar{T} = T/R_0$, it follows that $\bar{T} = \bar{R}_2\langle \bar{u} \rangle \times \bar{R}_1$ with $\bar{R}_2\langle \bar{u} \rangle$ of type M_{12} and $\bar{R}_1 \cong Z_2$. Thus $r(\bar{T}) \geq 5$, giving the same contradiction. Therefore $R_1 = \langle x \rangle$ and $u^2 = x$, as asserted.

Once again we reason as in Lemma 7.4. One checks that R_2 possesses a unique normal subgroup $Q \cong Q_8$ and so $Q \lhd T$. Setting $P = QQ^y$ and using the fact that T contains no subgroup isomorphic to $Q_8 \times Q_8$ and $\tilde{T}$ contains none isomorphic to Q_{16}, it follows as in Lemma 7.4 that $P = QC_P(Q)$ and either $P = Q$, $P \cong Q_8 \times Z_2$, or $|P| = 32$.

To analyze the last two possibilities, we first consider the structure of $Z = C_T(Q)$. Since $Q \lhd T$ and Q centralizes R_1, we have that $\tilde{Z} = C_{\tilde{T}}(\tilde{Q})$ and so $\tilde{Z} \cong Q_8$ with $\tilde{Z}' = \tilde{Q}'$. Examining the list of 2-groups of order 16 and using the fact that $Q' \times \langle x \rangle \subseteq Z$, it follows now that

$\Omega_1(Z) \cong Z_2 \times Z_2$, whence $\Omega_1(Z) = Q' \times \langle x \rangle$. Furthermore, as $P = QC_P(Q)$, we also have that $P \cong QZ$. But now if $P \cong Q_8 \times Z_2$, we see that $P = Q \times \langle x \rangle$.

We shall next argue that the case $|P| = 32$ is not possible. Observe first that any element a of QZ of order 4 with $\langle a^2 \rangle = Q'$ either lies in $Q \times \langle x \rangle$ or in Z. But y leaves Q' invariant and $Q^y \subseteq P \subseteq QZ$. Suppose $|P| = 32$, in which case certainly $Q^y \not\subseteq Q \times \langle x \rangle$. Since Q^y is generated by any two of its elements of order 4, no two such elements can lie in $Q \times \langle x \rangle$ and so by the preceding observation at least two of them must lie in Z, whence $Q^y \subseteq Z$. Since $|Z| = 16$ and $x \notin Q' = (Q')^y$, it follows then that $Z = Q^y \times \langle x \rangle \cong Q_8 \times Z_2$. But then $QZ \cong (Q_8 * Q_8) \times Z_2$ and so $r(QZ) = 5$, contrary to $r(G) \leq 4$. This proves the desired assertion. We therefore conclude that either $P = Q$ or $P = Q \times \langle x \rangle$.

Since y normalizes P and $P = Q$ or $Q \times \langle x \rangle$, it follows that y normalizes $QZ = QC_T(Q)$. But now if we set $V = QZ$ and $H = N_G(V)$ we conclude once again as in Lemma 7.4, using the structure of $N_M(V)$, that $R = T\langle y \rangle$ is a Sylow 2-subgroup of H and that R possesses a subgroup W of index 2 with $W \supset V$ such that $W = QW_1$, where again $W_1 = C_W(Q)$. Again we can choose y to lie in W_1, so that $W_1 = Z\langle y \rangle$. But by the structure of Z, we have that $Z/\langle x \rangle \cong Q_8$. Hence also $Z/\langle x^y \rangle \cong Q_8$. Since $x \neq x^y$, Z is thus a group of order 16 having two distinct quaternion homomorphic images. The only possibility is that $Z \cong Q_8 \times Z_2$, as is easily checked. Since $\tilde{Z}' = \tilde{Q}'$, Q' is not a direct factor of Z. Since $Q \cap Z = Q'$, we conclude that $r(QZ) \geq 5$, contrary to $r(G) \leq 4$ and the lemma is proved.

By Lemma 7.3, 7.5, and 7.6, $\bar{L} \not\cong M_{11}$, $L_3(3)$, or $U_3(3)$. Proposition 6.1 now yields that $\bar{L} \cong L_2(q)$, q odd, or A_7 and so Proposition 7.1 is proved.

8. <u>The noncyclic case</u>. By Propositions 6.1 and 7.1 we have that $\bar{L} \cong L_2(q)$, q odd, or A_7 and that $\bar{C} = \bar{R}_1 \cong R_1 \cong Z_2$ or $Z_2 \times Z_2$. Moreover, if the latter case occurs, then necessarily $\bar{L} \cong L_2(q)$, $q \equiv 3$, $5 \pmod 8$. Here R_1, R_2, T, y, and $\tilde{M}$ will have the same meaning as before. In this section we shall eliminate the second possibility. Thus we shall prove

<u>Proposition 8.1</u>. R_1 **is** of order 2.

Suppose false, in which case $R_1 \cong Z_2 \times Z_2$ and $\bar{L} \cong L_2(q)$, $q \equiv 3$, $5 \pmod 8$. Hence $R_2 \cong Z_2 \times Z_2$ and the Sylow 2-subgroup $\tilde{T}$ of $\tilde{M}$ is either equal to $\tilde{R}_2$ or else is isomorphic to D_8. We first prove

<u>Lemma 8.2</u>. We have $T \supset R_1R_2$.

<u>Proof</u>. Suppose false, in which case $T = R_1 \times R_2 \cong E_{16}$. We set $N = N_G(T)$ and $\bar{N} = N/C_G(T)$. Then $T\langle y\rangle \subseteq N$. Since $C_G(T) \subseteq M$, T is a Sylow 2-subgroup of $C_G(T)$ and so $C_G(T)$ is solvable. This implies that N is 2-constrained. Since G is an $\mathfrak{R}_3$-group, we conclude that N, and hence $\bar{N}$, is solvable.

Let u be any 2-element of $N - T$. We claim that $C_{R_1}(u) = 1$. Suppose false, in which case $H = C_G(C_{R_1}(u))$ is a 2-local subgroup of G containing $T\langle u\rangle \supset T$. But $C_M(C_{R_1}(u))$ covers $\bar{L}$ and so H is non-solvable, contrary to Lemma 5.1. This proves our assertion.

Observe next that by the structure of M, $N \cap M$ contains a 3-subgroup P which centralizes R_1 and normalizes, but does not centralize R_2.

Then $|\overline{P}| = 3$ and $N_{\overline{N}}(\overline{P})$ leaves R_1 invariant, this forces $|N_{\overline{N}}(\overline{P})|$ to be odd. Indeed, otherwise it would follow by the Frattini argument that $N - T$ contained a 2-element u which normalized R_1. But then $C_{R_1}(u) \neq 1$, contrary to the preceding paragraph.

We have that P centralizes a four subgroup of T, that $\overline{y}$ is an involution of $\overline{N}$, and that $|N_{\overline{N}}(\overline{P})|$ is odd. Hence if $O_2(\overline{N}) = 1$, we must have that $O(\overline{N}) \cong Z_3 \times Z_3$ and $\overline{N} = O(\overline{N})\langle\overline{y}\rangle$. Thus $T\langle y\rangle$ is a Sylow 2-subgroup of N. But as $C_{R_1}(y) = 1$ and R_1 is a four subgroup of T, it is immediate that $T\langle y\rangle \cong Z_2 \times Z_2 \int Z_2$. We see then that T is characteristic in $T\langle y\rangle$ and consequently $S = T\langle y\rangle$. But then G has a normal subgroup of index 2 by [27, Lemma 18], which is a contradiction. We conclude that $O_2(\overline{N}) \neq 1$.

If k is the order of a Sylow 2-subgroup of $\overline{N}$, then every involution of R_1 has at least k conjugates in T by the action of $\overline{N}$. Since T contains at most $12 = 15 - 3$ noncentral involutions, we conclude that $2 \leq k \leq 8$. Suppose $k = 8$. Every element of R_1 has exactly 8 conjugates in T, which is clearly impossible by the action of $\overline{P}$ on T. Thus $k \leq 4$.

Since $\overline{N}$ is isomorphic to a solvable subgroup of A_8 with $\overline{N}$ of even order and $N_{\overline{N}}(\overline{P})$ of odd order, there are precisely two possibilities for the structure of $\overline{N}$: namely, $\overline{N} \cong A_4$ or $\overline{N} \cong A_4 \times Z_3$ with $\overline{P} \neq O(\overline{N})$. To analyze these cases, we let R be a Sylow 2-subgroup of N containing $T\langle y\rangle$. Without loss, we can suppose that $R \subseteq S$. Our conditions imply that $\overline{RP} \cong A_4$.

We divide the proof into two cases according as R splits or does not split over T.

Case 1. R splits over T. Set $\tilde{N} = N/O(N) = N/O(C_G(T))$. If K denotes the inverse image of $\overline{RP}$ in N, it follows under the present assumption that $\tilde{K}$ also splits over $\tilde{T}$. Thus $\tilde{K}$ is isomorphic to a semidirect product of E_{16} by A_4 with A_4 acting faithfully on E_{16}. Now Lemma 2.2(iv) of Part II yields that $\tilde{R}$, and hence R, is either of type $L_3(4)$ or A_8. However, in the latter case, $\tilde{P}$ centralizes $Z(\tilde{R}) \cong Z_2$. But $C_{\tilde{R}}(\tilde{P}) = \tilde{R}_1$. Hence if $\langle r\rangle$ is the inverse image of $Z(\tilde{R})$ in R, then $r \in R_1$ and so $C_G(r)$ is nonsolvable, contrary to Lemma 5.1. We conclude that R is of type $L_3(4)$. Furthermore, if $R = S$, then S is of type $L_3(4)$, contrary to Proposition 4.2. Therefore also $R \subset S$.

Let v be an element of $N_S(R) - R$ with $v^2 \in R$. Since R is of type $L_3(4)$, R contains exactly one subgroup $T_1 \cong E_{16}$ distinct from T. Since R is a Sylow 2-subgroup of $N = N_G(T)$, v does not leave T invariant and so $T^v = T_1$. In addition, $R = TT_1$ and every involution of R lies in T or T_1. In particular, $\Omega_1(C_R(v)) \subseteq T \cup T_1$. Since v centralizes $C_R(v)$ and interchanges T and T_1, it follows that $\Omega_1(C_R(v)) \subseteq T \cap T_1 = Z(R) \cong Z_2 \times Z_2$. Thus $m(C_R(v)) \leq 2$ and so $m(C_{R\langle v\rangle}(v)) \leq 3$. The same conclusion holds by the same argument for any element v' of $R\langle v\rangle - R$ and so T, T_1 are the only elementary subgroups of order 16 in $R\langle v\rangle$. Thus $R = TT_1$ is characteristic in $R\langle v\rangle$. This forces $S = R\langle v\rangle$, otherwise some element of $N_S(R\langle v\rangle) - R$ would have to leave T invariant, contrary to the fact that $R = N_S(T)$. We therefore conclude that $S = R\langle v\rangle$ has order 2^7.

We argue next that $SCN_3(S)$ is empty. Suppose false, and let A be an elementary normal subgroup of S of order 8. By the structure of

R, $Z(R) \cong Z_2 \times Z_2$ is the unique normal four subgroup of R. Since $A \cap R \triangleleft S$ of order at least 4, we have $Z(R) \subseteq A$. But v interchanges T and T_1 and $T \cap T_1 = Z(R)$. Clearly then $A \not\subseteq R$ and so $S = RA$. Setting $\bar{S} = S/Z(R)$, it follows that $\bar{A} \cong Z_2$ and $\bar{A} \triangleleft \bar{S}$, so $\bar{A} \subseteq Z(\bar{S})$. But $\bar{S} = \overline{TT_1}\bar{A}$ with $\bar{T} \cap \bar{T}_1 = 1$ and $\bar{A}$ interchanging $\bar{T}$ and $\bar{T}_1$, so certainly $\bar{A} \not\subseteq Z(\bar{S})$. This contradiction establishes the desired assertion.

Finally as $\bar{P}$ normalizes $\bar{R}$, $N_N(R)$ contains a cyclic 3-subgroup P_1 which maps on $\bar{P}$. Since $\bar{P}$ centralizes R_1, so also does P_1. Since P_1 induces a group of automorphisms of R of order 3, it follows now from [16, Lemma 4.7] that P_1 acts nontrivially on $Z(R)$. Thus the involutions of $Z(R)$ are conjugate in G.

But $Z(R) \triangleleft S$ and as $Z(R) \cong Z_2 \times Z_2$, we have that $Z(R) \in U(S)$. However, this contradicts Proposition 4.3. This completes the proof of the lemma in Case 1.

<u>Case 2</u>. R <u>does not split over</u> T. Suppose R contains a second elementary subgroup $T_1 \neq T$ of order 16. By the action of R on T, we have that $C_T(r) \cong Z_2 \times Z_2$ for any r in $R - T$. Since T_1 centralizes $T \cap T_1$, it follows that $|T \cap T_1| \leq 4$, whence $R = TT_1$ and $|T \cap T_1| = 4$. But then a complement to $T \cap T_1$ in T_1 is a complement to T in R and so R splits over T, contrary to assumption. We conclude that T is the unique elementary subgroup of R of order 16. In particular, T is characteristic in R and as R is a Sylow 2-subgroup of $N = N_G(T)$, we see that $R = S$ is a Sylow 2-subgroup of G, whence also $|S| = 64$.

We argue next that $\overline{N} = \overline{SP} \cong A_4$. Indeed, in the contrary case, $\overline{N} = \overline{SP} \times O(\overline{N})$ with $O(\overline{N}) \cong Z_3$ and so S admits an automorphism group of type $Z_3 \times Z_3$. Since S has sectional 2-rank at most 4 and order 64 and contains $T \cong E_{16}$ with $C_S(T) = T$, it follows from Lemma 3.2 of Part I that S is of type $L_3(4)$. But then $S = R$ splits over T, which is not the case. This proves our assertion.

Now set $Z = Z(S)$. Let P_1 be a cyclic 3-subgroup of $N_G(S)$ which maps on $\overline{P}$. Since P_1 normalizes S, P_1 acts on Z. Furthermore, as $C_S(t) = T$ for t in $R_1^{\#}$, $Z \cap R_1 = 1$. But $C_T(P_1) = R_1$, so $T = Z \times R_1$ with $Z \cong Z_2 \times Z_2$ and P_1 acting nontrivially on Z. Since R_2 is the unique four subgroup of T normalized, but not centralized by P_1, we have, in fact $Z = R_2$.

Setting $P_0 = C_{P_1}(T)$, we have that $|P_1 : P_0| = 3$ and it is immediate that P_0 must centralize S. Hence P_1 induces a group of automorphisms of S of order 3. Setting $\tilde{S} = S/R_2$, we have that $|\tilde{S}| = 16$ and $C_{\tilde{S}}(P_1) = \tilde{R}_1 \cong Z_2 \times Z_2$. Hence either $\tilde{S} \cong E_{16}$ or $Q_8 \times Z_2$. However, in the latter case, $\tilde{S}' \subseteq \tilde{R}_1$ and it would follow at once that the involution of R_1 which maps on $\tilde{S}'$ lies in $Z(S) = R_2$, which is not the case. Hence $\tilde{S} \cong E_{16}$. This in turn implies that P_1 has no nontrivial fixed points on $W = [S, P_1]$ and that W has order 16. Since T is the unique elementary subgroup of S of order 16, this in turn forces $W \cong Z_4 \times Z_4$.

We have that $S = WR_1$ with $W \cap R_1 = 1$, so S is isomorphic to a split extension of $Z_4 \times Z_4$ by $Z_2 \times Z_2$. By Lemma 2.1(v) of Part II, the structure of S is uniquely determined. In fact, if we set $W = \langle a \rangle \times \langle b \rangle \cong Z_4 \times Z_4$ and $X = \langle e, d \mid e^2 = d^2 = 1 \rangle \cong D_8$ with $c = (ed)^2$ and with the action of X on W given by the relations

$$a^c = a^{-1},\ b^c = b^{-1},\ a^e = b,\ a^d = a^{-1}b^2,\ b^d = ba^2,$$

then we can identify R_1 with $\langle c,d\rangle$ and we have $S = \langle a,b,c,d\rangle$.

Finally we apply Thompson's transfer lemma to the maximal subgroup $S_1 = W\langle d\rangle$ and the involution c of $S - S_1$. Since c is a noncentral involution and since every involution of $W\langle d\rangle - W$ is conjugate to d in S, we conclude therefore that $c \sim d$ in G. But $C_S(c) = C_S(d) = T$ is a Sylow 2-subgroup of both $C_G(c)$ and $C_G(d)$ and consequently $c \sim d$ in $N = N_G(T)$. However, this clearly conflicts with the structure of N. Thus Case 2 also leads to a contradiction and the lemma is proved.

Thus to establish the proposition, it remains to treat the case $T \supset R_1R_2$. In this case $\tilde{T} = T/R_1 \cong D_8$ and so we have $|T:R_1R_2| = 2$. Thus $|T| = 32$ and $E = R_1R_2 \cong E_{16}$. It also follows that $T - R_1R_2$ contains an element u such that $u^2 \in R_1$. If u centralized R_1, then $Z(T) = R_1 \times (Z(T) \cap R_2)) \cong E_8$ and it would then follow that $C_{R_1}(y) \neq 1$. But then as in the preceding lemma, $C_G(C_{R_1}(y))$ would be a nonsolvable 2-local subgroup containing $R_1R_2\langle u,y\rangle = T\langle y\rangle \supset T$, contrary to Lemma 5.1. Thus $R_1\langle u\rangle$ is nonabelian and as its order is 8, we have $R_1\langle u\rangle \cong D_8$. Hence we can suppose that u is an involution, whence $T = \langle R_1 \times R_2\rangle\langle u\rangle$ with $R_i\langle u\rangle \cong D_8$, $i = 1,2$. We conclude that $T \cong Z_2 \times Z_2 \int Z_2$, which implies, in particular, that E, being the unique elementary abelian subgroup of T of order 16, is characteristic in T.

Hence if we set $N = N_G(E)$, we have that $T\langle y\rangle \subseteq N$. Since $C_G(E) \subseteq M$, we again have that N is 2-constrained and hence solvable. Likewise by the structure of M, $M \cap N$ contains a 3-subgroup P which centralizes R_1 and normalizes, but does not centralize R_2. Setting $\bar{N} = N/C_G(E)$, we have that $\bar{T}\langle\bar{y}\rangle = \langle\bar{u},\bar{y}\rangle$ is a subgroup of $\bar{N}$ of order 4. Moreover,

$|\bar{P}| = 3$ and we can choose P so that $\bar{P}$ is invariant under $\bar{u}$, whence $\bar{P}\langle\bar{u}\rangle \cong S_3$.

We claim that $|N_{\bar{N}}(\bar{P})|$ is not divisible by 4. Suppose false. Since $N_{\bar{N}}(\bar{P})$ leaves R_1 invariant, it follows by the Frattini argument that $N_N(R_1)$ contains a 2-subgroup V with $T \subset V$ such that $\bar{V}$ is a Sylow 2-subgroup of $N_{\bar{N}}(\bar{P})$ containing $\bar{T} = \langle\bar{u}\rangle$. But then $N_G(R_1)$ would be a nonsolvable 2-local subgroup of G containing V, contrary to Lemma 5.1. This proves our assertion.

Consider first the case that $O(\bar{N}) \cong Z_3 \times Z_3$. We have that $|\bar{N}|$ is divisible by 4, but $|N_{\bar{N}}(\bar{P})|$ is not divisible by 4. Since $\bar{N}$ is isomorphic to a subgroup of a Sylow 3-normalizer of A_8, we conclude that $|\bar{N}:O(\bar{N})| = 4$. Let V be a Sylow 2-subgroup of N such that $T \subset V$. As V splits over E, V is isomorphic to a subgroup of a 2-group of type A_{10} by Lemma 2.2(vii) of Part II. Thus $V \cong D_8 \times D_8$, $E_{16} \cdot Z_4$ or V is of type A_8. If $V \cong D_8 \times D_8$, then $Z(V) = Z(T) = \langle c, z\rangle$, where $\langle c\rangle = Z(R_1\langle u\rangle)$ and $\langle z\rangle = Z(R_2\langle u\rangle)$. But then Lemma 5.1 will be contradicted, as $C_G(c)$ is nonsolvable. If $V \cong E_{16} \cdot Z_4$ or V is of type A_8, then $S = V$, as E is the unique elementary abelian subgroup of order 16 of V. If $S \cong E_{16} \cdot Z_4$, then G can not be fusion-simple by [9]; so S is not of this form. Likewise S is not of type A_8 by Proposition 4.2.

Thus $O(\bar{N}) \not\cong Z_3 \times Z_3$ and now our conditions imply that $O_2(\bar{N}) \neq 1$. As in Lemma 8.2, the order of a Sylow 2-subgroup of $\bar{N}$ is at most 8. Also, as $\bar{u}$ inverts $\bar{P}$ and $|N_{\bar{N}}(\bar{P})|$ is not divisible by 4, we must have $O_2(\bar{N}) \cong Z_2 \times Z_2$ with $\bar{P}$ acting nontrivially on $O_2(\bar{N})$. Thus

$O_2(\overline{N})\overline{P}\langle\overline{u}\rangle \cong S_4$. This in turn implies that a Sylow 2-subgroup of $\overline{N}$ is isomorphic to D_8. Let V be a Sylow 2-subgroup of N containing $T\langle y\rangle$ and let R be the subgroup of index 2 in V such that $\overline{R} = O_2(\overline{N})$. Without loss, we can suppose that $V \subseteq S$ and that P normalizes R.

Let $r \in R$ be such that $\langle\overline{r}\rangle = Z(\overline{V})$. Then $C_{R_1}(r)$ is invariant under u as u normalizes R_1 and $\overline{u}$ centralizes $\overline{r}$. This forces $C_{R_1}(r) = 1$, otherwise, as in Lemma 8.2, $N_G(C_{R_1}(\langle r,u\rangle))$ would be a nonsolvable 2-local subgroup of G containing $E\langle u,r\rangle = T\langle r\rangle \supset T$, contrary to Lemma 5.1. It follows that $E\langle r\rangle \cong Z_2 \times Z_2 \int Z_2$.

As in Lemma 8.2, we have two cases to consider.

<u>Case 1</u>. R <u>splits over</u> E. Since $|Z(R)| \geq 4$, we conclude, by Lemma 2.2(iv) of Part II, that R is of type $L_3(4)$. Let E_1 be an elementary abelian subgroup of order 16 of R with $E_1 \neq E$. By the action of P on R, we have $E_1 = [R,P]$. Hence if we set $\tilde{N} = N/O(N)$, it follows that $\tilde{E}_1 \triangleleft \tilde{N}$. Moreover, $\tilde{P}$ acts fixed-point-free on $\tilde{E}_1$ and centralizes $\tilde{R}_1 \cong Z_2 \times Z_2$. We conclude now by the Frattini argument that $\tilde{N}$ is a split extension of $\tilde{E}_1$ by $N_{\tilde{N}}(\tilde{P})$. Since $\tilde{P}$ acts fixed-point-free on $\tilde{E}_1$, Lemma 2.2(viii) of Part II yields that $\tilde{V}$ and hence V is of type $\hat{A}_8$. Proposition 4.2 now yields that $S \supset V$. As V contains exactly two elementary abelian subgroups of order 16, $|N_S(V):V| = 2$. It is now easy to show, as in Case 1 of Lemma 8.2, that E and E_1 are the only two elementary abelian subgroups of order 16 in $N_S(V)$, that R is characteristic in $N_S(V)$, that $S = N_S(V)$, and that $SCN_3(S)$ is empty. But then as $Z(R) \in U(S)$ and all involutions of $Z(R)$ are conjugate, Proposition 4.3 again yields a contradiction.

<u>Case 2</u>. R <u>does not split over</u> E. As in Case 2 of Lemma 8.2, we

conclude that E is the unique elementary abelian subgroup of R of order 16 and that $\bar{N} = N/C_G(E) = \overline{VP} \cong S_4$. Likewise if P_1 denotes a cyclic 3-subgroup of $N_G(R)$ which maps on $\bar{P}$, it again follows that $R_2 = Z(R)$, that $|P_1:P_0| = 3$, where $P_0 = C_{P_1}(R)$, and that $W = [R,P_1] \cong Z_4 \times Z_4$. Thus $R = WR_1$ with $W \cap R_1 = 1$ and R has the structure of S in Case 2 of Lemma 8.2. Furthermore, we can choose P_1 to be inverted by the involution u. Hence $P_1R_1\langle u\rangle/P_0$ acts as a group of automorphisms of W. Thus the structure of $V = WR_1\langle u\rangle$ is uniquely determined by Lemma 2.1(v) of Part II, and so can be expressed in the form $V = \langle a,b,c,d,e\rangle$, where a,b,c,d,e satisfy the same relations as in Case 2 of Lemma 8.2. In particular, $W = \langle a,b\rangle$, $R_1 = \langle c,d\rangle$, $R = \langle a,b,c,d\rangle$, $\Omega_1(W) = R_2$, and $c = x$.

But now we check directly from the defining relations for V that E is the unique elementary abelian subgroup of V of order 16. Since $V = N_S(E)$, we therefore also conclude that $S = V$.

We have $C_S(d) = E$. We claim that E is a Sylow 2-subgroup of $C_G(d)$. Indeed, if not, then $|D:E| = 2$ for some subgroup D of $C_G(d)$ containing E. But then D normalizes E and so $D \subseteq N$. Setting $\tilde{N} = N/O(N)$, we have that $\tilde{D} \subseteq C_{\tilde{N}}(\tilde{d})$. However, it is immediate from the structure of $\tilde{N}$ that $C_{\tilde{N}}(\tilde{d}) = \tilde{E}\tilde{P}_1$. Since $|\tilde{D}:\tilde{E}| = 2$, this is a contradiction and establishes our assertion. Since $E = C_S(d)$ is a Sylow 2-subgroup of $C_G(d)$, it also follows that d is extremal in S.

Now we apply Thompson's transfer lemma to d and the maximal subgroup $S_1 = \langle a,b,de\rangle$ of S. Since $S_1/W \cong Z_4$ and $\langle c\rangle$ maps on $\Omega_1(S_1/W)$, it follows that $d \sim f$ in G for some involution f of $W\langle c\rangle = \langle a,b,c\rangle$. Since $\Omega_1(W) = R_2$ consists of central involutions, while $d \in R_1$ is noncentral, we have, in fact, $f \in \langle a,b,c\rangle - \langle a,b\rangle$. We may therefore assume that $f = c$ or ca. But in the first case, $f \in E$ (as $c = x$) and we conclude from the uniqueness of E that $d \sim f = c$ in $N = N_G(E)$, which is impossible by the structure of N. Thus we must have $f = ca$.

Since d is extremal in S, it follows now that $C_S(ca)^g \subseteq E = C_S(d)$ and $(ca)^g = d$ for some g in G. But $R_2 = \langle a^2,b^2\rangle \subseteq C_S(ca)$ and so $R_2^g \subseteq R_2$. On the other hand, considering the action of P on $E = R_1 \times R_2$, we see that the involutions of $E - R_2$ are all noncentral. Since the involutions of R_2 are central, we conclude that $R_2^g = R_2$.

Thus $g \in K = N_G(R_2)$. Since $R_2 = Z(R) \lhd S$, also $S \subseteq K$. Hence if we set $H = C_G(R_2)$, we have that $S \cap H = R$ is a Sylow 2-subgroup of C. We argue that H has a normal 2-complement. Indeed, $N \cap H = N_H(E) = O(N)R$ by the structure of N and so $N_H(E)$ has a normal 2-complement. However, it is immediate that R admits no nontrivial sutomorphisms of odd order which act trivially on E. Since E is characteristic in R, it follows now that also $N_H(R)$ has a normal 2-complement. Hence if we set $\tilde{H} = H/R_2$, we conclude that $N_{\tilde{H}}(\tilde{R})$ does as well. But $\tilde{R}$ is abelian and now Burnside's theorem yields that $\tilde{H}$ has a normal 2-complement, whence H does as well.

Finally our argument yields that $H = O(H)R$ and that $O(H) = O(K)$. Since E is characteristic in R, we thus have $K = O(K)N_K(E)$ by the Frattini argument. Hence N covers $K/O(K)$ and so the fusion $d \sim ca$ must occur in N, which is clearly not the case. This contradiction completes the proof of Proposition 8.1.

9. The structure of $O^2(\widetilde{M})$. By the preceding results, we are reduced to the case that $\overline{L} \cong L_2(q)$, q odd, or A_7 and $\overline{C} = \overline{R}_1 = \langle\overline{x}\rangle$. Thus $\widetilde{M} = \overline{M}/\overline{C} = \overline{M}\ \langle\overline{x}\rangle$ is correspondingly isomorphic to a subgroup of $P\Gamma L(2,q)$ or S_7. We continue the notation of the preceding sections.

In this section we shall prove

PROPOSITION 9.1. We have $O^{2'}(\widetilde{M}) \cong L_2(q)$, $PGL(2,q)$, $PGL^*(2,q)$, q odd, or A_7.

We assume false and proceed in a sequence of lemmas to derive a contradiction. Suppose first that $\overline{L} \cong L_2(q)$, where $q = p^k$, p a prime. Then $\widetilde{M}$ is isomorphic to a subgroup of $\mathrm{Aut}(\widetilde{L}) \cong P\Gamma L(2,q)$. We know that $\mathrm{Aut}(\widetilde{L})/\mathrm{Inn}(\widetilde{L})$ has the form $A_1 \times A_2$, where A_1 has order 2 with the inverse image of A_1 isomorphic to $PGL(2,q)$ and where A_2 is a cyclic group of order k whose elements are induced from the Galois group of $GF(q)$. It follows from this (and our assumption that Proposition 9.1 is false) that $O^{2'}(\widetilde{M})$ has one of the following two forms:

(a) $O^{2'}(\widetilde{M}) = \widetilde{K}\widetilde{W}$, where $\widetilde{K} \supseteq \widetilde{L}$, $\widetilde{K} \cong L_2(q)$ or $PGL(2,q)$, $\widetilde{K} \triangleleft \widetilde{M}$, $\widetilde{K} \cap \widetilde{W} = 1$, and $\widetilde{W}$ is a nontrivial cyclic 2-group whose elements are induced from the Galois group of $GF(q)$; or

(b) $O^{2'}(\widetilde{M}) = \widetilde{L}\langle\widetilde{a}\rangle$, where $\widetilde{a} = \widetilde{a}_1\widetilde{a}_2$ and $\widetilde{a}_1, \widetilde{a}_2$ map, respectively, on a generator of A_1 and a 2-element of A_2 of order at least 4. In particular, $O^2(\widetilde{M})/\widetilde{L}$ is cyclic of order at least 4 and $O^{2'}(\widetilde{M})$ contains a subgroup of index 2 of the form $\widetilde{L}\widetilde{W}$, where $\widetilde{W}$ is a nontrivial cyclic 2-group whose elements are induced from the Galois group of $GF(q)$. Moreover, 4 divides k. For uniformity, we set $\widetilde{K} = \widetilde{L}$ in this case.

In view of Lemma 2.2 of Part IV, we can assume without loss in both cases that $\widetilde{W} \subseteq \widetilde{T}$, that $\Omega_1(\widetilde{W})$ centralizes $\widetilde{R}_2$, and if $\widetilde{W} = \langle\widetilde{w}\rangle$, that

$C_{\tilde{R}_2}(\tilde{w}^i)$ is a Sylow 2-subgroup of $C_{\tilde{L}}(\tilde{w}^i)$ for all i. Likewise in case (b), if we consider the extension of $O^{2'}(\tilde{M})$ by $\tilde{a}_2$, we can suppose that $C_{\tilde{R}_2}(\tilde{a}_2)$ is a Sylow 2-subgroup of $C_{\tilde{L}}(\tilde{a}_2)$. But now it is easy to check in both cases that any element of $\tilde{T} - \tilde{R}_2\tilde{W}$ (if such exists) interchanges the two conjugacy classes of four groups in R_2.

On the other hand, if $\bar{L} \cong A_7$, we must have $O^{2'}(\bar{M}) \cong S_7$ as otherwise the proposition would hold. We can thus write $\tilde{M} = \tilde{L}\tilde{W}$, where $\tilde{W}$ has order 2 and is generated by an element which corresponds to a transposition of S_7. Again without loss we can assume that $\tilde{W} \subseteq \tilde{T}$ and that $\Omega_1(\tilde{W}) = \tilde{W}$ centralizes $\tilde{R}_2$. Moreover, by our choice of $\tilde{W}$, we have $C_{\tilde{L}}(\tilde{W}) \cong S_5$. Again for uniformity, we set $\bar{K} = \bar{L}$ and put this case under case (a).

We let $|\tilde{W}| = 2^m$. Then there exists an element $w \in T$ such that $w^{2^m} \in R_1$ and $\langle\tilde{w}\rangle = \tilde{W}$. We set $u = w^{2^{m-1}}$ and $W = \langle w,x\rangle$. Then $\langle\tilde{u}\rangle = \Omega_1(\tilde{W})$ and $u^2 \in R_1$, so that $u^2 = 1$ or x. Furthermore, $W/\langle x\rangle \cong \tilde{W}$ and as W is cyclic, W is abelian, whence $W \cong Z_{2^{m+1}}$ or $Z_{2^m} \times Z_2$. Setting $W_1 = \langle u,x\rangle$, we also have that W_1 centralizes R_2 and so $R_2W_1 \cong D_{2^n} \times Z_4$ or $D_{2^n} \times Z_2 \times Z_2$ according as $u^2 = x$ or 1. Moreover, as $W \neq 1$, we have that $n \geq 3$, whence $Z(R_2) \cong Z_2$. We set $\langle z\rangle = Z(R_2)$. Note also that in case (b), we have $n \geq 4$.

If $\bar{L} \cong PGL(2,q)$, then $\bar{T} \cap \bar{K} \cong D_{2^{n+1}}$ and we let t be an element of T such that $t^2 \in R_1$ and $\tilde{R}_2\langle\tilde{t}\rangle \cong D_{2^{n+1}}$. Again for uniformity, we set t = 1 if either $\bar{L} \cong L_2(q)$ and $\bar{K} = \bar{L}$ or if $\bar{L} \cong A_7$. Then either $T = R_2W\langle t\rangle$ and $R_2\langle x,t\rangle \triangleleft T$ or case (b) holds. Moreover, in the latter case, R_2W has index 2 in T. Furthermore, if t = 1, we check in each case that $T/R_2\langle x\rangle$ is cyclic and that $\Omega_1(T) = R_2\langle x\rangle$ or R_2W_1 according as $u^2 = x$ or 1. In the first case, the image of W_1 in $T/\Omega_1(T)$ is equal to $\Omega_1(T/\Omega_1(T))$ and so is characteristic in $T/\Omega_1(T)$. It follows therefore in either case that

$R_2W_1 = R_2\langle u,x\rangle$ is characteristic in T. Thus this is true whenever $t = 1$. Note also that when $t = 1$, $C_T(u) = R_2W$, so that either $u \in Z(T)$ or case (b) holds. On the other hand, if $t \neq 1$, we can assert only that $R_2W_1 \triangleleft T$. Finally we observe that any element of $T - R_2W$ interchanges the two conjugacy classes of four subgroups of R_2. In particular, $C_T(Y) \subseteq R_2W$ for any four subgroup Y of R_2 in all cases. These various properties of T will be used in our arguments.

We first prove

Lemma 9.2. The element u is an involution.

Proof. Suppose false, in which case $u^2 = x$. Consider first the case $t = 1$, whence $R = R_2W_1 = R_2\langle u\rangle$ is characteristic in T. Since $R \cong D_{2^n} \times Z_4$ in this case, it follows that $Z(R) = \langle z,u\rangle \cong Z_2 \times Z_4$. But then $\langle x\rangle = \mho^1(Z(R))$ is characteristic in R and hence in T. Since y normalizes T, this forces $x^y = x$, contrary to the fact that T is a Sylow 2-subgroup of $C_G(x)$ and $y \in S - T$. We thus conclude that $t \neq 1$, whence $\bar{L} \cong L_2(q)$ and case (a) holds with $\tilde{K} \supset \tilde{L}$.

By the structure of $\tilde{T}$, we have $\Omega_1(\tilde{T}) \subseteq \tilde{R}_2\langle\tilde{u}\rangle \cup \tilde{R}_2\langle\tilde{t}\rangle$. Since $u^2 = x$, it follows that $\Omega_1(T) \subseteq R_2\langle x,t\rangle$. Suppose $t^2 = x$, in which case $\Omega_1(T) = R_2 \times \langle x\rangle$. Then every elementary subgroup A of T of order 8 lies in $R_2 \times \langle x\rangle$ and considering the action of t on $R_2 \times \langle x\rangle$, we see that any elementary subgroup of T of order 8 is conjugate in T to A. Hence replacing y by yt' for suitable t' in T, we can assume without loss that y normalizes A. Since w centralizes a four subgroup of R_2 and since A was arbitrary, we can also assume without loss that $A \subseteq C_T(w)$, whence $C_T(A) = AW$. But y normalizes both T and A, and so y normalizes $C_T(A) = AW$. However, $AW = A\langle w\rangle$ is abelian with x the unique involution of $\mho^1(AW)$. Again $x^y = x$, giving the same contradiction. We thus conclude that $t^2 = 1$. Hence $\Omega_1(T) = \langle x\rangle \times R_2\langle t\rangle \cong Z_2 \times D_{2^{n+1}}$. Furthermore, if

A is as above, we reach the same contradiction if $A^{yt'} = A$ for some t' in T. Since $Z_2 \times D_{2^{n+1}}$ has precisely two conjugacy classes of E_8's, we see that all elementary subgroups of T of order 8 are conjugate in $T\langle y\rangle$ and that y normalizes no such subgroup of T.

We shall argue next that $S = T\langle y\rangle$. Indeed, let $U \in U(S)$. Since $u^2 = x$, x centralizes U and so $U \subseteq T$, whence $U \subseteq \Omega_1(T) \cong Z_2 \times D_{2^{n+1}}$. Since $n + 1 \geq 4$, $D_{2^{n+1}}$ contains no normal four subgroups and consequently $U = Z(\Omega_1(T)) = \langle x,z\rangle$ is the unique normal four subgroup of $\Omega_1(T)$. Thus $x \in U$ and so $|S:C_S(x)| = |S:T| \leq 2$, whence $S = T\langle y\rangle$, as asserted.

Observe next that y interchanges x and xz as y normalizes $U = \langle z,x\rangle$ and does not centralize x. Hence if y centralizes an involution $t' \neq z$ of T, then $t' \notin U$, $U\langle t'\rangle \cong E_8$, and y normalizes $U\langle t'\rangle$. However, we have shown above that y normalizes no elementary subgroup of T of order 8. Thus z is the unique involution of $C_T(y)$ and so $C_T(y)$ is cyclic or generalized quaternion.

We claim next that G has exactly two conjugacy classes of involutions. First of all, Thompson's transfer lemma implies that any involution of $S - T$ is conjugate to one in T, so we need only show that every involution of T is conjugate in G to z or x. By the structure of M, all involutions of $R_2 \times \langle x\rangle$ are conjugate in M to z, x, or xz. Since $x^y = xz$, all involutions of $R_2 \times \langle x\rangle$ are thus conjugate in G to z or x. Furthermore, if y normalized R_2, then considering the action of t on R_2, it would follow that yt' would normalize a four subgroup of R_2 for some t' in T, whence yt' would normalize an elementary subgroup of T of order 8. But then if we replaced y by

yt', this would contradict what we have shown above. Thus y does not normalize R_2. Since all involutions of R_2 are conjugate in M to z and neither x nor xz is a central involution, it follows that $r^y \in R_2\langle x,t\rangle - R_2\langle x\rangle$ for some involution r of R_2. Without loss we can assume that $t = r^y$, in which case all involutions of $R_2\langle t\rangle$ are conjugate in G to z. The only involutions of T not yet accounted for lie in the coset R_2tx and these are conjugate in T to either tx or ztx. Since $t = r^y$ and y interchanges x and zx, these are therefore all conjugate to rx or rzx and so are all conjugate in G to x. This establishes our assertion.

Suppose first that y can be taken to be an involution. Then by Thompson's transfer lemma, y is conjugate in G to x or z. Since x and z are each extremal in S, it follows that for some g in G, we have $y^g = x$ or z and correspondingly $C_S(y)^g \subsetneq T$ or S. Since S is not of maximal class, $|C_T(y)| \geq 4$. But $C_T(y)$ is cyclic or generalized quaternion with z as its unique involution. Hence $z = v^2$ for some v in $C_T(y)$. Then $v^g \in T$ or S, as the case may be. Suppose $v^g \in S - T$. Then just as with y, it follows that also $C_T(v^g)$ is cyclic or generalized quaternion with z as its unique involution. Since $z^g = (v^g)^2$ and $|S:T| = 2$, clearly $z^g \in C_T(v^g)$ and so $z^g = z$. On the other hand, if $v^g \in T$, then $z^g \in \Phi(T)$. However, we check easily from the structure of $T = \langle R_2,t,w\rangle$ that $\Phi(T)$ is abelian on two generators with $\Omega_1(\Phi(T)) = \langle z,x\rangle$. Since z is the only central involution of $\langle z,x\rangle$, we conclude in this case as well that $z^g = z$. Thus $y^g = x$ and $g \in C_G(z)$.

We now examine the structure of $K = C_G(z)$. Since G is an $\mathfrak{R}_3$-group, K is solvable. As usual, we set $\bar{K} = K/O(K)$ and $\tilde{K} = \bar{K}/\langle\bar{z}\rangle$.

Our conditions imply that $\bar{K} \supset \bar{S}$ and $O(\tilde{K}) = 1$. Set $Q = S \cap O_{2',2}(K)$, so that $\bar{Q} = O_2(\bar{K})$ and $\tilde{Q} = O_2(\tilde{K})$. Since $\tilde{x} \in Z(\tilde{S})$ and $\tilde{K}$ is solvable, $\tilde{x} \in \tilde{Q}$ and so $x \in Q$. Since $\bar{y}^{\bar{g}} = \bar{x}$ and $\bar{y}$ normalizes $\bar{Q}$, this implies that also $y \in Q$ and that $C_Q(y) \cong C_Q(x)$. Set $Q_1 = T \cap Q$. Since $S = T\langle y\rangle$, we have $Q = Q_1\langle y\rangle$. Since x centralizes Q_1 and does not centralize y, it follows that $|Q:C_Q(x)| = 2$, whence $|Q:C_Q(y)| = 2$. Thus $Q = C_Q(y)\langle x\rangle = C_{Q_1}(y)\langle y,x\rangle$. Since z is the unique involution of $C_{Q_1}(y)$ and $\langle y,x\rangle \cong D_8$ with $\langle z\rangle$ as its center, we see that $Q \cong C_{Q_1}(y) * D_8$ with $C_{Q_1}(y)$ cyclic or generalized quaternion. In particular, this implies that $|\bar{K}|$ is not divisible by 9 or 7.

Suppose 5 divides $|\bar{K}|$, in which case $C_{Q_1}(y) \cong Q_8$. Since K is solvable and $\mathrm{Aut}(Q_8*D_8)/\mathrm{Inn}(Q_8*D_8) \cong S_5$, S/Q is cyclic of order at most 4 and $|S| \leq 2^7$. This forces $T = \langle R_2,t,u\rangle$ with $n = 3$ and $|T| = 2^6$ (whence also $|S| = 2^7$). In particular, $T' \cong Z_4$ and $T' \subseteq R_2$. Thus y normalizes T' and as $t = r^y$ with $R_2 = T'\langle r\rangle$, it follows that y normalizes $R_2\langle t\rangle = \langle r,t\rangle$ and interchanges the generators r,t. Thus y inverts rt and we conclude that $R_2\langle t,y\rangle$ is dihedral of order $2^{n+2} = 2^5$. Hence $R_2\langle t,y\rangle/\langle z\rangle \cong D_{16}$. On the other hand, $\tilde{S}$ has order 2^6 and so is isomorphic to a split extension of E_{16} by Z_4. But then $\tilde{S}$ does not contain a subgroup isomorphic to D_{16}. This is a contradiction and shows that 5 does not divide $|\bar{K}|$.

Finally we have in any event that $Q = C_{Q_1}(y) * \langle y,x\rangle$ has 2-rank 2 as $C_{Q_1}(y)$ is cyclic or generalized quaternion. Since $m(S) \geq 3$, this forces $S \supset Q$ and so the only possibility is that $\bar{K}/\bar{Q} \cong S_3$, whence

$|S:Q| = 2$. Thus $|S:C_S(y)| \leq |S:C_{Q_1}(y)| \leq 4$. On the other hand, we again have that T' is cyclic of index 2 in R_2 and that y normalizes $R_2\langle t\rangle$ with $R_2\langle t,y\rangle$ being dihedral of order $2^{n+2} \geq 2^5$. But then certainly $|S:C_S(y)| \geq |R_2\langle t\rangle:C_{R_2\langle t\rangle}(y)| \geq 8$, a contradiction. We have therefore proved that y is not an involution.

To **treat** this case, we choose y in $S - T$ with $|y|$ minimal, and subject to this condition so that $|C_S(y)|$ is maximal. By [27, Lemma 16], y is conjugate to a in G for some a in T. In view of our choice of y, we can assume without loss that y or a is extremal in S. We have that $C_S(y) = \langle y\rangle * C_T(y)$ with z the unique involution of $Z(C_S(y))$ and $m(C_S(y)) = 2$. If y is extremal, then $a^g = y$ with $C_S(a)^g \subseteq C_S(y)$ for some g in G. Since $\langle x,z\rangle \subseteq C_S(a)$ and z is the only central involution of $\langle x,z\rangle$, we conclude at once that $z^g = z$. Thus $y \sim a$ in $C_G(z)$ in this case. In the contrary case, we have $y^g = a$ and $C_S(y)^g \subseteq C_S(a) \subseteq S$ for some g in G. As in the case that y was an involution, we have that $z = v^2$ for some v in $C_T(y)$ and it follows exactly as in that case that $z^g = z$. Thus $y \sim a$ in $C_G(z)$ in this case as well.

Finally let K, $\bar{K}$, $\tilde{K}$, and Q have the same meanings as before, so that $y \sim a$ in K and again $\bar{K} \supset \bar{S}$ and $x \in Q$. If $\langle\bar{x},\bar{z}\rangle$ is normal in $\bar{K}$, then $C_{\bar{K}}(\langle\bar{x},\bar{z}\rangle)$ is normal in $\bar{K}$ of index 2 with Sylow 2-subgroup $\bar{T}$. Clearly then the conjugacy of $\bar{y}$ and $\bar{a}$ in $\bar{K}$ is impossible. Thus $\langle\bar{x},\bar{z}\rangle$ is not normal in K. This implies that x is conjugate in K to an involution x' of $Q - \langle z,x\rangle$. In the present case $\Omega_1(S) \subseteq T$ and so $x' \in \Omega_1(T)$. But x' is a noncentral involution and so $x' \in R_2\langle t,x\rangle - R_2\langle t\rangle$.

Thus $x' = xt'$ with $t \in R_2\langle t\rangle$ and $t' \neq z$. Then $t' \in Q$ and so also $(t')^y \in Q$ as $Q \lhd S$. But as $R_2\langle t,y\rangle$ is dihedral, we see that $\langle t',(t')^y\rangle = R_2\langle t\rangle$, whence $R_2\langle t\rangle \subseteq Q$. We have $R_2\langle t\rangle \cong D_{2^{n+1}}$ with $n \geq 3$ and t' a noncentral involution of $R_2\langle t\rangle$. Since x centralizes $R_2\langle t\rangle$, we therefore conclude that $|Q:C_Q(x')| \geq 4$. On the other hand, $C_Q(x') \cong C_Q(x)$ as $\bar{x}$ and $\bar{x}'$ are conjugate in $\bar{K}$. But $|Q:C_Q(x)| \leq 2$ as $\langle z,x\rangle \lhd S$, so $|Q:C_Q(x')| \leq 2$, a contradiction. This completes the proof of the lemma.

In the balance of this section we analyze the case that $u^2 = 1$. In this case $W = \langle w\rangle \times \langle x\rangle$ with $W_1 = \langle u,x\rangle \cong Z_2 \times Z_2$.

We first prove

Lemma 9.3. The following conditions hold:

(i) Case (a) holds;

(ii) If $t = 1$, then $w = u$ and $\bar{L} \cong L_2(9)$.

Proof. Note that in case (b), we have $t = 1$ and $q = p^k$ with k divisible by 4. Hence certainly $\bar{L} \not\cong L_2(9)$ in this case. We see then that (ii) will imply (i).

Suppose then that $t = 1$, in which case $R = R_2W_1 = R_2\langle u,x\rangle$ is characteristic in T and $Z(R) = \langle z,x,u\rangle = E_8$. Then y normalizes $Z(R)$ and so y must therefore centralize some involution u_1 of $\langle x,u\rangle$. Since y does not centralize x, we have $u_1 = u$ or ux. But then $\tilde{u}_1 = \tilde{u}$ and so $C_{\bar{L}}(\bar{u}_1) \cong PGL(2,q_0)$ with $q_0^2 = q$ if $\bar{L} \cong L_2(q)$ and $C_{\bar{L}}(\bar{u}_1) \cong S_5$ if $\bar{L} \cong A_7$. Hence except in the first case with $q_0 = 3$, $C_{\bar{L}}(\bar{u}_1)$ is nonsolvable. Since $C_M(u_1)$ covers $C_{\bar{L}}(\bar{u}_1)$, it follows that also $C_G(u_1)$ is nonsolvable. Suppose first that case (b) does not hold, in which case $u_1 \in Z(T)$. But then $C_G(u_1) \supseteq T\langle y\rangle \supset T$ and we see that Lemma 5.1 is contradicted.

Hence if case (b) does not hold, we must have $q_0 = 3$ and $\bar{L} \cong L_2(9)$. This in turn forces $|\tilde{W}| = 2$, whence $w = u$. Thus all the conditions of (ii) are satisfied in this case.

Assume now that case (b) holds. Then $q_0 = p^{k_0}$, where $k_0 = \frac{1}{2}k$ is even. In particular, $C_{\bar{L}}(\bar{u}_1)$ is nonsolvable. This time we have $C_T(u_1) = R_2W$ is of index 2 in T and $T_1 = R_2W\langle y\rangle$ is a 2-subgroup of $C_G(u_1)$ with $|T_1| \geq |T|$. These conditions force $|T_1| = |T|$ and T_1 to be a Sylow 2-subgroup of $C_G(u_1)$, otherwise Lemma 5.1 again yields a contradiction.

The proof in this case requires a more subtle argument. First of all, it is immediate from the structure of T that $\Omega_1(T) = R_2 \times \langle u,x\rangle$ and that some element of $T - R_2W$ interchanges the two conjugacy classes of four subgroups of R_2. Hence T has only one conjugacy class of E_{16}'s and so, replacing y by ty' for suitable t' in T, we can assume without loss that y normalizes some subgroup A of T with $A \cong E_{16}$.

Next set $M_1 = C_G(u_1)$ and $\bar{M}_1 = M_1/O(M_1)$. Also let L_0 be a minimal R_2-invariant subgroup of M which maps onto $C_{\bar{L}}(\bar{u}_1)'$. Then L_0 is perfect, $R_2 \cap L_0$ is a Sylow 2-subgroup of L_0, $L_0/O(L_0) \cong L_2(q_0)$, and $L_0 \subseteq M_1$. Hence by Proposition 2.1 of Part III, $\bar{L}_0 \subseteq \bar{L}_1$, where $\bar{L}_1 = L(\bar{M}_1)$.

Since M_1 is nonsolvable and the Sylow 2-subgroup T_1 of M_1 has the same order as T, the results of Sections 5 and 6 apply to M_1 as well as to M. In particular, the structure of $\bar{L}_1$ and $\bar{C}_1 = C_{\bar{M}_1}(\bar{L}_1)$ is determined by Proposition 6.1. Since $\bar{L}_0 \subseteq C_{\bar{L}}(\bar{x})$ and $\bar{L}_0/O(\bar{L}_0) \cong L_2(q_0)$ with q_0 an odd square, clearly $\bar{L}_1 \not\cong A_7, L_3(3), M_{11}, U_3(3)$, or $L_2(q_1)$ with $q_1 \equiv 3,5 \pmod 8$. Hence the only possibility is that (I) of Proposition 6.1 holds with $\bar{L}_1 \cong L_2(q_1)$ for some $q_1 \not\equiv 3,5 \pmod 8$. In particular, it follows that $\bar{C}_1 = \langle\bar{u}_1\rangle$. This means that $\bar{x} \notin \bar{C}_1$ and so $\bar{x}$ does not centralize $\overline{L}_1$. We conclude at once that $q_1 = q_0^{\ 2}$ and that $\overline{x}$ induces a field automorphism of $\overline{L}_1$ of order 2 with $\overline{L}_0 = C_{\overline{L}_1}(\overline{x})'$. Thus $q_1 = q$. Furthermore, if we set $\overline{R}_2^* = \overline{T}_1 \cap \overline{L}_1$, then $\overline{R}_2^*$ is a Sylow 2-subgroup of $\overline{L}_1$ and $|\overline{R}_2^*:\overline{R}_2 \cap \overline{L}_0| = 2$. Let r be an involution of $R_2 - (R_2 \cap L_0)$. Since $<r,x>$ is a four group not containing u_1, it follows from Lemma 6.2(iv) of Part III, applied to $\overline{M}_1/<\overline{u}_1>$, that $\overline{r}^* \in \overline{R}_2^*$ for some element $\overline{r}^* \in <\overline{r},\overline{x},\overline{u}_1> - <\overline{x},\overline{u}_1>$. In particular, x centralizes R_2^* and $\overline{R_2^*W_1} = \overline{R}_2\overline{W}_1$.

Finally define w^*,u^*,W^*, and W_1^* analogously to w,u,W, and W_1 and let R_2^* be the inverse image of $\overline{R}_2^*$ in T_1. Then $W_1^* = <u^*,u_1>$, $|T_1:R_2^*W^*| \leq 2$, and any element of $T_1 - R_2^*W^*$ interchanges the two conjugacy classes of four subgroups of R_2^*. Note also that $R_2^* \cong D_{2^n} \cong R_2$ and that $n \geq 4$. Since $|R_2^*:R_2 \cap L_0| = 2$, we see that $<z> = Z(R_2^* \cap L_0)$ and hence that $<z> = Z(R_2^*)$. But now Lemma 2.2 (iv) and (viii) of Part IV, applied to $\overline{M}_1/<u_1>$, yields that $\Omega_1(C_{T_1}(R_2^*)) \subseteq \Omega_1(<z>W^*) \subseteq <z>W_1^*$. But x is an involution of T_1 which centralizes R_2^* by the preceding paragraph and so $x \in <z>W_1^*$. Since $x \notin <z,u_1>$, we conclude that $<z>W_1^* = <z,x,u_1> = <z>W_1$. Thus $R_2^*W_1^* = R_2^*W_1$, whence $R_2^*W_1^* = R_2W_1$, again by the preceding paragraph. Furthermore, as W^* centralizes both z and W_1^*, W^* centralizes x and consequently $R_2^*W^* \subseteq C_{T_1}(x)$.

But now as $R_2^*W_1^* \cong D_{2^n} \times Z_2 \times Z_2$ is clearly normal in T_1, our argument yields that no element of $T_1 - R_2^*W^*$ leaves invariant any elementary subgroup of $R_2^*W_1^*$ of order 16. However, $y \in T_1$ and y leaves A invariant by our choice of y. Since $A \cong E_{16}$ and $A \subseteq R_2W_1 = R_2^*W_1^*$, this forces y to lie in $R_2^*W^*$, whence y centralizes x, a contradiction. This completes the proof of the lemma.

As an immediate corollary, we have

Lemma 9.4. $\tilde{L}$ is not isomorphic to A_7.

Thus $\overline{L} \cong L_2(q)$ and case (a) holds. The proof in this case is quite delicate. We first argue that also $w = u$ when $t \neq 1$. However, we need a preliminary result.

<u>Lemma 9.5</u>. If $t \neq 1$, then all elementary abelian subgroups of T of order 16 are conjugate in T and lie in $R_2\langle x,u\rangle$.

<u>Proof</u>: If A is such a subgroup of T, then $x \in A$ as $m(T) \leq 4$. Thus $\tilde{A} \cong E_8$. Since $\bar{L} \cong L_2(q)$, q odd, and $\tilde{R}_2\langle\tilde{t}\rangle \cong D_{2^{n+1}}$, $\tilde{R}_2\langle\tilde{t}\tilde{u}\rangle \cong QD_{2^{n+1}}$, we have that $\tilde{A} \subseteq \tilde{R}_2\langle\tilde{u}\rangle$ and any other elementary abelian subgroup of $\tilde{T}$ of order 8 is conjugate in $\tilde{T}$ to $\tilde{A}$. From this, the lemma follows.

We can now prove

<u>Lemma 9.6</u>. We have $w = u$.

<u>Proof</u>: Suppose false. $C_{R_2}(w)$ contains a four subgroup B. We set $A = B\langle x,u\rangle$, so that $A \cong E_{16}$. By Lemma 9.3, we can suppose $t \neq 1$. Therefore by the preceding lemma any elementary abelian subgroup of T of order 16 is conjugate to A in T. Hence, as usual, we can assume without loss that y normalizes A. By the structure of T, we have that $C_T(A) = A\langle w\rangle \cong E_8 \times Z_{2^m}$. Since $m \geq 2$, this implies that y normalizes $\langle w^2\rangle$ and so y centralizes u.

Hence if we set $M_1 = C_G(u)$, we have that $T_1 = C_T(u)\langle y\rangle \subseteq M_1$. Since u does not centralize $t \in T$ and u centralizes $R_2\langle x,w\rangle$, we have that $|T:C_T(u)| = 2$ and hence that $|T_1| = T$. Since $m \geq 2$, we also have that $q > 9$, which implies that $C_{\bar{L}}(\bar{u})$ is nonsolvable. Thus M_1 is nonsolvable and as $T_1 \subseteq M_1$, a Sylow 2-subgroup of M_1 has order at least that of T. Our maximal choice of x and T now forces T_1 to be a Sylow 2-subgroup of M_1. Furthermore, as $|T_1| = |T|$, all the results we have established for M hold equally well for M_1.

Hence if we set $\bar{M}_1 = M_1/O(M_1)$, we have that $\bar{M}_1$ possesses a normal subgroup $\bar{L}_1 \cong L_2(q_1)$, q_1 odd, $q_1 \geq 5$, or A_7 and that $C_{\bar{M}_1}(\bar{L}_1) = \langle\bar{u}\rangle$. Setting $\tilde{M}_1 = \bar{M}_1/\langle\bar{u}\rangle$, it follows also that $\tilde{A}\langle\tilde{w}\rangle \cong E_8 \times Z_{2^{m-1}}$. In particular, $m(\tilde{M}_1) \geq 4$. However, as $\tilde{M}_1$ is isomorphic to a subgroup of $P\Gamma L(2,q_1)$, q_1 odd, or S_7, we have $m(\tilde{M}_1) \leq 3$. This contradiction establishes the lemma.

We have seen in the preceding lemma that when $t \neq 1$, we can choose y to normalize an elementary abelian subgroup of T of order 16. We prove a similar result when $t = 1$.

<u>Lemma 9.7</u>. If $t = 1$, then we can choose y to normalize each elementary abelian subgroup of T of order 16.

<u>Proof</u>: By Lemma 9.3, $\bar{L} \cong L_2(9)$ and so $R_2 = \langle a_1, a_2\rangle \cong D_8$ with $a_1^2 = a_2^2 = 1$ and $(a_1a_2)^2 = z$. Moreover, $T \cong D_8 \times Z_2 \times Z_2$ and $Z(T) = \langle z,u,x\rangle \cong E_8$. As in Lemma 9.3, y centralizes u or ux. Replacing u by ux, if necessary, we can assume without loss that y centralizes u. We also set $B_i = \langle a_i, z\rangle$ and $A_i = B_iZ(T)$, $i = 1,2$, so that A_1, A_2 are the only elementary abelian subgroups of T of order 16.

Suppose, by way of contradiction, that y does not normalize A_1 and A_2, whence $A_1^y = A_2$ and $A_2^y = A_1$. Consider first the case that $B_1^y = B_2$. Since $y \in H = C_G(u)$, B_1 and B_2 are thus conjugate in H. But as $C_{\bar{L}}(\bar{u}) \cong PGL(2,3)$ and contains $\bar{R}_2 = \bar{B}_1\bar{B}_2$, we see that $C_M(u)$ contains a 3-subgroup P which normalizes, but does not centralize B_1 or B_2, say B_1. Since $P \subseteq H$, all involutions of B_1 are thus conjugate in H. Since $R_2 = B_1B_2$, it follows that all involutions of R_2 are conjugate in H. But then H must be nonsolvable as $R_2 \cong D_8$. However, as $T\langle y\rangle \subseteq H$ and $T\langle y\rangle \supset T$, this contradicts Lemma 5.1. Hence $B_1^y \neq B_2$.

We set $N = N_G(A_2)$ and $\overline{N} = N/C_G(A_2)$ and study the structure of these groups. We have that $T \subseteq N$ as $A_2 < T$. If T is not a Sylow 2-subgroup of N, then $N_N(T)$ contains a 2-element y_1 with $y_1^2 \in T$. We have that y_1 normalizes A_2 as $y_1 \in N$ and so y_1 also normalizes A_1. Hence if we choose our Sylow 2-subgroup S to contain $T\langle y_1\rangle$ and take y_1 as y, the desired conclusion of the lemma holds. Therefore we can assume without loss that T is a Sylow 2-subgroup of N. But then $\overline{T} \cong Z_2$ and so $\overline{N}$ has a normal 2-complement.

Since P centralizes x as well as u, P normalizes A_1 and so $P^* = P^y$ normalizes A_2. Moreover, if we set $B^* = B_1^y \neq B_2$, B^* is a four subgroup of A_2 which is normalized, but not centralized by P^* and P^* centralizes A_2/B^*. Since $\overline{N}$ has a normal 2-complement and $|\overline{N}|$ is divisible by 6, $|\overline{N}|$ is not divisible by 7. Hence $|O(\overline{N})| = 3, 9$, or 15. However, in the last case, $O(\overline{N})$ would act regularly on A_2, contrary to the fact that $\overline{P}^* \subseteq O(\overline{N})$ and $\overline{P}^*$ does not act regularly on A_2. Thus $O(\overline{N}) \cong Z_3$ or $Z_3 \times Z_3$.

On the other hand, by Lemma 4.2(iii) and (iv) of Part V, $N \cap M$ contains a 3-subgroup Q which normalizes, but does not centralize B_2 and which centralizes $\langle uz, x\rangle$. (Note that by our choice of notation, $\tilde{B}_1 \subseteq C_{\tilde{L}}(\tilde{u})'$ and $\tilde{B}_2 \not\subseteq C_{\tilde{L}}(\tilde{u})'$). Thus, in fact, $O_3(\overline{N}) = \langle \overline{P}^*, \overline{Q}\rangle = \overline{P}^* \times \overline{Q} \cong Z_3 \times Z_3$. This implies that $\overline{P}^*$ leaves both $B_2 = [A_2, \overline{Q}]$ and $C_{A_2}(\overline{Q}) = \langle uz, x\rangle$ invariant. Since $B^* = B_1^y \neq B_2$, the only possibility is that $\overline{P}^*$ centralizes B_2 and $B^* = \langle uz, x\rangle$.

Our argument yields that $B^* \subseteq Z(T)$ and as $B^* = B_1^y$, all involutions of B^* are conjugate in G. But as B^* is of index 2 in $Z(T)$, y centralizes some involution b^* of B^*. Then $T\langle y\rangle \subseteq C_G(b^*)$. However,

$b^* \sim x$ in G as also $x \in B^*$. Since $T\langle y\rangle \supset T$ and T is a Sylow 2-subgroup of $C_G(x)$, this is clearly impossible; and the lemma is proved.

We assume henceforth that y is chosen to normalize an elementary abelian subgroup A of T of order 16. We set $N = N_G(A)$ and $\overline{N} = N/C_G(A)$. As usual, $C_G(A)$ has a normal 2-complement and N, $\overline{N}$ are both solvable. If $t \neq 1$, we have that $A \subseteq R_2\langle x,u\rangle$ by Lemma 9.5 while the same conclusion is obvious if $t = 1$ as then $T = R_2\langle x,u\rangle$. Thus $A = B \times \langle u,x\rangle$, where $B = A \cap R_2 \cong Z_2 \times Z_2$. Furthermore, again by Lemma 4.2(iii) and (iv) of Part V, $\overline{K} = \overline{N \cap M} \cong S_3$ and $\overline{P} = O_3(\overline{M \cap N})$ normalizes, but does not centralize B and centralizes $B^* = \langle u,x\rangle$ or $\langle uz,x\rangle$. We also set $k = |\overline{N} : \overline{K}|$ and fix all this notation.

Since $\overline{P}$ does not act regularly on A, it cannot centralize an element of $\overline{N}$ of order 5 and, as $\overline{N}$ is solvable, this implies that $|\overline{N}|$ is not divisible by 5. Since $y \in N$, $|\overline{N}|$ is divisible by 12 and again the solvability of $\overline{N}$ implies that $|\overline{N}|$ is not divisible by 7. Thus $|\overline{N}| = 2^2 3^b$. Furthermore, since $\overline{K} = C_{\overline{N}}(x) \cong S_3$, it also follows that a Sylow 2-subgroup of $\overline{N}$ has order at most 16. We therefore conclude that $k = 2,4,6,8$, or 12. We shall now analyze these various possibilities for k.

We first prove

Lemma 9.8. We have $k = 2,4$, or 8.

Proof: Suppose false, in which case $|\overline{N}|$ is divisible by 9. Let $\overline{Q}$ be a Sylow 3-subgroup of $\overline{N}$ containing $\overline{P}$, so that $\overline{Q} \cong Z_3 \times Z_3$. Then $\overline{Q}$ leaves both $B = [A,\overline{P}]$ and $B^* = C_A(\overline{P})$ invariant and so $\overline{Q} = \overline{P} \times \overline{P}^*$ with $B^* = [A,P^*]$ and $B = C_A(\overline{P}^*)$. We shall use these facts to derive a contradiction.

We set $V = R_2\langle u,x\rangle \cong D_{2^n} \times Z_2 \times Z_2$ and $Z = \langle z,u,x\rangle$. Then $Z = Z(V) \subseteq A$ and $Z \cong E_8$. Moreover, as $B^* = \langle u,x\rangle$ or $\langle uz,x\rangle$, $Z = \langle z\rangle \times B^*$ and so P^* leaves Z invariant. Furthermore, as V contains every elementary abelian subgroup of T of order 16 and V is generated by the totality of such subgroups, y leaves V invariant and so y leaves Z invariant. This implies that y centralizes an involution b^* of B^*. But $b^* \sim x$ as $x \in B^*$. Since $V\langle y\rangle \subseteq C_G(b^*)$ and $V\langle y\rangle \supset V$, V cannot be a Sylow 2-subgroup of $C_G(x)$. It follows that $T \supset V$ and hence that $t \neq 1$.

By the preceding paragraph $N_G(Z)$ contains a 3-subgroup P^* which centralizes z and normalizes, but does not centralize B^*. Since V is clearly a Sylow 2-subgroup of $C_G(Z)$, we can choose P^* to normalize V. Hence if we set $H = N_G(V)$ and $R = T\langle y\rangle$, we have that $\langle R,P^*\rangle \subseteq H$. Setting $\tilde{H} = H/C_G(V)V$, it follows that $|\tilde{H}|$ is divisible by 12. On the other hand, as R_2 is nonabelian, z is the unique involution of $Z \cap [V,V]$. Since $H \subseteq N_G(Z)$, this implies that z is isolated in H. Thus $\tilde{H}$ centralizes z. Since V is a Sylow 2-subgroup of $C_G(Z)$, also $\tilde{H}$ must act faithfully on $Z \cong E_8$. Hence the only possibilities are $\tilde{H} \cong A_4$ or S_4.

However, as z is isolated in H, x has 3 or 6 conjugates in H. Hence if $\tilde{H} \cong S_4$, we would have that $|C_{\tilde{H}}(x)|$ is divisible by 4, whence $|C_H(x)|$ is divisible by $4|V| = 2|T|$, contrary to the fact that T is a Sylow 2-subgroup of $C_G(x)$. Thus, in fact, $\tilde{H} \cong A_4$ and so $\tilde{R} = \langle \tilde{t},\tilde{y}\rangle$ is a normal Sylow 2-subgroup of $\tilde{H}$. Hence again we can suppose without loss that P^* normalizes R.

Observe finally that as $V/B^* \cong R_2 \cong D_{2^n}$ with $n \geq 3$, P^* centralizes V/B^*. Hence if we set $\bar{R} = R/Z$, we have that P^* centralizes

$\overline{V} = \overline{R}_2 \cong D_{2^{n-1}}$. Since $\overline{V} \lhd \overline{R}$, it follows that $\overline{V}$ also centralizes $[\overline{R}, P^*]$. But as $\overline{R}/\overline{V}$ is a four group not centralized by P^*, clearly $\overline{R} = \overline{V}[\overline{R}, P^*]$ and consequently $\overline{R} = \overline{V}C_{\overline{R}}(\overline{V}) = \overline{R}_2 C_{\overline{R}}(\overline{R}_2)$. In particular, $\overline{t} \in \overline{R}$ and we conclude now from the preceding factorization of $\overline{R}$ that $\overline{t}$ induces an inner automorphism of $\overline{R}_2$. However, this is a contradiction since $\overline{R}_2\langle\overline{t}\rangle \cong D_{2^n}$ inasmuch as $R_1R_2\langle t\rangle/R_1 \cong D_{2^{n+1}}$.

Lemma 9.9. We have $k \neq 8$.

Proof: suppose false, in which case $|\overline{N}| = 48$. If $\overline{P}$ is normal in $\overline{N}$, then $\overline{N}$ leaves $B^* = C_A(P)$ invariant. But then as $x \in B^*$, $|\overline{N}:\overline{K}| \leq 2$, where as above $\overline{K} = \overline{N \cap M}$, contrary to the fact that $k = 8$. Hence $\overline{P}$ is not normal in $\overline{N}$. Since $\overline{P}$ does not act regularly on A, we conclude now by considering the subgroups of A_8 of order 48 that $\overline{N} \cong Z_2 \times S_4$. Since $\overline{K} \cong S_3$, we have , in fact, $\overline{N} = O_2(\overline{N})\overline{K}$ with $O_2(\overline{N}) \cap \overline{K} = 1$. We let U be a $(T \cap N)$-invariant subgroup of the inverse image of $O_2(\overline{N})$ in N. Thus $|U| = 2^7$ and $C_U(x) = A$. By the Frattini argument there exists a cyclic 3-subgroup P in N which normalizes W, centralizes x, and maps on $\overline{P}$.

We set $Y = C_U(P)$ and let V be the inverse image of $[O_2(\overline{N}), \overline{P}]$ in U. We have that $A \subset U$ with $U/A \cong Z_2 \times Z_2$ and with W invariant under both P and $T \cap N$. Furthermore, as P is cyclic, it is immediate that $P/C_P(U) \cong Z_3$ and so Y maps onto $C_{\overline{N}}(\overline{P}) = Z(\overline{N}) \cong Z_2$. Since $C_A(P) = B^*$, Y is thus of order 8 and contains B^*. Since $C_W(x) = A$, $x \notin Z(Y)$ and so $Y \cong D_8$. Moreover, as $\overline{Y} = Z(\overline{N})$, $O_2(\overline{N}) = \overline{YV}$ and so $U = YV$.

Observe next that as $C_V(A) \subseteq A$, $Z(V) \subseteq A$. Consider first the case that $Z(V) \cong Z_2$. Then V is of type A_8 by Lemma 2.6. We shall now derive a contradiction by applying Lemma 2.7(ii). Indeed, by the structure of M, $C_G(A)$ has a normal 2-complement and so $C_G(A) = O(N)A$. Hence if we set $\tilde{N} = N/O(N)$, we have that $\tilde{V} \triangleleft \tilde{N}$, $C_{\tilde{N}}(\tilde{V}) \subseteq \tilde{V}$, and $|\tilde{P}| = 3$. Since $\tilde{V}$ is of type A_8, Lemma 2.7(ii) is applicable with $\tilde{N},\tilde{V}$ in the roles of X,Y respectively and yields that a Sylow 2-subgroup of $N_{\tilde{N}}(\tilde{P})$ has order at most 8. On the other hand, $\tilde{P}$ centralizes $\tilde{Y}$ and $|\tilde{Y}| = 8$ as $Y \cong D_8$. But $\bar{P}$ is inverted by an involution of $\bar{N}$ as $\bar{K} \cong S_3$. Hence $\tilde{P}$ is inverted by an involution of $\tilde{N}$ and so a Sylow 2-subgroup of $N_{\tilde{N}}(\tilde{P})$ has order at least 16. This contradiction shows that $Z(V)$ has order at least 4.

We have that $Z(V)$ is P-invariant and $Z(V) \subset A$. Considering the action of P on A, it follows that either $B \subseteq Z(V)$ or else $Z(V) \subseteq B^*$. However, in the latter case, $Z(V) = B^*$ as $|Z(V)| \geq 4$. Thus $x \in Z(V)$, contrary to $C_U(x) = A$. We therefore conclude that $B \subseteq Z(V)$. Since P has a nontrivial fixed point on V/B, we see now that $[V/B,P] \cong Z_2 \times Z_2$ or Q_8.

Suppose first that $[V/B,P] \cong Z_2 \times Z_2$, in which case $V_1 = [V,P] \cong E_{16}$ or $Z_4 \times Z_4$. In the first case, V splits over A and so V is of type $L_3(4)$ by [Part II, Lemma 2.2(iv)]. Since V and A are normal in T, and A,V_1 are the only two elementary abelian subgroups of V of order 16, we conclude that V_1 is normal in T. We also see that for a suitable choice of $\tilde{P}$, $\tilde{T}$ is a split extension of $N_{\tilde{T}}(\tilde{P})$ by $\tilde{V}_1$. On the other hand, $\tilde{P}$ corresponds to a 3-cycle in $A_8 \cong GL(4,2) \cong Aut(V_1)$ and $N_{A_8}(\langle 123 \rangle)$ has dihedral Sylow 2-subgroups of order 8. As $N_{\tilde{T}}(\tilde{P})$ is of order 16, some involution of $N_{\tilde{T}}(\tilde{P})$ must centralize $\tilde{V}_1$. This forces that $m(T) \geq 5$, contrary to $r(G) \leq 4$.

Assume then that $V_1 \cong Z_4 \times Z_4$. Since $R_2 \cong D_{2^n}$, $n \geq 3$, we have $N_{R_2}(A) \cong D_8$ and so $N_{R_2}(A)$ contains an involution r which centralizes B^* and such that $\bar{r}$ inverts $\bar{P}$. Since U is $(T \cap N)$-invariant by assumption, r leaves V_1 invariant and so $\langle r,x \rangle$ is a four group which acts on V_1. Clearly r and rx do not centralize $B = \Omega_1(V_1)$. Since $C_U(x) = A$, also x does not centralize V_1 and so $\langle r,x \rangle$ acts faithfully on V_1. But then by [Part II, Lemma 2.1(vi)] $V_2 = V_1\langle r,x \rangle$ is of type M_{12}. On the other hand, as $Y \times P$ acts on V_1 with $Y \cong D_8$ and $V_1 \cong Z_4 \times Z_4$, it is immediate that $Z(Y)$ centralizes V_1. But $Z(Y) \subseteq B^*$ as B^* is a normal four subgroup of Y. Since r centralizes B^*, $Z(Y)$ also centralizes $\langle r,x \rangle$ and so $Z(Y)$ centralizes V_2. However, $Z(Y) \not\subseteq V_2$ as then $Z(Y) \subseteq C_{V_2}(P) = \langle x \rangle$, whence $Z(Y) = \langle x \rangle$. But then $Y \subseteq C_W(x) = A$, which is not the case. Therefore $V_2 Z(Y) = V_2 \times Z(Y)$ and so $r(V_2 Z(Y)) = 5$, again a contradiction.

There thus remains the case $[V/B,P] \cong Q_8$. Again setting $V_1 = [V,P]$, we check that $V_1' \cong Z_2$ and that $V_1/V_1' \cong E_{16}$ or $Z_4 \times Z_4$. Clearly P centralizes V_1' and so $V_1' \subseteq Y$. Since $V_1' \triangleleft U$ and $Y \subseteq U$, we have, in fact, $V_1' = Z(Y)$.

We let r be as in the preceding paragraph. Since $\bar{r}$ inverts $\bar{P}$, we easily see that we can choose our cyclic subgroup P to be inverted by r. Hence if we set $Y_1 = Y\langle r \rangle$, it follows that Y_1 is a

Sylow 2-subgroup of $N_N(P)$ and is of order 16. We claim that $Y_1/Z(Y) \cong E_8$. Indeed, as $Y \cong D_8$ and r is an involution, $Y_1/Z(Y)$ is generated by involutions and so is either isomorphic to E_8 or D_8. But Y_1 does not split over $Z(Y)$ as $Y \subset Y_1$ with $Y \cong D_8$. Hence if $Y_1/Z(Y) \cong D_8$, then $Y_1 \cong D_{16}$ or QD_{16}, contrary to the fact that B^* is a normal four subgroup of Y_1. This proves our assertion.

Now set $H = V_1Y_1P$ and $\tilde{H} = H/Z(Y)$. Then $\tilde{V}_1 \triangleleft \tilde{H}$ and $\tilde{V}_1 \cong E_{16}$ or $Z_4 \times Z_4$ with $C_{\tilde{V}_1}(\tilde{P}) = 1$. The latter condition implies that $\tilde{V}_1 \cap \tilde{Y}_1 = 1$. Suppose $\tilde{V}_1 \cong E_{16}$. Since $r(\tilde{V}_1\tilde{Y}_1) \leq 4$ and $\tilde{V}_1 \cap \tilde{Y}_1 = 1$, $\tilde{Y}_1$ acts faithfully on $\tilde{V}_1$ and so $\tilde{Y}_1\tilde{P}$ is isomorphic to a subgroup of A_8. But $\tilde{Y}_1 \cong E_8$ by the preceding paragraph and $\tilde{Y}_1$ normalizes $\tilde{P}$. However, no subgroup of A_8 of order 3 has a normalizer which contains an elementary abelian subgroup of order 8.

We therefore conclude that $\tilde{V}_1 \cong Z_4 \times Z_4$. Since $\tilde{r}$ inverts $\tilde{P}$, $\tilde{r}$ does not centralize $\tilde{B} = \Omega_1(\tilde{V}_1)$ and hence neither does $\widetilde{rx}$. Furthermore, $\tilde{x}$ does not centralize $\tilde{V}_1$, since otherwise $|V_1 : C_{V_1}(x)| \leq 2$ as $\tilde{V}_1 = V_1/Z(Y)$ with $Z(Y) \cong Z_2$. But as $C_{V_1}(x) \subseteq A$, this would imply that $|V_1/V_1 \cap A| \leq 2$, which is not the case. Thus the four group $\langle\tilde{r},\tilde{x}\rangle$ acts faithfully on $\tilde{V}_1$ and consequently $\tilde{V}_2 = \tilde{V}_1\langle\tilde{r},\tilde{x}\rangle$ is of type M_{12}, again by [Part II, Lemma 2.1(vi)]. Moreover, as $\tilde{Y}_1 \cong E_8$ and $\tilde{Y}_1$ normalizes $\tilde{P}$, some involution $\tilde{y}_1$ of $\tilde{Y}_1$ centralizes $\tilde{V}_1$ by [Part II, Lemma 2.1(iv)]. But then $\tilde{Y}_1$ centralizes $\tilde{V}_2$ as it centralizes $\langle\tilde{r},\tilde{x}\rangle \subseteq \tilde{Y}_1$. Clearly $\tilde{Y}_1 = \langle\tilde{r},\tilde{x}\rangle \times \langle\tilde{y}_1\rangle$ and so $\tilde{V}_2\tilde{Y}_1 = \tilde{V}_2 \times \langle\tilde{Y}_1\rangle$. Since $\tilde{V}_2$ is of type M_{12}, we conclude now that $r(\tilde{V}_2\tilde{Y}_1) = 5$ and again we have a contradiction. This completes the proof of the lemma.

Lemma 9.12. We have $k \neq 4$.

Proof: Suppose false, in which case $|\overline{N}| = 24$. Since $\overline{P}$ does not act regularly on A, it corresponds in A_8 to a subgroup of A_8 generated by the product of two 3-cycles. Since the normalizer in A_8 of such a subgroup does not have order divisible by 8, $\overline{P}$ cannot be normal in $\overline{N}$. We conclude at once now the subgroup structure of A_8 and the fact that $\overline{K} \cong S_3$ that $\overline{N} \cong S_4$. Let V be a $(T \cap N)$-invariant Sylow 2-subgroup of the inverse image of $O_2(\overline{N})$ in N and also set $U = V(T \cap N)$, so that $|V| = 2^6$, $|U| = 2^7$, and W is a Sylow 2-subgroup of N. Again let P be a cyclic 3-subgroup of N which normalizes V, centralizes x, and maps on $\overline{P}$. Again we have that $P/C_P(V) \cong Z_3$. Likewise as in the preceding lemma there exists an involution r in $R_2 \cap N$ which centralizes B^* and for a suitable choice of P inverts P. Then $B^*\langle r\rangle \cong E_8$ and in the present case $Y = B^*\langle r\rangle$ is a Sylow 2-subgroup of $N_N(P)$.

Suppose first that $Z(V) \cong Z_2$. Again Lemma 2.6 yields that V is of type A_8. Moreover, by [Part II, Lemma 2.2(v)], $V_1 = [V,P] \cong Q_8 * Q_8$ with P acting fixed-point-free on $V_1/Z(V_1)$. Clearly, $Z(V_1) = V_1 \cap Y$ and so $|V_1Y| = 2^7$, whence $U = V_1Y$. Since $Y \cong E_8$, we see that U splits over V_1 and now Lemma 2.7(iii) yields that U is of type A_{10}.

Observe now that as $r(G) \leq 4$ and the centralizer of every central involution of G is solvable, the basic hypothesis of Lemmas 5.3 and 5.4 of Part II are satisfied. First of all, it follows from Lemma 5.3 of Part II that $|S| \leq 2^8$. If $|S| < 2^8$, then U must be a Sylow 2-subgroup of G as $|U| = 2^7$. But then S is of type A_{10}, contrary to Proposition 4.2. Hence $|S| = 2^8$ and we conclude now from Lemma 5.4 of Part II that S contains a maximal subgroup which is the central product of two subgroups isomorphic to either Q_{16} or QD_{16} which are interchanged under conjugation by an involution of S. In the first case, S is of type $PSp(4,p)$ for some $p \equiv 1,7 \pmod 8$, contrary to Proposition 4.2; while in the second case, Proposition 5.5 of Part II implies that G is not fusion-simple, which is again a contradiction.

Hence we have that $|Z(V)| \geq 4$. Again as in the preceding lemma, this implies that $B \subseteq Z(V)$ and if we set $V_1 = [V,P]$, then $V_1/B \cong Z_2 \times Z_2$ or Q_8. Suppose first that $V_1/B \cong Z_2 \times Z_2$, in which case $V_1 \cong E_{16}$ or $Z_4 \times Z_4$. In the first case, V splits over A and so by [Part II, Lemma 2.2(iv)], V is of type $L_3(4)$ as $|Z(V)| \geq 4$. Since $U = V\langle r\rangle$, it follows now from [Part II, Lemma 2.2(viii)] that U is of type $\hat{A}_8$. But r **centralizes** both B^* and $\langle z\rangle = Z(R_2)$, so $C_A(r) \cong E_8$. However, this is impossible by the structure of a Sylow 2-subgroup of type $\hat{A}_8$. Hence $V_1 \cong Z_4 \times Z_4$. Moreover, we also have that $V_1 \cap Y = 1$ and so $U = V_1Y$. Since $Y \cong E_8$, $C_Y(V_1) \neq 1$ by [Part II, Lemma 2.1(i)]. Since $Y = B^*\langle r\rangle$ and B^* stabilizes the chain $V_1 \supset B \supset 1$, we see that $C_Y(V_1) \subseteq B^*$ and so some involution b of B^* centralizes V_1. On the other hand, as usual, $\langle x,r\rangle$ acts faithfully on V_1 and so $V_2 = V_1\langle x,r\rangle$ is of type M_{12} by [Part II, Lemma 2.1(vi)]. But b centralizes both V_1 and $\langle x,r\rangle$ and so centralizes V_2, whence $r(V_2\langle b\rangle) \geq 5$, a contradiction.

We have therefore shown that $V_1/B \cong Q_8$. We argue next that U is a Sylow 2-subgroup of G. Suppose false and let S_1 be a Sylow 2-subgroup of G containing U. Then $N_{S_1}(U) - U$ contains an element s with $s^2 \in U$. Since U is a Sylow 2-subgroup of $N = N_G(A)$, we have that $A^s \neq A$. But $A^s \triangleleft U$ as s normalizes U and so $AA^s/A \triangleleft U/A$. In addition, $AA^s/A \neq 1$, as $A^s \neq A$. Since $U/A \cong D_8$ with $Z(U/A) \subseteq V/A$, it follows that $(AA^S/A) \cap V/A \neq 1$. Thus $AA^S \cap V \supset A$ and as AA^S is generated by it involutions, we conclude that $V - A$ contains an involution. Hence $V/B - A/B$ contains an involution. However, as $V_1/B \cong Q_8$, we see at once that $V/B = (V_1/B)(A/B) \cong Q_8 \times Z_2$ with $\Omega_1(V/B) = A/B$. This contradiction establishes our assertion.

Thus $|S| = |U| = 2^7$. We shall derive a contradiction in this case by means of Thompson's transfer lemma, but first we must set up the situation precisely. We have that $B \subseteq Z(V)$ and that $U = V\langle r\rangle$, where r is an involution of R_2, $B\langle r\rangle \cong D_8$, and r inverts P. Thus $\langle z\rangle = Z(R_2)$ centralizes both V and r, whence $z \in Z(U)$. Furthermore, $V_1 \lhd U$, $V_1/B \cong Q_8$, and $V_1' = \langle v_1\rangle \cong Z_2$. Since $V_1' \lhd U$, also $v_1 \in Z(U)$. Considering the action of P on $B\langle v_1\rangle$, it follows that every involution of $B\langle v_1\rangle$ is conjugate to one in $\langle z, v_1\rangle$ and so every involution of $B\langle v_1\rangle$ is a central involution. Observe next that as $V_1/B \cong Q_8$, we also have that $\Omega_1(V_1) = B\langle v_1\rangle = Z(V_1)$.

Our goal will be to show that every involution of $V_1\langle r\rangle$ is central. Since $v_1 \in Z(U)$, v_1 centralizes both x and $B\langle r\rangle$. Since $R_1 = \langle x\rangle \cong Z_2$, it follows that in $\overline{M}$, $\overline{v}_1$ induces a field automorphism of $\overline{L}$ of order 2. Hence by Lemma 6.2(iii) of Part III any involution of the form v_1r_1 with r_1 in $B\langle r\rangle$ is conjugate under the action of L to either v_1 or v_1z. Thus every involution of $V_1\langle r\rangle$ of this form is central.

Observe next that if we set $\tilde{N} = N/O(N)B$ we have that $\tilde{N} = \tilde{V}_1\tilde{P}\langle\tilde{r}\rangle \cong GL(2,3)$ and consequently every involution of $V_1\langle r\rangle$ is conjugate in N to either an involution of $Z(V_1)$ or to an involution of $Z(V_1)\langle r\rangle - Z(V_1)$. But as $Z(V_1) = B\langle v_1\rangle$, it follows in $\overline{M}$ that $Z(\overline{V}_1)\langle\overline{r}\rangle\overline{P} = \langle\overline{v}_1\rangle \times \overline{BP}\langle\overline{r}\rangle \cong Z_2 \times S_4$ and so every involution of $Z(V_1)\langle r\rangle - Z(V_1)$ is either contained in $B\langle r\rangle$ or has the form v_1r_1 for some r_1 in $B\langle r\rangle$. We know that all involutions of $Z(V_1)$ and all involutions of the form v_1r_1 are central. Since L has only one conjugacy class of involutions, so also are all involutions of $B\langle r\rangle$. Our argument therefore yields the desired conclusion that all involutions of $V_1\langle r\rangle$ are central.

Finally as x is noncentral, $x \notin V_1$ as all involutions of V_1 are central. But $x \in A \subseteq V$ and so $V = V_1\langle x\rangle$. Hence $U = V\langle r\rangle = V_1\langle x,r\rangle$. Thus $V_1\langle r\rangle$ is a maximal subgroup of W not containing x. Thompson's transfer lemma therefore forces x to be conjugate in G to some involution of $V_1\langle r\rangle$. However, this is impossible as all involutions of $V_1\langle r\rangle$ are central, while x is not. This completes the proof of the lemma.

Finally we prove

Lemma 9.11. We have $k \neq 2$.

Proof: Assume false, in which case $|\overline{N}| = 12$. Since $\overline{K} \cong S_3$, $\overline{N}$ is not isomorphic to S_4 and so $\overline{P} \lhd \overline{N}$. Since $B^* = C_A(\overline{P})$, it follows that $B^* \lhd N$.

We use this to prove that $t = 1$. Indeed if $t \neq 1$, then $Z(T) = \langle z,x\rangle$ and so y normalizes $\langle z,x\rangle$. But as $y \in N$ and y does not centralize x, this implies that x is conjugate to z or zx in N. Since $x \in B^*$ and $B^* \lhd N$, this forces z or zx to lie in B^*, which is not the case. Thus $t = 1$, as asserted.

By Lemma 9.3, $\overline{L} \cong L_2(9)$ and $T \cong R_2 \times \langle u,x\rangle = R_2 \times B^*$ with $R_2 \cong D_8$. In particular, $T \subseteq N$ and consequently $R = T\langle y\rangle$ is a Sylow 2-subgroup of N. We conclude that $B^* \lhd R$.

Finally we have that $T = AA_1$, where A_1 is a second elementary abelian subgroup of T of order 16. Since $T^y = T$ and $A^y = A$, y also normalizes A_1. Hence our entire analysis applies to A_1 as well as to A. Hence if we set $N_1 = N_G(A_1)$ and $\overline{N}_1 = N/C_G(A_1)$, $K_1 = M \cap N_1$, we must have that $|\overline{N}_1 : \overline{K}_1| = 2$ and that $\overline{K}_1 \cong S_3$. Again by Lemma 4.2(iii) and (iv) of Part V, $\overline{P}_1 = O_3(\overline{K}_1)$ normalizes, but does not centralize $B_1 = A_1 \cap R_2$ and centralizes $B_1^* = \langle u,x \rangle$ or $\langle uz,x \rangle$. Furthermore, we note that by the same lemma $B_1^* \neq B^*$. However, the first part of the argument of the present lemma yields that $B_1^* \triangleleft N_1$ and that $R = T\langle y \rangle$ is a Sylow 2-subgroup of N_1. In particular, we see that R normalizes both B^* and B_1^*, whence R normalizes $B^* \cap B_1^* = \langle x \rangle$. But then y centralizes x, which is not the case.

We have therefore shown that each possible value of k leads to a contradiction and so the proof of Proposition 9.1 is complete.

10. The structure of S. We are at last in a position to determine the structure of a Sylow 2-subgroup of G. We shall prove

Proposition 10.1. S is of type M_{12}.

We carry out the proof in a sequence of lemmas. By Proposition 8.1 and 9.1, $O^{2'}(\widetilde{M}) \cong L_2(q)$, $PGL(2,q)$, $PGL^*(2,q)$, q odd, or A_7 and $\overline{C} = \overline{R}_1 = \langle \overline{x} \rangle$. In particular, $\widetilde{T} \cong T/R_1 = T/\langle x \rangle$ is either dihedral or quasi-dihedral and $|\widetilde{T}:\widetilde{R}_2| \leq 2$.

We first prove

Lemma 10.2. T splits over R_1.

Proof: Suppose false. Since $R_1R_2 = R_1 \times R_2$, clearly we have

$T \supset R_1R_2$. This implies that $R_2 \cap Z(T) \cong Z_2$ and we set $\langle z\rangle = R_2 \cap Z(T)$. According as T is dihedral or quasi-dihedral, there exists an element t in $T - R_1R_2$ such that $t^2 \in R_1$ or $t^2 \in R_1\langle z\rangle - R_1$. If $t^2 \in R_2$, then $R_2\langle t\rangle$ is a group and $T = R_1 \times R_2\langle t\rangle$, whence T splits over R_1, contrary to assumption. Hence $t^2 \notin R_2$ and consequently we have correspondingly $t^2 = x$ or $t^2 = xz$. We also have that $Z(T) = \langle x,z\rangle$ and that $Z(T) \cap T' = \langle z\rangle$. Since y normalizes T, it follows that y centralizes z. Since $C_G(z) \supseteq T\langle y\rangle \supset T$, this implies that x is not conjugate in G to z. Since all involutions of R_2 are conjugate in M, x is, in fact, not conjugate in G to any element of R_2.

Our conditions imply that $\Omega_1(T) = R_1R_2$ and so every elementary abelian subgroup of T of order 8 lies in R_1R_2 and all are conjugate in T. Hence, as usual, we can assume without loss that y normalizes such a subgroup A of R_1R_2. We set $N = N_G(A)$ and $\bar{N} = N/C_G(A)$. We also set $B = A \cap R_2$, so that $B \cong Z_2 \times Z_2$ and $A = B \times \langle x\rangle$. In particular, $\langle x,z\rangle \subsetneq A$. If $|\bar{N}|$ were divisible by 7, then $x \sim z$ in N, contrary to the preceding paragraph. Hence this is not the case and so $\bar{N}$ is isomorphic to a $\{2,3\}$-subgroup of $L_3(2)$. As before, $\bar{K} = \overline{M \cap N} \cong S_3$. Since $y \in N$, we also have that $\bar{N} \supset \bar{K}$. We conclude at once therefore that $\bar{N} \cong S_4$, that x has exactly 4 conjugates in A under the action of N, and that N normalizes a four subgroup of A not containing x. Since x is not conjugate in G, to any involution in B by the preceding paragraph, this four group must be B. Therefore also $B \triangleleft N$. Note also that a Sylow 2-subgroup of N has order 2^6, while $|N_T(A)| = 2^4$.

Since z and t^2 lie in $\Phi(S)$, we have that $x \in \Phi(S)$ whether $t^2 = x$ or $t^2 = xz$. Hence if U is an element of $U(S)$, it follows that

$U \subseteq T$. We argue that $x \notin U$. Suppose false, in which case $|S:T| = 2$ and $S = T\langle y\rangle$. Clearly $N^g \cap S$ is a Sylow 2-subgroup of N^g for some g in G. Hence if we set $A_1 = A^g$, we have that $|N_S(A_1)| = 2^6$. If $A_1 \subseteq T$, then A_1 would be conjugate in T to A and as $|N_T(A)| = 2^4$, it would follow that $|N_T(A_1)| = 2^4$ whence $|N_S(A_1)| = 2^5$, which is not the case. Hence $A_1 \not\subseteq T$ and so $S = A_1T$. But now setting $B_1 = A_1 \cap T$, we conclude that $|N_T(B_1)| = 2^5$. This in turn implies that $|N_{\tilde{T}}(\tilde{B}_1)| \geq 2^4$. However, $\tilde{T}$ is dihedral or quasi-dihedral, so the only possibility is that $\tilde{B}_1 = \langle \tilde{z}\rangle = Z(\tilde{T})$ and that $x \in B_1$. But as A_1 is abelian, this gives $A_1 \subseteq C_S(x) = T$, which is not the case. Thus $x \notin U$, as asserted.

Since $U \subseteq T$, $U \triangleleft T$ and as $x \notin U$, it follows that $\tilde{U}$ is a normal four subgroup of $\tilde{T}$. Since $\tilde{T}$ is dihedral or quasi-dihedral with $\tilde{T} \supset \tilde{R}_2$, the only possibility is that $\tilde{T} \cong D_8$ and that $\tilde{R}_2 \cong R_2 \cong Z_2 \times Z_2$. In particular, this implies that $t^2 = x$ and $T = A\langle t\rangle \subseteq N$.

Let W be a Sylow 2-subgroup of N containing T and let V be the subgroup of index 2 in W such that $\bar{V} = O_2(\bar{N})$. Then $|W| = 2^6$, $\bar{V} \cong Z_2 \times Z_2$, and $B \subseteq Z(V)$. We also let P be a cyclic 3-subgroup of N which normalizes V, centralizes x, and maps on $O_3(\bar{K})$. We can choose P to be inverted by some element of $T - A$, which without loss we can take to be t. Our conditions imply that $[V/B,P] \cong Z_2 \times Z_2$ or Q_8. However, in the latter case one sees easily that $W/B \cong Q_{16}$ or QD_{16} with $A/B = Z(W/B)$. Then $\mho^1(W) \subseteq V$ with $\mho^1(W)B/B$ cyclic of order 4 and with $A \subseteq \mho^1(W)B$. Since $B \subseteq Z(V)$, B centralizes $\mho^1(W)$ and as $\mho^1(W)B/B$ is cyclic, it follows that $\mho^1(W)B$ is abelian. We conclude at once from this that $\mho^2(W)$ is of order 2 with $A = B \times \mho^2(W)$. But as $\mho^2(W) \triangleleft W$, we have $\mho^2(W) \subseteq Z(W)$, whence $A \subseteq Z(V)$. Thus V centralizes A, which is not the case.

We have therefore shown that $[V/B,P] \cong Z_2 \times Z_2$. Hence if we set $V_1 = [V,P]$, we conclude that $V_1 \cong E_{16}$ or $Z_4 \times Z_4$. Furthermore, $C_{V_1}(P) = 1$ and so $x \notin V_1$. Since $t^2 = x$, it follows that $V_1 \cap \langle t \rangle = 1$ and as $|W| = 2^6$, we see that $W = V_1\langle t \rangle$.

Suppose first that $V_1 \cong E_{16}$, in which case clearly $V_1\langle x \rangle \cong Z_2 \times Z_2 \int Z_2$ and $\Omega_1(W) = V_1\langle x \rangle$. It follows that V_1 is characteristic in W and that all involutions of $W - V_1$ are conjugate in W to x. Now let S_1 be a Sylow 2-subgroup of G containing W and suppose $S_1 \supset W$. Then $N_{S_1}(W) - W$ contains an element y_1 such that $y_1^2 \in W$. Since V_1 is characteristic in W, $x^{y_1} \in W - V_1$ and so $x^{y_1 w} = x$ for some w in W. But then $y_1 w$ centralizes x and as $T \subseteq W$ with T a Sylow 2-subgroup of $M = C_G(x)$, it follows that $y_1 w \in T$, whence $y_1 \in W$, which is not the case. We therefore conclude that $W = S_1$ is a Sylow 2-subgroup of G. But now [9] contradicts the simplicity of G.

Hence we must have $V_1 \cong Z_4 \times Z_4$. Since $C_G(A) \subseteq M$, we see at once that $C_G(A)$ has a normal 2-complement, which is clearly equal to $O(N)$. Setting $\tilde{N} = N/O(N)$, it follows that $\tilde{P} \cong Z_3$ and that $\tilde{P}\langle\tilde{t}\rangle$ is isomorphic to a subgroup of $\mathrm{Aut}(\tilde{V}_1)$. Furthermore, as $\tilde{x} = \tilde{t}^2$, it follows from the structure of $\mathrm{Aut}(Z_4 \times Z_4)$ that $\tilde{x}$ must invert $\tilde{V}_1$. We conclude therefore from the structure of $\tilde{N}$ that G is of type F_2 as this term is defined in Section 3. But now Propositions 3.6 and 4.2 together yield a contradiction. Thus T splits over R_1, as asserted.

By the preceding lemma, we have that $T = R_1 \times R_3$, where $R_3 \supseteq R_2$, $|R_3 : R_2| \leq 2$, and $R_3 \cong D_{2^n}$, $D_{2^{n+1}}$, or $QD_{2^{n+1}}$. If R_3 is nonabelian, we let z be the unique involution of $R_2 \cap Z(T)$; in the contrary case, $R_3 = R_2 \cong Z_2 \times Z_2$ and we let z be an arbitrary involution of R_2.

We next prove

Lemma 10.3. We can choose y to normalize an elementary abelian subgroup of R_1R_2 of order 8.

Proof: If $T = R_1R_2 \cong E_8$, the conclusion is obvious, while if $R_3 \cong QD_{2^{n+1}}$, then all elementary abelian subgroups of T lie in R_1R_2 and are conjugate in T and we obtain the desired conclusion by the usual argument. Hence we can suppose neither of these holds, in which case $R_3 \cong D_{2^m}$ for some $m \geq 3$. Thus $Z(T) = \langle z,x \rangle$ and $\langle z \rangle = Z(T) \cap T'$. Together our conditions imply that y centralizes z and $x^y = xz$. Since $Z(S) \subseteq T$, it follows that $\langle z \rangle = Z(S)$.

Now choose U in $U(S)$ and suppose first that $U \not\subseteq T$. Since $Z(S) \subset U$ and $U/Z(S) \subseteq Z(S/Z(S))$, we see that U normalizes every subgroup of T containing $Z(S)$ and, in particular, every elementary abelian subgroup of T of order 8. But then if we take y in $U - Z(S)$, the desired conclusion will hold. Hence we can assume without loss that $U \subseteq T$. If $x \notin U$, then $U\langle x \rangle \cong E_8$ and as $U\langle x \rangle = UZ(T)$, y normalizes $U\langle x \rangle$. But as $T = R_1 \times R_3$ with $R_3 \cong D_{2^m}$, $m \geq 3$, T has exactly two conjugacy classes of elementary abelian subgroups of order 8. Since y normalizes $U\langle x \rangle$, it follows that y leaves each class invariant under conjugation. Hence if A is an elementary abelian subgroup of R_1R_2 of order 8, $A^y \sim A$ in T and so $A^{yv} = A$ for some v in T. Replacing y by yv, we reach the desired conclusion. Therefore we can also suppose that $x \in U$, whence $|S:T| = 2$ and $S = T\langle y \rangle$.

We now set $R_3 = \langle a,b \rangle$ with a,b involutions and $b \in R_2$. Moreover, if $R_3 \supset R_2$, then clearly $a \notin R_2$. We have that $A = \langle z,x,a \rangle$ and $B = \langle z,x,b \rangle$ are representatives of the two conjugacy classes of elementary abelian subgroups of T of order 8. We can assume that y interchanges these two classes under conjugation or again we are done. Hence $B^y \sim A$ in T and so replacing y by yv for suitable v in T, we can suppose without loss that $B^y = A$. Since y normalizes $\langle x,z\rangle$, it follows that $b^y = ad$ for some d in $\langle z,x\rangle$.

We claim that d is in $\langle z\rangle$ if $R_3 = R_2$. Indeed, in this case, all involutions of R_2 are conjugate in M as are all involutions of the coset R_2x. Since $x^y = xz \in R_2x$, the second assertion implies that all involutions of $T - R_2$ are conjugate in G. Since x is noncentral and $z \in R_2$ is central, it follows now that no involution of R_2 is conjugate in G to x and hence to any involution of $T - R_2$. Since $b^y \in T$, we conclude, in particular, that $b^y \in R_2$, whence $b^y \in R_2 \cap A = \langle z,a\rangle$. Since $b^y = ad$ for some d in $\langle z,x\rangle$, this yields $d = 1$ or z and our assertion is proved.

Suppose, on the other hand, that $R_3 \supset R_2$. If $b^y = a$ or az, we again have $d \in \langle z\rangle$. However, if $b^y = ax$ or axz, we observe that also $T = R_1 \times \langle ax,b\rangle$ with $\langle ax,b\rangle \cong R_3$ and $\langle ax,b\rangle \supset R_2$. Hence we can replace a by ax, in which case we shall have $b^y = a$ or az. Thus we can suppose in all cases that $d \in \langle z\rangle$.

If $R_3 = R_2$, we have shown above that all involutions of R_3 are conjugate in G to z and all involutions of $T - R_3$ are conjugate in G to x. We claim that the same holds when $R_3 \supset R_2$. Indeed, since $b^y = ad$ with $d \in \langle z\rangle$, all involutions of R_2 are conjugate in G to b. Since all involutions of $R_3 - R_2$ are conjugate in R_3 to b, it follows that all involutions of R_3 are conjugate in G. Furthermore, as in the case $R_3 = R_2$ above, we have that all involutions of $R_1R_2 - R_2$ are conjugate in M to x. In particular, $bx \sim x$ in G. Hence $(bx)^y = (ad)(zx) \sim x$ in G. But every element of $T - (R_3 \cup R_1R_2)$ is conjugate in M to $adzx$. We thus conclude that all involutions of $T - R_3$ are conjugate in G to x. This proves our assertion.

Since x is noncentral and z is central, we conclude also that no involution of R_3 is conjugate in G to any involution of $T - R_3$.

We assume now that y is chosen in $S - T$ to have least possible order. We shall argue first that $y^2 \in R_3$. Suppose false, in which case certainly y is not an involution. Since T is maximal in S, we can apply [27, Lemma 16] to conclude that for some g in G, $y^g = w \in T$. By the structure of $T = R_1 \times R_3$, we have that $w^2 \in \langle (ab)^2 \rangle \subseteq R_3$. By assumption, $y^2 \in T - R_3$. However, as $y^2 \sim w^2 \in R_3$ and no involution of $T - R_3$ is conjugate to an involution of R_3, it follows now that y^2 is not an involution. Since $y^2 \in T$, this in turn implies that $y^2 \in \langle ab, x \rangle$. But then y^4 and w^4 lie in $\langle (ab)^2 \rangle$. Since $|y^4| = |w^4|$, this yields $\langle y^4 \rangle = \langle w^4 \rangle$ and that z is the unique involution of both $\langle y^4 \rangle$ and $\langle w^4 \rangle$. Since $y^g = w$, we conclude that $z^g = z$. Thus $g \in C_G(z)$ and so $y \sim w$ in $C_G(z)$.

As usual, we derive a contradiction by examining the structure of $K = C_G(z)$, $\overline{K} = K/O(K)$, and $\tilde{K} = \overline{K}/\langle \overline{z} \rangle$. Again we set $Q = S \cap O_{2',2}(K)$, so that $\overline{Q} = O_2(\overline{K})$ and, as usual, K is solvable and $x \in Q$. Likewise it follows that $\langle \overline{x}, \overline{z} \rangle$ is not normal in $\overline{K}$, otherwise $C_{\overline{K}}(\langle \overline{z}, \overline{x} \rangle)$ would be normal of index 2 in $\overline{K}$ with Sylow 2-subgroup $\overline{T}$, contradicting the conjugacy of $\overline{y}$ and $\overline{w}$ in $\overline{K}$. Hence x must be conjugate in K to an involution x' of $Q - \langle z, x \rangle$. By our choice of y, $S - T$ contains no involutions and so $x \in T = R_1 \times R_3$. Considering the way in which y acts on T, x' centralizes no element of $S - T$ and so $C_S(x') = \langle x', x, z\rangle \cong E_8$. But $|Q:C_Q(x)| \leq 2$ as $\langle x, z\rangle \triangleleft S$ and consequently the conjugacy of x and x' implies that also $|Q:C_Q(x')| \leq 2$. Hence either $Q = C_Q(x') \cong E_8$ or $|Q| = 16$. In the first case, y normalizes Q and $Q \subseteq T$, contrary to the fact that y normalizes no elementary subgroup of T of order 8 under our present assumptions. Thus $|Q| = 16$. If $Q \not\subseteq T$, we could have chosen y in Q and then y would normalize $\langle x', x, z\rangle \cong E_8$, giving the same contradiction. Hence $Q \subseteq T$ and now it is immediate from the structure of T that $Q \cong Z_2 \times D_8$. However, in this case Aut(Q) is a 2-group, whence $\bar{K} = \bar{S}$. But then $\langle \bar{x}, \bar{z}\rangle \triangleleft \bar{K}$, which is not the case. We have therefore shown that $y^2 \in R_3$.

Setting $R_4 = R_3\langle y\rangle$, our argument yields that R_4 is a maximal subgroup of S and that $S = R_4\langle x\rangle$. By Thompson's fusion lemma, x must be conjugate in G to an involution r of R_4. But as x is not conjugate to any involution of R_3, we must have $r \in R_4 - R_3$, whence $r \in S - T$. Thus $S - T$ contains an involution, and so by our minimal choice of y, we have that y is an involution.

If y normalized a four subgroup D of R_3 then y would normalize $\langle D, x\rangle \cong E_8$, contrary to our assumption that y normalizes no elementary abelian subgroup of T of order 8. We conclude at once from this that R_4 must be dihedral. Hence the unique maximal cyclic subgroup of R_4 is a normal subgroup of S of index 4. However, by [9], no simple group possesses a Sylow 2-subgroup of this form.

Henceforth we assume that y is chosen to normalize an elementary abelian subgroup A of R_1R_2 of order 8. As usual, we set $B = A \cap R_2$, $N = N_G(A)$, and $\bar{N} = N/C_G(A)$. In the present situation we have that $\bar{K} = \overline{M \cap N} \cong Z_3$ or S_3 according as $T = A$ is abelian or T is nonabelian. It follows now exactly as in the proof of Lemma 10.2 that $\overline{N}$ is correspondingly isomorphic to A_4 or S_4 and in both cases that $B \triangleleft N$ and that x has exactly four conjugates in A under the action of N; namely, the elements of $A - B$.

Likewise if V, P denote respectively a $(T \cap N)$-invariant 2-subgroup of N which maps an $O_2(\overline{N})$ and a 3-subgroup of N which normalizes V, centralizes x, and maps on $O_3(\overline{K})$, it follows that $[V/B, P] \cong Z_2 \times Z_2$ or Q_8. Furthermore, if $\overline{N} \cong S_4$, the argument of Lemma 10.2 shows that, in fact, $[V/B, P] \cong Z_2 \times Z_2$. We claim that the same holds when $\overline{N} \cong A_4$, in which case $T = A \cong E_8$. Indeed, if false, then V contains an element v such that $v^2 \in A - B$. But as all involutions of $A - B$ are conjugate in N to x, we can choose v so that $v^2 = x$. Then $v \in M = C_G(x)$ and so M contains an element of order 4, contrary to the fact that the Sylow 2-subgroup T of M is elementary abelian in this case. We therefore conclude in either case that $[V/B, P] \cong Z_2 \times Z_2$. Hence if we set $V_1 = [V, P]$, it follows now that $V_1 \cong E_{16}$ or $Z_4 \times Z_4$.

Finally we set $W = V(T \cap N)$, so that W is a Sylow 2-subgroup of N and $|W:V| \leq 2$. We fix all this notation for the balance of the proof.

We first eliminate the A_4 case by proving

Lemma 10.4. T is nonabelian.

Proof: Suppose false, in which case $\overline{N} \cong A_4$ and $W = V = V_1\langle x\rangle$ is a Sylow 2-subgroup of N. Suppose first that $V_1 \cong E_{16}$. Since $P \times \langle x\rangle$ acts on V_1 with P acting fixed-point-free, we have that $V \cong Z_2 \times Z_2 \int Z_2$, whence all involutions of $V - V_1$ are conjugate in V to x and V_1 is characteristic in V. It follows now exactly as in Lemma 10.2 that V is a Sylow 2-subgroup of G. But then G has a normal subgroup of index 2 by [27, Lemma 18], which is not the case.

Therefore $V_1 \not\cong E_{16}$ and so $V_1 \cong Z_4 \times Z_4$.

Clearly V is of type R_{2^5} as this term is defined in Section 3. Since G is simple, Proposition 3.7 shows that V cannot be a Sylow 2-subgroup of G. This in turn implies that A is not characteristic in V. In particular, $A \neq \Omega_1(V)$ and so $V - A$ contains an involution. Since $V = V_1\langle x\rangle$, this is possible only if x inverts some element of $V_1 - B$. But as P centralizes x and acts fixed-point-free on V_1, it follows that x inverts V_1.

Furthermore, as $C_G(A) \subseteq M$, it follows, as usual, from the structure of M that $C_G(A)$ has a normal 2-complement which is equal to $O(N)$. Thus $N = O(N)VP$ and we conclude now that G is of type F_1. Again Propositions 3.6 and 4.2 together yield a contradiction.

By the lemma, $\bar{N} \cong S_4$. Since $Z(N) = 1$ and $V_1 \cong E_{16}$ or $Z_4 \times Z_4$, Lemma 2.8 yields as a corollary:

<u>Lemma 10.5</u>. According as $V_1 \cong E_{16}$ or $Z_4 \times Z_4$, W is of type A_8 or M_{12}.

We next prove

<u>Lemma 10.6</u>. If W is a Sylow 2-subgroup of G, then S is of type M_{12}.

<u>Proof</u>: By the preceding lemma, either this lemma holds or S is of type A_8. However, Proposition 4.2 rules out the latter possibility.

By the lemma, Proposition 10.1 holds if W is a Sylow 2-subgroup of G, so we can assume henceforth that this is not the case and derive a contradiction. We first eliminate the M_{12} case.

<u>Lemma 10.7</u>. W is of type A_8.

<u>Proof</u>: Suppose false, in which case W is of type M_{12}. Then $V_1 \cong Z_4 \times Z_4$ and we conclude at once from the structure of $N/O(N)$ that

G is of type F_3. Once again Propositions 3.6 and 4.2 yield a contradiction.

By Lemmas 10.5 and 10.7, we have that $V_1 \cong E_{16}$. We set $C_1 = C_G(V_1)$, $N_1 = N_G(V_1)$, and $\bar{N}_1 = N_1/C$. Since W is of type A_8, V_1 is characteristic in W. Hence if S_1 is a Sylow 2-subgroup of G containing W, we have that $N_{S_1}(W) \subseteq N_1$. Since W is not a Sylow 2-subgroup of G, it follows that W is properly contained in a Sylow 2-subgroup W_1 of N_1. We fix all this notation as well.

Lemma 10.8. The following conditions hold:

(i) V_1 is a Sylow 2-subgroup of C_1;

(ii) $\bar{N}_1$ is isomorphic to a Sylow 3-normalizer in A_8;

(iii) W_1 is of type A_{10}.

Proof: Let Y_1 be a W-invariant Sylow 2-subgroup of C_1 and suppose $Y_1 \supset V_1$. Since x acts on Y_1/V_1, x centralizes some involution of Y_1/V_1 and so there exists an element y_1 in $Y_1 - V_1$ such that $[y_1, x] \in V_1$. But then as $V = V_1\langle x\rangle$, we see that $x^{y_1} \in V - V_1$. However, $V \cong Z_2 \times Z_2 \int Z_2$ and so all involutions of $V - V_1$ are conjugate to x by an element of V_1. Hence $x^{y_1 v} = x$ for some v in V_1 and so $y_1 v$ centralizes x. Since $y_1 v$ normalizes V_1, it also normalizes $B = C_{V_1}(x)$ and so $y_1 v$ normalizes $A = B \times \langle x\rangle$. Thus $y_1 v \in N$. Since W is a Sylow 2-subgroup of N and Y_1 is W-invariant, we conclude now that $y_1 v_1 \in W$. But as W is of type A_8, $C_W(V_1) = V$, whence $y_1 v_1 \in V_1$. Since $v_1 \in V_1$, it follows that $y_1 \in V_1$, which is not the case. This contradiction establishes (i).

Now set $T_1 = C_{W_1}(x)$. Then T_1 leaves $B = C_{V_1}(x)$ invariant and so T_1 normalizes A. As in the preceding paragraph, this implies that

$T_1 \subseteq W$. Thus $T_1 \subseteq C_W(x) = T \cap N$ and it follows that $T_1 = T \cap N$. Since $T = R_3 \times \langle x \rangle$ with R_3 nonabelian, we have that $T_1 = (R_3 \cap N) \times \langle x \rangle \cong D_8 \times Z_2$. Since V_1 is a Sylow 2-subgroup of C_1 and $B = V_1 \cap T_1$, we conclude that $\overline{T}_1 \cong Z_2 \times Z_2$.

On the other hand, as every involution of $V - V_1$ is conjugate to x by an element of V_1, it follows as in the preceding paragraph that the set Y_1 of elements y_1 of W_1 such that $[y_1, x] \in V_1$ has the form $Y_1 = V_1 T_1$. But clearly $\overline{Y}_1 = C_{\overline{W}_1}(\overline{x})$ and consequently $C_{\overline{W}_1}(\overline{x}) = \overline{T}_1 \cong Z_2 \times Z_2$. We conclude now that $\overline{W}_1$ is either dihedral or quasi-dihedral. Since W_1 is isomorphic to a subgroup of $GL(4,2)$, the only possibilities are that $\overline{W}_1 \cong D_8$ or $Z_2 \times Z_2$. However, in the latter case, $\overline{W}_1 = \overline{T}_1$, whence $W_1 = V_1 T_1 \subseteq W$, contrary to the fact that $W_1 \supset W$. Thus $\overline{W}_1$ is, in fact, dihedral of order 8.

Now we can establish (ii). Since V_1 is a Sylow 2-subgroup of $C_1 = C_G(V_1)$, C_1 has a normal 2-complement. This implies that N_1 is 2-constrained and as G is an $\mathfrak{R}_3$-group, it follows $\overline{N}_1$, and hence also N_1, is solvable. Observe next that $M \cap N_1$ leaves $C_{V_1}(x)$ invariant and so $K_1 = M \cap N_1$ normalizes A. Hence $K_1 \subseteq M \cap N$. As usual, we can choose P to be inverted by an involution r of R_3. Then $W = V_1\langle r, x\rangle$ and $P\langle r, x\rangle \subseteq N_1$. We conclude at once that $\overline{K}_1 = \overline{P}\langle \overline{r}, \overline{x}\rangle \cong Z_2 \times S_3$. Since a Sylow 3-subgroup of S_4 is not invariant under a four subgroup of S_4, it follows, in particular, that $\overline{N}_1 \ncong S_4$. But now examining the solvable subgroups of A_8 with Sylow 2-subgroups isomorphic to D_8 and containing a subgroup isomorphic to $Z_2 \times S_3$, we check that either $\overline{N}_1$ is isomorphic to a Sylow 3-normalizer in A_8 or else $O_3(\overline{N}_1) \cong Z_3$ with $O_3(\overline{N}_1)$ generated by an element which corresponds to a 3-cycle in A_8.

In the first case, we obtain (ii), so suppose then that $O_3(\overline{N}_1) \cong Z_3$. We claim first that $O_3(\overline{N}_1) \subseteq \overline{P}$. Indeed, in the contrary case, we would necessarily have $\tilde{N}_1 = \overline{N}_1/O_3(\overline{N}_1) \cong S_4$ with $\tilde{P} \neq 1$. However, as $\tilde{P}$ is invariant under the four subgroup $\langle \tilde{r}, \tilde{x} \rangle$ of $\tilde{N}_1$, this is impossible. Thus $\overline{P} = O_3(\overline{N}_1)$, as asserted. Again by the structure of A_8, $C_{\overline{N}_1}(\overline{P}) \cong Z_3 \times Z_2 \times Z_2$ or $Z_3 \times A_4$. In either case, $C_{\overline{N}_1}(\overline{P})$ is 2-closed and $O_2(C_{\overline{N}_1}(\overline{P})) = O_2(\overline{N}_1) \cong Z_2 \times Z_2$. In particular, it follows that $\overline{x} \in O_2(\overline{N}_1)$. Since $\overline{P}$ acts regularly on V_1 and $V_1\langle x \rangle \cong Z_2 \times Z_2 \int Z_2$, Lemma 2.9 now yields that the inverse image X_1 of $O_2(\overline{N}_1)$ in W_1 is of type $L_3(4)$. But then as $x \in X_1 - Z(X_1)$, we have that $C_{X_1}(x) \cong E_{16}$, contrary to the fact that T is a Sylow 2-subgroup of $C_G(x)$ and $m(T) = 3$.

Therefore (ii) holds and now (iii) is a consequence of [Part II, Lemma 2.2(v)].

Finally we prove

Lemma 10.9. W_1 is a Sylow 2-subgroup of G.

Proof: Again let S_1 be a Sylow 2-subgroup of G containing W_1. Also let T_1 be a conjugate of T which lies in S_1. If T_1 were maximal in S_1, then $T_1 \cap W_1$ would be normal in W_1 of index at most 2. However, as $T_1 \cong D_{2^m} \times Z_2$ or $QD_{2^m} \times Z_2$ for some m and W_1 is of type A_{10} (and hence isomorphic to $D_8 \int Z_2$), it is immediate that W_1 possesses no such normal subgroup. Hence T_1 is not a maximal subgroup of S_1. But now if U_1 is an element of $U(S_1)$, it follows that $|C_{S_1}(u_1)| > |T|$ for any u_1 in $U_1^{\#}$ and so we conclude from our maximal choice of x and T that $C_G(u_1)$ is solvable for each u_1 in $U_1^{\#}$.

On the other hand, as $W_1 = N_{S_1}(V_1)$, we have that $Z(S_1) \subseteq W_1$. But

$Z(W_1) \cong Z_2$ and $Z(W_1) \subseteq V_1$ as $W_1 \cong D_8 \int Z_2$. It follows therefore that $Z(S_1) = Z(W_1) \subseteq V_1$, whence also $U_1 \subseteq W_1$. Thus U_1 is, in fact, the unique normal four subgroup of W_1 and U_1 is contained in every elementary abelian subgroup of W_1 of order 16.

We have that $C_G(u_1)$ is solvable for each u_1 in $U_1^{\#}$ and that $N_G(V_1)/C_G(V_1)$ is isomorphic to a Sylow 3-normalizer in A_8. [Part I, Lemma 7.1] is therefore applicable to yield that $W_1 = S_1$ and hence that W_1 is a Sylow 2-subgroup of G.

But now Proposition 4.2 yields a final contradiction and the proof of Proposition 10.1 is complete. Now Proposition 4.4 yields Theorem A. Thus Theorem A holds and our Main Theorem is finally proved in its entirety.

Bibliography

1. Alperin, J.L., Sylow intersections and fusion, J. Alg.,6 (1967), 222-241.

2. Alperin, J.L. Brauer, R. and Gorenstein, D.,Finite groups with quasi-dihedral and wreathed Sylow 2- subgroups, Trans. Amer. Math. Soc., 151 (1970), 1-261.

3. __________, __________, and __________, Finite simple groups of 2-rank two (to appear).

4. Alperin, J.L. and Gorenstein, D.,The multiplicators of certain simple groups, Proc. Amer. Math. Soc.,17 (1966), 515-519.

5. Bender, H.,On groups with abelian Sylow 2-subgroups, Math. Z.,111 (1970) 164-176.

6. Brauer, R. and Fong, P.,A characterization of the Mathieu group M_{12}, Trans. Amer. Math. Soc.,84 (1966), 18-47.

7. Brauer, R. and Suzuki, M.,On finite groups of even order whose 2-Sylow group is a quaternion group, Proc. Nat. Acad. Sci. U.S.A.,45 (1959), 1757-1759.

8. Feit, W.,The current situation in the theory of finite simple groups, Actes Congress Intern. Math.,(1970) Tome 1, 55-93.

9. Fong, P.,Sylow 2-subgroups of small order I, II, (unpublished).

10. Frobenius, G.,Uber die charactere der mehrfach transitive gruppen S.-B. Preuss. Akad. Wiss.,(1904), 558-571.

11. Glauberman, G., Central elements of core-free groups, J. Alg., 4(1966), 403-420.

12. Goldschmidt, D., A conjugation family for finite groups, J. Alg., 16(1970), 138-142 .

13. Gorenstein, D.,Finite Groups, Harper and Row, New York , 1968.

14. ___________, On finite simple groups of characteristic 2 type, Inst. des Hautes Etudes Scient., 36(1969), 5-13.

15. ___________, Centralizers of involutions in finite simple groups, Chapter II of Simple Groups, Academic Press, New York, (1971).

16. Gorenstein, D. and Harada, K., A characterization of Janko's two new simple groups, J. Univ. Tokyo, 16(1970), 331-406.

17. ___________ and _________, On finite groups with Sylow 2-subgroups of type A_n, n = 8,9,10,11, Math. Z., 117(1970), 207-238.

18. ___________ and _________, On finite groups with Sylow 2-subgroups of type $\hat{A}_n$, n = 8,9,10,11, J. Alg., 19(1971), 185-227.

19. ___________ and _________, Finite groups of low 2-rank and the families $G_2(q)$, ${}^2D_4(q)$, q odd, Bull. Amer. Math. Soc., 77(1971), 829-862.

20. ___________ and _________, Finite groups whose Sylow 2-subgroups are the direct product of two dihedral groups, Ann. of Math., 95(1972), 1-54.

21. ___________ and _________, Finite groups with Sylow 2-subgroups of type PSp(4,q), q odd, (to appear).

22. Gorenstein, D. and Walter, J. H., The characterization of finite groups with dihedral Sylow 2-subgroups,J. Alg., 2(1965), I,II,II, 85-151, 218-270, 354-393.

23. ___________ and _________, The π-layer of a finite group, Ill. Jour. Math., 15(1971), 555-564.

24. ____________ and ____________, Centralizers of involutions in balanced groups, J. Alg., 20(1972), 284-319.

25. Griess, R., Schur multipliers of the known simple groups, Bull. Amer. Math. Soc., 178(1972), 68-71.

26. Harada, K., Groups with certain type of Sylow 2-subgroups, J. Math. Soc. Japan, 19(1967), 303-307.

27. ________, Finite simple groups with short chains of subgroups, J. Math. Soc. Japan, 20(1968), 655-672.

28. ________, Finite simple groups whose Sylow 2-subgroups are of order 2^7, J. Alg., 14(1970), 386-404.

29. ________, On some 2-groups of normal 2-rank 2, J. Alg., 20(1972), 90-93.

30. Higman, G., Suzuki 2-groups, Ill. Jour. Math., 7(1963), 79-96.

31. ________, Odd characterizations of finite groups, Mimeographed Notes, Oxford Univ.

32. Huppert, B., Endliche Gruppen I, Springer-Verlag, Berlin, Heidelberg, New York, 1967.

33. Janko, Z., Nonsolvable finite groups all of whose 2-local subgroups are solvable I, J. Alg., 21(1972), 458-517.

34. Janko, Z., and Thompson, J. G., On finite simple groups whose Sylow 2-subgroups have no normal elementary subgroups of order 8, Math. Z., 113(1970), 385-397.

35. MacWilliams, A., On 2-groups with no normal abelian subgroups of rank 3, and their occurrence as Sylow 2-subgroups of finite simple groups, Trans. Amer. Math. Soc., 150(1970) 345-408.

36. Mason, D., Groups with Sylow 2-subgroup dihedral wreath Z_2, (to appear).

37. ____________, Finite simple groups with Sylow 2-subgroups of type PSL(4,q), q odd, (to appear).

38. ____________, Finite groups with Sylow 2-subgroup the direct product of a dihedral and a wreathed group and related problems, (to appear).

38*. ____________, The characterization of certain finite simple groups of low 2-rank by their Sylow 2-subgroups, Ph.D. thesis, U. of Cambridge, 1972.

39. Sehgal, S., A generalization of a theorem of Z. Janko and J.G. Thompson, (to appear).

40. Smith, F., Finite groups whose Sylow 2-subgroups are the direct product of a dihedral and a semidihedral group, (to appear).

41. ____________, Finite groups whose Sylow 2-subgroups are the direct product of two semidihedral groups , (to appear).

42. Thompson, J.G., Nonsolvable finite groups all of whose local subgroups are solvable, Sections 1-6 Bull. Amer. Math. Soc., 74 (1968), 383-438; Sections 7-9, Pac. J. Math., 33 (1970), 451-536, (balance to appear).

43. Walter, J.H., The characterization of finite groups with abelian Sylow 2-subgroups, Ann. of Math., 89 (1969), 405-514.

44. Lundgren, R., A generalization of a result of J.G. Thompson, (to appear).

45. Syskin, Simple finite groups of 2-rank 3 with solvable centralizers of involutions, Algebra Logica, 10(97), 668-709.

46. Collins, M. and Solomon, R., A characterization of the simple groups $L_5(q)$ and $U_5(q)$, q odd, (to appear).

Rutgers University,
New Brunswick, New Jersey

Ohio State University
Columbus, Ohio